NITROGEN CERAMICS

NATO ADVANCED STUDY INSTITUTES SERIES

Proceedings of the Advanced Study Institute Programme, which aims at the dissemination of advanced knowledge and the formation of contacts among scientists from different countries.

The series is published by an international board of publishers in conjunction with NATO Scientific Affairs Division

| A | Life Sciences | Plenum Publishing Corporation |
| B | Physics | London and New York |

| C | Mathematical and Physical Sciences | D. Reidel Publishing Company Dordrecht and Boston |

| D | Behavioural and Social Sciences | Sijthoff International Publishing Company Leyden, The Neth. and Reading, Mass., USA |

| E | Applied Science | Noordhoff International Publishing Leyden, The Neth. and Reading, Mass., USA |

Series E: Applied Science — No. 23

NITROGEN CERAMICS

edited by

F. L. RILEY

Professor in the Department of Ceramics,
Houldsworth School of Applied Science,
The University of Leeds, U.K.

NOORDHOFF — LEYDEN — 1977

Proceedings of the NATO Advanced Study Institute
on Nitrogen Ceramics
University of Kent, Canterbury, U.K.
16-27 August 1976

ISBN-13:978-94-010-1300-0   e-ISBN-13:978-94-010-1298-0
DOI: 10.1007/978-94-010-1298-0

PREFACE

The suggestion that a NATO Advanced Study Institute would be an excellent forum for reviews and informed discussion on the broad subject of Nitrogen Ceramics, arose out of discussions with colleagues in the Department of Ceramics at the University of Leeds early in 1975. There was no doubt that such a meeting would be both very valuable and timely. Scientific and technological interest in the nitride ceramics and in silicon nitride in part-icular had been growing steadily during the 20-year period following 1955. The intensive five-year programme initiated by the Advanced Research Projects Agency of the U.S. Department of Defence, on the development of a design capability in brittle materials for high temperature applications, had been based principally on silicon nitride and silicon carbide ceramics, and was due to reach the end of its first stage in the autumn of 1976. It was clear that by then a considerable volume of information covering many aspects of silicon nitride would be available for presentation or review. Coincidentally, the same five-year period had seen the discovery, and increasingly detailed investigation, of ceramic materials based on the Al-Si-N-O and similar systems. Besides being of great interest for their crystal chemistry and structural relationships, some of these materials could be assumed potentially to be of equal importance to the silicon nitride ceramics. More recently progress had also been made in the sintering of covalent materials, as demonstrated for the case of silicon carbide. It seemed possible that the basic principles, as they were then understood, might be applied to silicon nitride to enable a high density material to be produced without the attendant problems of a less refractory grain boundary phase. The summer of 1976 was therefore seen as an appropriate time to take stock of developments, and to assess the best directions for future work.

The intention, in the selection of the broad theme of 'Nitrogen Ceramics', was to encourage the presentation of teaching contributions and reviews on all refractory inorganic materials containing nitrogen, so that a full perspective view of the subject area could be developed. Nonetheless the considerable

scientific and technological interest in silicon nitride ceramics
was well appreciated, and it seemed virtually inevitable that the
final Institute programme would have a strong bias in the direct-
tion of these particular materials. At the same time, and with
this in mind, deliberate steps were taken to draw into the
programme a number of participants experienced in other areas of
ceramics so as to make available their specialist knowledge, and
to facilitate the viewing of the nitrides in the context of
ceramic science and technology as a whole. The contributions made
by these participants to the Institute were particularly welcomed.

During the two weeks of the Institute, 18 invited review
lectures and some 60 shorter contributions of a teaching or
research review nature were presented. In order to keep the
published proceedings to a reasonable length, and price, not all
the shorter contributions have been included. Most of those
papers not included in full, and those not submitted for publication, are represented by their abstracts.

Two half-day sessions during the Institute were devoted to
the subject of 'Applications'. The first consisted of a series
of short reviews by nationals of the countries represented at the
Institute, and was intended to provide an overall picture of the
extent of interest in nitrogen ceramics. The second session consisted of a series of longer presentations, each dealing with a
specific topic. It has only been possible to include some of the
second series of presentations in this volume, but participants
at the Institute will remember the effectiveness of the afternoon
session, and will appreciate the efforts made by the speakers to
present informative, albeit necessarily brief, reviews of
programmes of work, and the current international interest being
shown at all levels, in these materials.

Many interesting discussions took place both within and outside the main meeting lecture theatre, and teams of reporters
worked diligently to record and summarise the flow of questions,
answers and discussion points. These reports have for the most
part been included in unamended form in this volume. Participants
will also recall the lively discussion sessions concerning the
representation of complex multicomponent systems, and the unresolved question of 'sialon' nomenclature. This latter question
again posed itself during the preparation of the index to this
volume. The answer chosen for this purpose has been that of a
division according to constituent element in strict alphabetical
order of metal followed by non-metal. This has the advantage of
allowing systems to be grouped very clearly, but of course does
not indicate exact composition or crystal structure (any more
than a description 'iron oxide' does).   Recognised phases
have been identified by the generally accepted coded descriptions
β', $0'$, x_4, for example. The provision of a more systematic and
universally acceptable nomenclature is clearly needed, particul-

arly so if the number of systems explored, and phases identified,
continues to increase at the rate of the last five years.

The outline programme for the Institute was drawn up at a
preliminary meeting held at the University of Limoges in the
spring of 1975. Professor M. Billy, Professor K.H. Jack and
Dr. F.F. Lange provided many valuable suggestions regarding both
the structure of the programme, and speakers able to set up the
framework of the Institute with reviews of important areas.
Dr. R.N. Katz, Professor G. Petzow and Professor A.L. Stuijts
also joined in the planning at an early stage. For a subject of
this nature, having both an academic, and also a strong technological content, close contact with relevant ongoing work in
Research Institutes, and University, Industrial and Government
laboratories was required if workers able to make important
contributions to the Institute programme were to be drawn in.
The planning group was able to ensure that this was done, thereby
further contributing through the uniformly high quality of the
presentations to the success of the programme.

Financial backing for the Institute was provided by the
Scientific Affairs Division of NATO, supported by generous
contributions from the National Science Foundation, Washington,
D.C., and the Office of Naval Research, Arlington. Virginia.

The Institute owed much to the quiet effectiveness of
Miss B. Harris and Mr. S.C. Ware, and the technical staff of the
Department of Physics at the University of Kent, in helping to
provide an ideal setting for the meeting. The invaluable
assistance of Miss B.M. Readman in the Institute Office at
Canterbury, and the extensive help of Miss S.H. Toon and
Mrs. T. Iredale in the Ceramics Department of the University of
Leeds, both before and after the Institute, is also gratefully
acknowledged.

F.L. Riley

Editor

CONTENTS

(R) indicates an invited Review Lecture

SECTION G

MICROSTRUCTURE

SECTION H

MECHANICAL PROPERTIES

SECTION I

ELECTRICAL PROPERTIES

SECTION J

OXIDATION AND CORROSION

The Perfect Participant

But al be that he was a philosophre,
Yet hadde he but litel gold in cofre;
But al that he myghte of his freendes hente,
On bookes and on lernynge he it spente,
And bisily gan for the soules preye
Of hem that yaf hym wherwith to scoleye.
Of studie took he moost cure and moost heede.
Noght o word spak he moore than was neede,
And that was seyd in forme and reverence,
And short and quyk and ful of hy sentence;
Sownynge in moral vertu was his speche,
And gladly wolde he lerne and gladly teche.

Geoffrey Chaucer (1343-1400)
The Canterbury Tales.

PARTICIPANTS

Organization Committee:

M. Billy	University of Limoges, France
K.H. Jack	University of Newcastle-upon-Tyne, U.K.
R.N. Katz	AMMRC, Watertown, U.S.A.
F.F. Lange	Rockwell International, Thousand Oaks, U.S.A.
G. Petzow	Max-Planck-Institut, Stuttgart, Germany
A.L. Stuijts	N.V. Philips, Eindhoven, Netherlands
F.L. Riley	University of Leeds, U.K. (Director)

Local Organization:

B. Harris and S.C. Ware, University of Kent.

Chairmen and Discussion Reporting Teams:

Session

A	August 16	C. Greskovich	J. Lang, D.P. Thompson
B	August 16	A. Mocellin	J.A. Leake, P.E.D. Morgan
C	August 17	T. Vasilos	J. Schlichting, S.C. Singhal
D	August 18	C. Jelacic	D.R. Clarke, T-Y. Tien
E	August 19	L.J. Gauckler	P. Goursat, A. Krauth, W.W. Smeltzer
F	August 20	P.E.D. Morgan	R.M. Cannon, F.F. Lange, A.J. Moulson
G	August 23	K.H. Jack	J.J. Brennan, J. Briggs
H	August 23	R.W. Davidge	P. Boch, J.D. Venables
H	August 24	A.G. Evans	W. Raüchle, J.E. Ritter, N.J. Tighe
H,I	August 25	M. Billy	D.R. Messier, J.E. Siebels
J	August 26	A.M. Diness	J.G. Desmaison, M.W. Lindley
K	August 26	D.J. Godfrey and R.N. Katz	
K	August 27	R.J. Brook	R.J. Lumby, A.N. Pick

Babini, Dr. G.N.   C.N.R. - Research Laboratory for Ceramic
    Technology, Via S. Nevolone 22, 48018 Faenza, Italy.
Becker, Mr. R.   Institut für Werkstoffkunde II, Universität
    Karlsruhe, 7500 Karlsruhe, W. Germany.
Billy, Prof. M.   Universite de Limoges, Centre de Récherches
    et d'Etudes Céramiques, 87100 Limoges, France.
Boch, Prof. P.   Université de Limoges, Allée André Maurois,
    87100 Limoges, France.
Bowen, Dr. L.J.   Department of Ceramics, The University of
    Leeds, Leeds LS2 9JT, U.K.
Boyer, Mr. S.M.   Department of Ceramics, The University of
    Leeds, Leeds LS2 9JT, U.K.
Brennan, Dr. J.J.   United Technologies Research Center,
    East Hartford, Connecticut 04108, U.S.A.
Briggs, Dr. J.   Battelle, Geneva Research Centre, 7 Route de Drize,
    1227 Carouge, Geneva, Switzerland.
Brook, Prof. R.J.   Department of Ceramics, The University of
    Leeds, Leeds LS2 9JT, U.K.
Cannon, Dr. R.M.   Department of Materials Science & Engineering,
    Massachusetts Institute of Technology, Cambridge, Mass., U.S.A.
Cetincelik, Dr. M.   P.O. Box 37, Bakanliklar, Ankara, Turkey.
Chowdhry, Dr. U.   Department of Materials Science & Engineering,
    Massachusetts Institute of Technology, Cambridge, Mass., U.S.A.
Clarke, Dr. D.R.   Department of Materials Science & Engineering,
    University of California, Berkeley, California 94720, U.S.A.
Cooke, Dr. R.G.   School of Materials Science, Bath University,
    Bath, BA2 7AY, U.K.
Dalgleish, Dr. B.J.   Imperial College of Science and Technology,
    University of London, London SW7, U.K.
Davidge, Dr. R.W.   Materials Development Division, AERE, Harwell,
    Didcot, Oxon. OX11 ORA, U.K.
Demit, Mr. J.R.   CERAVER, B.P. 113, 65001 Tarbes, France.
Desmaison, Dr. J.   Laboratoire de Chimie Minérale et Cinétique
    Hétérogène, Université de Limoges, 87100 Limoges, France.
Diness, Dr. A.M.   Office of Naval Research, Arlington,
    Virginia 22217, U.S.A.
Edmonds, Miss A.J.S.   Department of Metallurgy & Materials
    Science, University of Cambridge, Cambridge CB2 3QZ, U.K.
Engel, Mr. W.   Degussa Wolfgang, Forschung Metall., Abt. FM-7,
    Postfach 602, 6450 Hanau-Wolfgang, W. Germany.
Evans, Dr. A.G.   Rockwell International, Science Center,
    Thousand Oaks, California, 91360, U.S.A.
Gauckler, Dr. L.J.   Max-Planck-Institut für Metallforschung,
    Institut für Werkstoffwissenschaften, D-7000 Stuttgart 80,
    W. Germany.
Glaeser, Dr. W.   Hermann C. Starck Berlin Werk Goslar,
    3380 Goslar/Harz, Im Schleeke 78-91, W. Germany.
Godfrey, Dr. D.J.   Ministry of Defence (Procurement Executive),
    Admiralty Materials Laboratory, Poole, Dorset, U.K.

Goursat, Dr. P. Université de Limoges, Centre de Récherches et
    d'Etudes Céramiques, 87100 Limoges, France.
Greskovich, Dr. C. General Electric Co., Corporate Research and
    Development Center, Schenectady, New York 12301, U.S.A.
Grieveson, Prof. P. Department of Metallurgy, University of
    Strathclyde, 48 North Portland St., Glasgow G1 1XN, Scotland.
Gugel, Prof. E. Annawerk GmbH, Ceranox Division, D-8633 Roedental,
    Postfach 44, W. Germany.
Jack, Prof. K.H. Department of Metallurgy and Materials Science,
    The University, Newcastle-upon-Tyne, NE1 7RU, U.K.
Jelacic, Prof. C. Inorganic Chemistry Department, Tehnoloski
    Fakultet, Rudarska 75, 75000 Tuzla, Yugoslavia.
Johnson, Mr. R. A.M.E. Ltd., Dukesway, Team Valley, Gateshead,
    Tyne and Wear, U.K.
Katz, Dr. R.N. Materials Branch, AMMRC, Watertown, Mass., U.S.A.
Kizilyalli, Dr. H.M. Department of Physics, Middle East
    Technical University, Ankara, Turkey.
Klomp, Mr. J.T. N.V. Philips Research Laboratories, Kastanjelaan,
    Eindhoven, The Netherlands.
Knoch, Dr. H. DFVLR - Institut für Werkstoff-Forschung,
    Linder Höhe, 5000 Köln 90, W. Germany.
Krauth, Dr. A. Rosenthal Aktiengesellschaft, Institut für Werk-
    stofftechnik, 8672 Selb/Bayern, W. Germany.
Kriegesmann, Dr. J. Technische Universität Clausthal, Lehrstuhl
    für Glas und Keramik, Institut für Steine und Erden,
    3392 Clausthal-Zellerfeld, Zehntnerstr. 2A, W. Germany.
Kuwabara, Dr. M. Kyushu Institute of Technology, Sensui-cho,
    Tobata, Kitakyushu, Japan.
Lang, Prof. J. Université de Rennes, Laboratoire de Chimie
    Minérale C, 35031 Rennes Cedex, France.
Lange, Dr. F.F. Rockwell International, Science Center,
    Thousand Oaks, California, 91360, U.S.A.
Laurent, Dr. Y. Université de Rennes, Laboratoire de Chimie
    Minérale C, 35031 Rennes Cedex, France.
Leake, Dr. J.A. Department of Metallurgy & Materials Science,
    University of Cambridge, Cambridge CB2 3QZ, U.K.
Lee, Dr. J.G. Department of Materials Science and Engineering,
    University of Utah, Salt Lake City, Utah 84112, U.S.A.
Lindley, Dr. M.W. Ministry of Defence (Procurement Executive),
    Admiralty Materials Laboratory, Poole, Dorset, U.K.
Lumby, Dr. R.J. Lucas Industries Ltd., Group Research Centre,
    Shirley, Solihull, W. Midlands B90 4JJ, U.K.
Major, Mr. L.D. Department of Ceramics, Case Western Reserve
    University, Cleveland, Ohio 44106, U.S.A.
Mangels, Mr. J.A. Ford Motor Company, Turbine Development
    Department, 20,000 Rotunda Drive, Dearborn, Michigan, U.S.A.
Martinengo, Dr. P.C. Fiat Central Research Laboratories,
    Turin, Italy.
Messier, Dr. D.R. AMMRC, Watertown, Mass. 02172, U.S.A.

Mocellin, Dr. A.   Centre des Matériaux, Ecole des Mines,
    B.P. 87, 91003 Evry Cedex, France.
Morgan, Dr. P.E.D.   Franklin Institute Research Laboratories,
    Benjamin Franklin Parkway, Philadelphia, P.A. 19103, U.S.A.
Moulson, Dr. A.J.   Department of Ceramics, The University of
    Leeds, Leeds LS2 9JT, U.K.
Nicol, Dr. A.W.   Department of Minerals Engineering, The University
    of Birmingham, P.O. Box 363, Birmingham B15 2TT, U.K.
Niwano, Mr. K.   Industrial Research Institute of Hyogo Prefecture,
    Yukihira-cho, 3-Chome, Suma-ku, Kobe, Japan.
Oda, Mr. I.   Research & Development Laboratory, NGK Insulators
    Ltd., Suda 2-56, Mizuho-Ku, Nagoya, Japan.
Olapinski, Dr. H.   Feldmühle AG, Postfach 1149, 7310 Plochingen,
    W. Germany.
Petzow, Prof.Dr. G.   Max-Planck-Institut, Stuttgart, W. Germany.
Pick, Dr. A.N.   The Dyson Group, Research & Development
    Laboratory, Owler Bar, Sheffield S17 3BJ, U.K.
Plant, Mr. J.H.   Associated Engineering Developments Ltd.,
    Engineering Ceramics Unit, Rugby, Warwickshire CV22 7DB, U.K.
Popper, Mr. P.   British Ceramic Research Association, Queen's Road,
    Penkhull, Stoke-on-Trent, Staffs., U.K.
Raúchle, Dr. W.   Daimler-Benz AG, Abt. ZWE, Postfach 202,
    D-7000 Stuttgart 60, W. Germany.
Riley, Dr. F.L.   Department of Ceramics, The University of Leeds,
    Leeds LS2 9JT, U.K.
Ritter, Dr. J.E. Jr.   Mechanical Engineering Department,
    University of Massachusetts, Amherst, Mass. 01002, U.S.A.
Schaeffer, Dr. H.A.   Institut für Workstoffwissenschaften III,
    Universität Erlangen-Nürnberg, 852 Erlangen, W. Germany.
Schlichting, Dr. J.   Institut für Chemische Technik, Universität
    Karlsruhe, 75 Karlsruhe, Kaiserstrasse 12, W. Germany.
Schwetz, Dr. K-A.   Elektroschmeltzwerk Kempten GmbH,
    D-896 Kempten, Postfach 1526, W. Germany.
Siebels, Mr. J.E.   Volkswagenwerk AG, 3180 Wolfsburg 1, W. Germany.
Singhal, Dr. S.C.   Westinghouse Electric Corporation, Research &
    Development Center, Pittsburgh, Pa. 15235, U.S.A.
Smeltzer, Prof. W.W.   Department of Metallurgy and Materials
    Science, McMaster University, Hamilton, Ontario, Canada.
Stuijts, Prof. A.L.   N.V. Philips Research Laboratories,
    Eindhoven, The Netherlands.
Thompson, Dr. D.P.   Crystallography Laboratory, The University,
    Newcastle-upon-Tyne, NE1 7RU, U.K.
Tien, Prof. T-Y.   Department of Materials and Metallurgical
    Engineering, University of Michigan, Ann Arbor,
    Michigan, 48103, U.S.A.
Tighe, Dr. N.J.   Inorganic Materials Division, National Bureau
    of Standards, Washington D.C., U.S.A.
Torre, Mr. J-P.D.   Centre des Matériaux, B.P. 87, 91003 Evry-Cedex,
    France.

Trontelj, Dr. M. Institute J. Stefan, University of Ljubljana,
    Jamova 39, 61000 Ljubljana, Yugoslavia.
Umebayashi, Mr. S. National Industrial Research Institute of
    Kyushu, 841 Shuku machi, Tosu City, Saga Pref., Japan.
Vasilos, Dr. T. AVCO Corporation, Systems Division, Lowell,
    Massachusetts, 01851, U.S.A.
Venables, Dr. J.D. Martin Marietta Laboratories,
    Baltimore, Maryland, 21227, U.S.A.
Vincenzini, Prof. P. C.N.R. - Laboratorio di Ricerche Tecnolo-
    giche per la Ceramica, Via S. Nevolone 22, 48018 Faenza, Italy.

GENERAL INTRODUCTION

INTRODUCTORY LECTURE

P. Popper

The British Ceramic Research Association,
Queens Road, Penkhull, Stoke-on-Trent, ST4 7LQ,
England.

1. INTRODUCTION

When Dr.Riley first approached me about giving the Introductory Lecture of this Seminar on Nitrogen Ceramics he said something like this: "You have probably been longer in this field than anybody else, you covered a wider field than silicon-nitrogen ceramics; it would be interesting to have your views about the past, present and future on a personal basis." I said: "O.K.", because in the midst of the winter a fortnight in Canterbury in August in the company of the most important people working on nitrogen ceramics seemed a very attractive proposition.

When I saw for the first time the list of attendees and the subject of their contributions I wondered whether I was wise in accepting. I remembered a discussion with my former director on the subject of nervousness when giving a paper at a Conference. He said he didn't think that any contributor ought to feel apprehensive because it was most likely that he had worked on the particular subject of his discourse for several years and knew more about it than anybody else in the audience. Well, when looking through the list I became convinced that on almost anything I might say there would be many people here who knew a lot more about it than I do.

Since I was asked to give my personal views I think it might be best to give you some of my background. I studied Applied Physics at the Technische Hochschule in Vienna and changed to Electric Engineering when I came to England in 1939. My first job apart from minor engagements (as glass blower, tennis instructor and weight checker) was as Assistant Lecturer in Radio & Physics

at Reading University. I am mentioning all this because I think
one's education and past experience strongly influences one's
professional outlook. Mine is, I believe, the scientific approach
of a *physicist* - this, by the way, is a useful way of drawing attention to the lack of training in basic chemistry and my total
ignorance and therefore suspicion of the usefulness of chemical
thermodynamics. Secondly as an *engineer* I have a strong desire to
see that something useful comes out of my own efforts and equally
of those of other people. It is in connection with the latter that
I am often very critical of some so-called "academic" research
projects which, it seems to me, neither lead to a "useful" product
nor provide a particularly good subject for training in research;
by *useful* product I mean a product which somebody wants at a price
for which it can be produced *profitably* by somebody else. I will
refer to this later on several occasions.

Although I enjoyed my time at Reading University immensely I
was convinced that as an Engineer I must get experience of industry,
and I joined the Materials Research Laboratory of Philips in this
country. As a physicist I got involved through the measurement
side with the very interesting developments that were then taking
place in ceramics, i.e. the ferroelectric titanates and magnetic
ferrites. This was my first contact with ceramic technology and
led eventually to my joining the British Ceramic Research Association,
which was then largely engaged in traditional, i.e. clay-based,
ceramic materials. My new colleagues at the B.Ceram.R.A. started
asking me such questions as "Why do you call titanates, ferrites,
nitrides, etc. ceramics? There is no clay in them." This made me
think about a definition of ceramics and since the title of this
Seminar is Nitrogen Ceramics it might be worthwhile to spend a few
minutes on the meaning of the word ceramics. I frequently start my
lectures by saying that I find it increasingly difficult to give a
definition of ceramics, but I will attempt to say what, in my
opinion, *ceramic materials* are and what a *ceramic article* is.

According to the Materials Scientist all materials are either
metals, organics or ceramics. This means that *ceramic materials*
are, from a chemical point of view, compounds between metals and
non-metals, i.e. oxygen, nitrogen, carbon, silicon, boron, etc. In
other words ceramic materials are oxides, nitrides, carbides,
silicides, borides, etc. A *ceramic article* is one that has been
manufactured by *ceramic technology*, i.e. shaped from a material in
a finely divided state and consolidated into a monolithic body by
a heat-treatment called "sintering". Usually *ceramic articles* are
made by applying *ceramic technology* to *ceramic materials*.

When I joined the B.Ceram.R.A. the first project I worked on
was entitled "New Electrical Ceramics". It was a Ministry of Supply
contract - it was the type of contract that doesn't exist these
days; it was entirely open-ended and lasted eventually 12 years,

a situation any member of a contract research organization will
only meet in his wildest dreams. When I asked my sponsors what
they meant by "new electrical ceramics" all I could get from them
was the following statement: "Presently all the electrical ceramics,
with the exception of silicon carbide, are oxides, and this is only
because our atmosphere contains oxygen. In a different environment
there may be some quite different materials with interesting properties," and indeed the only natural nitrides have been found in
meteorites, e.g. osbornite,[1] a titanium nitride TiN and sinoite[2]
a silicon oxynitride Si_2ON_2. I questioned them further about what
they meant by "interesting properties". It seemed to me a waste of
time to find an expensive alternative to an electrical insulator
with the insulating properties of aluminium oxides, or an electrical
conductor like Cu. The only answer I could get was "look at any
material with a large energy gap between the filled and empty
energy bands." I was not very satisfied with this statement as it
seemed to me a very unrewarding task to synthesize a vast number
of new materials in the hope of finding a material with an "interesting property". I did however come to think that whilst finding
a material that excelled in a *single property*, was - from an engineer's
point of view - not very useful, a material with the right combination of properties could be. Thus a material with the low
electrical conductivity of Al_2O_3 but coupled with a substantially
higher thermal conductivity or lower thermal expansion or lower
hardness could be of advantage, i.e. such a material could have
the same electrical insulating properties as alumina but would have
a better thermal stress resistance or be more readily machined.

I have always found the "suck it and see" approach a very
undignified way for a scientist to set about things, and I therefore asked Dr.Ruddlesden, who collaborated with me on this project,
to look into the relations between structure and properties so that
we could apply whatever knowledge was then available to the selection of materials worth studying. I myself surveyed the literature
on all simple binary non-oxide compounds, i.e. nitrides, phosphides,
sulphides, carbides, borides, etc. I recorded in the form of the
Periodic Table the information that I was able to extract from the
literature in 1955.[3] Most of this came from investigations of basic
inorganic chemistry which flourished in Germany at the turn of the
last century; the information was mostly limited to chemical
formula, melting or decomposition point, crystal structure (when
available), appearance (crystallinity, colour), solubility in water
or acids. We discarded thermally unstable and water soluble compounds and from general ideas about the relationship between
structure and properties we selected from the nitrides, boron
nitride, aluminium nitride and silicon nitride for further study.
It may be worth while here to quote from Dr.Ruddlesden's paper
presented at the first "Special Ceramics Symposium" in 1959.[4]
"When looking for materials which will be electrical insulators,
both electronic and ionic conduction must be avoided." This suggests

that the investigation should be restricted to compounds containing elements of Groups V and VI. Also, since increased atomic weight gives increased ionic polarizability and increased conduction, attention should be confined to materials of low atomic weight. In this category we have the Group III-V compounds, which resemble the Group IV elements in crystal structure and are also semiconductors but have much larger gaps between the filled and empty allowed energy levels and show even less intrinsic conduction. Also in this category lie the oxides and sulphides, which are well known as good insulators, and the nitrides and phosphides, most of which have received little attention and some of which have not yet been prepared.

These nitrides and phosphides are now being studied at the British Ceramic Research Association. Silicon nitride, which has already been mentioned on account of its hardness and strength, is a very good insulator and so is aluminium nitride, which has the wurtzite structure. Their T_e values (temperature at which the resistivity is 1 Megohm cm) are 1050°C and 650°C respectively. Boron nitride is electrically very different from graphite; the different charges on its boron and nitrogen atoms make it a very good insulator, T_e values as high as 1300°C being obtained for the pure material."

Since this is a Seminar on nitrogen ceramics and not on silicon-nitrogen ceramics, which is the subject of most papers to be given, I will give in my lecture relatively more emphasis to nitrogen ceramics not containing silicon. But first a few general comments on nitrogen ceramics. Notwithstanding the suggestions by the contract sponsors mentioned above, we do live in an oxygen-containing atmosphere and in general the oxide is·thermodynamically more stable than the corresponding nitride and therefore most nitrides will oxidize at high temperatures. The rate of the oxidation depends on the stability and texture of the oxide film formed and is low when a coherent SiO_2 film forms, as with silicon nitride, and high with BN where the B_2O_3 layer has a low melting point and high volatility.

There is of course the possibility of a non-oxidizing environment, e.g. high vacuum in electronic devices or an inert atmosphere in metallurgical applications; here the limiting factor will be dissociation, which is sometimes more severe than with the corresponding oxide, e.g. the equilibrium pressure of 1 atm. N_2 is reached with Si_3N_4 at about 1900°C. Thus silicon nitride has its limitations in very high temperature applications, particularly when the partial pressure of nitrogen is low and where there is a fast gas flow (e.g. in rocket nozzles).

2. BORON NITRIDE

This material attracted me because of the isoelectronic relationship between B-N and C-C. The hexagonal form of BN is sometimes described as "white graphite" and it held out a promise of being an easily machinable high temperature electrical insulator. We set out trying the standard ceramic technology, i.e. first prepare a powder which would then be sintered to a monolithic body. There are a great number of ways in which BN powder can be produced; they are all typical of the production of nitride powders in general. I mention only a few:

> (1) nitridation of the element (B);
> (2) reduction of the oxide (B_2O_3) by carbon and nitridation;
> (3) reaction of boric oxide or acid with NH_3;
> (4) vapour reaction of halide (BCl_3) with ammonia.

(1) requires an expensive starting material and unless the powder is very fine the reaction is slow and usually incomplete. With (2) the problem is to obtain a product free of C. We mostly used method (3). Because of the low melting point of B_2O_3 one has to use an inert filler, (e.g. $Ca_3(PO_4)_2$ or $CaCO_3$) to prevent fusion; the filler is removed by acid treatment and any unreacted B_2O_3 by extraction with hot alcohol. Once one has produced some boron nitride this can be used as fusion-preventing filler which does not require to be removed. Similarly, one can use AlN or Si_3N_4 as a filler when nitriding Al or Si above their melting point.

When one tries to sinter the BN powder produced as above no appreciable sintering takes place and one has to resort to hot-pressing as a method of consolidation. As so often, one finds that the purer the BN powder, i.e. the closer its nitrogen content approaches the theoretical value (56.5%), the lower is the hot-pressed density. No thorough study has been carried out on this but it appears that low temperature nitrided powder contains an appreciable amount of oxygen, probably in the form of a poorly crystallized boron oxy- (or hydroxy-) nitride, which is insoluble in alcohol but aids densification by hot pressing. I believe that the presence of amorphous phases is not uncommon in other nitrogen ceramics; these are often neglected as they are not shown up by X-ray diffraction and there is no accurate method for determining their content. Similarly there is no rapid and convenient method of determining chemically the nitrogen and oxygen content. I am looking forward to hearing the paper describing the formation of BN from CaB_6 and wonder what its purity and hot pressing characteristics are.

Method (4), the vapour reaction, gives the purest and best crystallized powder (thin plates) if the reaction takes place at a very high temperature. Such a powder will not densify by pressure

sintering but will provide the ideal powder medium for very high
temperature (e.g. for quasi-isostatic hot pressing) and is an
excellent material for a die wash in hot pressing.

The vapour method, when associated with a hot graphite substrate,
provides a means of coating graphite with an electrically insulating
layer. If built up to substantial thickness and finally freed from
the graphite this method can be used to manufacture crucibles for
use in crystal growing and substrates for microwave circuitry.

Pyrolytic, or C.V.D. (chemically vapour deposited) boron
nitride, like pyrolytic graphite, is a highly anisotropic material.
The hexagonal layers of the lattice are parallel to the substrate,
resulting in a high thermal conductivity within the layer but 100
times less in the direction through the thickness of a deposit.[5]

The B-N/C-C analogy always made one speculate whether BN with
the diamond-type cubic structure exists. I remember the considerable
excitement I felt when, shortly after we started work on the hexagonal type of BN, Wentorf[6] succeeded in producing the cubic allotrope called "Borozon". This, like the formation of synthetic
diamond, required the very high temperature and pressure available
in the "Belt" apparatus of the G.E. Company. The material was
heralded under the fanfares of "harder than diamond", a claim
which I believe was never substantiated. However, Borazon is being
used today as an abrasive[7] in preference to diamond for the dry
grinding of certain steels and there are numerous Russian references
to its being used as a machining tool, either as a single point
tool or in sintered form. Sintering to density 3.35 g/cm^3 (theoretical 3.45) can be performed at pressures of 60-75 kbar at 1800°C
in graphite capsules.

There is an occasional reference to "isotropic pyrolytic boron
nitride" first announced by the Raytheon Co.[8] This is rather a
misnomer as it refers to BN of the hexagonal structure which has
been deposited in a manner leading to a randomly oriented, rather
poorly crystallized material. Although of considerably lower
density than theoretical this material was said to be vacuum-tight,
isotropic and have a very high thermal conductivity. The latter
has, however, not been substantiated.[5]

Hot-pressed hexagonal boron nitride has not yet found any
very large-scale application. It is attractive as an electrical
insulator because of the ease with which it can be machined, and
for this reason is a most useful material to have around in laboratories engaged on high temperature research. It does not
readily react with graphite at very high temperatures and thus is
a useful electrical insulator in contact with graphite heaters.
It is also used at more moderate temperatures as an electrical
insulator of high thermal conductivity in high voltage semiconductor

devices. An interesting point to mention is that its thermal
conductivity does not drop as rapidly with increasing temperature
as other electrical insulators (e.g. BeO). It has also - even at
very low density in the form of a pressed compact - a remarkably
high electrical breakdown voltage, which I have ascribed to a very
long breakdown path resulting from its platy powder particles. It
is inert with respect to many highly reactive chemical substances,
e.g. molten cryolite.

Nobody who handles boron nitride can fail to notice its talc-
like appearance and apparent low friction, which again results
from its layer structure. The commercial literature suggests its
use as a bearing material; however, because of its softness it is
only usable at the very lowest contact pressures. When drawing
on the analogy with graphite it must also be remembered that the
low friction of graphite is due to an adsorbed layer of moisture;
graphite bearings are known to seize in a high vacuum. It has
been shown[9] that, whilst water vapour does not act in the same way
with boron nitride, certain organic vapours lower the friction of
degassed boron nitride, e.g. silicone fluids have been used in
boron nitride lubricating suspensions.

What do I see as the main problems of boron nitride? As far
as the hexagonal allotrope is concerned nobody has succeeded in
making a high purity and high density body for a reasonable price.
Hot pressing as a production route is not as objectionable with
boron nitride as with other ceramics because of its easy machin-
ability. Production by pyrolysis is very expensive and the very
marked anisotropy of many properties of the product is limiting.

There was a time when I, rather naively, extended the BN/CC
analogy to a consideration of organic materials, thinking that
there may be a BN analogue for all carbon-based polymers and that
this might provide a way of obtaining materials with the mechanical
properties of organic polymers but with a higher thermal stability.
Within limitations this approach works; there is a BN analogue of
the benzene ring called "borozole" made up of a B-N hexagonal ring.
In our early work on the pyrolytic deposition of boron nitride we
used TCB (trichloroborazole) as the source of B and N, just as we
use now MTS (methyltrichlorosilane) for the deposition of silicon
carbide, believing then that there was a great virtue in using a
precursor which contains the required elements in the correct
proportion.

3. ALUMINIUM NITRIDE

If we recall to ourselves the middle section of the top of
the Periodic Table, i.e.

```
...  B   C   N  ...
...  Al  Si  P  ...
```

we see that there is a close relation between the III-V compounds
AlN and BP, and SiC. We looked at both. Because of the close
relation with SiC one would expect these materials to be of
interest as refractories, abrasives or high energy gap semiconductors. I will only mention AlN.

We used as preparation route the nitridation of Al powder,
probably not a very good route because of the low melting point of
aluminium ($\sim 660°C$). Although this must be almost 20 years ago I
remember the rather fascinating reaction products we observed under
the microscope. They often consisted of hollow, almost spherical
shells and if one looked closely there was a small hole in the
shell. Some of the larger broken shells looked just like the egg
shells left by a hatched pullet. We concluded that the aluminium
had melted and the skin was nitrided - but where was the chicken,
or the egg? One must assume that either the aluminium had run
out as a liquid or evaporated as a gas. But why? We did not
succeed in sintering AlN to a reasonably high density but were
able to hot-press it. Nowadays various ways of sintering AlN to
almost theoretical density are known. The following approaches
have been successful: reaction-sintering of AlN + Al in a nitriding atmosphere; grinding to micron and submicron particle size;
additions of Ni, Co or Fe or Y_2O_3 to AlN.

Electrically conducting boats are made commercially by hot
pressing a mixture of BN, AlN and TiB_2. They are used for the
continuous evaporation of Al. The components are adjusted to get
such a resistivity that they can be heated by passing a suitable
current through them.

In spite of quite useful properties and a very good high
temperature stability I am not aware of any large scale commercial
application using substantially pure aluminium nitride. I believe
that this is because the properties of AlN are so much like those
of SiC, except perhaps for some specific chemical compatibility,
and SiC powder is produced mostly for use as an abrasive in very
large quantities and is thus available at very low cost.

4.   INTERSTITIAL NITRIDES AND OTHERS

These are usually considered as "hard metals". They have the
cubic NaCl-type structure, are very hard and have melting points
in the range 2000-3000°C, and are closely related to the interstitial carbides. The following mononitrides belong to this category:
ScN, TiN, VN, ZrN, CbN, ThN, and UN. Their electrical conductivity
is metallic and their thermal conductivity has an electronic

contribution.

Despite their excellent high temperature properties, good corrosion resistance and high hardness, the interstitial nitrides have not found any extensive use. Thin films of TiN on tungsten carbide tool tips have been found advantageous for certain machining operations and are being used increasingly; also thin films of reactively sputtered nitrides are of interest in electronic circuits.

Because of their possible use as nuclear fuels the nitrides of uranium and plutonium have been fairly extensively investigated. UN looks attractive because of its high thermal conductivity, high fissile atom density and dimensional stability at high temperatures.

A new type of reactor employing UN has recently been proposed. It is called the Actinide Nitride-Fuelled (ANF) reactor.[10] It is suggested that it can recycle nuclear wastes, thus reducing the present problems of environmental safety and radioactive waste storage, and it is self-correcting. Its principle is as follows:

A graphite-lined core contains uranium dissolved in molten tin under light nitrogen pressure. UN will precipitate and settle on the bottom of the reactor vessel. When the amount of UN reaches a critical mass fission begins and the reactor is in operation. However, if the temperature rises above the planned operational level UN will dissociate, putting uranium back into solution and thus decreasing the size of the UN mass. This equilibrium is said to provide the ANF reactor automatically with stability against runaway conditions. It is also said that spent fuel can be reprocessed and new fuel added without closing down the reactor, thus substantially raising its economic performance.

One of the nitrides that always intrigued me - though hardly falling within my definition of nitrogen ceramics - is polysulphur nitride $(SN)_x$. This polymer has been known since 1920 but only recently have experiments indicated that it is an anisotropic metal with a superconducting transition temperature near 0.3K. Crystals of the polymer are typically needle-shaped, 5 to 10mm long and 0.5mm wide, with a distorted hexagonal cross-section. Some crystals have optically-smooth surfaces and have been used for specular reflectance studies. SEM investigation has shown that all of the crystals are composed of fine bundles of fibers running parallel to the chain or needle direction. The typical fibre structure observed indicates that the diameter of the individual fibres is sometimes less than 200 Å.

5.   COMMENTS ON PRODUCTION METHODS OF POWDERS AND COMPONENTS

Before I address myself specifically to silicon-containing
ceramics I should like to make some general comments regarding
the production of powder and components of nitrides in general.

I have already indicated when discussing boron nitride that
a variety of methods is available. It seems to me that workers
have in general stuck to one particular method for each material,
e.g. formation of Si_3N_4 by nitridation of Si, formation of BN by
the nitridation of the oxide. It is for this reason that I am
glad to see the paper by Prof.Cutler on the production of Si_3N_4
powder from silica. I would be interested to hear what happens
when you start with a fine sand, Aerosil or ground vitreous silica.
In this connection, let me say that I deplore that people rarely
quote their failures; by not doing so many people will repeat
the unsuccessful routes. I think unsuccessful attempts are worth
quoting, even if no explanation is at hand, but particularly so if
there is an explanation. Let me also say that young people
neglect too much work done more than, say, 10 years ago by simply
saying it's "ancient", and furthermore that the most surprising
observations are well worth taking note of, even if there is no
plausible explanation at the time. Observations are usually
correct, explanations rarely so. I have always thought that the
observation of Evans & Chatterji[12] that the rate of nitridation of
silicon in argon containing 0.2 v.% N_2 as impurity was considerably
faster and more uniform than in atmospheric nitrogen was a very
important one. It is these rather unexpected observations which
need following up rather than dismissing with the word "rubbish".
It is only now, after almost 20 years, that an explanation in
terms of nucleation rate is plausible.

To return to the production of powders. I would like to see
many more chemical approaches, both of the more refined "bucket
chemistry" type and high temperature methods like formation of
powders in a plasma. The behaviour of these powders, when well
characterized, in subsequent fabrication might teach us a lot
more than using the same powder with a pinch of this or that added.

6.   REACTION SINTERED SILICON NITRIDE (RSSN)

I have already used the term reaction sintering in connection
with forming an AlN body. I prefer the term reaction sintering to
reaction bonding - although I may be in a minority judging from
the titles by the various authors - because it fits in with my
"definition" of ceramics. I prefer to use the term bonding for
the joining of two dissimilar materials, e.g. metal/ceramic bonding.
I came across reaction sintered silicon nitride for the first
time - and so probably did many other people - when trying to form

Si_3N_4 powder by nitriding silicon powder. In spite of packing the powder very loosely the result was a quite coherent "mono-lithic" replica of the powder volume. Thus the powder sintered and it did this by reaction from Si to Si_3N_4 when it gained by about 60% (theoretical 66%) in weight.

I was immediately fascinated by it and came to the conclusion that this was the best way of producing ceramic components and my thoughts suggested that alumina components should be made by compacting and oxidizing aluminium powder, titania components by oxidizing titanium, etc. Most of my attempts turned out complete failures; the pressed metal compacts expanded when being oxidized and developed little strength. My only success was reaction-sintered SiC, by attempting to treat a mixture of SiC + C in a vapour of silicon.[13] The *observation* of a strong impervious body of self-bonded SiC with some free Si was *correct*; the *explanation* was *wrong* - it was not silicon vapour but liquid silicon which infiltrated the compact by capillary reaction.[14] When mentioning self-bonded silicon carbide it might be worth making another point: I used first lampblack as the source of carbon but the method didn't work. The sample expanded, was weak and contained a lot of free silicon. Change of the source of carbon to a colloidal type of graphite did the trick and it has worked well ever since. The lesson to be learned: different powders can give very different results. The results can probably be explained but often one doesn't bother but pursues the success, and rightly so.

After a fairly long infatuation with reaction-sintering I have come down to earth and wonder whether it is such a good process after all. What are its disadvantages?

(1) If one of the reactants is a gas the product will necessarily be a porous and thus relatively weak material.

(2) The combination of reaction and sintering in one process leads to difficulties (exothermic heat of reaction, thickness limitation).

(3) Any flaws present in the unfired body will not heal during sintering when there is no shrinkage.

The only advantage is that the dimensional change during densifi-cation is very small but this I consider now almost as an accident - but more of this later.

In 1959 at the first B.Ceram.R.A. Special Ceramics Symposium Mr.Parr and his colleagues from the Admiralty Materials Labs. gave their first paper on Silicon Nitride, an Engineering Ceramic;[15]

I gave, in an Appendix, a brief survey of our developments at the B.Ceram.R.A. When I look at the photographs of components which were then produced and what is being done now, one wonders what has happened in the intervening 17 years. Four years before 1959 Mr.Parr and I had probably never heard of silicon nitride and of reaction sintering.

Let me give now my personal views why developments have moved much more slowly than might have been expected.

6.1. Silicon nitride has been oversold.

It was the one and only engineering ceramic. It was said to be cheap because Si was the most abundant element of the earth's crust and N the most abundant element in our atmosphere. It was not said that a lot of energy was required to extract the silicon from the minerals and the nitrogen from the atmosphere and that even more energy was required to put the two together. I am therefore very interested in Dr.Davidge's and Mr.Osborne's cost and energetics analysis. A quite important person in the field, when asked what he considered to be the potential of reaction sintered silicon nitride, replied: "You might just as well ask what is the potential of steel". When an engineer tried it in an application and the component fractured, then he was blamed for not knowing how to design with brittle materials. I was asked by a member of the U.S. Defence Department after all the praises of Si_3N_4 had been sung to him whether I believed all that he had been told, and I replied: "Of course not, but even if only a quarter is true it is still a material well worth considering". For hot applications its low thermal expansion and high creep resistance leaves it almost without a competitor. I think both varieties, RSSN and HPSN, are here to stay.

6.2. The technical problems of manufacture have been underestimated.

There was a time when it was considered that every engineering firm would have a little shop tucked away in an odd corner where somebody would prepare compacted silicon powder "billets"; these would then be machined in the machine shop and sent back for nitridation. I have been saying for many years: "Any fool can make silicon nitride but to produce silicon nitride consistently with properties close to the best obtainable is probably the most difficult job in ceramics". I have already stated that flaws will not heal; similarly any low density area in a green compact will remain low density after sintering. This means that the shaping process must provide a perfect specimen. Let me state further difficulties all connected with the reaction sintering process:

(a) The reaction sintering process is extremely sensitive to impurities in the raw materials. The catalytic effect of parts per million of Fe on the kinetics of nitridation is well known.

At the last Science of Ceramics meeting at Cambridge, last September, Dr.Atkinson presented a joint paper with Dr.Moulson on the reaction between silicon powder and nitrogen.[16] I got the impression that for the first time somebody understood the mechanism, as far as pure silicon was concerned, and that they were well on the way to understanding the effect of iron. I am therefore looking forward to Dr.Moulson's paper on the effect of iron impurities.

A lot of experiments have been performed and sometimes heated arguments have been had on the issue whether α and β are two allotropes if Si_3N_4 or whether $\alpha-Si_3N_4$ is an oxynitride. My view on this has been that α and β are obviously very closely related structures which are energetically not very different; therefore small impurities can have a considerable effect on which crystal structure forms. There is evidence that in the presence of oxygen $\alpha-Si_3N_4$ is favoured, whilst Al favours the β structure. What is perhaps far more important to know is whether textural differences due to $\alpha-Si_3N_4$ and $\beta-Si_3N_4$ exist and do they affect the properties. Are α and β Si_3N_4 formed by a different reaction mechanism, e.g. reaction of nitrogen with a vapour (SiO) or with solid or molten silicon?

(b) I have referred to the negligible change in dimension during sintering as an accident. This means that the pores in a silicon compact are being filled with silicon nitride. I would have thought that this must therefore involve a vapour transport reaction but rarely does the work on kinetics take this into account. (I find all the attempts of curve fitting based on idealized conditions irrelevant. Similarly, I find the attempts of which physical chemists are so fond, of determining activation energies of a temperature-dependent process, a complete waste of time. Very frequently they are based on the temperature dependence over a very short range, and drawing a straight line through three points on a log vs. $\frac{1}{T}$ plot is not a great achievement. The value of the activation energy derived does not help in proposing a mechanism for a working hypothesis).

(c) The reaction silicon plus nitrogen is strongly exothermic and the local temperature rises can be considerable, leading to melting or sintering of the silicon, thus changing drastically the available surface area and rate of reaction. In this

connection I have frequently put forward the view that the
widely practiced 2 or 3-step nitridation schedule is almost
the worst possible choice. At the beginning of a step there
is very rapid take-up of nitrogen and the danger of exothermic
melting is considerable; for the rest of the time, take-up of
nitrogen is very slow and maintaining the temperature constant
seems a waste of time and energy. I would have thought that
the optimum temperature/time schedule would be a continuous
rise in temperature at a rate which keeps the exothermic temperature rise inside the compact within safe limits.

This brings me to another point – the often complained-about
lack of reproducibility. The variable furnace loading in
manufacture must present considerable problems – it may vary
in total weight, it may be part powder, part compacts, and
the compacts will vary in size. I believe that reproducibility
will be dramatically improved only when uniform loads will
go at a constant rate through the nitriding furnace.

(d) Thinking of the problems of large-scale production, I don't
quite know what to make of the equilibrium partial pressures
of the order of 10^{-20} atm. which are being quoted from the
thermodynamic data. It is obviously impossible to control
impurities in a production furnace, which may contain ordinary
refractories, to that order. However, any of the reactions
predicted by thermodynamics must be limited by the total
amount of impurities available.

(e) I believe that in the early days there was a gross neglect of
ceramic technology. All shaping applications are well known
ceramic processes, e.g. slip casting, die and isostatic
pressing, extrusion, flame spraying, plastic forming (injection
moulding) machining of unfired or prefired compacts. A large
number of patent specifications on silicon nitride cover well-
known ceramic processes and apply them to Si powder. Particle
packing and therefore particle size distribution has been a
rather neglected subject.

6.3. Applications are coming along much more slowly than was expected

With very few exceptions there is no ready "replacement"
market for silicon nitride. Because of its price it must do something substantially better than existing materials can do. This may
mean a longer component life with consequent low replacement rate,
or it must do something which could not be done before. This sort
of innovation takes time and if one thinks about an entirely new
type of engine with a highly uprated performance the development
cost is very high and few firms can afford that sort of expenditure;
thus government support is needed. This is now forthcoming in a
large measure in the U.S. and in Germany, which is bound to lead

to a shortening of the development process.

7.   HOT-PRESSED SILICON NITRIDE (HPSN)

When it comes to trying to find suitable additives to aid densification by pressure sintering the "undignified" approach of adding a bit of this and that seems the most successful because the process is not well understood. There was a time when one thought that a structural transformation was an important link in the densification mechanism; one starts with a powder high in $\alpha-Si_3N_4$ content and finishes with predominantly $\beta-Si_3N_4$. In SiC it is the other way round; one goes from the cubic $\beta-SiC$ to hexagonal $\alpha-SiC$. But in a fairly recent but rather neglected paper[17] it was shown that the $\alpha-\beta$ transformation had nothing to do with densification but was important for the development of strength.

The low creep resistance at high temperatures has provided considerable problems with hot-pressed materials. It always seemed to me that densification by hot-pressing is a creep or deformation process and that any additives which aid densification will lower the high temperature creep resistance. I can only think of two ways in which this can be overcome:

(i)   by raising the hot-pressing temperature and thus widening the gap between processing and utilization temperature - but here one runs immediately into a problem because of dissociation of silicon nitride. (Perhaps some of the more complex oxy-nitrides have a higher dissociation temperature);

(ii)   by using a glass which has a low viscosity during hot-pressing but which devitrifies at a certain holding temperature.

Hot-pressing is an expensive process; it requires an expensive powder and components will require at least some if not a considerable amount of diamond machining. Nevertheless one has to keep in mind that HPSN is probably the strongest available ceramic material and as such will always find some application - the most likely field being in metal forming.

8.   CONCLUDING REMARKS

At this rather late stage of preparing the paper I thought it would be a good idea if I looked at all the abstracts of the papers to be presented and made some comments. The first thing immediately obvious is that most papers are concerned with Si- and N-containing ceramics. One is therefore tempted to ask: why? Is it just fashionable to do research on silicon nitride and related substances?

Whilst there may be some of this I am convinced that, apart from SiC and perhaps AlN, the silicon-based nitrogen ceramics are the most likely materials that will allow engines to operate at high temperatures, thus being more efficient, saving fuel and decreasing the toxicity of exhaust gases.

The other immediate response is: How can this large activity of very many research laboratories be absorbed by the few firms who are producing silicon nitride? Work on oxynitrides has shown us that nitrogen can be introduced into silicate structures; this opens up a vast field of new materials of considerable complexity. It would seem to me to be more profitable to elucidate some basic principles of atomic substitutions than to investigate more and more complex systems. Thus I would like to know what does increasing z in $Si_{6-z}Al_zN_{8-z}O_z$ do to the properties important for a high temperature engineering ceramic, i.e. thermal and chemical stability, thermal expansion and conductivity, etc? The possibility of component manufacture from complex oxynitrides by sintering in nitrogen or controlled atmospheres at $1750^{\circ}C$ is certainly an exciting prospect. However, there are no furnaces available to do this, certainly not on a large scale. Who will supply the considerable funds that are necessary, and who will be capable of selecting which of the lines that are being researched now should be developed onto a production scale?

I have raised many questions in this Introductory Lecture and I hope to receive replies to some of them during the next fortnight - but I am equally certain that at the end of this Seminar more new unanswered questions will have been posed. This is progress.

REFERENCES

1.   F.A.Bannister, Mineral.Mag. 26, 36, 1941.
2.   C.A.Anderson, K.Keil & B.Mason, Science, 146, 256, 1964.
3.   P.Popper, Non-oxide Ceramic Dielectrics, Progress in Dielectrice, Vol.1, Ed.J.B.Birks & J.H.Schulman, Heywood & Co., 1959.
4.   S.N.Ruddlesden, Relations Between Properties & Structures, Special Ceramics, p.3., Ed.P.Popper, Heywood & Co., 1960.
5.   A.Simpson & A.D.Stuckes, J.Phys.C: Solid St.Phys. 4, 1710, 1971, J.Phys.D: Appl.Phys. 9, 621, 1976.
6.   R.H.Wentorf, J.Chem.Phys. 26, 956, 1957.
7.   D.M.Busch & P.J.Strachan, Eng.Dig. 36(3), 39, 1975.
8.   S.R.Steele & J.Pappis, 8th Conf. on Tube Techniques, I.E.E.E., 1966.
9.   G.W.Rowe, Wear, 3, 274, 1960.
10. R.R.Jones, Industrial Res. 18(5), 55, 1976.
11. R.H.Geiss & H.Arnal, Industrial Res. 18(5), 83, 1976.
12. J.W.Evans & S.K.Chatterji, J.Phys.Chem. 62, 1064, 1958.
13. P.Popper, Preparation of Dense Self-bonded Silicon Carbide,

Special Ceramics, p.209, Ed.P.Popper, Heywood & Co., 1960.

14. C.W.Forrest, P.Kennedy & J.V.Shennan, Special Ceramics 5, Ed.P.Popper, p.99, B.Ceram.R.A., 1972.

15. N.L.Parr, G.F.Martin & E.R.W.May, Preparation, Micro-structure, and Mechanical Properties of Silicon Nitride, Special Ceramics, p.102, Ed.P.Popper, Heywood & Co., 1960.

16. A.Atkinson & A.J.Moulson, Science of Ceramics 8, p.111, British Ceramic Society, 1976.

17. R.J.Weston & T.G.Carruthers, Proc.Brit.Ceram.Soc. <u>22</u>, 197, 1973.

ACKNOWLEDGMENTS

The author would like to thank Mr.A.Dinsdale, O.B.E., Director of Research, B.Ceram.R.A. for permission to publish this paper.

Section A

BONDING AND PHASE RELATIONSHIPS

BONDING IN NITROGEN CERAMICS

P.E.D. Morgan

The Franklin Institute Research Laboratories
Philadelphia, Pa. 19103.

ABSTRACT. Nitrides auger well as systems for fundamental bonding studies. In relating the properties of nitrides to those of oxides it is suggested that the latter are more covalent than formerly appreciated. It is not anticipated that nitrides will violate the tried and tested general principles on which oxide bonding and crystal studies are so well founded.

1. INTRODUCTION

Bonding is extremely complex in detail and I want, therefore, to concentrate on relatively simple chemical and structural evidence to discuss bonding in nitrides. We currently seem to be at the advent of a major reappraisal of bonding engendered by the recent advances in experimental techniques for looking more directly at bonding electrons; photo electron spectroscopy, X-ray emission shifts, K absorption edge studies and so on. This is encouraging more MO calculations to be attempted but, as yet, the results are often inconclusive.

Most structure types considered typical of oxides are represented also in nitrides thus: wurtzite – AlN & $BeSiN_2$, rock salt YN, Ba_2NF, Mg_3NF_3, distorted rutile TaON, fluorite UN_2, "spinel Al_3O_3N", intermediate wurtzite – rock salt Mg_2NF, phenacite – $\beta-Si_3N_4$ etc.

Other nitrides are more alloy-like, e.g. TaN, Th_3N_4 & Th_2N_2O,

---

*This work has been supported in part by U.S. Office of Naval Res.

UVN$_2$ (isotypic with UMoC$_2$). More complex types are known, e.g. hexagonal Li$_2$ZrN$_2$, Li$_2$ThN$_2$ and Li$_2$UN$_2$ similar to the β–Mg$_3$As$_2$ type and the various silicate analogues, melilite, apatite, yttrium aluminate, as well as unique structures such as α–Si$_3$N$_4$, Si$_2$N$_2$O, etc.

As comparison with the oxides is inevitable, we must examine (very cursorily) what is known about the bonding in these before proceeding. As the electronegativity of nitrogen is substantially lower than oxygen and very considerably lower than fluorine it might seem at first sight surprising that generally similar structures are formed. Two ideas suggest themselves if nitrides are not highly ionic (as we will discuss): one, that oxides are not as ionic as formerly thought and, two, maybe the structure formed is not as dependent on the relative ionic/covalent ratio as was originally believed.

Classically the oxides were treated as if they were almost entirely ionic but, recently, diverse evidence has appeared to suggest otherwise. This is not to suggest, of course, that, empirically, the concepts of ionic radius and radius ratios, "electrostatic balance" and field structure maps do not work - manifestly they do. However, measurements of radii of "ions" by precise X-ray techniques do not give the same answers as crystallographically estimated radii and the use of "x-ray radii" does not lead to as good agreement with Pauling's first rule. Moreover an elegant study of bond length–bond strength relationships (1) (using Pauling's second rule) leads to smooth curves from oxides considered to be largely ionic through to oxides generally considered to be much more covalent. This confirms what Pauling suspected much earlier (2) that, while the nature of the second rule was originally thought to be electrostatic (3), it works as well when valence rather than charge is considered, i.e. a "valence balance rule". Moreover, as Johnson has discussed (4), there are features of the Born–Haber–Mayer cycle, that are self-confirming, particularly to do with the choice of the bond length that must be empirically measured and inserted in the calculation, that itself changes in a complex way with the type of bonding.

The relative ionic/covalent content of a bond, measured by the electronegativity difference (an old concept) between the atoms concerned, was first quantified by Pauling (5). The excess energy of a bond between unlike atoms, over and above the mean (arithmetic, or better, geometric), was ascribed to the extra resonant energy of ionic forms. The ionic energy contribution was then scaleable as the electronegativity difference between the atoms raised to the second power, i.e.:

$$30(x_A - x_B)^2 = E(A-B) - [E(A-A).E(B-B)]^{\frac{1}{2}}$$

Electronegativities were assigned (H = 2.1 arbitrarily), and remarkable agreements between different bonds obtained. Even now a good first estimate of bond energy for an unknown bond can be obtained by this method.

Pauling carefully pointed out that only single bonds could be compared with each other (other rules covered the comparison of multiple bonds). Two problems immediately appeared: it was never clear how the electronegativity difference in an isolated molecular single bond, and the associated magnitude of the ionic contribution, could be transferred to multiple bonds in the solid state. In the 1939 edition of "The Nature of the Chemical Bond" (5) lithium iodide was treated by assuming that the covalent part of the Li-F isolated bond, 57%, resonates between the six equivalent octahedral rock salt positions and that the balance for each bond is ionic making each bond 90% ionic. Later it became apparent from model systems that charge is "smeared out" and Pauling postulated the "Principle of Electroneutrality" (6). For solid CsF he now allowed that each bond have 9% covalent character (as found for the isolated gaseous molecule) and added these up (instead of dividing among the bonds as earlier) to give a total covalency 54% and a partial change on Cs and F of approximately +1/2 and -1/2.

Later editions of "The Nature of the Chemical Bond", e.g. 1960 (2), contain the Electroneutrality Principle yet still give estimates of bonds in solids as highly ionic, i.e. Si-O, 50% (charge on Si would be +2), BeO, 60% (Be + 1.2) B-O, 44% (B + 1.3) violating the principle.

The problem of translating the ionic contribution in an isolated bond to the case of a solid is now seen to be due either to the change of hybridisation on going from the simple molecule to the solid (i.e. one-ended sp^3 or similar → multi-centered p^3 for rock salt) or to the effects of lone pairs. The effect of lone pairs may be seen in both the isolated molecule and the solid (sometimes in different ways) and are very important in many materials of interest to ceramists, mineralogists and glass technologists. In many isolated molecules such as BF_3, $BeCl_2$, $SiCl_4$ etc. the assumption that the bonds are single is incorrect; many chemical indications (bond length, bond energies, acid-base reactions, etc.) now make it clear that $(p \rightarrow p)\pi$ or $(p \rightarrow d)\pi$ bonds may be present in these molecules. Therefore, Pauling's electronegativity principle cannot be applied in its simple form to estimate the ionic content as some of the extra energy, believed originally to come from ionic resonance in fact comes from multiple bonding and the electropositivity of the cation is overestimated (especially apparently for Be, B and Si) (7).

Not only is the ionic contribution exaggerated in the isolated molecule but, in the solid (i.e. BeO, B_2O_3 and borates, SiO_2 and silicates), the same lone pair donation may further reduce the partial charge at the cation. Pauling was not (characteristically) unaware of this inconsistency and did propose (8) (for example) $(p \rightarrow d)\pi$ back bonding from oxygen to silicon as a mechanism of reducing the formal + 2 charge at silicon in silicates.

Pauling's electroneutrality principle does seem to be substantially correct so that the net charge on cations in many ceramics is quite small especially for silicon, boron and beryllium. Structurally, as we know a great deal about these elements in important ceramic materials, I will concentrate on them.

2. BORON COMPOUNDS

In some ways boron is an example of the most extreme case where the potential for the effects of π bonding was originally neglected. Good reviews exist (9-11) and I don't wish to duplicate these but only to add some pertinent data of use in the solid state. A study of bond lengths and energies in boron compounds is quite revealing. The desire for boron to use the empty p orbital, even when lone pairs are not available at the ligand, is clearly shown by the formation of multi-centered bonds such as in diborane, B_2H_6. It is reasonable to suspect that the net effect of this is to bring extra negative charge into the region of the boron atoms thus reducing the overall ionicity.

Following an idea of Johnson (1) it is interesting to compare the bond energies* E(B-X), where X does, or does not, have lone pair orbitals, with similar carbon and nitrogen bonds: for X without lone pairs, i.e. H, C, or B: B-H, 85, C-H, 99.5, N-H, 84.3; B-C 80, C-C 82.6, N-C 54.7; B-B, 74, C-B, 88, N-B, 40, kcal mole^{-1}; the carbon bonds are the strongest. If the ligand X contains a lone pair, the order changes: B-N, 82, C-N, 73, N-N, 38; B-O, 125, C-O, 86, N-O, 39; B-F, 154, C-F, 117, N-F, 67; B-Cl, 105, C-Cl, 78, N-Cl, 45, B-Br, 88, C-Br, 65, etc. so that the boron bonds become considerably stronger than those of carbon or nitrogen. The simple explanation is that $(p \rightarrow p)\pi$ bonding between the lone pair p orbitals on X and the empty p orbital of boron strengthens the bond which then has multiple bond character. The change in E(B-C) with groups on the carbon capable of more or less π-donation to the empty p orbital of boron has also been studied (14,15). So we have (with the π-contribution following the oblique stroke, all values being in kcal mole^{-1}): ϕBCl_2 116/28, $\phi_2 BCl$ 112/24, ϕBBr_2 115/27, $\phi_2 BBr$ 110/22, $\phi_3 B$ 106/18 and $(C_6 H_{11})_3 B$ 88/0. Recently the same workers have compared E(B-C) (16) for o- and p-substituted tolyldichloroboranes; the o- (116 kcal mole^{-1}) produces a weaker bond than the p-substitution (125 kcal mole^{-1}) due to the twisting of the ϕ-B bond because of steric interference in the o-case reducing the π component.

---

*Unless otherwise noted, bond energies have been calculated from data in (13) or from (9-11).

A study of bond lengths is equally revealing[*]. In solids bond lengths usually change with co-ordination (this leads to tables of "ionic radii" with adjustments for co-ordination). This does not mean that bond lengths change <u>because</u> of coordination change but rather that both bond lengths and coordination are functions of the hybridisation in the orbitals being used. The orbitals used in higher coordinations, i.e. double ended p^3 for rock salt or sp^3d^2 etc. are larger and more diffuse than the orbitals of tetrahedral bonding, sp^3, or triangular sp^2 etc.

That this is so can be seen in some simple cases, e.g. N-H in NH_3 is 1.015Å, in $(NH_4)^+$ ~1.035Å, the coordination changes, the hybridisation does not. The opposite may be seen for the tetrahedral carbon atom attached to another carbon atom (18), i.e.

$$-\overset{|}{\underset{|}{C}}-\overset{|}{\underset{|}{C}}\underset{\diagdown}{\diagup} \quad , \quad -\overset{|}{\underset{|}{C}}-\underset{\diagdown}{\overset{|}{C}}\diagdown \quad , \quad -\overset{|}{\underset{|}{C}}-C\equiv$$

the C-C lengths are 1.542, 1.501, 1.459Å respectively as the change from $sp^3 \rightarrow sp^2 \rightarrow sp$ occurs on the attached atom (the hybridisation or coordination at the central carbon atom under consideration does not change). The increasing s character on the attached atom leads to a shortening of the bond even though the coordination on the central atom does not change[**]. In C-N also: sp^3, 1.473Å, sp^2, 1.470Å, sp 1.427Å. For boron too, the coordination may change with little or no change in bond length, e.g. B-C in triangular $(CH_3)_3B$ 1.56Å, in tetrahedral BH_3CO, 1.57Å, tetrahedral $[(CH_3)_2HB]_2$ ~1.59Å. The differences are less than within triangular types, e.g. B-C in $(CH_3)_2BF$, 1.55Å, in CH_3BF_2, 1.60Å. If the change with coordination is larger than this order of change we should look for additional effects such as bonding as being responsible for the excess change when lone pairs are present. Where π bonding between carbon and boron is possible, the bond is, of course, considerably shorter, e.g. $\phi-BCl_2$, 1.52Å. This is apparent also in the B-F bond in triangular BF_3, 1.29Å, where π bonds can occur and in tetrahedral, (BF_4) 1.42Å, where no π bond occurs and, in a (presumably) less ionic case, $(CH_3)_2O.BF_3$, 1.43Å. The large change is attributable mainly to π bonding rather than coordination per se. It has been appreciated for many years that boron halogen trigonal bonds are not single but contain an appreciable π component, which was apparent also from studies of the energy of adduct formation in various boron compounds (19). Of more ceramic interest, for the B-O bond, where oxygen has lone pair p orbitals, the large difference in bond length between triangular sp^2 B-O, e.g. 1.36Å in $B(OH)_3$, 1.38Å in $(B_3O_6)^{3-}$ rings, 1.39Å in $B(OCH_3)_3$, and tetrahedral sp^3 B-O, 1.48 ± 0.06Å in a variety of solids, must be due to π bond effects in the former absent in the latter. The analogous potential for π bonding in the B-N case is evidenced by the change in bond

---

[*]Unless otherwise noted, values are from (17).

[**]The older notion of hyper-conjugation is now played down.

length; B-N single bond $NH_3.BF_3$ 1.58Å, $NH_3.BH_3$ 1.56Å, $(NH_3)_2BH_2$
1.60Å (a weak bond), B-N bond order ~1-1/2, as in borazole,
$B_3N_3H_6$, 1.44Å, or trichloroborazole 1.41Å, B=N as in $BH_2=NCl_2$,
1.38Å (20), agreeing nicely with calculated values for $NH_2=BH_2$ (21).
The truly single B-N bond distance is remarkably constant when
very different ligands are attached to the molecule, e.g. for
$(CH_3)_3N.BH_3$, 1.638Å and for $(CH_3)_3N.BF_3$ 1.636Å (22). The $(p{\rightarrow}p)\pi$
bonding effect can be manifest as a bridging effect across ring
systems (23) as in "boratran" where the close boron-nitrogen
approach is associated with an increased bonding energy of the
nitrogen lone pair p as determined by photo-electron spectroscopy.

Pauling calculated a value of 2.0 (H=2.1) for the electronegativity of boron but inevitably π bonding was built into the system
by the use of boron halides and oxides. The ionic resonance was
thereby overestimated and boron made too electropositive. Following a suggestion of Quane (7) for silicon we can recalculate x_B
using only <u>single</u> bonds E(B-C), E(B-H) and E(B-B) (24), now more
accurately assessed, and obtain the value of 2.1($\pm$ 0.05) using
Pauling's GM form. It would be folly to attempt to assess the
relative merits of these values based on Mulliken's method; a
discussion of the problems associated with real orbital electronegativity values has been presented by Hinze and Jaffé (25).
Using the value 2.1, the B-O and B-N isolated <u>single</u> bonds become
39% and 20% ionic instead of 44% and 22% as originally determined
(2). Moreover the partial donation of extra electrons to boron
either by $(p{\rightarrow}p)\pi$ bonding or by lone pair "dative bonding" will
further reduce the ionicity of these bonds (as measured by partial
charge on the atoms) to some much lower value agreeing with the
electroneutrality principle (6).

Direct evidence that the partial charge at boron in boric
oxide, may be small comes from a calculation of E(B-O) (three triangular bonds per boron) using ΔH_f for boric oxide and known (albeit roughly for boron) heats of atomisation of boron and oxygen.
E(B-O) is 125 kcal mole^{-1}; initially, assuming covalent type
bonds, E(B-O) for $(CH_3O)_3B$ is 127 kcal mole^{-1}. There appears to
be no excess ionic crystal energy in B_2O_3 over and above the
slight ionic contribution to each individual bond such as may
exist in trimethylborate where ionic effects between molecules
are insignificant (bond energies of molecules are calculated for
the gaseous phase). The usual amorphous state of boric oxide is
explicable (at least partly) if there is little ionic crystal
energy to be gained from the ordering of small partial charges.
Boric oxide appears to show simple covalent bond additivity just
as found in organic compounds.

3. BERYLLIUM AND ALUMINUM

Beryllium has the same tendency as boron to use empty p orbitals in bonding. In the simplest cases, when lone pairs exist in

the ligand this is achieved as in BeF_2 and BeO by dative bonding.
When lone pairs are not available on the ligand the formation of
multi-centered bonds occurs, thus BeH_2 and $Be(CH_3)_2$ are polymers,
with roughly tetrahedral hybridisation at beryllium. As for boron,
one must conclude that this effect produces a net donation of negative charge to the beryllium atom reducing the partial charge. The
(Be-O) bond in bromellite is then not 60% ionic (2) but some considerably lower value and the Be-N bonds in Be_3N_2 and $BeSiN_2$ (tetrahedral beryllium in each case) would be largely covalent. Aluminum
shows similar behaviour in model compounds, e.g. Al_2Cl_6, $Al_2(CH_3)_6$
etc. Again the net effect of donation to empty p orbitals will be
to reduce the partial positive charge at the aluminum atom. In solid
AlN, the partial charge at Al that might be calculated from 3 single
Al-N σ bonds (using, for example, a Pauling estimate) will be reduced by the development of the fourth "dative bond" whereby the
nitrogen lone pair donates to the unfilled aluminum p orbital, i.e.

AlN is, therefore, considerably more covalent than estimated from
a simple use of Pauling's electronegativity equation.

Occasionally model compounds exist, allowing a fruitful comparison with simple oxides: for example, basic beryllium acetate,
$(CH_3CO_2)_6Be_4O$ has 4 beryllium atoms, at the corners of a tetrahedron,
tetrahedrally coordinated to oxygen, an oxygen atom at the centre of
the tetrahedron coordinates to the 4 beryllium atoms at the corners,
acetate groups join the edges. The Be-O central bonds have lengths
of 1.624Å very close to Be-O in bromellite at 1.652Å; if there was
a large ionic contribution in bromellite, Madelung effects might be
expected to change the bond length by more than is actually found.

4. SILICON AND $(p \rightarrow d)\pi$ BONDING IN THE Si-N AND Si-O BONDS

It has been recognised for a long time (26) that π-bonding was
potentially possible between the p type lone pair on a nitrogen
atom and empty d orbitals on silicon. Early attempts were even
made to quantify the number and magnitude of π-bonds in this type
of situation. It was believed, (26) for example, that for a silicon
atom bonded tetrahedrally by sp^3 bonds it would be possible to form
two strong π-bonds, one weak π-bond and one non-π-bonded bond with
4 tetrahedrally placed suitable substituents. From group theory
(27) it was understood that of the 8 maximum possible π-bonds**,
the central atom, in this case silicon, could form a maximum of
5 π-bonds using all the d orbitals. In general, if the substituents
are identical then the π-bonding would be "smeared out" over all

*All bonds in AlN are, of course, equivalent sp^3 hybrids; the pic-
 ture view is merely a mental aid helping to link bonding in solids
 with the conventions used to depict bonding in simple model compounds.
**If the 4 attached atoms had 2 lone pair p orbitals each, i.e.
 $SiCl_4$ of SiO_2

the bonds by resonance effects. However the $(p \rightarrow d)\pi$-bond remained "elusive" (28).

The pros and cons of $(p \rightarrow d)\pi$-bonding in Si-N is summarized (29) in an excellent review article (104 pages and 556 references). It is interesting that at the time, while nearly all chemical evidence (reaction kinetics, compound stability, dipole moment, ionization constant, bond lengths, bond angles, etc.) could all be rather neatly explained by $(p \rightarrow d)\pi$-bonding, more direct physical evidence (NMR, ESR, IR, UV spectra, etc.) did not give unambiguous support for the theory. No better theory, however, has explained what is observed.

Using the normal covalent radii of silicon and nitrogen the Si-N bond length is calculated to be ~1.81Å. Normally in amine compounds, such as $(CH_3)_3N$, nitrogen is sp^3 coordinated with the nitrogen at the pyramidal apex with C-N-C bond angles of ~ 108°. It was with some excitement, therefore, that when the silicon analogue, trisilylamine, $(SiH_3)_3N$, was analyzed (30) it was found to contain a Si-N bond length of 1.738Å and a Si-N-Si bond angle of 119° 36' ± 30'; the molecule, within experimental error, was planar. Similarly, it was found that in H_3SiNCO the Si-N bond length was even shorter at 1.699Å and the SiNC unit was linear, whereas in H_3CNCO the C-N-C angle is 125 ± 5° (near the 120° expected for sp^2 hybrids).

In the case of trisilylamine, the planarity was reasoned (29) to be due to preferred sp^2 hybridization at nitrogen allowing better overlap of the lone pair p orbital with vacant d orbitals on silicon. For H_3SiNCO the nitrogen was sp hybridized allowing better overlap between the lone pair p and empty d's and with the bond even further shortened by the additional $(\pi \rightarrow d)\pi$ bond from the π orbital formed between nitrogen and carbon.

All attempts to make H_3SiOCN, silyl cyanate, have failed, always the isocyanate is produced instead. Most chemists' intuition would suggest the cyanate with an Si-O would be preferred to the isocyanate with a Si-N bond. There is obviously a stabilizing influence for the iso compound. Much "chemical evidence" suggests that $(p \rightarrow d)\pi$ bonding is stronger for the Si-N link than for Si-O. The low basicity of silylamines (29) was taken to demonstrate that the nitrogen lone pair p was not readily available (as in carbon amines) for dative (or Lewis base) bonding. Other interesting examples that have been analyzed are: $[(CH_3Si)_2N]_2Be$ (31), Si-N, 1.726Å, Si-N-Si<, 129° 19'; $(CH_3Si)_2NH$ (32), Si-N, 1.735Å, Si-N-Si <, 127° 54'; $(H_3Si)_2NBF_2$ (32), Si-N, 1.737Å, Si-N-Si <, 125° 36'; $[(CH_3)_3SiNSi(CH_3)_2]_2$ (33), Si-N, 1.719Å with a planar 4 membered ring. All these compounds have planar nitrogen units and all suggest strong $(p \rightarrow d)\pi$ bonding.

In silylammonium complexes no lone pair p-orbital would exist at nitrogen allowing π overlap. These complexes were found to be extremely unstable (a consequence possibly of lack of stabilization by π-bonding). Recently, a silylammonium complex has been prepared

(34) but as yet no determination made on the Si-N bond length.
For the same reason only the normal cyanide H_3SiCN is known. In
the isocyanide the nitrogen lone pair would be involved in bonding
to carbon and less available for $(p{\rightarrow}d)\pi$ bonding to silicon and
this seems to make the compound unstable. Some recent evidence has
tended strongly to support the $(p{\rightarrow}d)\pi$ -bonding mechanism in several respects. Photoelectron spectroscopy (35) on a range of
silicon compounds including silicon nitride has directly shown the
effect to be important and, moreover, to indicate a bond order for
Si-N of ~ 1.2. This agrees (36) with an estimate of 1.26 based
upon bond shortening. Several other photoelectron studies have
supported $(p{\rightarrow}d)\pi$ bonding in silicon compounds, particularly (37)
in silyl and germyl halides, both the shape and the presence of
the Si-Cl vibrational progression in the spectrum of SiH_3Cl indicated $(p{\rightarrow}d)\pi$ from Cl—Si. This was extended with later evidence
(38) on vinyl silanes (39,40) and allyl silanes (40). Studies
indicated through conjugation (though empty d's). Evidence was
found (41) for $(p{\rightarrow}d)\pi$ bonding in phenyl-silane. Calculations (42)
supported the findings. The case of fluorosilanes was less clear
(43). However, studies of silylphosphines (44) do not show the bond
length, bond angle anomalies of the nitrogen compounds. Studies of
oxyanions of several of the second row elements has seemed to
support the theory (45). In exhaustive reviews Gibbs et al.(46,47)
treated the covalent interactions (including the use of Si 3d orbitals) in silicates. Gibbs has frequently pointed out the imposs-
ibility of Si^{2+} being a stable species in solids because of the
large ionization energy*. The inclusion of d orbital basis sets
seemed to improve the agreement of MO calculations with experiment
in some studies (48) but other workers (49,50) found that data
could be rationalized without recourse to 3d involvement. A photoelectron study of "Silatrans" (23) indicated an increase in the
binding energy of the nitrogen p lone pair, difficult to rational-
ize without $(p{\rightarrow}d)\pi$ bonding, the Si-N distance is ~ 2.1Å with the
Si 5 coordinate.

One of the major earlier problems was that the measured Si-N
bond energy had agreed (29) with that predicted by considerations
of electronegativity theory; the invoking of π-bonding was thought
unnecessary. It has been pointed out (7) that Pauling's original
concept of electronegativity took no account of the type of bonding,
σ or π, involved and, in fact, the electronegativity of silicon
had been calculated using compounds where π-bonding was already
present: the argument was circular and the result, therefore, to
be expected. Using only situations where π-bonding is impossible
(7), e.g. Si-C, Si-H, recalculating electronegativities and applying the new data to the Si-N bond produces a bond one-third stronger
than predicted by σ bonding alone. Moreover, the recalculated
electronegativity of silicon, 2,3, is in much better agreement

*The second ionization of Si requires 677 kcal mole^{-1}, greater
than the first ionization of neon at 499 kcal mole^{-1}.

with the Mulliken value than the original Pauling estimate (25),
and correlates better with chemical inductive effects in organic
compounds. The bulk of the evidence currently seems to support
the notion that π-bonding is important in the Si-N bond. Moreover,
the bond order is about 1.3, implying that for nitrogen, bonded to
three silicons, the lone pair is almost entirely involved in π-
bonding.

While the characteristics of a single Si-N bond are becoming
understood, an area that remains largely unresolved is the effect
of multiple π bond substituents at a single tetrahedrally coordinated central atom; this should be investigated. Some curious effects
have already been noted, e.g. the Si-F bond length in SiF_4 is shorter than in H_3SiF, implying that increasing competition for the d
orbitals has not weakened each individual bond (d orbital shrinkage
may occur!). The effect of spacial disposition of multiple p lone
pairs of nitrogen around silicon is not known, but an attempt has
been made (51) to explain different $(p \rightarrow d)\pi$ overlaps in the cases
of α and β quartz. With some plausibility it was suggested that
different d orbitals do not interact with lone pair p orbitals to
the same degree. In the special case of solid silicon nitrides
the lone pair p orbitals on nitrogen are held by the 3-dimensional
structures in definite orientations in space with respect, not
only to the central silicon atom, but also, with each other.

4.1 Tetrahedrally, sp^3, bonded N

Several mixed crystalline silicon nitrides are known where the
nitrogen is tetrahedrally sp^3 hybridized. In this configuration
nitrogen cannot π bond to silicon. Si-N bond lengths (Å) are:
$LiSi_2N_3$ 1.750 (52), $BeSiN_2$ 1.755 (53), $MnSiN_2$ 1.845 (54) and
$MgSiN_2$ 1.873 (55). The expected length is 1.81Å for a single covalent σ bond. Confirmation is awaited from bond length and bond
angle measurements in the only presently known silylammonium complex (34). The bonds seem quite normal and the expected ionic
content, if we use Quane's (7) value of 2.3 for silicon, is 11%.
(The expectation using the traditional Pauling value would be 30%.)
Using $\kappa = 2.3$, the Si-O single bond (with O sp^3 coordinate) would
be 30% ionic (cf. 50% earlier) and SiC (as in silicon carbide)
would be 1% (cf. 12%). We can, incidentally, use Pauling's GM
relation directly upon the bonds in diamond, silicon and silicon
carbide and calculate a value of $\sim$ 4% ionic for the Si-C bond.
This method would overestimate the ionic content because the small
excess ionic crystal energy is assumed to be ionic resonance
within each bond. The agreement is therefore gratifying.

4.2 Trigonal, sp^2, bonded N

In α and β-Si_3N_4 and in Si_2N_2O the nitrogens are trigonal (almost planar) sp^2 bonded. In Si_2N_2O (56) the Si-O bond length, 1.623Å
and <Si-O-Si 147°42' are quite normal for silicates, the Si-N bond
lengths are 1.692, 1.716 and 1.750Å, the <N-Si-N's, 106° 24',

110° 0' and 110° 6' close to optimal tetrahedral, and the
<Si-N-Si's, 116° 7', 116° 25' and 125° 20' putting the nitrogen
only 0.1Å out of the plane of 3 silicons. Inspection of the crystal
and crystal radii does not make it obvious why nitrogen should be
in the planar position, there is room in the crystal for pyramidal
nitrogen. One must assume that, as for trisilylamine, $(p \rightarrow d)\pi$
bonding is responsible for the planarity and the short Si-N bond
length. More details of this and other structures are contained
in (57).

For α-Si_3N_4 (58) the average Si-N bond length is 1.738Å; there
are 8 distinct Si-N bonds from 1.707Å - 1.779Å. The <N-Si-N's are
close to tetrahedral. The nitrogens are out of plane by 0.04, 0.10,
0.16 and 0.27Å. The disposition of lone pair N p orbitals around
Si is very similar for 4 types of nitrogens in α-Si_3N_4 and for
nitrogens in Si_2N_2O.

For β-Si_3N_4 the average Si-N bond length is 1.736Å (individually 1.730(2), 1.740, 1.745Å). The <N-Si-N's are slightly closer
to tetrahedral than in α and the two nitrogens are out of plane by
0.00Å and 0.07Å. The combination of these features (giving β a
slightly higher density, 3.19g/cc over α, 3.18g/cc) led to the
prediction that β was the more stable phase (57); and that, in
combination with other observations, α was probably always a metastable phase. The disposition of N lone pair p orbitals is also
suggestive of a "special arrangement" dictated by better $(p \rightarrow d)\pi$
overlap.

The range of observed Si-N bond lengths, from 1.873-1.692
(nearly 11%), is larger than encountered with Si-0 bonds; these
range from ~1.679-1.563 (7%)*. Al-0 bonds in silicates vary even
less. It appears that π bonding effects may be larger for Si-N
than for Si-0 and this would be a fruitful area for calculations.
If we assume that silicon nitrides and oxides are covalent and
calculate mean bond energies, surprising agreement with simple
model compounds is obtained. From Si_3N_4 E(Si-N) is ~80 kcal mole^{-1}.
For $[(CH_3)_3Si]_2$ NH, 76 kcal mole^{-1}, and for (CH_3) $SiN(C_2H_5)_2$ 87
kcal mole^{-1} (29). Even better agreement can be found for E(Si-0),
111 kcal mole^{-1} for quartz and 108 kcal mole^{-1} for $(CH_3O)_4Si$.
There appears to be no extra ionic energy in amorphous quartz
over and above that in the isolated $(CH_3O)_4Si$ molecule.

The notion of covalent bond additivity can apparently be extended to mixed crystals, e.g. Be_2SiO_4, phenacite. E(Be-0), from
bromellite is some 71 kcal mole^{-1}. E(Si-0) from α-quartz is 111
kcal mole^{-1}. For phenacite 4E(Be-0) + 4E(Si-0), using heats of
atomization and the known heat of formation of phenacite, is 956
kcal mole^{-1}. Adding the separately calculated values gives 1012
kcal mole^{-1}. The summation of the separate bonds, as if they were
covalent bonds, therefore only overestimates the total bond energy

---

*See: Miscellaneous Refs. Pant and Cruickshank.

of phenacite by 56 kcal mole^{-1}. The error is in the direction expected if in fact small partial charges do reside at Be and Si because BeO and SiO_2 are more densely packed (more Madelung contribution) than phenacite, which contains large open channels.

5. GERMANIUM NITRIDES

It is generally accepted (29) that $(p{\to}d)\pi$ bonding is less important in germanium chemistry. Photo-electron spectroscopy has tended to confirm this.

The much more limited work in Ge systems supports the ideas presented earlier. The structure of Ge_2N_2O has not been determined but is expected shortly (59). All the lengths and angles have not been calculated as before; some of the available data, e.g. (60), look quite uncertain. Average lengths Ge-N in Å: $CaGeN_2$ (61) 1.853, $ZnGeN_2$ (62) 1.94, $MgGeN_2$ (55) 1.953, $MnGeN_2$ (54) 1.97, α-Ge_3N_4 (60) 1.84, β-Ge_3N_4 (60) 1.82, covalent radii 1.88. The total variation is 8%, less than for Si-N. The Ge-N bond lengths are shorter in the π-bonded α and β-Ge_3N_4, shortest in the β, as for Si_3N_4. Ge-N-Ge < 's are near to the normal tetrahedral.

In β-Ge_3N_4 one N atom is planar and the other only slightly non-planar, as in β-Si_3N_4. In α-Ge_3N_4, however, two nitrogens are markedly less planar in environment than in α-Si_3N_4; being 0.57Å out of the plane of respective silicons. To be pyramidal the out-of-plane distance in this case would be 0.61Å. These atoms, as far as has been determined, are nearer to pyramidal positions and not tending towards planarity at all. It is tempting to conclude that this is a result of less $(p{\to}d)\pi$ bonding. The other nitrogens seems to be fairly similar in α-Ge_3N_4 to α-Si_3N_4. No great trust is placed in the available structure determinations and we await further data. Ge has a slightly larger covalent radius than Si; on steric considerations, one might expect the nitrogen to be more planar in α-Ge_3N_4 than in α-Si_3N_4, not less.

The β-configuration again appears to be the lower energy type.

6. ESTIMATES OF PARTIAL CHARGE

It has been insinuated here that the partial charge at atoms in crystals, even in the most ionic cases, is less than most investigators have imagined. The study of nitrides will highlight this effect and may well provide good model crystal situations for theoretical studies. It is only possible here to list recent studies and reviews that seem to endorse this thinking (63-68). In particular new physical techniques and calculations show promise of illuminating this subject in the near future. One of the most promising treatments (69) leads to calculated partial changes in <u>molecules</u> of $\sim$ +0.55 for Be in BeF_2 and $BeCl_2$, +0.65 for Mg in MgF_2 and $MgCl_2$. The authors noted some problems with beryllium halides believed due to $(p{\to}p)\pi$ interactions and also specifically avoided speculating on Si and Ge due to obvious extra π bonding effects not implicit in their original assumptions.

7. SOME FUTURE IMPLICATIONS IN NITRIDES

From bonding (and other considerations, phase rule etc.) it was predicted (70) (before any other publications) that the β'-sialon series would be an essentially defectless series of compositions which are most simply written as $Si_{3-x}Al_xN_{4-x}O_x$. Moreover, it can be anticipated that there may be a preferential association of aluminum and oxygen in the structure; oxygen has one extra electron than nitrogen and aluminum one less than silicon, by associating Al-O achieves an isoelectronic state to Si-N and this may lower the crystal energy. This will be interesting to investigate. The preference of nitrogen for the planar configuration in Si_2N_2O suggests that X (or J) phase may well contain this configuration and that sheets similar to those in Si_2N_2O be cross linked through octahedral aluminum (probably). The idea that nitrogen and oxygen may be rather randomly disposed in new oxynitride crystals (including silicate types) is already entering the literature. However, this is extremely unlikely. In Si_2N_2O, for example, the nitrogen and oxygen atoms are quite rigidly disposed in special sites. In analogous minerals where oxygen and fluorine (or hydroxide) ions co-exist, e.g. apatite, topaz, there is ordered disposition and the mixed nitride fluorides e.g. Mg_2NF (71) also show this feature very nicely.

It is anticipated that new types will tend to obey Pauling's second crystal rule (valence balancing) and this requires that nitrogen and oxygen occupy different sites with respect to the cations so that the bond strength reaching nitrogen be ~ 3 and oxygen $\sim 2^*$. Attempts to introduce "new rules" to govern the new structures must be resisted unless absolutely unavoidable.

REFERENCES

1. I.D. Brown and R.D. Shannon, 'Empirical Bond Strength–Bond Length Curves for Oxides', Acta Cryst. A29, 266 (1973).
2. L. Pauling, Nature of the Chemical Bond, 547 (1960).
3. L. Pauling, 'Principles Determining the Structure of Complex Ionic Crystals', J. Amer. Chem. Soc. 51, 1010 (1929).
4. D.A. Johnson, 'Some Thermodynamic Aspects of Inorganic Chemistry', Camb. Univ. Press (1968).
5. L. Pauling, The Nature of the Chemical Bond, Cornell Univ. Press (1939).
6. L. Pauling, 'The Modern Theory of Valency', J. Chem. Soc. 1461 (1948).
7. D. Quane, 'Electronegativity and π-Bonding Strengths of Silicon and Germanium', J. Inorg. Nucl. Chem. 33, 2722 (1971).
8. L. Pauling, 'Interatomic Distances and Bond Character in the Oxygen Acids and Related Substances', J. Phys. Chem. 56, 361–365 (1952).

*This is an alternative view of why, in β'-sialon, on average, aluminum will have one oxygen as a nearest neighbour.

36

9.   The Chemistry of Boron and its Compounds, ed. E.L. Mutterties
      (Wiley, New York, 1967).
10.   T. Onak, Organoborane Chemistry, Acad. Press (1975).
11.   Progress in Boron Chemistry - series especially Vol. 3.
      Ed. R.J. Brotherton and H. Steinberg, Pergamon Press (1970).
12.   D.A. Johnson, Some Thermodynamic Aspects of Inorganic
      Chemistry, Camb. Univ. Press (1968).
13.   National Bureau of Standards Technical Note 270-3 & 270-6
      (1971). Government Printing Office, Washington D.C. (1968).
14.   A. Finch, P.J. Gardner, E.J. Pearn and G.B. Watts,
      'Thermochemistry of Triphenylboron, Tricyclohexylboron and
      some Phenylboron Halides', 63, 1880-8 (1967).
15.   A. Finch, P.J. Gardner and G.B. Watts, 'The Standard Enthalpy
      of Formation of Diphenylborinic Acid', Chem. Comm., 1054-55
      (1967).
16.   A. Finch, P.J. Gardner, N. Hill and K.S. Hussian, 'The
      Thermochemistry of o- and p-Tolyldichloroboranes and the
      Boron-to-Carbon Bond Strength', J. Chem. Soc. 2543-45 (1973).
17.   Tables of Interatomic Distances and Configuration in Molecules
      and Ions, Special Publications No. 11 and 18, Chem. Soc.
      London (1958) and (1965).
18.   W.J. Orville-Thomas, 'The Structure of Small Molecules',
      Principles of Modern Chemistry 1, Elsevier (1966).
19.   F.A. Cotton and J.R. Leto, 'Acceptor Properties, Reorganizat-
      ion Energies, and π Bonding in the Boron and Aluminum Halides',
      J. Chem. Phys. 30, 993-998 (1959).
20.   F.B. Clippard, Jr. and L.S. Bartell, 'Electron Diffraction
      Determination of the Molecular Structure of Dimethylamino-
      dichloroborane', Inorg. Chem. 9, 2439 (1970).
21.   O. Gropen and H.M. Seip, 'Rotational Barrier and Structure
      of Aminoborane Determined by Ab Initio Calculations', Chem.
      Phys. Let. 25, 206-8 (1974).
22.   P. Cassoux, R.L. Kuczkowski, P.S. Bryan and R.C. Taylor,
      'Microwave Spectra of Trimethylamine-Borane. The Boron-
      Nitrogen Distance and Molecular Dipole Moment', Inorg. Chem.
      14, 126-129 (1975).
23.   S. Cradock, E.A.V. Ebsworth and I.B. Muiry, 'Photoelectron
      Spectra and Bonding of (N-B)-2,8,9-Trioxa-5-aza-1-bora-bicyclo
      (3.3.3) undecane (Boratran) and some 2,8,8-Trioxa-5-aza-1-
      silabicyclo (3.3.3) undecanes (Silatrans)', J. Chem. Soc.,
      Dalton Trans., 25-29 (1975).
24.   A. Finch, P.J. Gardner and A.F. Webb, 'The Standard Enthalpy
      of Formation of Tetramethoxydiboron', J. Chem. Thermodynamics
      4, 495-498 (1972).
25.   J. Hinze and J.H. Jaffe, 'Electronegativity. I. Orbital Elec-
      tronegativity of Neutral Atoms', J. Am. Chem. Soc. 84, 540
      (1962).
26.   G.E. Kimball, 'Directed Valence', J. Chem. Phys., 8, 188-198
      (1940).

27. F.A. Cotton, Chemical Applications of Group Theory, Wiley-Interscience, 212 (1971).

28. E.W. Randall and J.J. Zuckerman, 'The Elusive (p→d)π Bond', Chem. Comm. 20, 732-33 (1966).

29. E.A.V. Ebsworth, 'd-Orbitals and the Formation of Internal π-bonds', The Bond to Carbon, Part I, ed. Alan G. MacDiarmid, 1, 22-104 (1968).

30. K. Hedburg, 'The Molecular Structure of Trisilyamine (SiH₃)₃N', J. Amer. Chem. Soc. 77, 6491 (1955).

31. A.H. Clark and A. Haaland, 'The Molecular Structure of Bis [di(trimethylsilyl)amino]-beryllium', Chem. Comm. 16, 912-13, (1969).
    A.H. Clark and A. Haaland, 'On the Molecular Structure of Bis [di(trimethylsilyl)amino]-beryllium, [((CH₃)₃Si)₂N]₂Be', Acta Chem. Scand. 24, 3024-30 (1970).

32. A.G. Robiette, G.M. Sheldrick and W.S. Sheldrick, 'The Conformation of Three Disilazanes', Chem. Comm. 15, 909-10 (1968).
    A.G. Robiette, G.M. Sheldrick and W.S. Sheldrick, 'Determination of the Molecular Structure of N-Difluoroboryl-Disilazane by Electron Diffraction', J. of Molec. Struc. 5, 423-31 (1970).

33. P.J. Wheatley, 'An X-ray Diffraction Determination of the Crystal and Molecular Structure of Tetramethyl-NN'-bistrimethylsilylcyclodisilasane', J. Chem. Soc., 1721-24 (1962).

34. R.E. Highsmith, J.R. Bergerud, and A.G. MacDiarmid, 'Reaction of the Protonic Acids, HCo(PF₃)₄ with NN-Dimethyl-(trimethylsilyl)amine to form Silylammonium Compounds', Chem. Comm. 1, 48-9 (1971).

35. R. Nordberg, H. Brecht, R.G. Albridge, A. Fahlman and J.R. van Wazer, 'Binding Energy of the "2p" Electrons of Silicon in Various Compounds', Inorg. Chem., 9, 2469-74 (1970).

36. U. Wannagat, 'The Chemistry of Silicon-Nitrogen Compounds', Advances in Inorganic Chemistry and Radiochemistry 6, 225-78, (1964).

37. S. Cradock and E.A.V. Ebsworth, 'Photo-electron Spectra of Silyl and Germyl Halides and (p→d)π Bonding', Chem. Comm. 1, 57-59 (1971).

38. D.C. Frost, F.G. Herring, A. Katrib and R.A.N. McLean, '(p→d)π Bonding in Halosilanes; Evidence from Photoelectron Spectroscopy', Chem. Phys. Let. 10, 347-50 (1971).

39. U. Weidner and A. Schweig, 'Theory and Application of Photoelectron Spectroscopy VII. Through-conjugation through a tetrahedral silicon atom', J. Organ. Chem. 37, C29-30 (1972).

40. U. Weidner and A. Schweig, 'Theory and Application of Photoelectron Spectroscopy. V. The Nature of Bonding in Vinyl and Allylsilanes', J. Organometallic Chem. 39, 261-66 (1972).

41. R.A. Neill McLean, 'The Bonding of a Silicon Atom with a Phenyl Ring: the Photoelectron Spectrum of Phenylsilane', J. Canadian Chimie, 51, 2089-91 (1973).

42. W.B. Perry and W.L. Jolly, 'The Use of X-ray Photoelectron Spectroscopy to Assess d Orbital Participation in Compounds of Silicon and Germanium', Chem. Phys. Let. 17, 611-13 (1972).

38

43. S. Cradock, E.A.V. Ebsworth and R. Alastair Whiteford, 'Photoelectron Spectra of Some Simple Fluorosilanes', J. Chem. Soc. 22, 2401-4 (1973).

44. C. Glidewell, P.M. Pinder, A.G. Robiette and G.M. Sheldrick, 'Molecular Structures of Silylphosphine, Silylmethylphosphine and Silyldimethylphosphine: an Electron Diffraction Study', J. Chem. Soc. 1402 (1972).

45. D.S. Urch, 'Direct Evidence for 3d-2p π-bonding in Oxy-anions', J. Chem. Soc. (A) 3026 (1969).

46. P.H. Ribbe, M.W. Phillips, G.V. Gibbs, 'Tetrahedral Bond Strength Variations in Feldspars', NATO Conf. Felspars, Univ. of Manchester Press, 25 (1972).

47. G.V. Gibbs, M.M. Hamil and S.J. Louisnathan, L.S. Bartell, and Hsiukang Yow, 'Correlations Between Si-O Bond Length, Si-O-Si Angle and Bond Overlap Populations Calculated Using Extended Huckel Molecular Orbital Theory', Am. Mineralogist, 57, 1578 (1972).

48. G.A.D. Collins, D.W.J. Cruickshank and A. Breeze, 'Ab Initio Calculations on the Silicate Ion, Orthosilicic Acid and Their $L_{2,3}$ X-ray Spectra', J. Chem. Soc., Faraday Trans. II, 68, 1189-1195 (1972).

49. W.B. Perry and W.L. Jolly, 'Valence Electron Binding Energies of Some Silicon Compounds from X-ray Photoelectron Spectroscopy', J. Electron Spectro & Rel. Phenom. 4, 219 (1974).

50. W.B. Perry and W.L. Jolly, 'Correlation of Core Electron Binding Energies with Charge Distributions for Compounds of Carbon, Silicon and Germanium', Inorg. Chem. 13, 1211 (1974).

51. D.W.J. Cruickshank, 'The Role of 3d-Orbitals in π-Bonds between (a) Silicon, Phosphorus, Sulphur, or Chlorine and (b) Oxygen or Nitrogen', J. Chem. Soc., 5486 (1961).

52. J. David, Y. Laurent, J. Pierre Charlot and J. Lang, 'Étude cristallographique d'un nitrure I 4_2 5_3. La structure tétra-édrique type wurtzite de $LiSi_2N_3$', Bull. Soc. fr. Minéral. Cristallogr. 96, 21-24 (1973).

53. Von P. Eckerlin, 'Zur Kenntnis des Systems $Be_3N_2-Si_3N_4$, IV,' Z. anorg. allg. Chem. 353, 225-336 (1967).

54. M. Maunaye, R. Marchand, J. Guyader, Y. Laurent, 'Structure de $MnSiN_2$ et $MnGeN_2$', Bull. Soc. fr. Minéral. Cristallogr. 94, 561-4 (1971).

55. J. David, Y. Laurent and J. Lang, 'Structure de MgSiN et $MgGeN_2$', Bull. Soc. fr. Minéral. Cristallogr. 93, 153 (1970).

56. I. Idrestdt and C. Brosset, 'Structure of Si_2N_2O', Acta Chemica Scand. 18, 1879 (1964).

57. P.E.D. Morgan, 'Study of π-bonding in Silicon Nitride and Related Compounds', ONR, Arlington, Va., Report F-C2429-06, August 1, 1974, NTIS, AD-7849979GA.

58. R. Marchand, Y. Laurent and J. Lang, 'Le Structure de Nitrure de silicium α', Acta Cryst. B25, 2157 (1969).

59. J.C. Labbe and M. Billy, 'Préparation et caractérisation d'un oxynitrure de germanium', Comptes Rendus Acad. Sc. Paris, 277, 1137 (1973).

60. S. Wild, P. Grieveson and K.H. Jack, 'The Crystal Structures of Alpha and Beta Silicon and Germanium Nitrides', Special Ceramics 5, 385-95, Brit. Ceram. Soc. (1970).

61. M. Maunaye, J. Guyader, Y. Laurent and J. Lang, 'Étude structurale de $CaGeN_2$ et $Ca_{1-x}GeN_2$', Bull. Soc. fr. Minéral. Cristallogr. 94, 347-52 (1971).

62. M. Maunaye, P. L'Haridon, Y. Laurent and J. Lang, 'Étude structurale de $ZnGeN_2$', Bull. Soc. fr. Minéral. Cristallogr. 94, 3-7 (1971).

63. A.G. Revesz, 'π-Bonding and Delocalisation Effects in SiO_2 Polymorphs', Phys. Rev. Let. 27, 1578 (1971).

64. B.G. Cocksey, J.H.D. Eland and C.J. Danby, 'Photoelectron Spectra of the Zinc and Cadmium Halides', J. Chem. Soc., Faraday Trans. 2, 69, 1558 (1973).

65. A.F. Povey and P.M.A. Sherwood, 'Covalent Character of Lithium Compounds Studied by X-ray Photoelectron Spectroscopy', J. Chem. Soc., Faraday Trans. 2, 70, 1240 (1974).

66. J.A. McGinnety, 'Crystal Forces in Lithium Tetrafluoroberyllate', J. Chem. Phys. 59, 3442 (1973).

67. D.E. Parry and M.J. Tricker, 'Covalent Character of Alkali Metal Halides from X-ray Photoelectron Spectroscopy', Chem. Phys. Let. 20, 124 (1973).

68. J.E. Huheey, J.C. Watts, 'Halogen Electronegativity and Isomer Shifts of Tin Compounds. Another Example of the Importance of Charge Capacity', J. Inorg. Chem. 10, 1553 (1971).

69. R.S. Evans and J.E. Huheey, 'Electronegativity, Acids, and Bases – III. Calculation of Energies Associated with Some Hard and Soft Acid-Base Interactions', J. Inorg. Nucl. Chem. 32, 777 (1970).

70. P.E.D. Morgan, 'Amorphous Silicon Nitride and a Classification of Sialon', Bull. Amer. Ceram. Soc. 53, 392, 4-S2-74 (1974).

71. S. Andersson, 'Magnesium Nitride Fluorides', J. Sol. St. Chem. 1, 306 (1970).

Miscellaneous References

C.A. Coulson, 'd-Electrons in Chemical Bonding', The Robert A. Welch Foundation Conference on Chemical Research XVI, Theoretical Chemistry, Houston, Texas (1972).

K.A.R. Mitchell, 'The Use of Outer d Orbitals in Bonding'. Chem. Revs. 69, 157 (1969).

J.C. Phillips, 'The Chemical Bond and Solid-State Physics', Phys. Today, 23, 23 (1970).

A.P. Mortola, H. Basch and J.W. Moskowitz, 'An Ab Initio Study of Permanganate Ion', Internat. J. Quan. Chem. VII, 727 (1973).

B.C. Tolfield, 'The Study of Electron Distributions in Inorganic Solids', Prog. Inorg. Chem. 20, 153 (1976).

A.K. Pant and D.W.J. Cruickshank, 'A Reconsideration of the Structure of Datolite, $CaBSiO_4(OH)$'. Zeit. Krist. 125, 286-297 (1967).

DISCUSSION (Lang, Thompson)

Brook: A charge at the Be site of 1.2-x positive is quoted for BeO. However, Cline has observed that BeO is an ionic conductor with the Be ion migrating with its formal charge of +2, a finding reinforced by application of the Nernst-Einstein relation to beryllium diffusion data. Similarly Mills and Kröger suggest normally charged (2-) oxygen interstitial ions as a conducting species in pure SiO_2, which is considered even more 'covalent'. Can you explain this discrepancy, which seems to limit the utility of 'covalent-ionic' concepts in treating transport processes in ceramics?

Morgan: Just as when NaCl dissolves in water and the Na^+ charge spreads out over the sphere of solvation, I imagine the charge at a vacancy just as at an atom spreads to its neighbours. Of course, there is a net residual charge otherwise a field would have no effect. The net spread-out charge does behave as though it were at a point just as a charged sphere behaves (to the outside) as a net point charge, but inside the sphere where the bond effects occur the distribution of the charge is very important.

Schaeffer: There is evidence that changes in covalent bonding have an effect on the charge, which can be noticed in electrical conductivity data. For example, in a $CaO-Al_2O_3-SiO_2$ melt, oxygen possesses the highest mobility, yet it is still not responsible for the charge transport (i.e. the Nernst-Einstein equation does not hold) due to the high extent of covalency in the Si-O bond. Concerning experimental tools for determining effective charges in crystals and glasses, the positron annihilation technique appears to be rather promising.

Groskovich: Do you believe that much delocalization of electrons exists in the Si_3N_4 structures?

Morgan: It does appear that electron density spreads back from nitrogen to silicon via some sort of π bonding. This does not mean delocalization in the sense of aromatic type effects but rather to the extent that may occur, for example in graphite-type boron nitride. All electron orbitals are of course delocalized to a greater or lesser extent.

Riley: What is the nature of the bonding in metallic nitrides?

Morgan: Normally in the rock salt structure, multi-centred p^3 bonding is involved. This could be the state for nitrogen in transition metal compounds such as TiN, VN. The metal can use p and d orbitals and there are difficulties in deciding between $(p \rightarrow p)\sigma$ overlap and $(p \rightarrow p)\pi$ overlap which may occur together with the same p orbital. In addition $(p \rightarrow d)\pi$ overlap inevitably is present just as for ReO_3, TiO and other metallic oxides.

REPRESENTATION OF MULTICOMPONENT SILICON NITRIDE BASED SYSTEMS

L.J. Gauckler and G. Petzow
Max-Planck-Institut für Metallforschung
Institut für Werkstoffwossenschaften, PML
7 Stuttgart-80, W.-Germany

1. INTRODUCTION

In the last decade, the compound silicon nitride (Si_3N_4) has
attracted much attention because of its potential application as
a high temperature engineering material. Despite its excellent
properties it is still not being used extensively today because
there are difficulties in fabricating silicon nitride ceramics.
Strong silicon nitride ceramics can only be made with the addition of sintering aids. The effective sintering aids were found
to be Al_2O_3, MgO, Y_2O_3, etc. The studies of these sintering
aids have led to the discovery of a new family of materials – the
solid solution of metal oxides in metal nitrides. Effort has been
concentrated on systems with silicon nitride as one of the compounds. However, other metal-nitride/metal-oxide systems should
also be investigated. Compounds such as oxinitrides, borides,
carbides of metals and semi-metals are also potential high temperature materials. In general, the relationship of these materials
can be expressed by Fig. 1 (1). Nitrogen, oxygen and carbon, as
well as silicon, boron and one of the other metals, e.g. aluminum,
are drawn in a six component system. The six components result
in 63 geometric elements: the six corners (1-component systems),
15 edges (the binary systems), 20 planes, 15 three- and 6 four-
dimensional zones and 1 five-dimensional zone. The bond characteristics in such compounds change over a wide range. This is
expressed by the circles around the six cornered system which represent different bond characteristics. For the above considered
systems the bond characteristics range from the covalent to metallic and ionic. Therefore, a large variety of properties and
different fabrication possibilities can be expected.

42

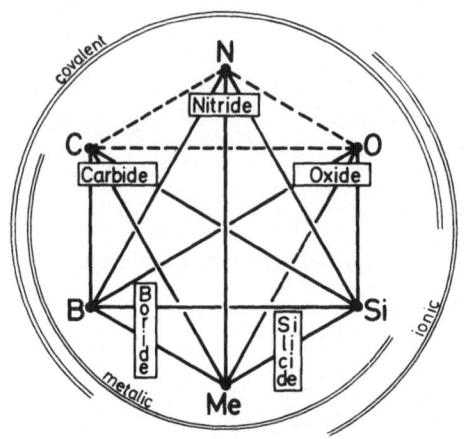

Fig. 1:

Systems with non-metal high temperature materials after Petzow (1).

2. LITERATURE SURVEY

The first ß-Si$_3$N$_4$ solid solution was reported to exist in the system Si$_3$N$_4$-Al$_2$O$_3$, by Oyama and Kamigaito (2) in Japan, and Jack and Wilson (3) in Great Britain. Both of these publications reported extensive solid solution formation of Al$_2$O$_3$ in ß-Si$_3$N$_4$. In later publications, the solid solution forming regions were given in ternary systems by these authors. Jack (4) used the ternary system Si$_3$N$_4$-SiO$_2$-Al$_2$O$_3$, and Oyama (5), on the other hand, used the system Si$_3$N$_4$-AlN-Al$_2$O$_3$. The phase diagrams of these systems are presented in Fig. 2. As shown in these diagrams, there are wide ternary solid solution regions in both of these systems. The diagrams in Fig. 2a and 2b are isothermal sections of the corresponding systems. Jack (4) reported the 1700° C section, and Oyama (5) prepared his samples at 1760° C. The solid solution formation along the join Si$_3$N$_4$-Al$_2$O$_3$ requires the existence of non-metal interstitials or metal site vacancies in the ß-Si$_3$N$_4$ lattice. Oyama (5) has suggested the existence of silicon vacancies as an explanation of his results. The single-phase ß-Si$_3$N$_4$ solid solution forming regions in the system Si$_3$N$_4$-Al$_2$O$_3$-Be$_2$SiO$_4$, Si$_3$N$_4$-Al$_2$O$_3$-Li$_2$O and Si$_3$N$_4$-Al$_2$O$_3$-MgO were reported by Jack (4). The 1700°C isothermal phase diagrams of these systems are given in Fig. 2c, 2d and 2e respectively. The isothermal diagram in the system Si$_3$N$_4$-Al$_2$O$_3$-Ga$_2$O$_3$, reported by Oyama (6), is presented in 2f. As shown in these diagrams, extensive ternary solid solutions exist in all of these systems.

In the system Si$_3$N$_4$-Al$_2$O$_3$, at high temperature the reaction

$$Si_3N_4 + 2\ Al_2O_3 \rightleftarrows 3\ SiO_2 + 4\ AlN \dots\dots\dots\dots 1$$

might occur, which is a double exchange reaction. Therefore, the results of our equilibrium studies have been presented in terms of reciprocal salt systems.

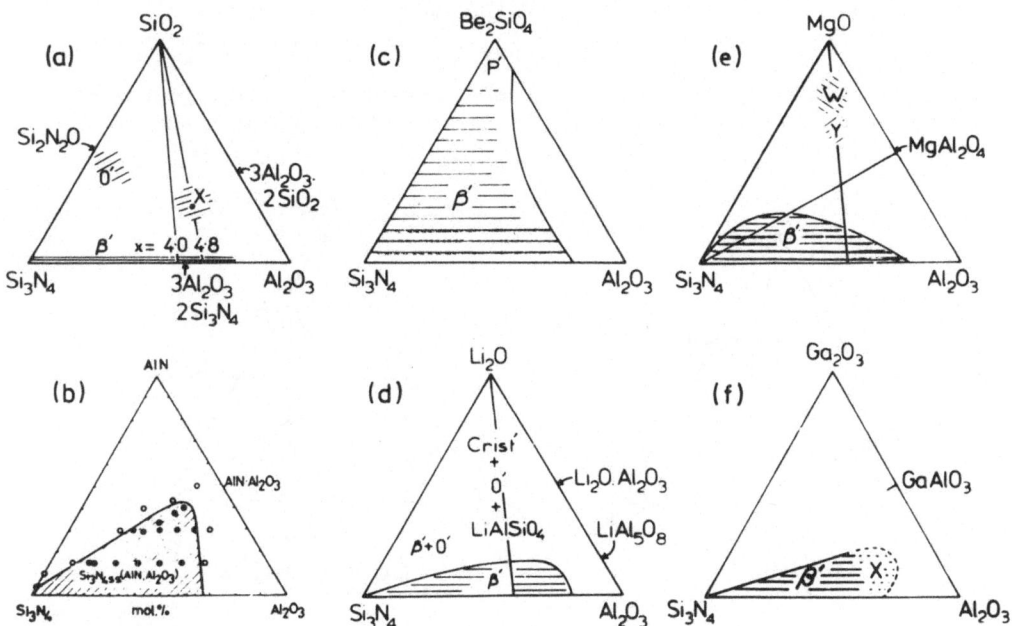

Fig. 2:

Isothermal phase diagrams of Si_3N_4 systems. The shaded area indicates solid solution region.

a. The system Si_3N_4-SiO_2-Al_2O_3 at $1700°$ C, after Jack (4)
b. The system Si_3N_4-AlN-Al_2O_3 at $1760°$ C, after Oyama (5)
c. The system Si_3N_4-Be_2SiO_4-Al_2O_3 at $1700°$ C, after Jack (4)
d. The system Si_3N_4-Li_2O-Al_2O_3 at 1500-$1700°$C, after Jack (4)
e. The system Si_3N_4-MgO-Al_2O_3 at $1700°$ C, after Jack (4)
f. The system Si_3N_4-Al_2O_3-Ga_2O_3 at $1730°$ C, after Oyama (5).

3. REPRESENTATION OF RECIPROCAL SALT SYSTEMS

The graphical representation of any four component system, under isobaric conditions, requires three concentration variables and the temperature. For the representation of the three concentrations at a constant temperature, the most common figure used is the equilateral tetrahedron with the four components at the corners. In Fig. 3 such a system is shown with Si, Al, O and N. Reciprocal salt systems are mixtures of components containing at least two cations (A, B) and two anions (X, Y). The binary components are located on the joins between the metals and the non-metals. The concentration variables in Fig. 3a being the atomic-percentage, and in Fig. 3b the equivalent-percentage. In Fig. 3a the components are located on a plane which transforms to a square in the case of equivalent-percentage in Fig. 3b. In this system, at temperatures below 1800° C, condensed phases are only known to exist under the Si^{4+}, Al^{3+}, O^{2-} and N^{3-} valency states. With the restriction of electroneutrality all possible combinations of the four elements are located in the Si_3N_4, AlN, Al_2O_3, and SiO_2 plane.

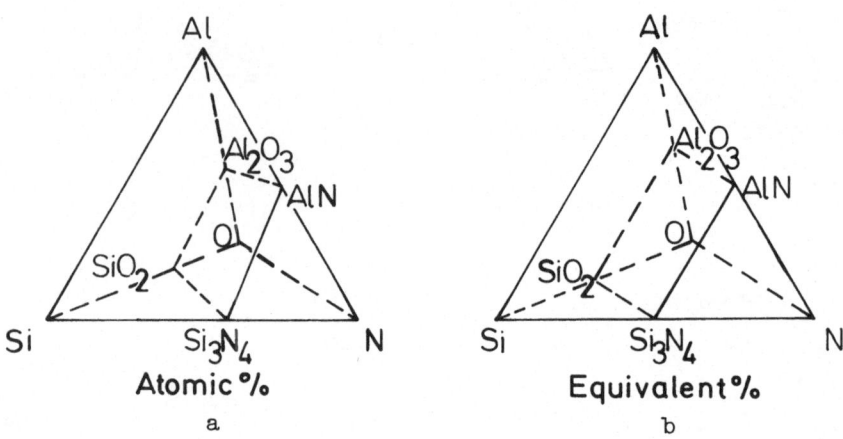

Atomic %   Equivalent %

a   b

Fig. 3:

The quaternary system Si-Al-O-N with the quasiternary system Si_3N_4-AlN-Al_2O_3-SiO_2.

This is valid whether atomic or equivalent-percentages are used. The restriction to the above mentioned valencies and electroneutrality reduces the number of the three concentration variables to two. In this system the highest number of phases in equilibrium is four, that is in the case of nonvariance. Because of the different valency states of the Si_3N_4-AlN-Al_2O_3-SiO_2 system components it is useful to express their concentrations in equivalent percentage (eq.-%).

$$\text{eq.-\% Al} = \frac{\text{at.-\% Al} \cdot 3}{\text{at.-\% Al} \cdot 3 + \text{at.-\% Si} \cdot 4} \cdot 100 \quad \ldots\ldots\ldots\ldots 2$$

$$\text{eq.-\% Si} = 100 \% - \text{eq.-\% Al} \ldots\ldots\ldots\ldots\ldots\ldots\ldots\ldots\ldots\ldots\ldots\ldots\ldots 3$$

$$\text{eq.-\% O} = \frac{\text{at.-\% O} \cdot 2}{\text{at.-\% O} \cdot 2 + \text{at.-\% N} \cdot 3} \cdot 100 \ldots\ldots\ldots\ldots 4$$

$$\text{eq.-\% N} = 100 \% - \text{eq.-\% O} \ldots\ldots\ldots\ldots\ldots\ldots\ldots\ldots\ldots\ldots\ldots\ldots\ldots 5$$

From equivalent-percentages can be calculated to the atomic-percentages with

$$\text{at.-}\% \text{ Al} = \frac{4 \text{ eq.-}\% \text{ Al}}{3 \text{ eq.-}\% \text{ Si} + 4 \text{ eq.-}\% \text{ Al} + 6 \text{ eq.-}\% \text{ O} + 4 \text{ eq.-}\%\text{N}} \cdot 100 \qquad \ldots 6$$

$$\text{at.-}\% \text{ Si} = \frac{3 \text{ eq.-}\% \text{ Si}}{3 \text{ eq.-}\% \text{ Si} + 4 \text{ eq.-}\% \text{ Al} + 6 \text{ eq.-}\% \text{ O} + 4 \text{ eq.-}\%\text{N}} \cdot 100 \qquad \ldots 7$$

$$\text{at.-}\% \text{ O} = \frac{6 \text{ eq.-}\% \text{ O}}{3 \text{ eq.-}\% \text{ Si} + 4 \text{ eq.-}\% \text{ Al} + 6 \text{ eq.-}\% \text{ O} + 6 \text{ eq.-}\%\text{N}} \cdot 100 \qquad \ldots 8$$

$$\text{at.-}\% \text{ N} = \frac{4 \text{ eq.-}\% \text{ N}}{3 \text{ eq.-}\% \text{ Si} + 4 \text{ eq.-}\% \text{ Al} + 6 \text{ eq.-}\% \text{ O} + 6 \text{ eq.-}\%\text{N}} \cdot 100 \qquad \ldots 9$$

Such a representation of a four-component system was used by Jänecke in 1908 (7) to describe the equilibria of reciprocal salt pairs in water. A similar method was already used by Löwenherz in 1894 (8). More detailed descriptions and examples are discussed by Vogel (9) as well as Ricci (10). For Si_3N_4 based systems this method was first used by Gauckler and Huseby (11). The restriction to a constant valency state is not always fulfilled regarding the A,B/X,Y system. A system may exist in which two or more components are located on a join of metal to non-metal. The simplest example of such a system may be illustrated by considering a binary compound with a metal component possessing a varying valency. Restricting to a system with A (4+), B(3+)$\rightarrow$B(4+)/X(3-), Y (2-) and the compounds A_3X_4, AY_2, BY_2, BX$\rightarrow B_3X_4$, the system shown in Fig. 4 results. Such a system can no longer be treated as a reciprocal salt system as illustrated by the example Si, Zr/N, O, with the binary compounds ZrN and Zr_3N_4.

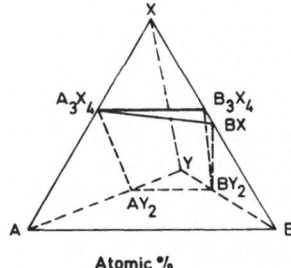

Atomic %

Fig. 4:

A system of the type A,B/X,Y in which one metal has a varying valency state in equilibrium with one non-metal.

3.1 Five component systems

Multiple component systems (n > 4) may be treated in a similar way to reciprocal salt systems regarding non varying valencies and electroneutrality. For example, the five component condensed system Si-Al-Be-O-N can be treated as a quasiquaternary reciprocal salt system. In Fig. 5 a projection of the four dimensional concentration zone is shown. The binary compounds are located on the metal/non-metal joins.

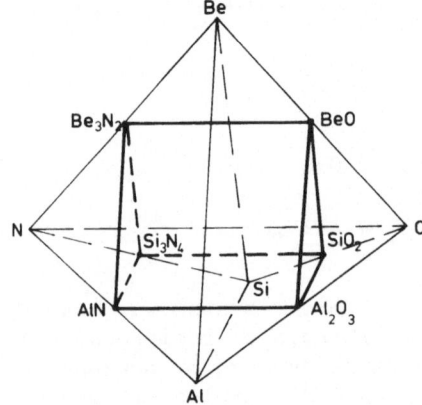

Fig. 5:

The quinary system Si-Al-Be-O-N with the quasiquaternary system Si_3N_4-AlN-Be_3N_2-SiO_2-Al_2O_3-BeO (12).

With the above mentioned restrictions, the number of variables is lowered by 1. The system can be presented in a three-dimensional space as a quasiquaternary system. Taking the compounds of interest as components, the system becomes a regular triangular prism. This system contains six components, nine binary subsystems, two ternary subsystems and three reciprocal ternary systems. Compounds with common non-metals forming end triangles and the distance between these end systems measures the concentration of the non-metal. This type of system was called "double reciprocal" by Jänecke, in early work (13). First results of quasiquaternary systems based on Si_3N_4 were published in (14).

4. THE SYSTEM Si-Al-Be-O-N BASED ON EXPERIMENTAL RESULTS

Specimens in this system were prepared either by hot-pressing or pressureless sintering. Starting materials used were mixtures of appropriate proportions of α-Si_3N_4, β-Be_3N_2, AlN, SiO_2, BeO and αAl_2O_3. No more than three of those compounds were used for single composition. The hot-pressing was carried out at 1750° C to 1900° C in BN-coated graphite dies at pressures of 28 to 30 MN/m^2 for 1 to 2 h. The cylindrical samples were 10 to 35 mm in diameter and weighed about 3 to 20 g. The temperatures were measured by Pt6Rh-Pt30Rh thermocouple and compared with an optical pyrometer. The furnace temperature was controlled by the optical pyrometer which varies the current through the die. The temperature was accurate within $\pm$ 15° C during equilibration. The pressureless sintering process will be described later. The metal concentration in some specimens were analyzed by atomic absorption (15) and the non-metals by hot-gas-extraction. Phases were identified by x-ray diffraction using monochromated CuKα-radiation

in a diffractometer. Some of the phases were identified using
light- and SE-microscopy.

4.1 The system $Si_3N_4-AlN-Al_2O_3-SiO_2$

The first results of an isothermal section at $1760°$ C of this
system were published by the authors in 1974 (11,14,16). The system shown in Fig. 6 is based on the results of 106 different compositions, hot pressed at 1760 to $1780°$ C. These compositions
can be found in the thesis (12). In the SiO_2 rich part of the
system, at $1760°$ C, six phases, Si_3N_4, Si_2N_2O, SiO_2 rich melt,
$3 Al_2O_3 \cdot 2 SiO_2$, Al_2O_3 and X_1 ($Si_4Al_4O_{11}N_2$) were observed. The
oxide solubility in $ß-Si_3N_4$ and in Si_2N_2O, is less than 3 eq.-%.
The equilibria in the binary $SiO_2-Al_2O_3$ at $1760°$ C were in good
agreement with those reported by Aramaki and Roy (17). The solid
solution formation in the compounds Si_2N_2O, $3 Al_2O_3 \cdot 2SiO_2$ and
X_1 are poor. The diffraction lines of X_1 were in good agreement
with those given by Oyama (18). However, the composition is
slightly different from those reported previously (4,19). The density measured by Archimedes principle is 3.05 g/cm^3. The melting
temperature was measured in a high temperature DTA at argon pressure as low as 2.6 x 10^{-3}, and was found to be between 1740 and
$1780°$ C. After melting, a drastic increase of the gas pressure in
the vacuum system occurred, due to the decomposition of the specimen. At $1760°$ C in the SiO_2 rich side of the $ß-Si_3N_4$ solid solution a 2-phase region ($ß$ + liquid) exists. As the temperature
decreases, a reaction takes place where $ß$ + liquid eventually
form X_1. This was also indicated by high temperature DTA measurements. For this reason, the dashed section of the diagram
may be valid at a temperature slightly $<1750°$ C.

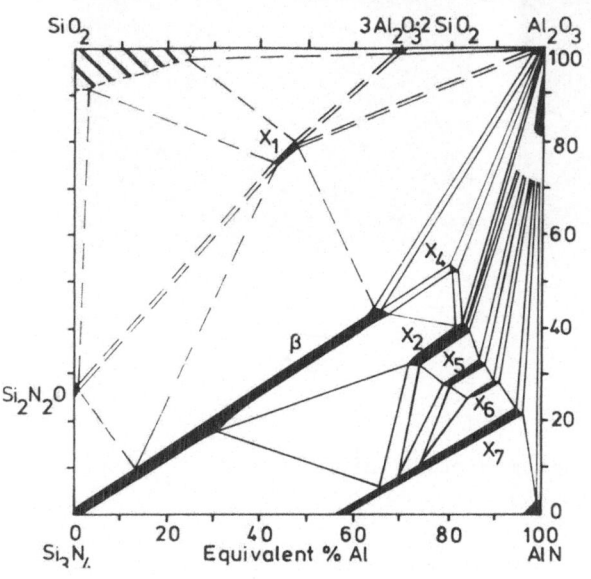

Fig. 6:

Isothermal phase diagram
of the system
$Si_3N_4-AlN-Al_2O_3-SiO_2$ at
$1760 °C$ (11).

48

As found by Oyama (5) and Jack (4), single-phase specimens should be obtained on the diagonal line Si_3N_4-Al_2O_3. However, such specimens were found to contain three phases: ß-Si_3N_4 ss, X_1 phase and liquid. This was not only indicated by x-ray diffraction but was also shown by light microscopy (Fig. 7). This examination procedure is absolutely necessary for specimens with low X_1 concentration and poorly crystallized third phase. In Fig. 7a, three phases are detectable. In the light micrographs the bright phase is elemental silicon. The elongated grains with higher reflectivity are the ß-phase and X_1. From their morphology it is here not possible to distinguish X_1 from ß. Material between the grains with low reflectivity is glass. The grains often show lamellae of varying orientation. This probably is due to twinning and growth imperfections.

In the AlN rich part of the system, a large ß-Si_3N_4 solid solution was found as well as a series of AlN rich phases X_2, X_4, X_5, X_6, and X_7. In the binary system Si_3N_4-AlN the solubilities in Si_3N_4 and AlN are less than 2 to 3 eq.-%. Between Si_3N_4 and AlN at 57 eq.-% Al a new phase (X_7) was found with a metal/non-metal ratio of 8:9 and the composition of $Si_3Al_5N_9$. In the subsystem AlN-Al_2O_3 good agreement with Lejus (20) was obtained. Single phase ß-Si_3N_4 region was determined very carefully.

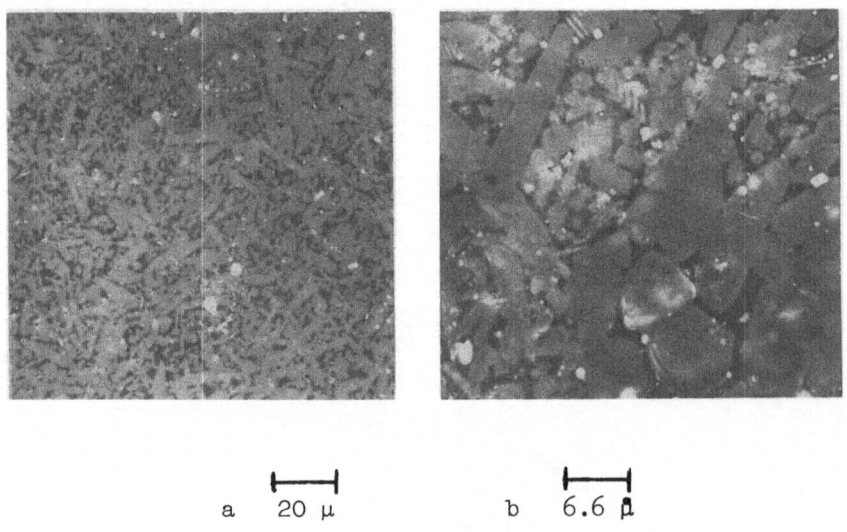

a   20 µ   b   6.6 µ

Fig. 7:

The microstructure of a specimen containing X_1 + ß-Si_3N_4 ss and liquid (light micrographs)

As reported earlier (11, 12, 21) the composition is described by:

$$Si_{6-x}Al_xO_xN_{8-x} \cdots\cdots\cdots\cdots\cdots\cdots\cdots\cdots 10$$

where x = 0 to 4.2 at 1760° C, which is different from

$$Si_{6-0.75x}Al_{0.67x}O_xN_{8-x} \cdots\cdots\cdots\cdots\cdots 11$$

where x = 0 to 8 as reported by Jack et.al. (4) and implied by Oyama (5).

This does not require the existence of non-metal interstitials or metal site vacancies in the Si_3N_4 lattice. Oyama (5) has suggested the existence of silicon vacancies as an explanation of his result, which are not in agreement with ours. Along the line of the metal/non-metal ratio (3:4) there is a linear increase of the lattice parameters of ß-Si_3N_4 with increasing Al and O concentration.

In Fig. 8 the microstructures of a hot-pressed ß-Si_3N_4 solid solution with x = 2.7 and a hot-pressed ß-Si_3N_4 specimen containing 3 wt.-% MgO are shown. The average grain size of the solid solution is larger as compared to the MgO containing material. The grains are less elongated. The grain size of the ß-Si_3N_4 solid solution increases with increasing aluminum and oxygen content. This influences some of the properties of these materials. In the specimens with increasing AlN concentration, 5 new phases (X_2, X_4, X_5, X_6, and X_7) were identified. Except for the phase X_4 the single phase fields are extensive along lines of constant metal/non-metal ratios. The structures of these phases were previously proposed by Jack (22) to be for X_4 (8 H), X_2 (15 R), X_5 (12 H), X_6 (21 R) and for X_7 (27 R) with the cell dimensions given in Table 2.

Phase	Ramsdall notation	hexagonal unit cell dimensions in Å	
X_4	8H	a = 2.988	c = 23.03
X_2	15R	a = 3.01	c = 41.81
X_5	12H	a = 3.029	c = 32.91
X_6	21R	a = 3.048	c = 57.19
X_7	27R	a = 3.059	c = 71.98

Table 2: Phases indentified by their x-ray patterns in the system Si_3N_4-AlN-Al_2O_3-SiO_2 (11) and the structures proposed by Jack (22).

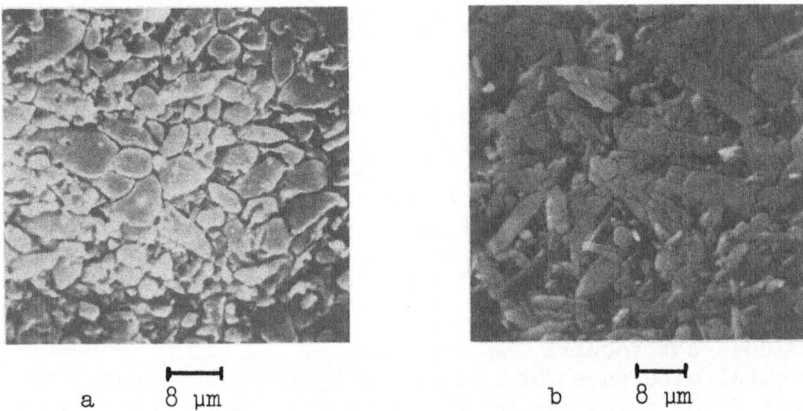

a   8 μm   b   8 μm

Fig. 8:

SEM-microstructures of hot-pressed β-Si$_3$N$_4$ at 1780°C, 1 h, 30 MN/m^2.
a: β-Si$_3$N$_4$ solid solution (x = 2.7)
b: β-Si$_3$N$_4$ with 3 wt.-% MgO.

    For the phases X$_4$ (8H) and X$_2$ (15R) the observed and the cal-
culated d-spacings were in good agreement. For X$_5$, X$_6$, and X$_7$,
however, they were not in good agreement for all compositions in
the solid solution ranges. This led to the assumption that these
phases probably belong to another, but similar set of AlN-polytype structures.

    The microstructures of these different AlN-polytypes are simi-
lar. The microstructure of the X$_2$ (15R) phase is shown in Fig. 9.

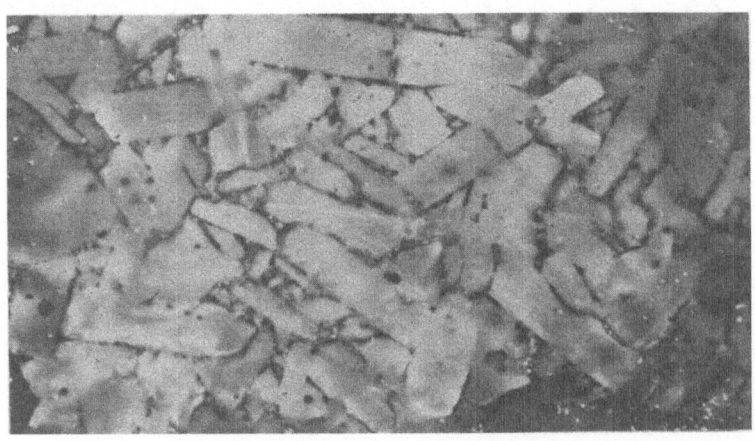

Fig. 9:

Light-micrograph of a X$_2$-phase, hot pressed at 1780°C, 1 h,
30 MN/m^2 (80 eq.-% Al, 30 eq.-% O).

The average grain size is 3 times larger compared to the β-Si_3N_4 ss and the grains being more elongated. The compounds X_2 to X_7 were found to be stable up to 1950°C. At this temperature the solid solution of X_6 and X_7 extend to the binary system AlN-Al_2O_3. The d-spacings of the phases X_6 and X_7 are in good agreement with those reported by Lejus (20) for the phase 'X' in the binary system AlN-Al_2O_3. This suggests that the phase 'X' consists of two phases as shown in a revised AlN-Al_2O_3 diagram in Fig. 10.

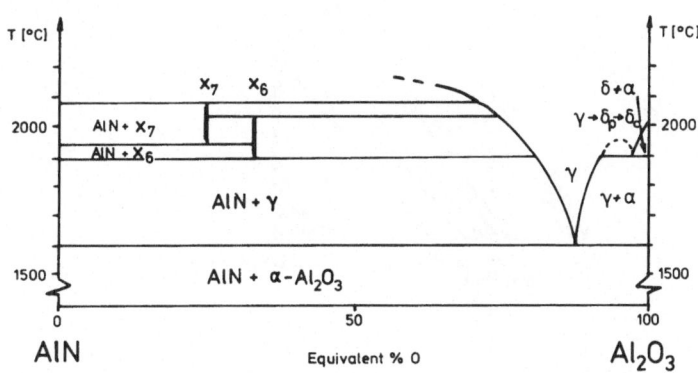

Fig. 10:

Proposed phase diagram AlN-Al_2O_3 with the AlN-polytypes X_6 and X_7 at temperatures above 1920°C (11, 12).

4.2. The system Si_3N_4-Be_3N_2-BeO-SiO_2 at 1780°C

The system Si-Be-O-N was reported by Huseby et.al. (23) in earlier work. For all phases investigated in this system, the solid solubilities extend in the directions in which the metal to non-metal ratios remain constant. For example, the large solubility range in β-Si_3N_4 extends towards Be_2SiO_4, as a large solubility range in $BeSiN_2$ extends towards BeO. The exchange of Si by Be and N by O decreases the unit cell volume.

In the Be_3N_2 rich part of the diagram a series of 8 new compounds were found: $Be_6Si_3N_8$, $Be_{11}Si_5N_{14}$, $Be_5Si_2N_6$, $Be_9Si_3N_{10}$, $Be_8SiO_4N_4$, $Be_6O_3N_2$, $Be_8O_5N_2$, and $Be_9O_6N_2$. In Fig. 11 it is shown for these phases, denoted with Y, that in these polytypes the metal/non-metal ratio extends from 1:1 ($BeSiN_2$) to 3:2 (Be_3N_2) i.e. from 1 to 1.5 while in the Si-Al-O-N system the ratio ranges from 3:4 (X_4) to 1:1 (AlN) i.e. from 0.75 to 1. The structures of the Be containing phases are reported to be very similar to those of the AlN polytypes (26).

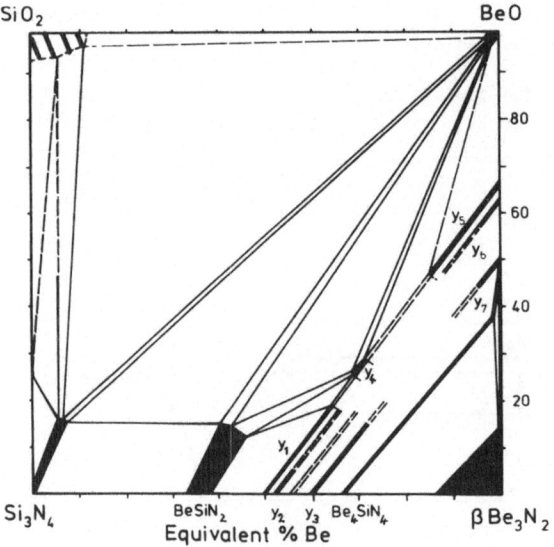

Fig. 11:

Isothermal section at 1780°C of the system $Si_3N_4-SiO_2-BeO-Be_3N_2$ (23) with the solid solution $\beta-Si_{6-x}Be_xO_{2x}N_{8-x}$.

4.3. The systems $Si_3N_4-AlN-Be_3N_2$ and $SiO_2-Al_2O_3-BeO$

The condensed system Si-Al-Be-N as well as the oxygen containing system Si-Al-Be-O are parts of the five component system shown in Fig. 5. Whereas the phase equilibria in the Si-Al-Be-O system are already well established (25, 26), no data are available for the Si-Al-Be-N system. Therefore hot-pressing experiments were necessary in the Si-Al-Be-N system using $\alpha-Si_3N_4$, $\beta-Be_3N_2$ and AlN as starting powders. The compositions of the specimens and the phase equilibria are shown in Fig. 12.

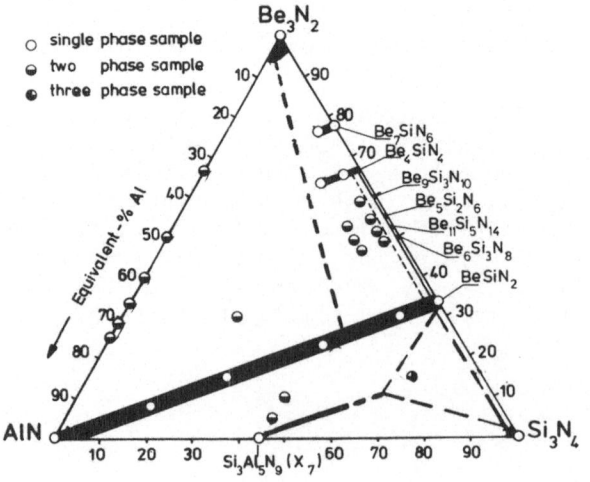

Fig. 12:

The system Si-Al-Be-N at 1780°C (27).

For ease of comparison equivalent-% were used as concentration units.

The subsystem Si_3N_4-Be_3N_2, except the phase with a constant metal to non-metal ratio of 4:3 at 77.5 eq.-% Be, was reported by Huseby (23). This phase was predicted by Thompson (24) and was afterwards identified in the specimens. No other phases were detected in the binary system Be_3N_2-AlN. In the quasiternary system a complete solid solubility exists from $BeSiN_2$ to AlN. In Fig. 13 the changes of the lattice parameters of the wurtzit-type struc-. ture are shown. A linear increase of the cell volume occurs with increasing Al concentration. From the phases in the Si_3N_4-Be_3N_2 subsystem only the phase Be_3N_4 showed a solubility greater than 5 eq.-% Al.

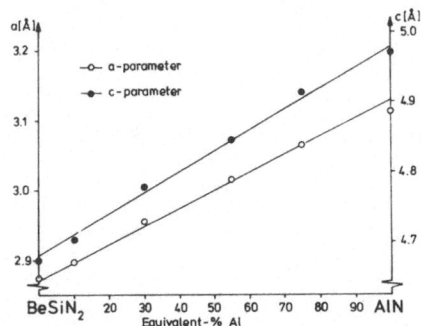

Fig. 13:

Change of lattice parameters in the solid solution AlN-$BeSiN_2$ at 1780°C (27).

4.4. The β-Si_3N_4 solid solution in the system Si_3N_4-AlN-Al_2O_3-SiO_2-BeO-Be_3N_2

Inside the quasiquaternary system, the single phase solid solution with β-Si_3N_4 structure restricted to the plane connecting points Si_3N_4, Be_2SiO_4, $BeAl_2O_4$ and AlN:Al_2O_3 was found, as shown in Fig. 14. These four points being coplanar in the quasiquaternary diagram. All composition points on this plane have a constant metal to non-metal ratio of 3:4. This plane is shown in Fig. 14b. A large area of single-phase region with the β-Si_3N_4 structure is located on this plane. No single-phase solid solution was found either above or below it. The lattice parameters changes in this solid solution are shown in Fig. 15. In the Al free solid solution (Fig. 15a) the a- and c-axis decrease with increasing Be concentration. In the Be free solid solution, the lattice parameters increase (Fig. 15b) with increasing Al concentration. Fig. 15c shows series of solid solutions with a constant concentration of 65 eq.-% Al, the lattice parameters decrease again with increasing Be content. A similar behaviour was found in the systems Si-Al-Mg-O-N and Si-Be-Mg-O-N (28).

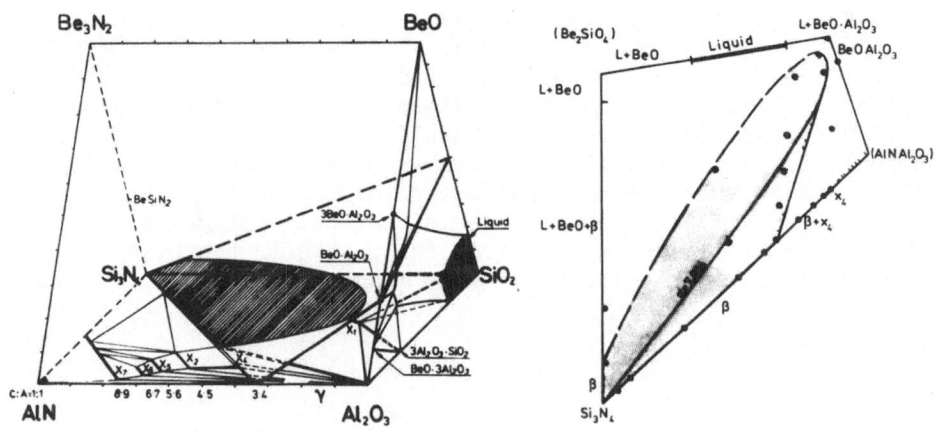

Fig. 14:

Phase diagram of the quaternary reciprocal salt system Si_3N_4-SiO_2-AlN-Al_2O_3-Be_3N_2-BeO at 1760°C (12).

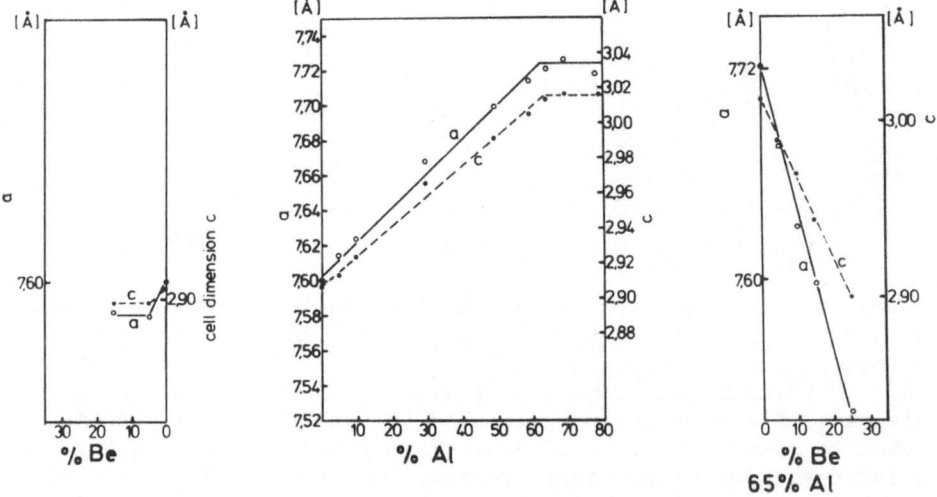

Fig. 15:

Unit cell dimensions of the β-Si_3N_4 solid solution at 1760°C in the subsystem Si_3N_4-AlN:Al_2O_3-BeO·Al_2O_3-Be_2SiO_4 (12).

4.5. Misfit Factor

The lattice of ß-Si$_3$N$_4$ expands when larger aluminum atoms dissolve
in the crystal and contracts when smaller beryllium atoms dissolve
in it. Naturally, oxygen atoms enter the lattice accompanying the
metal atoms during solid solution formation. The covalent radii
of oxygen is smaller than that of nitrogen. Therefore, oxygen
should also effect lattice parameters. When there are more than
one kind of metal atoms entering the solid solution, the lattice
parameters of ß-Si$_3$N$_4$ change in a more complex manner. From these
data, one tends to believe that there is an upper limit and a
lower limit of lattice distortion that the ß-Si$_3$N$_4$ can tolerate.
It would be very helpful if one could set up a rule using the tabulated values of atomic size to predict the solid solubility
limits of different elements in the ß-Si$_3$N$_4$. The following is an
attempt to do so. Of course, this method can only be applied to
substitutional ß-Si$_3$N$_4$ solid solutions containing no extrinsic
vacancy or interstitial. This seems to be true from the results
of Gauckler (11, 12) and Huseby (23) in the aluminum and beryllium
containing silicon nitride systems. One can assume that the
strongest factors influencing the solid solution formation would
be, among others, the size and charge differences between the
foreign atoms and the host. One can write an expression to account
for the lattice distortion of the host crystal caused by accommo-
dating foreign atoms. The following formula is used to express
these differences. Thereby 't' is defined as the "Misfit Factor".
The misfit factor 't' given below includes the influences from
both the size and charge from the foreign atoms in the lattice.
The size is in the formula as r_i and r_j, however, the charge of
the foreign atoms is accounted for by the factor n_j. n_j has
different values for the same value of n_i when the charge of the
foreign atoms are different. This difference comes from the
different number of oxygen atoms needed to balance the total charge
brought about by substituting different charged metal atoms for
silicon in solid solution. For instance, one oxygen is needed
for every aluminum atom and two oxygen atoms are needed for every
beryllium atom to form the stoichiometric solid solution of 3:4.

$$ t = \frac{\dfrac{3\Sigma n_i r_i}{r_{Si}\Sigma n_i} + \dfrac{4\Sigma n_j r_j}{r_N \Sigma n_j}}{7} \quad \ldots\ldots\ldots\ldots\ldots\ldots 12 $$

where n_i = fraction of the foreign metal atoms

     n_j = fraction of the foreign non-metal atoms

and r_j = radii of the foreign non-metal atoms.

When the misfit factor is plotted against the number of foreign atoms introduced, a map can be drawn to outline the tolerance
of the ß-Si$_3$N$_4$ lattice. The amount of foreign atoms in the mixture

56

can be described by the function:

$$f = \frac{\dfrac{3\Sigma n_i}{n_{Si}+\Sigma n_i} + \dfrac{4\Sigma n_j}{n_N+\Sigma n_j}}{7} \quad \dots\dots\dots\dots\dots \; 13$$

where n_{Si} and n_N are fractions of Si and N atoms respectively. The logarithm of the Misfit Factor 't' of some compositions in the systems Si, Al, Be/N, O and Si, Be, Mg/N, O were computed and are plotted against the function 'f' in Fig. 16. The single-phase composition were drawn as solid squares and triangles, and the multiphase compositions were drawn as open squares and triangles. Pauling's tetrahedral covalent radii were used for this computation. It is clearly shown in Fig. 16 that the single-phase region is confined in the center part of the plot where the misfit is smallest.

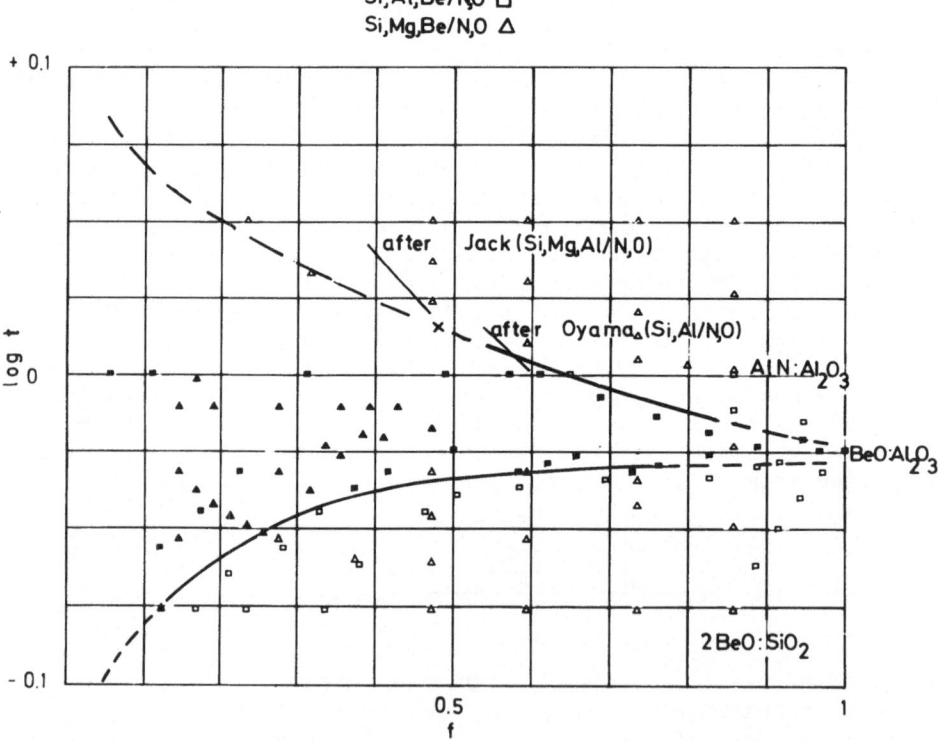

Fig. 16:

Logarithm of the misfit factor 't' of foreign atoms in β-Si$_3$N$_4$ metal oxides solid solutions.

Substitantial amount of foreign atoms can be accomodated in the
β-Si$_3$N$_4$ lattice where the misfit is small. When the misfit becomes larger, only small amounts of foreign atoms enter the lattice. A curve is drawn in the plot to outline the solubility
limit.

It should be noted that all of the data points were taken
from isothermal phase diagrams at about 1700 to 1800°C. The misfit
factors were calculated for the compositions laying on the solubility limits in the system Si$_3$N$_4$-Al$_2$O$_3$-MgO, reported by Jack (4),
and the system Si$_3$N$_4$-AlN-Al$_2$O$_3$, reported by Oyama (2). These
points were plotted in Fig. 16. It can be seen that these points
fit the curve quite well. It should be noted that only the composition having a metal to non-metal ratio of 3:4 from the results
of Jack (4) and Oyama (2) were used. This solid solution formation
restricted to the 3:4 plane indicates that the lattice β-Si$_3$N$_4$
does not accomodate high concentrations of extrinsic defects. This
being acceptable since one does not expect extrinsic defects in
excessive quantities to exist in highly covalent bonded compounds
such as silicon nitride.

More systems should be studied to test the usefulness of the
misfit factor. Additional compositions should be choosen among
the oxides and nitrides whose metal atoms normally fit in the tetrahedral sites. When we restrict our discussions to the compositions having a constant metal to non-metal ratio of 3:4, and
not assuming lattice extrinsic vacancy or interstitial, the compositions should have general chemical formula as given in Fig.
17. The formula are different for different charged cations. When
the concept of the reciprocal salt systems is used, all compositions should lay on the appropriate joins.

A combination of three ternary reciprocal salt systems will
be a quaternary reciprocal salt system. Nine of these systems are
given in Fig. 17 as examples. Joins of constant metal to non-
metal ratio of 3:4 are drawn in. When these joins are connected,
planes are formed. Generalized formulas of these solid solutions
are given underneath the corresponding diagrams. Our hypothesis
suggests that compositions on these planes having 't' and 'f'
factors inside the boundary in Fig. 17 will give single phase
solid solutions with β-Si$_3$N$_4$.

The misfit factor should be used with caution. There are
other factors which will affect the solid solubility. For example, the character of the chemical bonds in the β-Si$_3$N$_4$ metal
oxides solid solutions is an unknown factor which may have strong
influence. Solid solubility is also affected by the shape and relative positions of the minima on the free energy composition
curves of both of the phases in equilibrium at high temperature.
That is, the solubility limit of β-Si$_3$N$_4$ solid solution is also
affected by the free energy of the phases in equilibrium with it.

58

Therefore, the calculation proposed in this paper is only the
first approximation. However, this calculation will provide a
guide to approximate the solubility limits, when the value of
the atomic radii can be estimated.

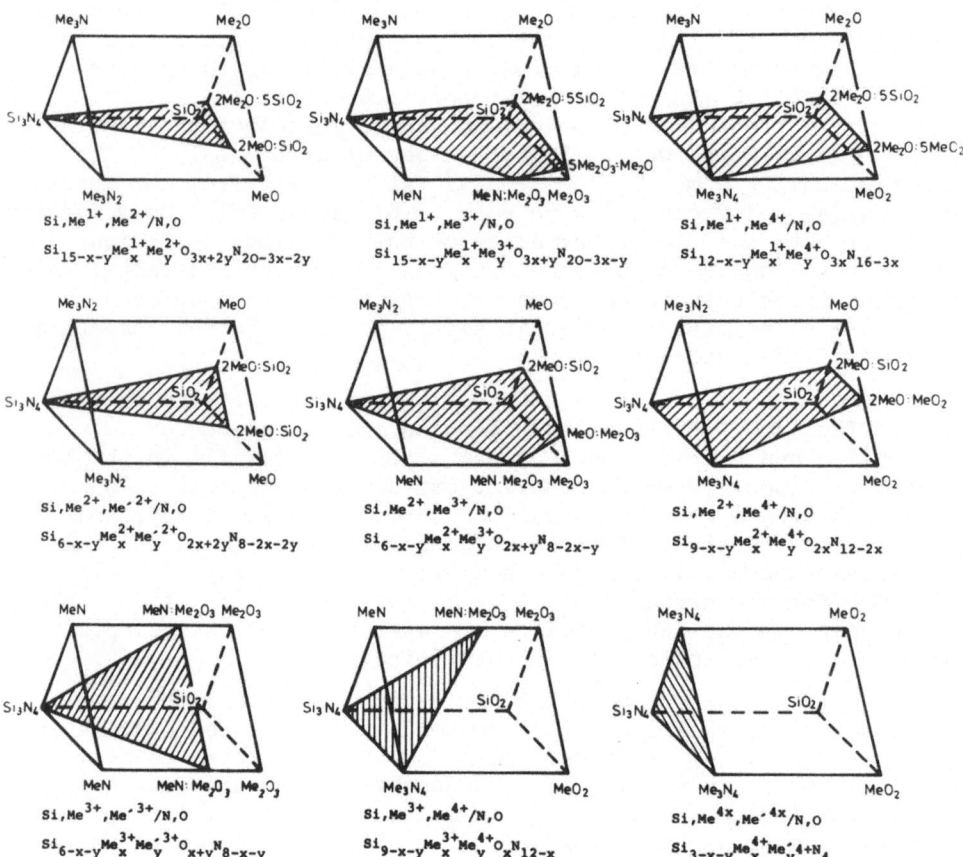

Fig. 17:

Phase diagrams of quaternary reciprocal salt systems showing the
planes of the metal to non-metal ratio of 3:4 for different
valencies of metal atoms.

5.   SUMMARY

The contradicting data on compositions and properties of Si_3N_4
based solid solutions are due mainly to uncertain and incorrect
phase diagrams in the past. The phase equilibria in the systems

Si-Al-O-N, Si-Be-O-N and Si-Be-Al-N were experimentally determined. The presentation of such metal-nitride/metal-oxide systems is possible in terms of reciprocal salt systems if the reaction:

$$12Si_3N_4 + 6Me_2^{(y+)}O_y \rightleftharpoons 4Me_3^{(y+)}N_y + 3ySiO_2 \dots\dots\dots\dots 14$$

might occur.

The knowledge, obtained from the studies in the system Si-Al-Be-O-N could be applied to other systems with similar metal and non-metal components. β-Si_3N_4 solid solutions could be predicted in other systems from a misfit factor, which accounts for the size and charge of the substituting atoms.

ACKNOWLEDGEMENTS

The work has been carried out with support from the Bundesministerium für Forschung und Technologie, BMFT, Bundesrepublik Deutschland, which is gratefully acknowledged. Thanks are due to T.Y. Tien, University of Michigan, Ann Arbor, U.S.A., and H.L. Lukas, Max-Planck-Institut für Metallforschung, Stuttgart, for helpful discussions.

REFERENCES

1. G. Petzow, F.E. Buresch; Z. f. Werkstofftechnik/J. of Mat. Techn. 3, Nr.5 (1972) 243-250.

2. Y. Oyama, O. Kamigaito; Japan J. Appl. Phys. 10, 1937 (1971).

3. K.H. Jack, W.J. Wilson; Nature Physical Science, 283, 28 (1972).

4. K.H. Jack; Trans. & K. Brit. Ceram. Soc. 72 (1973) 376-384.

5. Y. Oyama; Japan J. Appl. Phys. 11 760 (1972).

6. Y. Oyama; Japan J. Appl. Phys. 11 1572 (1972).

7. E. Jänecke; Z. phys. Chem. 51 (1908) 132; 71 (1911) 1.

8. Löwenherz; Z. phys. Chem. 13 (1894) 459.

9. R. Vogel; "Die heterogenen Gleichgewichte", Leipzig 1959.

10. J.E. Ricci; in "Phase Diagram of Fused Salts", Molten Salt Chemistry, M. Blander, Ed. Interscience Publ., N.Y., 1964

11. L.J. Gauckler, J. Huseby; presented at the Second Powder Metallurgical Seminar, Max-Planck-Institut, Stuttgart, 1974 see J. Amer. Ceram. Soc. 58 (1975) 346 and 58 (1975) 377.

60

12. L.J. Gauckler; "Gleichgewichtsuntersuchungen in den Systemen Si-Al-O-N und Si-Al-Be-O-N" Ph.D.thesis, Univ. Stuttgart 1976, 83.

13. E. Jänecke; Z. phys. Chem., 82, 1 (1913).

14. Am. Ceram. Soc. Fall Meeting Sept. 1975, for abstract see Am. Ceram. Soc. Bull. August 1975 p. 739, (18-BE-75F) see also reference 12.

15. L. Kotz, G. Kaiser, P. Tschoepel and G. Tölg; Fresenius Z. Anal. Chem., 260, 3 (1972) 207-209.

16. L.J.Gauckler, H.L. Lukas and T.Y. Tien; Mat. Res. Bull. 11 (1976) 503-312.

17. S. Amaraki, R. Roy; J. Am. Ceram. Soc. 42, 12 (1959) 644.

18. Y. Oyama; Yogyo-Kyokai-Shi, 82, 7 (1974) 351-357.

19. Y. Oyama, O. Kamigaito; Yogyo-Kyokai-Shi, 80, 8 (1972) 327-336.

20. A.M. Lejus; Rev. Hautes Tempér. et. Réfract. 1 (1964) 53-59.

21. R.L. Lumby, B. North, A.J. Taylor; in Special Cer. No. 6", P. Popper, Ed., Brit. Cer. Res. Ass. (1975) 283-308.

22. K.H. Jack; J. Mat. Sc. 11 (1976) 1135-1158.

23. J. Huseby, H.L. Lukas and G. Petzow; J. Am. Cer. Soc. 58 (1975) 377.

24. D.P. Thompson; J. Mat. Sc. (1976) 1377-1380.

25. S.M. Lang, C.L. Fillmore, L.H. Maxwell; J. Res. Nat. Cur. Stand. 48, 4 (1952) 301.

26. R.A. Morgan, F.A. Hummel; J. Am. Ceram. Soc. 32, 8 (1949) 255.

27. G. Schneider, L.J. Gauckler; (to be published).

28. L.J. Gauckler; (to be published).

DISCUSSION (Lang, Thompson)

Greskovich: In phase equilibrium studies on compositions which exhibit large weight losses at high temperature, how do you determine the final composition of the specimen?

Gauckler: If there are no free-metal inclusions, the following procedure is carried out: The specimens are dissolved in HF/HNO₃ at 180°C under 100 atm, and the metal ions afterwards are analysed by atomic absorption. Smaller amounts of N, and O, are analysed by hot-gas extraction. Larger amounts of N are analysed by the Kjeldal method.

Lumby: By using TEM and EPMA we have been able to positively identify the intergranular phase in the Si_3N_4-Al_2O_3 system as a silicate glass.

Cannon: Can you predict the partition of cation impurities between the various crystalline phases and the amorphous phase? This would be valuable for knowing the properties of amorphous phases which may exist with various crystalline phases, and the phases which may give better high temperature properties, or less sensitivity to impurities.

Gauckler: Not yet.

Brook: An important assumption in the misfit factor is that the phases preserve stoichiometric ratios of cation:anion. Since departures from stoichiometry are known in these systems (for example 3% solubility of O in Si_3N_4), how accurate do you believe the assumption to be? Is there experimental evidence for similar departures in the sialons, such as β'-sialons?

Gauckler: The misfit factor has only been calculated for stoichiometric Si_3N_4 solid solutions and takes into account the geometric differences in size and differences in charges of foreign atoms compared to the hosts. No calculation has been carried out for α-Si_3N_4 or other Si-Al-O-N phases except the β-Si_3N_4 ss. As the tolerance factor is only one of the factors which will influence the incorporation of foreign atoms in a lattice for very low and very high concentrations of dissolved atoms it is not very precise.

Lange: Solid solutions are known to decrease the thermal conductivity of materials. Thus, their study in Si_3N_4 systems is important for avoiding such systems for structural ceramics, and for finding them for insulating materials.

The solid solutions discussed for Si_3N_4 are substitutional. What is the reason for this?

Gauckler: Interstitials or lattice vacancies may both form the basis of deviations of the stoichiometry in a solid solution. In a four co-ordinated, highly covalently bonded material, these possibilities are not very likely. The bonds of an inserted atom must be saturated by other bonding orbitals coming from the lattice - this would make it necessary to change the strong bonds already satisfied from the host lattice. Vacancies in the lattice require opening a covalent bond. In a four co-ordinated structure this is not possible without changing its atomic arrangement to incorporate vacancies on lattice sites. The structure would no longer be the original, it would collapse and another phase would appear.

Morgan: You talk of using covalent radii but are you not in fact using crystal radii?

Gauckler: We used covalent radii, published by Pauling in The Nature of Chemical Bond.

A SIMPLE THERMOCHEMICAL MODEL FOR THE Si-Al-O-N SYSTEM

J.P. Torre and A. Mocellin

ABSTRACT : A thermochemical model has been constructed, for the pseudoquaternary system Si_3N_4-AlN-Al_2O_3-SiO_2, based on a limited number of simplifying assumptions. Among these was that experimental coexistence diagrams for the condensed phases could be considered true equilibrium diagrams at 1750°C. Also, when homogeneity ranges exist in the condensed phases, the latter are taken as ideal pseudo-binary or-ternary associations of the end members : Si_3N_4, Al_2O_3, SiO_2, AlN, Si_2N_2O.
The compositions of the gas phase in equilibrium with one or several others, could be computed and the various results where shown to be all compatible with each other. The calculations also clearly suggest that the atmosphere plays a dominant role in the formation of other phases and that no satisfactory understanding of the corresponding mechanisms can be expected, unless potential vapour transfers are properly taken into account. It has also been possible to estimate the effects of slight temperature or total pressure changes on the equilibrium gas-phase composition and on the stability fields.

1. INTRODUCTION

Chemical reactivity and phase relationship studies in the Si_3N_4-Al_2O_3 system have revealed the existence of a range of joint solid solubilities of aluminium and oxygen in the β-Si_3N_4 lattice. These so-called "β-Sialon" solutions were at first thought to be quasi binaries with an Al/O ratio of 2/3 |1-4|. It was subsequently shown that AlN could also be accommodated by silicon nitride, together with Al_2O_3 |5-6|. But the presence of a second phase (labelled "X") |2,7,9|, in sintered AlN-weak mixtures later suggested that the

homogeneity range of the β' solution lay primarily along the
Si₃N₄-(Al₂O₃.AlN) join of the pseudo ternary with a chemical formula close to :

$$Si_{3-x}Al_xO_xN_{4-x}(x \leqslant 2.25) \qquad\qquad |\ 7-9\ |$$

Other phases containing all four elements have also been observed
in the Si-Al-O-N system | 7-14 |. But the reaction mechanism leading
to any of these are still incompletely understood. The formation
of "X" during the sintering of Si₃N₄-Al₂O₃ mixtures has been claimed
to be related to the presence of SiO₂ layers on the particles of the
starting Si₃N₄ powders |7,10| but quite sizable amounts of "X" have
been occasionally found in some samples | 4-6| suggesting that sur-
face silica may not be the only cause for "X" formation. Weight losses also have been noted during sintering. So far, they have only
been empiricaly controlled |4,10,11| , and the phases observed after
such sintering treatments on compacts of blended Si₃N₄, Al₂O₃, AlN
and SiO₂ powders, have been more or less explicitely assumed to
result from reactions between condensed phases only, i.e. with no
participation of the atmosphere in the reaction mechanisms. This is
probably an oversimplification and in view of the available experimental evidence, it would seem more realistic a priori to allow for
gas-solid interactions in discussions about phase stability in the
Si-Al-O-N system. In the present paper then, a thermochemical model
based on a few simplifying assumptions, will be proposed. The phase
stability conditions will thence be theoretically established with
the presence of the vapor phase being explicitly taken into account.

2. ELEMENTS OF THE MODEL

Systematic X-ray diffraction analysis of the phases present in reactively sintered Si₃N₄-SiO₂-Al₂O₃-AlN blends has allowed experimental pseudoquaterny isothermal coexistence diagrams for the
various condensed phases |8,9,12| found in the Si-Al-O-N system
to be drawn. The most recent of these diagrams |12| is shown on
figure 1. They are constructed in the Si₃N₄-SiO₂-Al₂O₃-AlN plane
section of the basic Si-Al-O-N tetrahedron. A regular square is
obtained by expressing the concentrations in equivalents just as
in a classical reciprocal salt system |15|. Other phases, which do
no appear on figure 1, have recently been observed in the AlN rich
part of the square. Information on these being however still quite
limited, |7,8,9,12,14|, they have not been taken into account in
the present study, which is thus only concerned with the silica
rich part of the pseudo-quaternary. It is believed that such a restriction should not be of much consequence in practice since interest in this class of materials has been based primarily on their
potential use in oxidizing environments. Given such a framework,
the computational results to be presented below, are based on the
following four assumptions.

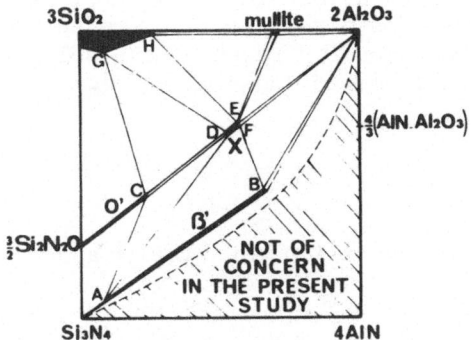

Fig. 1. The Si_3N_4-SiO_2-Al_2O_3-AlN system |12|

1st assumption

The experimental diagram shown on figure 1 will be taken as a true
equilibrium diagram at 1750°C.

2nd assumption

The small amount of nitrogen which, according to figure 1, can be
incorporated in the liquid SiO_2-Al_2O_3 solution |16| will be neglec-
ted, and the latter will be considered an ideal binary solution
|17|, and labelled : $SiO_2s(Al_2O_3)$.

3rd assumption

Figure 1 and the available experimental information about atomic
arrangements in the Si-Al-O-N phases |4,5,12,14| suggest that both
0' and X can be considered as binary associations of $\{Si_2N_2O\}$ and
$\{Al_2O_3\}$ groupings. It will moreover be assumed that 0' is an ideal
solid solution. But, such an assumption cannot hold for phase "X" :
the ideality of $SiO_2s(Al_2O_3)$ would not be consistent with the presumably very small homogeneity range of "X" (Appendix I).
In view of this small range however, the activities of $\{Si_2N_2O\}$
and $\{Al_2O_3\}$ in "X" will merely be assumed to be linear functions
of the corresponding concentrations.

4th assumption

The homogeneity range of β' will be assumed to lie along the
Si_3N_4-Al_3O_3N join, and $\{Si_3N_4\}$, $\{Al_2O_3\}$ and $\{AlN\}$ groupings to remain
undissociated with the solution; $\{AlN\}$ and $\{Al_2O_3\}$ groupings will

66

thus be considered to be in equal concentrations in the solution.
It will then be assumed that the three dissolved species have an
ideal behaviour on some part of the homogeneity range of the solution, the extent of which has to be determined and will be represented by x and y such that :

$$x \leqslant c_{Al_2O_3}^{\beta'} = c_{AlN}^{\beta'} \leqslant y$$

where $c_{Al_2O_3}^{\beta'}$ and $c_{AlN}^{\beta'}$ are the molar concentrations of {Al$_2$O$_3$} and
{AlN} in β'.
Thermochemical activity plots for the various dissolved species,
readily follow from the above assumptions and are given on figure 2.
Also indicated are the compositions corresponding to particular special compositions labelled A,B,C... in the reference diagram
(figure 1). Numerical values are given in Table I, as they were
read on the diagram. There are three independant
components in this system which therefore has two degrees of freedom when two condensed phases and the vapour are in equilibrium
with each other. If the temperature is fixed (1750°C in what follows) there is only one degree of freedom left and so the system
will be fully determined once one concentration has been assigned
a value. Equilibrium conditions between two of the condensed phases can thus be calculated as functions of any one concentration
or activity. But in practice, computations can be made only if
explicit numerical values are given to the parameters of the model,
which appear on figure 2. Some "analytical constraints" inferred
from Assumption 1, at once reduce the number of these parameters
(Table II). It may be noted that, as well as the phase rule, they
would hold true for any system exhibiting a phase diagram similar
to that of figure 1.

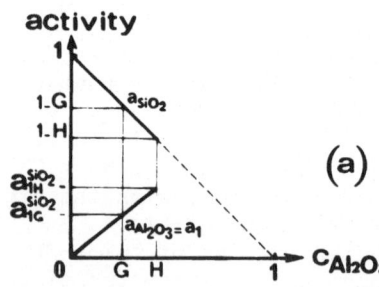

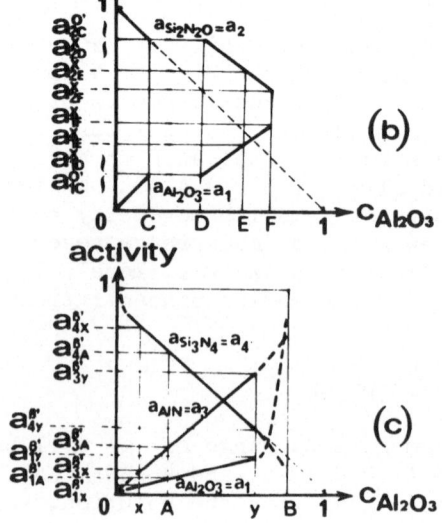

Fig.2. Thermochemical activity
plots :
a) SiO$_2$ s(Al$_2$O$_3$)
b) O' and X
c) β'

Table 1 : Compositions at some special points in the reference diagram (Figure 1)

Point of the diagram	Phase	molar concentration %				
		SiO_2	Al_2O_3	Si_2N_2O	AlN	Si_3N_4
A	β'	–	5	–	5	90
B	β'	–	36	–	36	28
C	O'	–	29	71	–	–
D	X	–	58	42	–	–
E	X	–	62	38	–	–
F	X	–	63	37	–	–
G	$SiO_2S(Al_2O_3)$	96	4	–	–	–
H	$SiO_2S(Al_2O_3)$	80	20	–	–	–

Table II : Analytical constraints internal to the model

Phases in equilibrium	Representative points on Figure 1	Relations between parameters of the model
O' and X	C,D	$a_{1C}^{O'} = a_{1D}^{X}$ (1) $a_{2C}^{O'} = a_{2D}^{X} = 0.71$ (2)
O', X and $SiO_2S(Al_2O_3)$	C, D, G	$a_{1C}^{O'} = a_{1D}^{X} = a_{1G}^{SiO_2}(=\frac{G}{H}a_{1H}^{SiO_2})$ (3)
X mullite and $SiO_2S(Al_2O_3)$	E, H	$a_{1E}^{X} = a_{1H}^{SiO_2}$ (4)
X and Al_2O_3	F	$a_{1F}^{\mathbf{X}} = 1$ (5)
O', X and β'	C, D, A	$a_{1A}^{\beta'} = a_{1C}^{O'} = a_{1D}^{X}$ (6)
β' and Al_2O_3	B	$a_{1y}^{\beta'} = 1$ if y = 1 (7)

3. COMPUTATION METHOD

The purpose of the computations is to specify the limits within which the parameters that are not fixed by the analytical cons-traints of Table II must be chosen in order that sensible numerical values can be calculated for $\{Al_2O_3\}$, $\{AlN\}$, $\{Si_3N_4\}$, $\{Si_2N_2O\}$, and $\{SiO_2\}$ concentrations and the composition of the vapour phase be reached for each individual equilibrium. From equations (1) to (6) (Table II) and from figure 2, it follows that :

$$a_{1A}^{\beta'} = a_{1C}^{o'} = a_{1D}^{x} = a_{1G}^{SiO_2} = 0.14$$

$$a_{1E}^{X} = a_{1H}^{SiO_2} = 0.86$$

Then the only adjustable parameters left are :

$$x, \; y, \; a_{3y}^{\beta'}, \; a_{4x}^{\beta'} \; a_{2F}^{X}$$

These will be referred to as p_k ; $k = 1,\ldots,5$

Considering the relative stabilities of the possible vapour spe-cies, only four of them were assumed to be possibly present at 1750°C : N_2, O_2, SiO and Al_2O |18-22|. Then for any equilibrium between two condensed phases involving n different condensed che-mical species, the equilibrium parameters can be calculated as functions of a given concentration (which will be P_{N_2}) thanks to $(n + 4) - C = n + 1$ equilibrium equations (C : number of indepen-dant components i.e. 3). The resolution was carried in three suc-cessive steps :

1st step

For each equilibrium (labelled : "j") between two condensed phases, the equation giving each equilibrium parameter x_j^i as a functin of P_{N_2} and of the adjustable parameters of the model $\{p_k\}$ has been written.

2nd step

Values of each p_k warranting the existence of some continuous P_{N_2} range for the definition of $x_j^i = f_j^i(P_{N_2})$ functions have then been computed by trial and error. Let these values be $\xi_j(p_k)$.

3rd step

For each p_k, the intersection :

$$\xi(p_k) = \bigcap_j \xi_j(p_k)$$

has finally been determined and checked not to be empty.

Consider, as an example, the equilibrium between "β'" and "X". The various condensed species involved in this equilibrium are :

$\{AlN\}$, $\{Al_2O_3\}$, $\{Si_3N_4\}$ in β' ; $\{Al_2O_3\}$ and $\{Si_2N_2O\}$ in X.

If P_{N_2} is given, then 13 unknowns remain (3 partial pressures in the vapour phase, 3 activites and 3 concentrations in X).

The necessary 13 independant equations must comprise 6 equilibrium equations to be chosen among all the linear cominations of :

$$Si_3N_4(\beta') + \frac{3}{2} O_2 = 3 SiO + N_2 \qquad (8)$$

$$2 Si_3N_4(\beta') + \frac{3}{2} O_2 = 3 Si_2N_2O(X) + N_2 \qquad (9)$$

$$Al_2O_3 (X) \qquad\quad = Al_2O_3(\beta') \qquad (10)$$

$$Al_2O_3 (X) + N_2 \quad = 2 AlN(\beta') + \frac{3}{2} O_2 \qquad (11)$$

$$Al_2O_3 (X) \qquad\quad = Al_2O + O_2 \qquad (12)$$

plus the equilibrium equations between condensed species within each solution. The corresponding mass action laws are deduced from figure 3 and from equations (1) to (6) :

$$a_{Si_2N_2O}^X = \frac{0.14\,(1-a_{2F}^X)}{0.86} - \frac{0.71-a_{2F}^X}{0.86} \cdot a_{Al_2O_3}^X \qquad (13)$$

$$a_{Al_2O_3}^{\beta'} = \frac{2.80}{a_{3y}^{\beta'}} \cdot a_{AlN}^{\beta'} \cdot y \qquad (14)$$

$$a_{Si_3N_4}^{\beta'} = \frac{a_{4x}^{\beta'}}{2.80(x-1)} \cdot a_{Al_2O_3}^{\beta'} - \frac{a_{4x}^{\beta'}}{x-1} \qquad (15)$$

$$a_{Si_3N_4}^{\beta'} = \frac{a_{4x}^{\beta'}}{x-1} \cdot \frac{y}{a_{3y}^{\beta'}} \cdot a_{AlN}^{\beta'} - \frac{a_{4x}^{\beta'}}{x-1} \qquad (16)$$

Relations between activities and concentrations, i.e. plots on figure 3, must be expressed analytically :

$$a_{Al_2O_3}^X = \frac{0.86}{F-D} \cdot c_{Al_2O_3}^X + \frac{0.14 \cdot F-D}{F-D} \tag{17}$$

$$a_{Si_2N_2O}^X = \frac{0.71-a_{2F}^X}{D-F} \cdot c_{Al_2O_3}^X + \frac{a_{2F}^X \times D - 0.71 \cdot F}{D-F} \tag{18}$$

$$a_{Si_3N_4}^{\beta'} = \frac{a_{4x}^{\beta'}}{x-1} \cdot c_{Al_2O_3}^{\beta'} - \frac{a_{4x}^{\beta'}}{x-1} \tag{19}$$

$$a_{AlN}^{\beta'} = \frac{a_{3y}^{\beta'}}{y} \cdot c_{Al_2O_3}^{\beta'} \tag{20}$$

$$a_{Al_2O_3}^{\beta'} = 2.80 \cdot c_{Al_2O_3}^{\beta'} \tag{21}$$

Mass balance equations finally, must also be taken into account :

$$c_{Al_2O_3}^X + c_{Si_2N_2O}^X = 1 \tag{22}$$

$$c_{Al_2O_3}^{\beta'} + c_{AlN}^{\beta'} + c_{Si_3N_4}^{\beta'} = 1 \tag{23}$$

$$c_{AlN}^{\beta'} = c_{Al_2O_3}^{\beta'} \tag{24}$$

$$P_{O_2} + P_{N_2} + P_{Al_2O} + P_{SiO} = P_{total} \quad (= 1atm) \tag{25}$$

Computation has been practically conducted from equations (8) to (13), (17, (17), (19), (21), (22) to (25), the choice of which is further discussed in Appendix I. Numerical values fo the standard reaction free energies used throughout this work are summarized in Table III.

<u>Table III</u> : Standard Gibbs free energies

Reaction	$\Delta G°$ (cal/mole)	Reference
$3\ Si + 2\ N_2 = Si_3N_4$	$-209000 + 96.8T$	(20)
$Si_3N_4 + \frac{3}{2} O_2 = Si_3N_4$ $3\ SiO + 2\ N_2$	$-107000 - 147.8T$	(20),(21)
$2\ Si_3N_4 + \frac{3}{2} O_2 =$ $3\ Si_2N_2O + N_2$	$-287000 + 25.4T$	(20)
$Al_2O_3 + N_2 =$ $2\ AlN + \frac{3}{2} O_2$	$-246400 - 25T$	(19),(21)
$Al_2O_3 = Al_2O + O_2$	$-371200 - 94T$	(22)
$Si_2N_2O + \frac{3}{2} O_2 =$ $2\ SiO_2 + N_2$	$-220400 + 24.4T$	(20),(21)
$2\ SiO_2 + 3\ Al_2O_3 =$ $3\ Al_2O_3 \cdot 2\ SiO_2$	$- 3600 + 15.6T$	(19),(21)

It appeared on the final results that SiO is the prevailing gaseous species besides nitrogen. The variation ranges of the adjustable parameters of the model on the other hand, were shown to be restricted to :

$$0.70 \geqslant a_{2F}^{X} \geqslant 0.65$$

$$x = 0.02; \ y = 1$$

$$0.002 \geqslant a_{4x}^{\beta'} \geqslant 0.001$$

$$0.05 \geqslant a_{3y}^{\beta'} \geqslant 0.01$$

4. RESULTS AND DISCUSSION

4.1. Equilibrium diagram

Results of the computations show that the whole problem can be
solved by choosing the $\{p_k\}$ values as indicated above. Oxygen
partial pressure variations versus P_{N_2} are given in figure 3 for
the various equilibria between two condensed phases. For each curves the uncertainty related to the ranges of possible values of
the adjustable parameters of the model has also been represented.
Two dimensional fields in such a diagram correspond to the stability for each individual condensed phase, in equilibrium with the
gas phase only. Within each of those fields, the system has two
degrees of freedom and P_{O_2} cannot be calculated as a function of
P_{N_2} unless another concentration has been fixed. For example,
iso-$C^{\beta'}_{Al_2O_3}$ curves have been drawn in the single-phased β' part of
the diagram. Similarly, P_{Al_2O} versus P_{N_2} curves could in principle
be drawn to represent the various equilibria, but they would have
little interest in practice. Generally speaking within the P_{N_2}
range of interest (i.e. 5×10^{-2} atm $\leqslant P_{N_2} \leqslant 5 \times 10^{-1}$ atm) the
following relations were confirmed :

$$10^{-8} \text{ atm} \leqslant P_{Al_2O} \leqslant 10^{-4} \text{ atm}$$

$$P_{N_2} + P_{SiO} \simeq 1 \text{ atm}$$

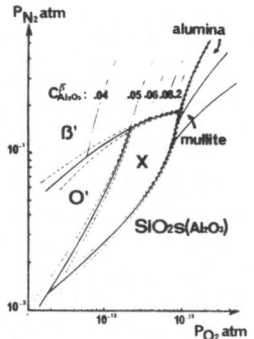

Fig. 3. Equilibrium diagram 1750°C, 1 atm

4.2. Analytical validity of results

The first point to be emphasized is that the following discussion
only may have any significance in as much as the reference diagram
(Figure 1) is correct. On the other hand there is no a priori reason to expect a possible extension of the model to the AlN rich part
of this diagram. Neither have the existence of condensed SiO (still
a matter of controversy) or of alpha-Si_3N_4 been taken into account
(Appendix I). It is however possible to roughly estimate the maximum error which may result from the simplifying assumptions of the
model. For example, in the case of the β'-X equilibrium, neglecting
uncertainties on the thermochemical data (Table III) from equation
(12) :

$$\frac{\Delta P_{Al_2O}}{P_{Al_2O}} + \frac{\Delta P_{O_2}}{P_{O_2}} = \frac{\Delta a_{Al_2O_3}}{a_{Al_2O_3}} \qquad (26)$$

The variation range of $a_{Al_2O_3}^{\beta'}$ is limited by the "analytical cons-
traints" (Table II) and Assumption 1. The minimum value of $a_{Al_2O_3}^{\beta'}$
(0.14 as mentionned before) can be calculated remembering that
$SiO_2 s(Al_2O_3)$ is an ideal solution (Assumption 2) and that $a_{Al_2O_3}^X$
approximately is a linear function of $c_{Al_2O_3}^X$ (certainly the
less questionable assumption of the model in view of the available
experimental data about the SiO_2-Al_2O_3 binary |16-17|, and the
small homogeneity range of "X"). As activity versus concentration
variations can only be monotonous in homogeneous solutions it
follows that the maximum value of $\dfrac{\Delta a_{Al_2O_3}}{a_{Al_2O_3}}$ is about 10. Hence

$\dfrac{\Delta P_{Al_2O}}{P_{Al_2O}}$ and $\dfrac{P_{O_2}}{P_{O_2}}$ both are smaller than one order of magnitude,

which is insignificant given the absolute values of P_{Al_2O} and P_{O_2}.
The maximum absolute error made by assuming that $P_{N_2} + P_{SiO} \simeq 1$ atm
is equal to the maximum value of $P_{O_2} + P_{Al_2O}$ i.e. 10^{-4} atm. The
gas phase composition is thus believed to be adequately known for
practical purposes.

The foregoing argument about the β'-X equilibrium may be extended
to any other two phase equilibrium since Equation (12) can always
be invoked. It should however be pointed out that iso-concentration
profiles in the condensed phases such as those for β', shown on
figure 3, are less precisely known than the field boundaries. They
should only be taken as qualitative estimates.

It may finally be of practical interest to use the present model
and computation scheme for investigating the influence of slight
temperature and total pressure changes on the relative stabilities
of the various phases in the system. Adding to the basic hypothe-
ses a further assumption that the diagram on figure 1 is not substantially modified by a ± 25°C temperature variation and/or a
0.1 atm total pressure increase, then computations similar to those
previously discussed can be carried out. The corresponding average
results are shown on figures 4 to 6. It appears among others that,
in accord with experimental observations |23|, a reduction in the
extent of the "X" phase field is predicted when temperature is
decreased or pressure increases.

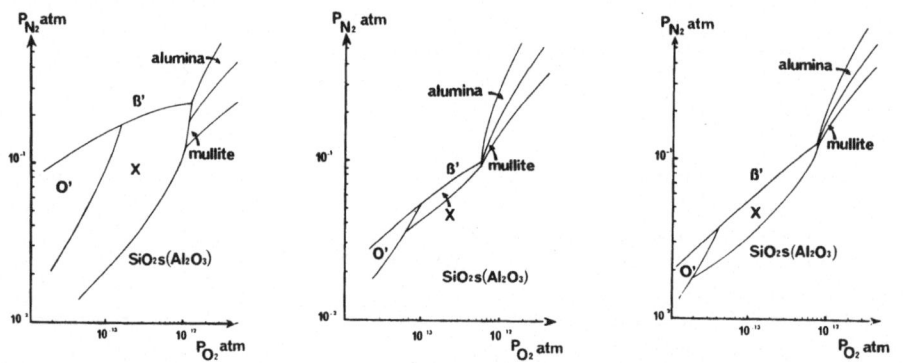

Fig. 4. 1775°C, 1 atm   Fig. 5. 1725°C, 1 atm   Fig. 6. 1750°C, 1.1 atm

5. CONCLUSION

The present work is believed to show that :

(1) Equilibrium conditions between condensed phases in a sizable
portion of the Si-Al-O-N system can theoretically be established,
starting from already available experimental information.

(2) A computation scheme based on a limited number of simplifying
assumptions has been established and applied to the Si_3N_4-AlN-Al_2O_3-
SiO_2 planar section of the Si-Al-O-N tetrahedron, except for the
AlN rich part of this section where experimental information is
still insufficient.

(3) Results of the computation confirm the importance of atmosphere
compositions in setting the existence and stability conditions of
individual condensed phases in the system. It follows that gas
transfers within the compacts or exchanges with the environment are
more than likely to take place during the practical reaction and
sintering treatments that are usually being carried.

(4) Slight changes of the system temperature or total pressure substantially modify the stability conditions of condensed phases for

a given atmosphere composition, as is suggested by the diagrammatic
representation of the theoretical computation results.

Acknowledgments:

An essential ingredient of this work : the reference diagram of
Figure 1 was most kindly made available by Professor K.H. JACK,
prior to publication. Very useful discussions with Dr. G. LESOULT
of this laboratory are also gratefully acknowledged. Financial
support was provided by D.R.M.E. under Contract No. 74/531.

APPENDIX I - COMPLEMENTS ABOUT THE MODEL AND COMPUTATIONS

A/ Thermochemical description of phase "X"

If phase "X" is assumed to be an ideal solution, then equation (5)
leads to:

$$a^X_{1E} = \frac{E}{F} \text{ and } a^X_{1D} = \frac{D}{F}$$

Hence, using equations (3) and (4) :

$$a^X_{1D} = \frac{D}{F} = \frac{G}{H} \cdot a^{SiO_2}_{1H} = \frac{G}{H} \cdot \frac{E}{F}$$

which is not possible according to Table I.

Making $\frac{D}{F} = \frac{G}{H} \cdot \frac{E}{F}$ become true would imply that everything else
being the same, $\frac{D}{F}$ is reduced, i.e. the homogeneity range of "X"
is much greater than reported in Figure 1. This would be in con-
tradiction with the most recent experimental observations |8,12,13|.

B/ Complements about the β'-X equilibrium

Any other activity versus concentration equation than used in the
text can be deduced from equations (17) to (21) using equations
(22) to (24). Fifteen equations (6 equilibrium equations plus
equations (17) to (25) are thus available in order to solve for the
thirteen unknowns of the problem. However these equations are not
independent: the set of equilibrium equations must involve the
literature data, which are represented by equations (8) to (12);
as 6 equations are necessary, one of the equations (13) to (16),
which only are consequences of the assumptions of the model, must
be used. Using equations (22) to (24), it can be shown then that
one of the equations (17) to (21) becomes redundant. For instance:
choosing equation (13) i.e. the equilibrium equation between Al_2O_3

and Si_2N_2O within phase "X", equation (17) can be obtained from (13), (18) and (22); conversely equation (18) can be obtained from (13), (17) and (22). The total number of equations is thus reduced to 14. The existence of the second extra equation must be related to assumption 4 concerning phase β' which leads to the equilibrium equation (15) between $\{Al_2O_3\}$ and $\{AlN\}$ within β'. This last equation must be compatible with the following one:

$$Al_2O_3(\beta') + N_2 = 2 \; AlN \; (\beta') + \frac{3}{2} O_2$$

which is obtained from (10) and (11). All redundancies therefore are only apparent.

C/ Condensed SiO, and $\alpha-Si_3N_4$

The existence of the equilibrium:

SiO (condensed) = SiO (gas)

has been claimed by several authors |30-32|, with a corresponding constant less than 0.1 at 1750°C |21,32|. Hence condensed SiO should appear as soon as P_{SiO} becomes higher than 0.1 atm. As the present computations show that:

$P_{N_2} + P_{SiO} \simeq 1$ atm, it follows from figure 3 that none of the condensed phases under study could exist alone, i.e. as a stable compound in the absence of condensed SiO. As a consequence, the observed conditions of coexistence for the phases mentioned on figure 1 would not correspond to equilibrium states unless the presence of condensed SiO were properly taken into account. But as this latter compound only has been reported in an amorphous state, its stability remains questionable, and it can safely be excluded from discussions on stability diagrams.

According to some authors |24-27| α and $\beta-Si_3N_4$ are respectively low and high temperature crystallographic modifications of the same compound. In such a case, $\alpha-Si_3N_4$ is not of concern here. Other authors |28,29| have suggested however that $\alpha-Si_3N_4$ was an oxynitride having a defect structure and a small range of homogeneity ($Si_{11.4}N_{15}O_{0.3}$ - $Si_{11.5}N_{15}O_{0.5}$). According to this latter assumption the equilibrium conditions between α and o' on the one hand and between β and O' on the other hand could have been computed using the method proposed in the present work. The results however would have been highly imprecise. Consider for instance the following equilibrium equation, which would necessarily have entered the computations:

$$\frac{76}{3} \beta \; Si_2N_4 + O_2 = \frac{20}{3} Si_{11.4}N_{15}O_{0.3} + \frac{2}{3} N_2$$

The corresponding equilibrium constant is:

$$K = \exp \left(\frac{4 \times \frac{20}{3} \Delta G°_\alpha - \frac{76}{3} \Delta G°_\beta}{RT} \right)$$

with ΔG_α° = standard free enthalpy of formation of $\frac{1}{4}$ $Si_{11.4}N_{15}O_{0.3}$
and ΔG_β° = standard free enthalpy of formation of β-Si_3N_4, both known
|28,29| with the same uncertainty ($\pm$ 5 kcal). It can then be shown
that the relative uncertainty for K at 1750°C is about 6500% !

REFERENCES

1. K.H. Jack, W.I. Wilson, Nat. Phys. Sci. 238, 1972, 28.
2. K.H. Jack, Trans. & J. Brit. Ceram. Soc. 72, 1973, 376.
3. Y. Oyama, O. Kamigaito, Yogyo-Kyokai-Shi, 80, 1972, 327.
4. Y. Oyama, Japan J. Appl. Phys. 12, 1973, 500.
5. Y. Oyama, Japan. J. Appl. Phys. 11, 1972, 760.
6. Y. Oyama, Yogyo-Kyokai-Shi, 82, 1974, 7.
7. J. Lumby, Special Ceramics 6, BCRA, P. Popper ed., 1974.
8. L.J. Gauckler, H.L. Lukas, G. Petzow, J.Am.Ceram.Soc. 58, 1975, 347.
9. K.H. Jack, Journées d'Etude sur les Nitrures II, U.E.R. des Sciences, Limoges, 1975.
10. F.F. Lange, Contract N0019-73-C-0208, Naval Air Systems Command, Final Report, 1974.
11. J. Demit, Contrat D.R.M.E. 74/154, rapport d'avancement, 1975.
12. K.H. Jack, J. Mat. Sci. 11, 1976, 1135.
13. L.J. Gauckler, Mat. Res. Bull. 11, 1976, 503.
14. D.P. Thompson (Ref. 9).
15. J. Zernike, Chem. Phase Theory (Kluwer, Deventer, Netherlands), 1955.
16. Greig. Am. J. Science, 13, 1927, 1 et 133.
17. M. Rey, Cahiers du CESSID, IRSID, 1954.
18. L. Brewer, A.W. Searcy, J. Am. Chem. Soc. 73, 1951, 5408.
19. Janaf Thermochemical Tables, 1965.
20. K. Blegen, Special Ceramics 6, BCRA, P. Popper ed., 1974.
21. O. Kubaschewski, Ll.E. Evans, C.B. Alcock, Metallurgical Thermochemistry, Pergamon Press, 1967.
22. J. Drowart, G. de Maria, R.P. Burns, M.B. Inghram, J. Chem. Phys. 32, 1960, 1366.
23. J.P. Torre, A. Mocellin, to be published.
24. H.F. Priest, F.C. Burns, G.L. Priest, E.C. Skaar, J. Am. Ceram. Soc. 56, 1973, 385.
25. A.J. Edwards, D.P. Elias, M.W. Lindley, A. Atkinson, A.J.Moulson, J. Mat. Sci. 9, 1974, 516.
26. I. Kohatsu, J.W. McCauley, Mat. Res. Bull. 9, 1974, 917.
27. C.M.B. Henderson, D. Taylor, Trans. Brit. Cer. Soc. 74, 1975, 49.
28. S. Wild, P. Grieveson, K.H. Jack, Special Ceramics 5, BCRA, P. Popper ed., 1970.
29. I. Colquhoun, S. Wild, P. Grieveson, K.H. Jack, Proc. Brit. Cer. Soc., D.J. Godfrey ed., 1973.
30. W.A. Krivsky, R. Schuhmann, Trans.Met.Soc. AIME, 221, 1961, 898.
31. G.I. Distler, E. Tokmakova, Fiz.Khim.Probl.Krist. 2, 1971, 172.
32. S.I. Nagai, K. Niwa, M. Shinmei, T. Yokokawa, J. Chem. Soc., Faraday Trans. I, 69, 1973, 1628.

DISCUSSION (Lang, Thompson)

Jack: At first sight, your representation is a little strange because it suggests that at a constant oxygen partial pressure an increase in p_{N_2} causes a change from mullite to alumina. This, of course, is because the total pressure is one atmosphere and an increase in p_{N_2} means a decrease of p_{SiO}.

1) What assumptions do you make about the partial pressure of Al_2O?

2) From your results can you say something about the relative stabilities of β'-sialon and β-Si_3N_4?

3) Are kinetics and not thermodynamics the decisive factor?

Torre: 1. As indicated, the gas phase was taken to contain no other species than N_2, SiO, O_2 and Al_2O, such that

$$p_{N_2} + p_{SiO} + p_{O_2} + p_{Al_2O} = p_{Total} \text{ (1 atm in this closed system)}.$$

The working simplifying assumption made to carry the calculations through was that $p_{Total} \approx (p_{N_2} + p_{SiO})$.

The final results are in quite good agreement with this approximation and p_{Al_2O} is negligible with respect to p_{N_2} and/or p_{SiO}.

2. One of the adjustable parameters of the model is the maximum activity of $\{Si_3N_4\}$ in the β' solution. Acceptable results (in terms of the internal consistency of the overall model and the basic assumptions) could only be obtained for quite small values of this activity, corresponding to an Al_2O_3N-weak solution (2% Al_2O_3 approximately). Following that line of argument then, it must be concluded that the activity of Si_3N_4 rapidly decreases with the degree of Al and O substitution, and that pure β-Si_3N_4 is less stable than a β' solution.

3. With respect to practical ease of formation, one may invoke the usual flux equation: $J = -M\nabla\mu$ and argue that, although the present calculations may at best bring some indications on $\nabla\mu$, experimental information (i.e. diffusion results) on M   still seems unavailable for a clear identification of the decisive factor (kinetics versus thermodynamics).

Grieveson: With regard to the assumption of ideal behaviour in the Si_3N_4-Al_2O_3 system β' phase, I feel that an alternative assumption used in carbide systems might be useful. The assumption is that the heat of formation of the β' solid solution is zero and the entropy can be calculated using a statistical mechanics approach. The concept is of random replacement of silicon by aluminium and nitrogen by oxygen. Such assumptions have proved to be highly successful in predicting activities in solid carbide solutions such as Fe_3C-Mn_3C, $Fe_5C_2-Mn_5C_2$ and $Fe_7C_3-Mn_7C_3$.

Torre: In principle we agree that the statistical mechanics approach could provide another meaningful route. It seems to us however that, in order to remain consistent with the available structural information on these Si-Al-O-N phases, Al-O and Si-N couples would be taken as mutually substituting, rather than single atoms, so that each of the two sublattices could not be assumed to be quite randomly populated.

Brook: The assumption made in drawing up the diagrams is that complete equilibrium is reached in the Si-Al-O-N system. An alternative assumption, which I think may be closer to the situation usually encountered, is that the cation ratio stays relatively constant but that equilibration on the anion lattice occurs. Have you attempted to compose any diagrams based on this second approach?

Torre: No, it has not been done that way, but the corresponding p_{N_2} versus p_{O_2} diagram which would be constructed on that basis would look the same as ours.

SOLID-STATE REACTIONS IN THE SiAlON SYSTEM

P. Goursat

Lab. Chimie Minérale et Cinétique Hétérogène,
E.R.A. au C.N.R.S. n° 539, U.E.R. des Sciences,
123, Rue Albert Thomas, 87100 LIMOGES, FRANCE.

ABSTRACT. Sialons ceramics have high vapour pressures and are not stable at high temperatures. By removal of silicon, oxygen and nitrogen, solid-solutions $\beta'(SiAlON)$ obtained with Si_2N_2O-Al_2O_3 or Si_3N_4-Al_2O_3 mixtures give at 1700°C and in an argon atmosphere a non-stoichiometric compound : $\delta(SiAlON)$ with an hexagonal lattice. This last one by the same process, change to an another non-stoichiometric phase : $\delta'(SiAlON)$ and the complete loss of silicon leads to γ-aluminum oxynitride.

The chemical analysis of reacted compounds and the determination of weight losses, show that with solid-state reactions, non-stoichiometric domains are more extended than by hot-pressing technique and for a structure the ratio cations/anions vary within wide limits.

The SiAlON system has been studied up to now mainly using hot-press sintering /1,2,3,4,5/. From these results we have been led to studying this system by a more classical method, for solid-state reaction in order to obtain a better thermodynamic equilibrium and also to avoid glassy phases.

It is well known that at high temperatures SiAlON materials have a high vapour pressure of silicon monoxide, silicon, oxygen, and nitrogen. It is precisely this property that we have used to study the quaternary diagram and in intention to determine the extent of the non-stoichiometric domains.

1. REACTIONS $Si_2N_2O-\alpha Al_2O_3$. FORMATION OF SiAlON PHASES

Silicon oxynitride and alumina in the powdered form are intimately dry mixed (molecular ratio $Si_2N_2O/Al_2O_3 = 1$), then submitted to a pressure of 3×10^3 bars. The pellets (m = 1,5 g) undergo an isothermal heat treatment under argon atmosphere at atmospheric pressure. The results obtained are summerised in table I.

Table I

Temperature °C	Heat treatment (hours)	Reaction Products
1300	12	Mullite traces
1450	48	Mullite + Si_3N_4 traces
1550	6	Mullite traces + β'(SiAlON)
1600-1650	4	β'(SiAlON)
1650-1700	4	δ(SiAlON)
1700-1750	4	δ'(SiAlON)

They show that silicon oxynitride reacts with α-alumina below 1500°C to form mullite and silicon nitride. At high temperatures, the nitride formed by the decomposition of silicon oxynitride /6/ leads to a solid-solution β'(SiAlON). This phase also disappears at 1650-1700°C to form a new quaternary phase called δ /7/ and which in its turn changes to δ'(SiAlON) /8/.

2. SOLID-SOLUTION β'(SiAlON)

Following the same experimental method, we have analysed the solid-solution β'(SiAlON) derived from mixtures $Si_2N_2O-\alpha Al_2O_3$ and $Si_3N_4-\alpha Al_2O_3$. The analysis for silicon and aluminium are obtained by atomic absorption method, nitrogen by volumetric analysis and the results are presented in table II.

The solid-solutions form by the insertion of alumina into the nitride structure. However, in contrast to reaction by hot-pressing, there is a considerable loss of silicon, oxygen, nitrogen during the reaction as can be seen from the values of $\Delta m/m_o$ (m_o = initial weight, Δm = loss) (table III). During the reaction a local equilibrium is set up between the solid and gazous phases and this equilibrium is continually displaced by the loss of the volatile species.

The ratio of cations/anions is no longer constant and equal to 6/8, nor are the amounts of aluminium and silicon simply related to the quantity of oxygen and nitrogen, as is the case for hot-pressing. The region of non-stoichiometry is much extended,

Table II

Initial mixture weight %	β'(SiAlON)				
	Si weight %	Al weight %	O weight %	N weight %	Formula
Si_2N_2O : 71.5 Al_2O_3 : 28.5	22.2	36.79	18.61	22.38	β'_I $Si_{2.26}Al_{3.87}O_{3.37}N_{4.63}$
Si_3N_4 : 75 Al_2O_3 : 25	41.67	16.09	12.31	29.93	β'_{II} $Si_{4.21}Al_{1.68}O_{2.12}N_{5.88}$
Si_3N_4 : 40 Al_2O_3 : 60	32.80	25.00	17.20	25.00	β'_{III} $Si_{3.35}Al_{2.65}O_3N_5$

Table III

β'(SiAlON)	$\frac{\Delta m}{m_o}$ %	Si	Al	O	N
I		77.18	0	69.21	54.04
II		23.60	0	13.68	17.69
III		26.71	ϵ	26.27	15.96

as it is represented figure I, where ours datas an compared with hot-pressing results.

3. δ(SiAlON) PHASE

Specimens (1,5 g) of solid-solutions β'(SiAlON) initially compacted under 2×10^3 bars have been decomposed under argon at 1700°C. X-ray patterns have shown the presence of a new phase δ(SiAlON), for which the chemical analysis is presented in table IV.
From the atomic ratios, the formula for this phase is

$$SiAl_4O_2N_4.$$

This phase is hexagonal with lattice parameters a = 3.011(3) Å ; c = 41.80(2) Å and is identical to the phase 15 R postulated by JACK and THOMPSON /9/.
The relative weight losses during the reaction of 3 specimens (weight % : Si = 28·6 ; Al = 27·6 ; O = 23·2 ; N = 20·5) are given in table V, to understand the problem of determining the exact composition.
The ratio $\Delta m/m_o$ shown that there is an important loss of material. Though the region of non-stoichiometry has not been completely determined (fig. 1) newertheless it appears that the region

Table IV

Reaction products	Si weight %	Al weight %	O weight %	N weight %	Al/N	Al/O	Al/Si
δ_1 (SiAlON)	12.3	47.20	15.9	24.6	0.99	1.76	3.98
δ_2 "	12.2	48.20	15.0	24.6	1.02	1.90	4.20
δ_2 "	12.8	47.8	14.6	24.8	1.00	1.94	3.87
δ_4 "	12.5	48.0	14.5	25.0	1.00	1.96	3.98
$SiAl_4O_2N_4$	12.5	48.2	14.3	25.0	1	2	4

Table V

Reaction products	$\frac{\Delta m}{m_o}$ weight %			
	Si	Al	O	N
δ_2	75.8	3.5	64.4	32.4
δ_3	75.0	3.4	64.8	32.7
δ_4	76.3	3.2	64.0	33.4

is extended and that reaction products have not always exactly the same atomic ratio of Al/N = 1, and theirs composition change with the initials compounds.

4. δ'(SiAlON) PHASE

During the decomposition of δ(SiAlON) an important loss of silicon, oxygen, and nitrogen occurs once again. The resulting product consists of aluminum nitride and δ'(SiAlON), as shown by X-ray diffraction. This mixture in loosing all of its silicon becomes γaluminum oxynitride and aluminum nitride. The aluminum nitride can be removed from this mixture using a solution of NaOH(M) at boiling point for 1/2 hour. Analysis of the specimens carried out as previously described, are presented in table VI.

The different ratios Al/N, N/O, Al/Si show that non-stoichiometric compounds are formed ; the ratios cations/anions are not constant. It can be seen that complete removal of silicon leads to $Al_{12}O_{12}N_4$, which corresponds to a limit of the γ(AlON) region. The extent of the δ'(SiAlON) non-stoichiometric region has not been fully characterised (fig. 1).

Table VI

| Weight % | | | | Al/N | N/O | Al/Si | Compounds |
Si	Al	O	N				
10.6	50.2	14.4	24.8	1.05	1.97	4.91	δ'
10.6	50.9	12.1	26.4	1.00	2.5	5.00	$Si_{2.29}Al_{11.43}O_{4.57}N_{11.43}$
6.4	53.2	20.7	19.7	1.40	1.09	8.62	δ'
6.4	53.2	20.5	19.9	1.39	1.11	8.60	$Si_{1.36}Al_{11.66}O_{7.59}N_{8.41}$
5.8	54.0	19.9	20.3	1.38	1.17	9.66	δ
5.4	54.4	20.2	20.0	1.41	1.13	10.45	δ'
5.4	54.3	20.5	19.8	1.42	1.10	10.43	δ'
5.1	53.9	23.2	17.8	1.57	0.88	11.00	$Si_{1.07}Al_{11.73}O_{8.53}N_{7.47}$

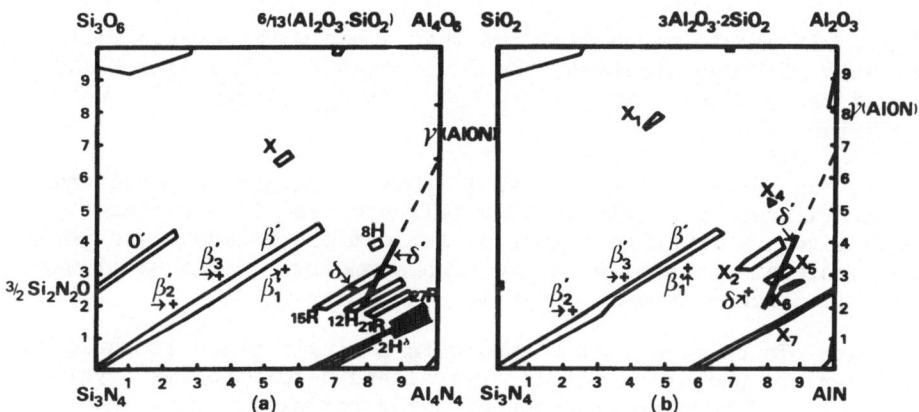

Fig. 1 : Isothermal section Si_3N_4–AlN–$Al_2O_3SiO_2$ of the system SiAlON (a) by Jack and (b) by Petzow.

It can be concluded that using solid-state reactions, the non-stoichiometric domains of SiAlON compounds become more apparent in the sense that the ratio of cations/anions need no longer be constant. By contrast, work with hot-pressing has tended to emphasize the existence of a constant ratio cations/anions.

86

REFERENCES

1. Y. Oyama, Jap. J. Appl. Phys., 1971, 11, 760.

2. K.H. Jack, J. Brit. Ceram. Soc., 1973, 72, 376.

3. P. Drew and M.H. Lewis, J. Mat. Sci., 1974, 11, 1833.

4. E. Gugel, I. Petzenhauser and A. Fickel, Pow. Met. Int., 1975, 7, 66.

5. L.J. Gauckler, H.L. Lukas and G. Petzow, J. Amer. Ceram. Soc., 1975, 58, 346.

6. P. Lortholary and M. Billy, C.R. Acad. Sci., 1974, 278, 1343.

7. D. Brachet, P. Goursat and M. Billy, C.R. Acad. Sci., 1975, 280, 1207.

8. D. Brachet, P. Goursat and M. Billy, C.R. Acad. Sci., impublished results.

9. D. Thompson and K.H. Jack, private communication.

DISCUSSION (Lang, Thompson)

Lange: It should be pointed out that, when working with reciprocal salt Si-M-O-N diagrams, the four corners of the square diagram should be labelled: Si_3N_4, $^b/x\, M_2O_x$, $3SiO_2$ and $^4/x\, M_3N_x$, where x = metal valency.

Jack: How can you be sure that δ'-phase is a single phase and not a mixture? The orthorhombic cell proposed is very large and would give a very large number of reflections unless there were many systematic absences due to high symmetry, or unless it was a superlattice of a smaller unit.

Goursat: To be sure that δ'-phase is a single phase it would be necessary to prepare a single crystal and determine its crystal structure. I have indexed powder X-ray patterns. From the results obtained using the well-known Ito's method, an orthorhombic cell seems to be probable.

Grieveson: Our observations on the volatilisation of silica have shown that the volatilisation rate is proportional to $\frac{1}{p^2}$. It would seem from Dr. Goursat's results that these materials will be of little value under low pressures.

Goursat: All experiments were done under argon at atmospheric pressure, and extensions to reduced pressures were not made.

Section B

CRYSTAL CHEMISTRY

SOME ASPECTS OF THE STRUCTURE AND CRYSTAL CHEMISTRY OF NITRIDES

J. LANG

Laboratoire de Chimie Minérale C, Université de Rennes,
France.

ABSTRACT. All metals, with the exception of the alkali metals but
one and the noble metals, form nitrides. The crystal chemistry of
these compounds is so vast that we have studied only certain aspects
of them, by a consideration of their cation coordination number,
which consideration facilitates the classification of more complicated compounds. With the exception of some low coordination number
structures which one meets with elements of groups I and II and with
some transition elements, the majority of structures are of well-
known types and correspond to a cation environment of 4 or 6
nitrogen atoms. The different types of structure are discussed
from the point of view of their inter-relationships, and the properties of the compounds. Certain characteristics peculiar to
nitrides, as opposed to those found in oxides, sulphides, etc., are
presented. Analogies between nitrides and oxynitrides are pointed
out, as is the question of the polymorphism of certain compounds.

INTRODUCTION

With its low electronegativity and its very stable molecule,
nitrogen cannot react easily with a number of other elements or it
does so at a high temperature. However, it gives a large group of
nitrides. Their structures are of many different types and it is
difficult to give a comprehensive view of them. Their bonds are less
ionic than those in oxides though many of them have also a metallic
bond character.

Ordinarily, according to the type of their bond, nitrides are
classified into ionic, covalent, semi-metallic or molecular compounds. The last will be left out here. Ionic nitrides given by
the more electropositive elements are readily hydrolysed and have a
basic character, as opposed to the nitrides of the second group

which are mainly acidic and are more stable towards aqueous solutions. Some of them, however, dissociate easily with increasing temperature. The semi-metallic nitrides are given by transition elements and they often have a range of homogeneity. This simple classification is especially useful for binary compounds. In ternary nitrides different types of bonding exist in the same compound, nitrogen atoms forming polyhedra round the more covalent metal and charge balance being maintained by the insertion of the appropriate numbers of the more ionic metal between polyhedra.

As are the structures of many oxides, those of many nitrides are formed by a close packed lattice in which holes of different types are completely or partially occupied. It seems useful to consider the coordination number of cations which is very often equal to 4 or 6. In some compounds the coordination number has a lower value and in some lanthanum compounds values as high as 7 are found. Structural types are thus classified according to this view-point in this paper.

STRUCTURES WITH 4 COORDINATED CATIONS

In most of the cases where a cation is four coordinated by nitrogen atoms, the environment is not of the square-planar type but of the tetrahedral one. A very important group of compounds includes those where nitrogen atoms are also tetrahedrally co-ordinated. These compounds have then a tetrahedral structure.

Tetrahedral structures

A type of structure is said to be tetrahedral when every atom in this structure is tetrahedrally coordinated (1). Each atom needs 4 tetrahedral orbitals which are occupied by 8 valency electrons. Every orbital being common to two atoms, every atom has a valency electron concentration (v.e.c.) equal to 4. This is a necessary condition for such a compound to exist. This is naturally done for the elements of Group IV in their diamond structure. For the elements of other groups there is an excess or a lack of valency electrons, but if one considers two groups at equal distances from group IV, the excess of valency electrons of one group can compensate for the lack of the other. Thus a v.e.c. of four can be maintained by combining Zn with S, Ga with P, or Cu and Cl (Grimm-Sommerfeld compounds), namely 2+6, 3+5, 1+7, each element being represented by the number of its valency electrons. By progressive and transverse substitutions, new formulae can be found for possible tetrahedral compounds as indicated by Figure 1. For binary compounds $A_m B_n$, with x and y representing the number of valency electrons, one must have $\dfrac{mx + ny}{m + n} = 4$, m and n being integers.

This relation has 10 solutions: 4 corresponding to Grimm-Sommerfeld compounds and also the formulae $3_2 6$, $3_3 7$, 15_3, $1_2 6_3$, $2_3 7_2$ and 25_2.

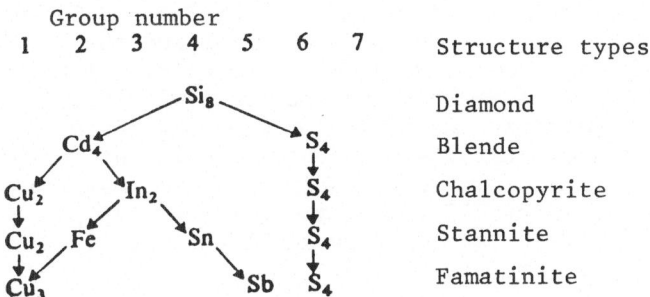

Figure 1. Transverse substitutions in tetrahedral structures.

The only known binary compounds correspond to the last of these 6 formulae: e.g. ZnP_2, but by combining conveniently all of them, more complicated compounds with v.e.c. = 4 can be found. Of course those new complicated formulae have to be in accordance with the chemical properties and valency of the elements. Here are some examples.

$$1\,5_3 + (1_2\,6_3)_4 \longrightarrow (1_3\,5\,6_4)_3 \qquad Cu_3SbS_4$$

$$1_2\,6_3 + 2\,6 + 4 \longrightarrow 1_2\,2\,4\,6_4 \qquad Cu_2FeSnS_4$$

$$1\,5_3 + 4_2 \longrightarrow 1\,4_2\,5_3 \qquad LiSi_2N_3$$

In many cases, tetrahedral compounds have a structure related to the blende or wurtzite type. Nitrogen atoms form a cubic or hexagonal close-packed lattice in which half the tetrahedral holes are occupied by atoms of an electropositive element.

The binary nitrides with a normal tetrahedral structure are 35 compounds, i.e. the nitrides of group III. The ionic character increases from boron to thallium. Boron is a special case because of the polymorphism of BN (2). This nitride is isoelectronic with C_2 and as with carbon the stable structure is of the graphite type. The interatomic distance (1.44 Å) is not very different from C-C (1.42 Å) but the delocalisation of electrons is very low, thus the resistivity of BN is very high ($\rho > 10^{12}\Omega\,cm$). The dipolar interactions between sheets are weak but sufficient to explain their relative displacement and the superposition along the c axis of nitrogen and boron atoms. This nitride is not very reactive chemically and not moisture sensitive when it is very pure (3). BN can give a cubic form. This transformation needs simultaneously a high temperature (1800°C) and a high pressure (65 kb) (4) to distort the sheets and to make the free electron pairs of nitrogen atoms interact with boron orbitals. Recently it has been shown that cubic BN could be obtained at 1000°C and 32 kb in the presence of alloys acting as catalysts (5) and in other different conditions (5a). According to Vickery (6) a high pressure is not even needed. This cubic nitride is isostructural with blende and has some properties in common with diamond, especially its exceptional hardness. Cubic and hexagonal forms of BN have about the same energies and the stability of the type of structure is related to the entropy term which depends upon the number of defects and increases with the temperature. At a low temperature and a high pressure, a boron nitride with the wurtzite structure has been obtained (7) but recent work has shown that this form is probably always metastable (8). All other nitrides of group III belong to this last type, even though a cubic form of GaN stabilised by small amounts of oxygen has recently been described (9,10,11).

In the case of ternary nitrides, only 245_2 and 1425_3 compounds have a v.e.c. equal to 4. Their structure is tetrahedral only if the cations M^I or M^{II} are ordinarily tetrahedrally coordinated. Compounds are known for Li, Be, Mg, Zn and Mn. The energies of the $M^{IV}-N$ bonds decrease rapidly down group IV, and hence only silicon and germanium are involved in ternary nitride formation. The carbon compounds are cyanamides, the structure of which are quite different. The first known example of a ternary nitride with a tetrahedral structure was $BeSiN_2$ (12). Other phases are known in the system $Be_3N_2-Si_3N_4$ (13) but they do not have a v.e.c. = 4. Other compounds ABN_2 have been prepared (A = Mg, Zn, Mn; B = Si, Ge) and with the exception of $ZnGeN_2$ which crystallises in the monoclinic system (space group $P2_1$) all belong to the orthorhombic system with the same space group as $BeSiN_2$ ($Pna2_1$), and give an ordered structure (14-20). The nitrogen atoms form a slightly distorted hexagonal close-packed lattice in which the M^{IV} and M^{II} metal atoms each

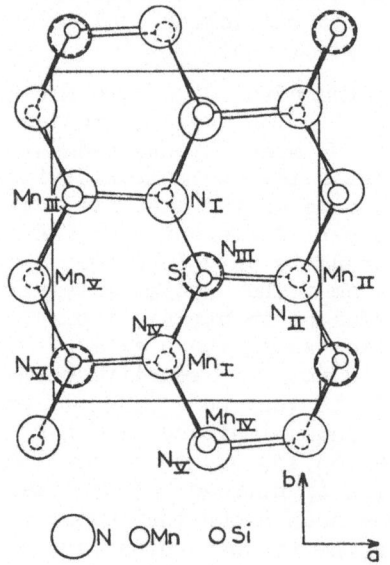

Figure 2. The structure of $MnSiN_2$ projected on the (001) plane.

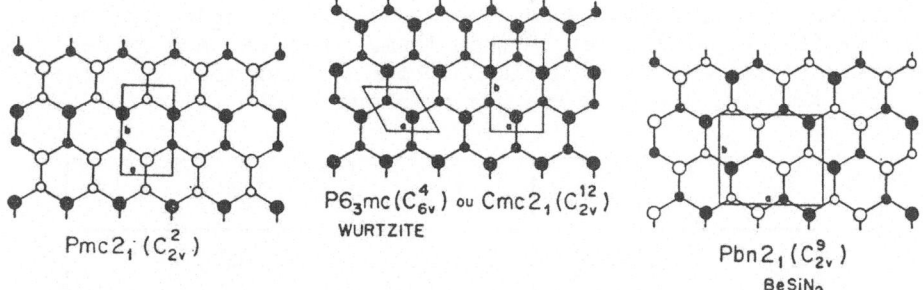

Figure 3. The wurtzite structure and the two possible ordered
structures for a ternary compound.

occupy $\frac{1}{4}$ of the tetrahedral holes. Figure 2 shows the structure
of $MnSiN_2$ projected on the (001) plane. $ZnGeN_2$ is also related to
the wurtzite type and is ordered as shown by neutron diffraction
(19).

A wurtzite-type ternary compound with equal numbers of the two
metals A and B can order in two ways as shown in Figure 3. Here,
only metal atoms are shown, and types A and B are distinguished by
open and closed circles. Small circles represent the same atom at
a z-coordinate of 0.5. These two structures correspond to two types
of cell; the one of larger volume has been found for all the ternary
nitrides whilst the other has not been observed for any ABC_2 compound.

The other ternary tetrahedral structures are those of LiB_2N_3
compounds with B = Si or Ge. They are isotypic and analogous to
Cu_2SiS_3 (21). They crystallise in the orthorhombic system with
space group $Cmc2_1$. The nitrogen atoms form a distorted hexagonal
lattice and are surrounded either by 2 Li + 2 Si or by 1 Li + 3 Si.
Figure 4 shows a perspective view of the cell. As previously
indicated, two types of ordered ternary structures AB_2C_3 can be
obtained from the ZnS type and in both the long cell dimension is
three times that of ZnS. It has to be noted that one of them is
still unknown in the present range of AB_2C_3 compounds. These
examples give an interesting result.

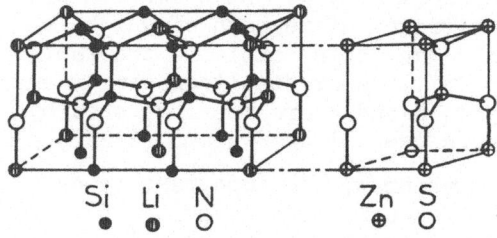

Figure 4. Perspective view of the cell of $LiSi_2N_3$ compared to
the cell of wurtzite.

94

If we compare the ABX_2 compounds such as 145_2 and 136_2 on the one hand and the AB_2X_3 compounds such as 14_25_3 and 1_246_3 on the other, we find a parallelism between them. Nitrides and oxides have the wurtzite structure while sulphides, phosphides and arsenides have the zinc blende structure as shown in the following table:

	245_2	136_2	14_25_3	1_246_3
W type	$BeSiN_2$ $MgSi(Ge)N_2$ $MnSi(Ge)N_2$ $ZnGeN_2$	$\beta LiGaO_2$ $\beta NaAlO_2$ $NaGaO_2$ $\beta NaFeO_2$	$LiSi_2N_3$ $LiGe_2N_3$	Li_2SiO_3 Na_2SiO_3 Cu_2GeS_3 (h.t.)
B type	$ZnSi(Ge)P_2$ $CdSi(Ge)P_2$ $MgSiP_2$ $CdSnP_2$	$CuBSe_2$ $CuAlS_2$ $AgGaSe_2$ $AgFeS_2$	$CuSi_2P_3$ $CuGe_2P_3$ $CuGe_2As_3$	Cu_2SiS_3 (l.t.) Cu_2GeS_3 Cu_2GeTe_3 Cu_2SnTe_3

However, there is one exception with $CaGeN_2$. The v.e.c. for this compound is 4 but calcium ions as a rule are not tetracoordinated except in Ca_3N_2. $CaGeN_2$ is tetragonal and its structure is intermediate between those of cristobalite and chalcopyrite (22) as shown in Figure 5. The calcium ion is in a distorted 12-fold coordination but four of these nitrogens are much closer than the rest and form a distorted tetrahedron. In this way, the structure can be related to that of chalcopyrite.

We shall deal very little with the question of the group IV nitrides. Many papers will be given on this topic in this conference. These nitrides have a v.e.c. = 4.57 and can be considered as defect tetrahedral structures whatever the form α or β. Each silicon or germanium atom is surrounded by 4 nitrogen atoms each of which is common to 3 (M-4N) tetrahedra. This last point explains the fact that, unlike silicon oxide, silicon nitride crystallises easily and does not give a glass. The two forms α and β are distinguished by the stacking of the planes containing the nitrogen atoms (23-25).

When the compounds are not tetrahedral, two important types of structures are found which have the cubic close-packing of nitrogen atoms. No such structure related to the hexagonal close packing is known. If the cations occupy $\frac{3}{4}$ of the tetrahedral positions, we get the M_3N_2 structure of group II nitrides; if all

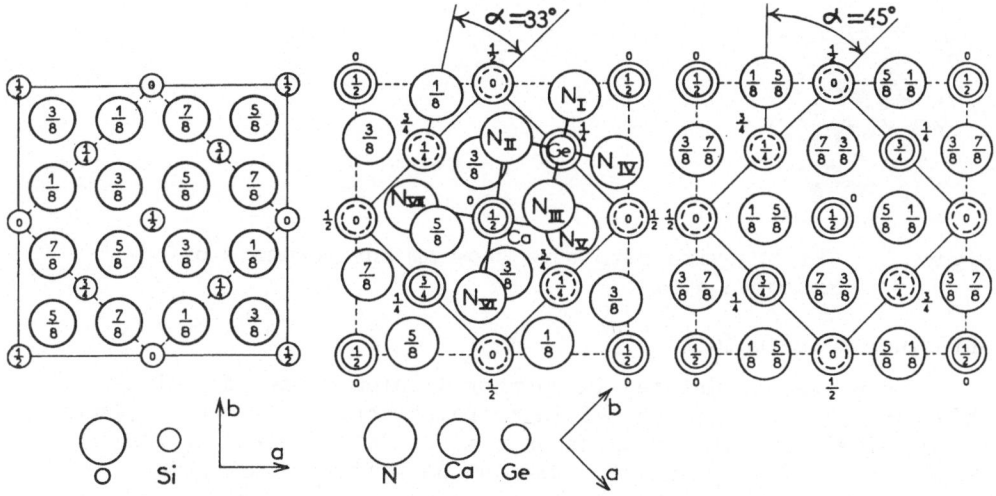

Figure 5. Projection on the (001) plane of structures from left to right: cristobalite, $CaGeN_2$, chalcopyrite.

tetrahedral positions are occupied we get an antifluorite structure.

Antibixbyite structures

M_3N_2 nitrides are isotypic except those of the heavier elements of subgroups IIa and IIb (26-28). They crystallise with the antibixbyite structure (anti-Mn_2O_3) and space group Ia3. Cations are located in an ordered fashion in $\frac{3}{4}$ of the possible tetrahedral sites, thus the nitrogen atoms are surrounded by 6 cations distributed on 6 of the corners of a slightly distorted cube as indicated in Figure 6. It is very surprising that the Ca^{++} has such an environment which is practically never encountered with calcium compounds. The interatomic distances in Ca_3N_2 are rather large and because of the repulsive interactions between nitrogen atoms surrounding vacant sites, the holes are rather large and the density has a low value. Many other crystalline forms have been described for calcium nitrides (29) but the crystal structures are not always known. Beryllium nitride which is not so ionic as the calcium compound also has two types of structure. Beside the cubic, a new hexagonal form has been described, but the crystal structure is not as yet completely established (30). The anti-Mn_2O_3 structure is also found in some rare earth compounds. Lanthanide subnitrides probably do not exist, but the mononitrides are well known; under a nitrogen pressure, compounds LnN_x ($x \neq 1$, Ln = Tb, Dy, Nd, Tm, Lu) have been obtained and give X-ray diagrams identical with Mn_2O_3-type oxides (31). Uranium nitride αU_2N_{3+x} ($0.20 < x < 0.48$) also belongs to this class (32-34).

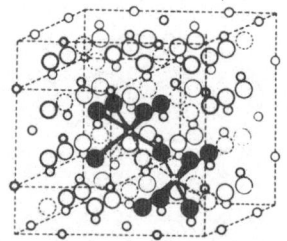

Figure 6. The bixbyite structure showing environments of
            nitrogen atoms.

Antifluorite structure

These are encountered in ternary lithium compounds, where
this element is always in 4-fold coordination. In these compounds,
Li is joined to elements which have an oxidation state ranging
from 2 to 6. The general formula can be written $Li_{2x-3}M^xN_{x-1}$ (M
= Mg, Zn, Al, Ga, Ti, Si, Ge, V, Mn, Cr). In the antifluorite
cell, cations are distributed on tetrahedral positions either at
random, as in LiMgN, or in an ordered fashion in other compounds.
The other cations are distributed statistically in 16-fold special
positions as indicated in Figure 7 (35-38). For all compounds
where M/Li $\neq$ 1 the cell paramater must be doubled. In the case of
Li_5SiN_3 to take account of the stoichiometry, one third of the 16-
fold positions are also occupied by lithium atoms (39). The
structure of Li_9CrN_5 has not been determined and is probably related to the same type (40). It is surprising that in such a compound the chromium atom keeps a value of its oxidation state as
high as 6, since in the nitride it has formally a value of 3. As
a rule such high oxidation states are kept only in oxygen compounds.
The same remark applies to the case of Li_7VN_4. For Mn, only the
oxidation state V is kept in Li_7MnN_4.

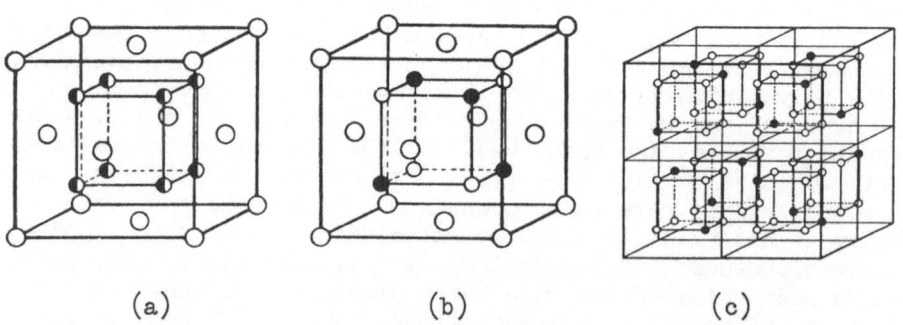

Figure 7. Ordered and disordered anti-fluorite structure.

    (a) LiMgN   (b) LiZnN   (c) Li_3AlN_2

These high oxidation states were verified by measurements of the nitrogen absorption during the reaction, by redox titration and total analysis. The partial pressure of nitrogen over the ternary compound Li_9CrN_5 is lower than over CrN. None of the highly charged atoms gives distinct anionic structural group such as those formed in metallates. The structure would then have been much more complicated and not so closely related to a simple ionic structure. However, M-4N tetrahedra differ appreciably from Li-4N tetrahedra and are more closely packed (Al-N = 1.85 Å, Ga-N = 1.95 Å, Li-N = 2.15 Å).

These substances can react with lithium oxide to give solid solutions, e.g. $Li_5SiN_3.2Li_2O$, which keeps the antifluorite structure. The cell parameter decreases with increasing oxide content. Similar compounds to those described above have been prepared using metals of higher atomic number (e.g. LiBaN, Li_7NbN_3 or Li_9MoN_5). These compounds have different structures but can assume an anti-fluorite arrangement by incorporating small quantities of lithia as for example LiSrN, $0.2\ Li_2O$ or LiBaN, $0.2\ Li_2O$.

Anti-La_2O_3 structures

The final structure-type in which lithium atoms occupy tetrahedral holes is found in compounds of transition metals of formula Li_2MN_2 where M = Zr, Th, U, Hf. The structures of the Zr, Th and U compounds are related to the anti-La_2O_3 type, but the structure of the Zr compound is different from that of the other two. The nitrogen atoms are 6 or 7 coordinated if we consider that they are surrounded by 3 transition elements and 3 or 4 Li atoms (41) as indicated in Figure 8 representing the Li_2ZrN_2 structure. The La_2O_3 structure is observed in the rare earth dinitrides, LnN_2, obtained by Ettmayer (31) with Ce, Pr, and Nd under a pressure of nitrogen as high as 300 atmospheres. The X-ray diagrams are identical to those of oxides of La_2O_3 type and the structure must have a defect cationic lattice in which atoms are probably distributed at random.

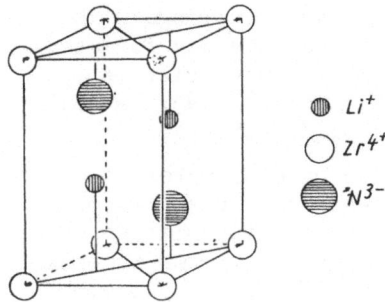

      Li^+

      Zr^{4+}

      N^{3-}

Figure 8. Structure of Li_2ZrN_2.

STRUCTURES WITH 6 COORDINATED CATIONS

The transition metal nitrides we are now dealing with often obey the Hägg rules (50). The structure is determined by the radius ratio $k = r_N/r_M$ where r are the atomic radii. If k is less than 0.59, the metal atoms form very simple structures (hcp or fcc); if k is greater than 0.59 the structure is complicated. For k < 0.59, the nitrogen is accommodated in the largest interstitial sites of the simple host structure. The site has to be small enough to give a sufficient bonding between metal and non-metal and large enough to maintain the bonding between metal atoms of the host lattice. The interstitial sites are tetrahedral, octahedral or trigonal prismatic. The first are too small and as k values for transition elements and lanthanides nitrides are less than 0.59, nitrogen is accommodated either in an octahedron or trigonal prism.

Structures related to cubic close packing

The most simple is the NaCl type structure to which belong the interstitial mononitrides. Cations are formally in the oxidation state +3 and all octahedral holes are occupied. However, some compounds have a range of homogeneity and the phase diagrams are not well known. Titanium can dissolve nitrogen up to 22 N at. % (42,44) and the phase TiN_x with the NaCl structure appears for x > 0.86 with a defect anionic lattice (43). Sometimes the compound is stoichiometric (HfN), but both sublattices are defective (12.3% vacancies) (45). In such a compound physico-chemical studies show that N atoms are in an oxidation state ranging from 0 to -3. Some electrons of the metal, not accepted by N, are inserted in the conduction band. The same character is found in Ti or ZrN_x with x > 1. A large proportion of N evolves as molecular N_2 when the nitride is dissolved in acids. Besides N≡, it is supposed that N is also present in an atomic state, the two sublattices having a defect structure (65). The study of these phases, especially LnN, is very complicated. It is difficult to know if an homogeneity range exists for LnN, if so, it is narrow (0.05 or 0.1 a.u.). Measurements of the cell parameter are ineffective because parameters vary weakly with composition and also vary in range with oxygen impurity content (46,47). These nitrides readily form solid solutions with other mononitrides or carbides: one series is CrN-TiN (48), and UN with Zr, Hf, Cr, or Nd nitride (49); a lack of miscibility can occur, e.g. TiN-UN (49) or ZrN-CrN (48).

Not being Hägg compounds, some nitrides have, however, a structure related to the NaCl type. Ca_2N and Sr_2N have a rhombohedral cell in which half the planes (III) containing nitrogen are vacant. The structure is formed of sheets where nitrogen is octahedrally accommodated and M is tricoordinate (51,52). If there are more vacancies and only one quarter of the nitrogen sites are occupied we have positions in an ordered fashion. Then the octahedra are sharing only corners.

Structures related to hexagonal close packing

With an hexagonal lattice, we get the anti-NiAs structure, found in δ'NbN. The metal atoms are accommodated in trigonal prisms. A small percentage of octahedra are vacant because of a stoichiometry deviation. It is also possible to have a mixed packing of metal atoms, giving not ABA but AABBA. With all the octahedral holes filled with nitrogen we have the structure of NbN which is of the anti-TiP type. If octahedral sites are incompletely filled by nitrogen atoms, we have new types related to the anti-NiAs structure. If the octahedral sites are half-filled at random, we have a W_2C type structure, found for example for Nb_2N and Ta_2N. The corners of the trigonal prism surrounding the metal atoms are partially occupied. The ordered structure is mainly found with carbides. Partial occupation of octahedral holes is observed in nickel nitride, Ni_3N, where only one third of the holes are filled with nitrogen. The Ni atoms form a h.c. packing and N a single hexagonal packing (53-56).

The last type of structure is formed by a simple hexagonal packing for the metal atoms and for nitrogen (WC structure). Figure 9 shows this, and the NiAs structures. All atoms are accommodated in trigonal prisms. This structure allows the atoms to be in closer contact with one another than they are in NaCl. M-M distances are 1.3 M-N instead of 1.414 M-N as in NaCl (57).

Perovskite compounds

Nowotny and co-workers (58,59) have studied many compounds in which the metal substructure is no longer close packed. Non-metal atoms, however, are accommodated in octahedral interstitial sites. The most important of these structures is the perovskite type. This structure is commonly found for oxides ABO_3. The charge of oxygen may be balanced by exactly two cations such as 1+5, 2+4 or 3+3 or even more complicated combinations. The nitrogen compounds are MT_3N where M is a non-transition metal, T a transition metal

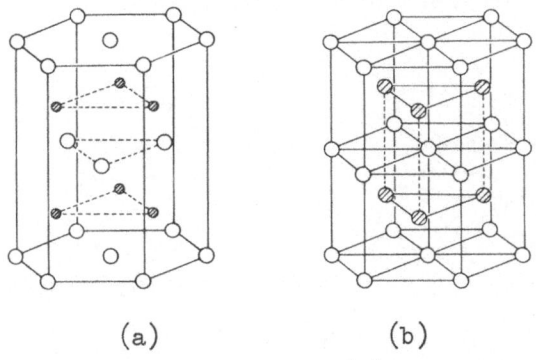

(a)   (b)

Figure 9. NiAs structure (a) and WC structure (b).

or lanthanide. The nitrogen atoms can be replaced by boron or carbon atoms and solid solutions can be obtained as $M_4N_{1-x}C_x$. The following table gives a list of such compounds which are of interest for their magnetic properties.

T/M	Al	Mn	Fe	Co	Ni	Cu	Zn	Ga	Ge	As	Rh	Pd	Ag	In	Sn	Ir	Pt	Au	Hg
Cr					+	+	+	+	+						+	+	+		
Mn		+	+	+	+	+	+	+	+	+	+	+	+	+	+	+	+	+	+
Fe	+		+		+	+	+		+	+		+	+		+	+		+	
Co					+	+	+								+	+			
Nd		+						+							+	+			

In the cubic cell, nitrogen takes the place of cation B and is octahedrally coordinated by T atoms. The N-6T octahedra share corners and are surrounded by M atoms arranged in a simple cubic structure; the T atoms are surrounded octahedrally by 2T and 4M.

We can consider that nitrogen perovskites are related to either the T_4N or MT_3 structures (Figure 10). In the T_4N group of nitrides, two examples are known, Fe_4N and Mn_4N (60). The metal atoms T occupy two independent positions 000 and $\frac{1}{2}0\frac{1}{2}$. The first can be substituted in an ordered fashion by M atoms giving a simple cubic structure.

In the MT_3 alloys of ordered $AuCu_3$ type, nitrogen atoms are inserted in the octahedral site $\frac{1}{2}\frac{1}{2}\frac{1}{2}$. These ordered alloys are rare while there are many perovskites. The insertion of nitrogen stabilises the ordered structure particularly in manganese-containing compounds. In the system MMn_3-MMn_3N, the progressive insertion of N gives phases such as MMn_3N_x but the solid solution is not a continuous series. The ordered MMn_3 structure is kept up to ~ 0.1. Beyond $x \sim 0.5$ the perovskite appears, but for $x \sim 0.2$ an hexagonal phase appears. MMn_3 planes of the structure form h.c. packing with $c/a = 0.816$ and nitrogen atoms are inserted in a disordered manner in the octahedral holes coordinated entirely by Mn atoms. Octahedra 2M, 4Mn are always empty (Figure 11).

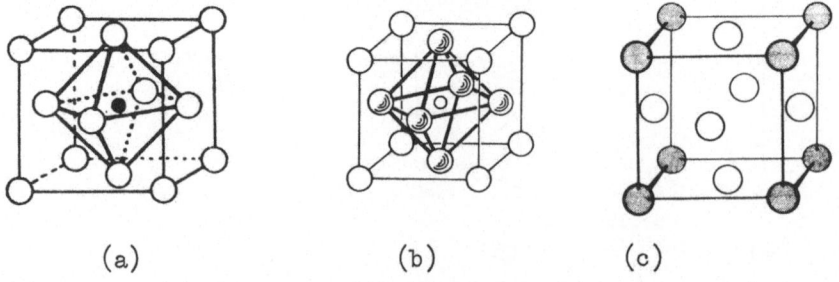

Figure 10. Structures of (a) Fe_4N, (b) perovskite, (c) MT_3 alloy.

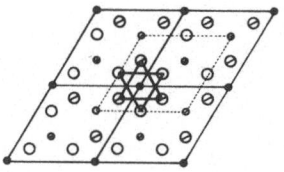

Figure 11. The hexagonal structure of MMn_3N_x.

With increasing temperature for a given compound, or according to the nature of M and T, the simple perovskite structure becomes more complicated. For $SnMn_3N$ between 4.2 and 237 K or $CuMn_3N$ at 293 K, we have a tetragonal cell (T_1) due to a slight deformation of the cubic one. The value of c/a may be smaller (M = Cu, Sn) or larger (M = Sn, Sb) than 1 (61).

Another tetragonal cell (T_4) has been found (63) with Cr_3AsN, which is four times larger than the primitive cubic cell ($a = a'\sqrt{2}$ and $c = 2a'$, a' = cub. param.). This is the result of an alternated arrangement of T_6N octahedra along the c axis. These octahedra are rotated in one direction and then in the other by an angle of about 15°, as indicated in Figure 12. The M atoms are then surrounded by only 4 close neighbours, the interatomic distances being reduced by the rotation of octahedra.

A third type, Cr_3GeN, intermediate between these first two has been described by Boller (64). The tetragonal cell is twice the primitive cubic with $a = a'\sqrt{2}$ and $c \sim a'$, a' = cub. param. The T_6N octahedra are always rotated in the same way but they are distorted because Ge atoms are located at $\frac{1}{2} \pm u$ and 4Cr at $z = u'$. The interatomic distances are modified and this type can be related to the U_3Si structure. Thus the coordination number of Ge atoms increases up to 10. Two of the eight neighbours have very large bond lengths, but 4 chromium atoms at about the same interatomic separation enter the coordination polyhedron. Figure 13 enables

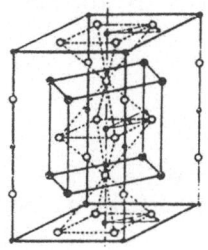

Figure 12. The Cr_3AsN structure.

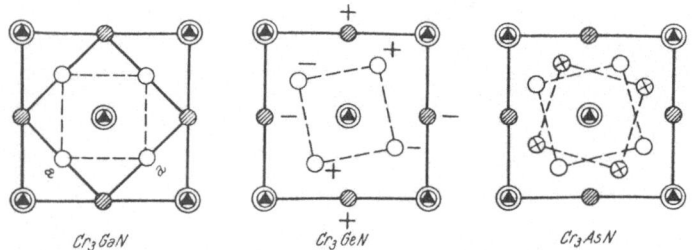

Cr_3GaN Cr_3GeN Cr_3AsN

Figure 13. Perovskite related structures.

a comparison of the three structures to be made.

The most complicated structure is an orthorhombic cell (O_8) eight times larger found by Barberon (61) for $As_{1-x}Sb_xMn_3N$. This structure is intermediate between the T_1 and T_4 cells and can transform into either of these two modifications. As we can get T_1 by a deformation of a cubic cell, we have O_8 by a slight deformation of the tetragonal cell. The similarity with T_4 is due to alternate rotation of distorted T_6N octahedra along the c axis.

The comparative study of these perovskites shows that the magnetic moment of Mn increases with the distance Mn-N. Different schemes, assuming either covalency (66) or electron transfer (67-69) have been proposed. A study of the N electronic state shows that this element is in an oxidation state of 0 or -1. The electron transfer is not important and N does not seem to play the donor role which is attributed to it to explain the magnetic properties. The distortion of cells can be related to the number, n, of valency electrons of M, as indicated by the following scheme:

T_1 c/a >1	cubic	T_1 c/a >1	T_4
			$\longrightarrow$ n
($CuMn_3N$) 1.3	($ZnMn_3N$) 3.5	($SbMn_3N$) 3.8	($GeMn_3N$)

A geometrical factor can interfere and the distortion decreases with increasing r_M, e.g. $GeMn_3N$ (T_4) and $SnMn_3N$ (T_1), $CuMn_3N$ (T_1) and $AgMn_3N$ (cub.). The two deformation processes would be independent. T_1 would be due to crystal field effects and T_4 to the covalent character of the Mn-M bonds.

STRUCTURES WITH 2, 3 OR 5 COORDINATED CATIONS

Anti-ReO_3 structure

If there is hole at $\frac{111}{222}$ in a perovskite cell, we have the anti-ReO_3 structure of Cu_3N. The different kinds of atoms do not have the same location as in perovskite. The N atoms form a simple cubic structure and the metal atoms are in the middle of the edges. Thus the monovalent copper has a c.n. of 2 and N is octahedrally surrounded (70). This nitride is not very stable and dissociates at 350°C.

 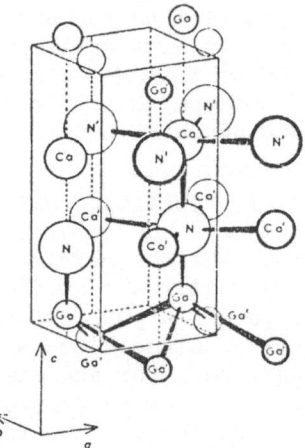

Figure 14. Li$_3$N structure.　　Figure 15.　CaGaN structure.

Structure of Li$_3$N

This structure is rather uncommon. In almost all of its compounds, Li is tetrahedrally coordinated but in the binary nitride such a coordination number cannot be achieved. In fact if Li had the c.n. 4, N atoms would be surrounded by 12 Li and it is impossible to get more than 8 tetrahedra with a common corner. Li$_3$N crystallises in the hexagonal system (62). The N atoms form a simple packing and cations are located at $00\frac{1}{2}$, $\frac{1}{3}\frac{2}{3}0$ and $\frac{2}{3}\frac{1}{3}0$ (Figure 14). Li ions are surrounded either by 2 or 3 N and each N has 8 neighbours at the corners of a hexagonal bipyramid. Solid solutions Li$_{3-x}$T$_x$N with Cu, Co and Ni can be obtained and only the Li on the edges can be substituted at random and in a low proportion (17-20 at %) (71). These nitrides are readily hydrolysed and are not refractory compounds though Li can give some refractory nitrides when reacted with suitable elements.

NbCrN structure

A few compounds have this structure, which is encountered either in stoichiometric compounds (72,74) or in solid solutions such as Ta$_{1-x}$Cr$_{1+x}$N (73). The cell is tetragonal and the structure can be considered as a distortion of a b.c. cubic arrangement of metal atoms with a Nb-Nb-Cr-Cr sequence in which N is inserted in one sixth of the octahedral holes. However, the co-ordination polyhedron is quite different from one to another as can easily be seen in Figure 15, which shows the structure of CaGaN, which has this structure (75). The nitrogen atoms are surrounded octahedrally

by 5 Ca and 1 Ga, each Ca is coordinated by 5 N and each Ga by 4
Ga and 1 N. One interesting feature is the layer disposition of
gallium atoms which can give metallic interactions. The calcium
is probably ionised as the Ca-N distances are very similar to those
found (28). This compound has metallic character and is in fact a
semi-conductor, the conductivity of which is as good as that of a
pure metal.

The question of oxynitrides will be developed elsewhere in
this conference but some characteristics may be pointed out. The
main question arises whether nitrides are really nitrides or not.
In fact it is always extremely difficult to avoid contamination
by oxygen when preparing a pure compound. In some cases the properties of very pure substances are significantly different from
those commonly described, for example in their stability towards
hydrolysis. In other cases as previously seen, the oxygen contamination does not allow one to know if a given compound is stoichiometric or not.

However oxides and nitrides can give solid solutions or form
new ternary or quaternary compounds. The most promising ones as
refractory materials are the "sialons". Even when there is a
narrow range of solid solution, it is often possible to get a
partial replacement of oxygen by nitrogen in many oxides or vice
versa in nitrides. However it seems that the partial replacement
of oxygen by nitrogen with a suitable modification of cations is
not as easy to achieve in some compounds such as silicates, as is
the reverse substitution in nitrides.

REFERENCES

1. E. Parthé, Cristallochimie des structures tétraédriques,
    Gordon and Breach, Paris, 1972.
2. R.S. Pease, Nature, 167, 722, 1950. Acta Cryst., 5, 356, 1952.
3. M.D. Lyutaya, I.G. Chemysh, O.A. Frenkel, Porosh Met., 10,
    86, 1970.
4. R.H. Wentorf, J. Chem. Phys., 26, 956, 1957.
5. H. Saito, M. Ushio, S. Nagao, Yogyo Kyokai Shi, 78, 1, 1970.
5a. K. Ichinose, T. Aoki, Y. Maeda, Proc. Int. Conf. High Pressure
    4th, 436, edited by J. Osugi. Phys-Chem. Soc. Japan, Kyoto,
    1975.
6. R.C. Vickery, Nature, 184, 268, 1959.
7. I.N. Frantsevich, T.R. Balan, A.B. Bochko, Doklady Akad. Nauk.
    SSSR, 218, 951, 1974.
8. F.R. Corrigan, F.P. Bundy, J. Chem. Phys., 63, 3812, 1975.
9. J.I. Pankove, J. Luminescence, 7, 114, 1973.
10. W. Seifert, A. Tempel, Krist. Techn., 9, 1213, 1974.
11. P. Verdier, R. Marchand, Rev. Chim. Minér., 13, 145, 1976.
12. A. Rabenau, P. Eckerlin, Special Ceramics, Brit. Ceram. Res.
    Ass., Stoke-on-Trent, 1, 136, 1960. Naturwiss., 46, 106, 1959.

I.C. Huseby, H.L. Lukas, G. Petzow. J. Am. Ceram. Soc., 58, 377, 1975.

P. Eckerlin, Zeit. Anorg. Chem., 353, 225, 1967.

P. Eckerlin, A. Rabenau, H. Nortmann, Zeit. Anorg. Chem., 353, 113, 1967.

J. David, Y. Laurent, J. Lang, Bull. Soc. Fr. Mineral. Cristall., 93, 153, 1970.

S. Wild, P. Grieveson, K.H. Jack, Special Ceramics 5, Edited by P. Popper, B.Ceram.R.A., Stoke-on-Trent, p.289, 1972.

M. Maunaye, P. L'Haridon, Y. Laurent, J. Lang, Bull. Soc. Fr. Mineral. Cristall., 94, 3, 1971.

M. Wintenberger, M. Maunaye, Y. Laurent, Mat. Res. Bull., 9, 1049, 1973.

M. Maunaye, R. Marchand, J. Guyader, Y. Laurent, J. Lang, Bull. Soc. Fr. Mineral. Cristall., 94, 561, 1971.

J. David, Y. Laurent, J.P. Charlot, J. Lang, Bull. Soc. Fr. Mineral. Cristall., 96, 21, 1973.

M. Maunaye, J. Guyader, Y. Laurent, J. Lang, Bull. Soc. Fr. Mineral. Cristall., 94, 347, 1971.

O. Borgen, H.M. Seip, Acta Chem. Scand., 15, 1789, 1961.

R. Marchand, Y. Laurent, J. Lang, Acta Cryst. B, 25, 2157, 1969.

I. Kohatsu, J.W. McCauley, Mat. Res. Bull., 9, 917, 1974.

M. Von Stackelberg, R. Paulus, Z. Phys. Chem. B, 22, 305, 1933.

J. David, Y. Laurent, J. Lang, Bull. Soc. Fr. Mineral Cristall., 94, 340, 1971.

Y. Laurent, J. Lang, M.T. Le Bihan, Acta Cryst. B, 24, 494, 1968.

Y. Laurent, J. Lang, M.T. Le Bihan, Acta Cryst. B, 25, 199, 1969.

P. Eckerlin, A. Rabenau, Zeit. Anorg. Chem., 304, 218, 1960.

D. Hall, G.E. Guzz, G.A. Jeffrey, Zeit. Anorg. Chem. 369, 108, 1969.

R. Kieffer, P. Ettmayer, S. Pajakoff, Monatsh. Chem., 103, 1285, 1972.

N. Masaki, H. Tagawa, J. Nucl. Materials, 57, 187, 1975.

T.E. Rundle, N.C. Baezinger, A.S. Wilson, R.A. McDonald, J. Am. Chem. Soc., 70, 99, 1948.

J. Tobisch, W. Hase, Phys. Stat. Sol., 21, K11, 1967.

R. Juza, F. Hund, Zeit. Anorg. Chem., 257, 1, 1948.

R. Juza, F. Hund, Zeit. Anorg. Chem., 257, 13, 1948.

R. Juza, W. Gieren, J. Haug, Zeit. Anorg. Chem., 300, 61, 1959.

R. Juza, K. Langer, K. Von Benda, Ang. Chem. Internat. Edit., 7, 360, 1968.

R. Juza, H.H. Weber, E. Meyer-Simon, Zeit. Anorg. Chem., 273, 48, 1955.

R. Juza, J. Haug, Zeit. Anorg. Chem., 309, 276, 1961.

A.P. Palisaar, R. Juza, Zeit. Anorg. Chem., 384, 1, 1971.

106

42. M.B. Arbuzov, B.V. Khaenko, E.T. Kachkovskaya, Porosh Met., 13, 69, 1973.
43. F.W. Wood, O.G. Paasche, Microstructural Science, 2, 101, Edited by G.P. Fritzke, J.H. Richardson, J.L. McCall, American Elsevier Publ. Co. Inc., 1974.
44. B. Holmberg, Acta Chem. Scand., 16, 1255, 1962.
45. M.E. Straumanis, C.A. Faunce, Zeit. Anorg. Chem., 353, 329, 1967.
46. R.C. Brown, N.J. Clark, J. Inorg. Nucl. Chem., 36, 1777, 2287, 2507, 1974.
47. N. Lorenzelli, J. Melamed, J.P. Marcon, Colloques Internationaux du C.N.R.S. No. 180, 1, 375, Paris, 1970.
48. R. Kieffer, P. Ettmayer, F. Petter, Monatsh. Chem., 102, 1182, 1971.
49. H. Holleck, E. Smailos, F. Thuemmler, Monatsh. Chem. 99, 985, 1968.
50. G. Hägg, Zeit. Phys. Chem. B, 12, 33, 1931.
51. E.T. Keve, A.C. Skapski, Inorg. Chem., 7, 1757, 1968.
52. J. Gaudé, P. L'Haridon, Y. Laurent, J. Lang, Bull. Soc. Fr. Mineral. Cristall., 95, 56, 1972.
53. T. Terao, Japan J. Appl. Phys., 4, 353, 1965.
54. L.H. Butorina, Z.G. Pinsker, Sov. Phys. Crystallog. 5, 560, 1960.
55. R. Juza, Die Chemie, 58, 25, 1945.
56. N. Schönberg, Acta Chem. Scand., 8, 199, 1954.
57. G. Brauer, E. Mohr-Rosenbaum, Monatsh. Chem., 102, 1311, 1971.
58. H. Nowotny, F. Benesovsky, Phase Stability in Metals and Alloys, Edited by P.S. Rudman, J. Stringer, R.I. Jaffe, 319, McGraw Hill, New York, 1967.
59. H. Nowotny, W. Jeitschko, F. Benesovsky, Planseeber. Pulvermet. 12, 31, 1964.
60. K.H. Jack, Acta. Cryst., 5, 404, 1952.
61. M. Barberon, These, Paris, 1973.
62. E. Zintl, G. Brauer, Zeit. Electrochem., 41, 102, 1935.
63. H. Boller, Monatsh. Chem., 99, 2444, 1968.
64. H. Boller, Monatsh. Chem., 100, 1471, 1969.
65. M.E. Straumanis, C.A. Faunce, W.J. James, Acta. Metall., 15, 65, 1967. Inorg. Chem., 5, 1027, 1966.
66. J. Goodenough, A. Wold, R.J. Arnott, J. Appl. Phys., 31, 3425, 1960.
67. C.W. Wiener, J.A. Berger, J. Metals, 7, 360, 1955.
68. R. Juza, H. Puff, F. Waganknecht, Z. Elektrochem., 61, 804 and 810, 1957.
69. Stadelmaier, Developments in the Structural Chemistry of Alloy Phases, Plenum Press, 1969.
70. R. Juza, H. Hahn, Zeit. Anorg. Chem., 244, 133, 1940, and 248, 118, 1941.
71. W. Sachsze, R. Juza, Zeit. Anorg. Chem., 259, 121, 1949.
72. D.H. Jack, K.H. Jack, J. Iron St. Inst., 210, 790, 1972.
73. P. Ettmayer, Monatsh. Chem., 102, 858, 1971.
74. H. Hughes, J. Iron St. Inst., 205, 775, 1967.
75. P. Verdier, P. L'Haridon, M. Maunaye, R. Marchand, Acta. Cryst. B, 30, 226, 1974.

DISCUSSION (Leake, Morgan)

Popper: I indicated in my lecture that we were able to select BN, AlN, Si_3N_4 as the binary compounds of greatest practical interest. Could you select from the great numbers of ternary nitrides those which are likely to have high temperature stability, low thermal expansion and high thermal conductivity. What principles would you employ?

Lang: Many of the nitrides whose structure has been described were not studied as refractory materials; for example those with a perovskite structure are interesting for their magnetic properties. The refractory ternary nitrides seem to be those obtained with lithium or beryllium nitrides combined with stable covalent nitrides. $LiSi_2N_3$ has been patented for refractory purposes. A great many ternary nitrides react readily with O_2 at high temperature ($< 1000°C$) and the measurement of their thermal expansion and conductivity seems not to have been made. To select suitable compounds, it would probably be useful to consider the properties of binary nitrides giving a ternary one and the affinity of the elements for N_2 and O_2 respectively.

Jack: May I comment on the different structures shown by "NbN". These are solely a result of oxygen impurity. There are about as many "Nb-N" phase diagrams as there have been investigators because actually each has worked across a different section of the Nb-N-O system! ε- and δ-NbN accommodate different oxygen concentrations and in any one crystal you can get regions of ε in a matrix of δ because of oxygen segregation. The difference between ε and δ is just a difference of stacking close-packed Nb planes. The effect of minute amounts of O on the structures of interstitial nitrides can be very great.

We have just prepared what I believe is the first reported quaternary nitride $MgSiAlN_3$. This is isostructural with $LiSi_2N_3$ with Mg occupying 4-fold Li sites and Si+Al occupying Si 8-fold Si sites in a random manner.

Mocellin: Is there any information about the structure evolution of LnN_2-type compounds when, after being formed under high N_2 pressure, the conditions are brought back to normal?

Lang: After the pressure has been released, a lot of nitrogen is evolved and the initial structure of lanthanide nitride is probably restored.

THE CRYSTAL CHEMISTRY OF THE SIALONS AND RELATED NITROGEN CERAMICS

K.H. Jack

The Wolfson Research Group for High-Strength Materials,
Crystallography Laboratory, The University,
Newcastle upon Tyne, England.

ABSTRACT. The acronym "sialon" was originally given to new
compounds derived from silicon nitride and oxynitride by
simultaneous replacement of silicon and nitrogen by aluminium and
oxygen. Similar replacements are possible in other structures and
other metals such as lithium, beryllium and magnesium can also be
incorporated. Sialons are built up of one-, two-, and three-
dimensional arrangements of $(Si,Al)(O,N)_4$ and $(M,Si)(O,N)_4$ tetrahedra in the same way that the structural unit in the silicates
is the SiO_4 tetrahedron. The new materials show structure types
based not only on α and β silicon nitrides and silicon oxynitride
but also on silicon carbide and aluminium nitride, eucryptite,
spinel, melilite, apatite, wollastonite and other naturally
occurring silicates and aluminates. They include glasses as well
as crystalline phases and are being explored for their thermal,
mechanical, chemical and electrical properties. Phase relationships in the yttrium and magnesium sialon systems are discussed
to illustrate the technological advantages offered by "ceramic
alloying".

1. THE SIALONS

In each of its structural modifications silicon nitride is a
typically covalent solid built up of SiN_4 tetrahedra joined in a
three-dimensional network by sharing corners; each nitrogen corner
is common to three tetrahedra. Indeed (see Fig. 1), the atomic
arrangement of $\beta-Si_3N_4$ is identical with that of phenacite, the
beryllium silicate Be_2SiO_4 built up of BeO_4 and SiO_4 tetrahedra.
In the mineral silicates the SiO_4 tetrahedral units, each carrying
four negative charges, may occur separately or may be joined by

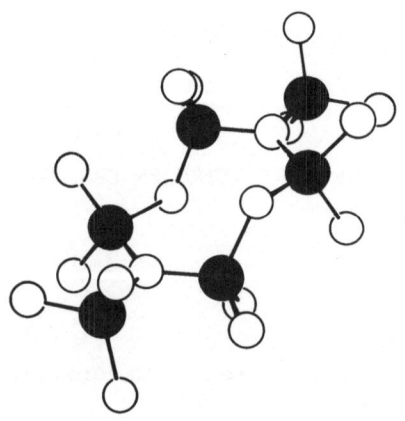 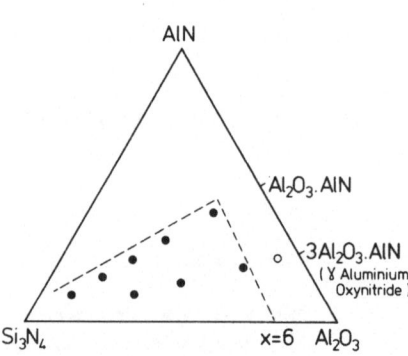

Fig. 1. The β-Si_3N_4 structure.   Fig. 2. Early (incorrect) phase diagram for Si_3N_4-Al_2O_3-AlN

sharing oxygen corners into rings, chains, two-dimensional sheets, or three dimensional networks. Aluminium plays a special role because the AlO_4 tetrahedron - with five negative charges - is about the same size as SiO_4 and can replace it in the rings, chains and networks provided that valency compensation is made elsewhere in the structure. By analogy, it seemed possible to replace N^{3-} by O^{2-} in silicon nitride if at the same time Si^{4+} was replaced by Al^{3+}. Charge compensation might also be feasible by introducing other metal atoms such as Mg^{2+} and Li^{1+}. In fact, it was suggested (1) that a variety of materials, vitreous as well as crystalline, could be obtained built up of $(Si,Al)(O,N)_4$ tetrahedra in the same way that the almost infinite range of silicates is built up of $(Si,Al)O_4$ units. We reported the first of these new phases, β'-"sialon" (2) at about the same time as Oyama & Kamigaito (3) and Tsuge et al. (4) in Japan. The British and Japanese work has developed in similar directions and has even made the same mistakes.

1.1 Early experimental results

The X-ray diffraction patterns of the products obtained by reacting together silicon nitride and alumina at 1750-2000°C were identical with that of β-Si_3N_4 except that reflexions move to slightly lower angles as the Al_2O_3 content increases up to about 70 w/o. The homogeneous sialon product was therefore termed β'. The following alternative compositions for β' were considered:

(1)   β · $\mathrm{Si}_6^{24}\,\mathrm{N}_8^{\overline{24}}$

$\downarrow$

β' $\mathrm{Si}_{6-0.75x}^{24-3x}\,\mathrm{Al}_{0.67x}^{2x}\,\mathrm{O}_x^{\overline{2x}}\,\mathrm{N}_{8-x}^{\overline{24-3x}}$

$\mathrm{Al}_{5.33}^{16}\,\mathrm{O}_8^{\overline{16}} \equiv \mathrm{Al}_2\mathrm{O}_3$

(2)   β $\mathrm{Si}_6^{24}\,\mathrm{N}_8^{\overline{24}}$

$\downarrow$

β' $\mathrm{Si}_{6-z}^{24-4z}\,\mathrm{Al}_z^{3z}\,\mathrm{O}_z^{\overline{2z}}\,\mathrm{N}_{8-z}^{\overline{24-3z}}$

$\mathrm{Al}_6^{18}\,\mathrm{O}_6^{\overline{12}}\,\mathrm{N}_2^{\overline{6}} \equiv \mathrm{Al}_2\mathrm{O}_3 \cdot \mathrm{AlN}$

and on the basis of limited chemical analysis and the observation of only one single-phase crystalline product, it was concluded that the sequence (1) represented the reaction between silicon nitride and alumina. It was not appreciated that products represented by (2) might be obtained with (i) volatilisation of silicon monoxide and nitrogen or, in the reducing environment of graphite in the hot press, (ii) volatilisation of silicon and carbon monoxides, or (iii) with simultaneous formation of a silica-rich glass. For example, possible reactions to produce β' with $z = 4$ are:

$$4\mathrm{Si}_3\mathrm{N}_4 + 6\mathrm{Al}_2\mathrm{O}_3 \quad = \quad 3\mathrm{Si}_2\mathrm{Al}_4\mathrm{O}_4\mathrm{N}_4 + 6\mathrm{SiO} + 2\mathrm{N}_2 \quad \ldots \text{ (i)}$$

$$\mathrm{Si}_3\mathrm{N}_4 + 2\mathrm{Al}_2\mathrm{O}_3 + \mathrm{C} = \mathrm{Si}_2\mathrm{Al}_4\mathrm{O}_4\mathrm{N}_4 + \mathrm{SiO} + \mathrm{CO} \quad \ldots \text{ (ii)}$$

$$\mathrm{Si}_3\mathrm{N}_4 + 2\mathrm{Al}_2\mathrm{O}_3 = \mathrm{Si}_2\mathrm{Al}_4\mathrm{O}_4\mathrm{N}_4 + \mathrm{SiO}_2 \quad \ldots \text{ (iii)}$$

Similar β'-sialon phases were obtained by reacting silicon nitride with lithium-aluminium spinel, $\mathrm{LiAl}_5\mathrm{O}_8$, and also with magnesium-aluminium spinel, $\mathrm{MgAl}_2\mathrm{O}_4$. The metal:non-metal atom-ratio 3M:4X in the reacting spinels is, of course, the same as that in the silicon nitride. Subsequently, $\mathrm{Si}_3\mathrm{N}_4$ was reacted with equi-molecular mixtures of $\mathrm{Al}_2\mathrm{O}_3$ and AlN (that is, the equivalent of the spinel $\mathrm{Al}_3\mathrm{O}_3\mathrm{N} \equiv \mathrm{Al}_2\mathrm{O}_3 \cdot \mathrm{AlN}$) and then with varying ratios of the oxide and nitride. From the results it was concluded that the β'-phase field extended not only along the joins $\mathrm{Si}_3\mathrm{N}_4\text{-}\mathrm{Al}_2\mathrm{O}_3$ and $\mathrm{Si}_3\mathrm{N}_4\text{-}\mathrm{Al}_3\mathrm{O}_3\mathrm{N}$ but also covered the region between these limits. The phase diagram deduced for the $\mathrm{Si}_3\mathrm{N}_4\text{-}\mathrm{Al}_2\mathrm{O}_3\text{-}\mathrm{AlN}$ system at above 1700°C is shown by Fig. 2 and was identical with that proposed by Oyama (5); both are now known to be incorrect!

Doubts about this β' homogeneity range were raised by Lumby et al. (6) who suggested from creep measurements that sialons with overall compositions along the Si_3N_4-Al_2O_3 join contained glassy phases whereas compositions with 3M:4X showed minimum high-temperature creep. Since then, preparative work at Stuttgart (7) and Newcastle (8) using mixtures of different reactants (e.g. from Si_3N_4, Al_2O_3, AlN, Si_2N_2O, SiO_2) to obtain the same product has shown that β' extends along the join Si_3N_4-Al_3O_3N from z = 0 to about z = 4.

2. THE Si-Al-O-N SYSTEM

Results of hot-pressing, usually at 1750°C but sometimes at higher and lower temperatures (1550-2000°C), are shown in Fig. 3 which is an idealised behaviour diagram and does not necessarily represent thermodynamic equilibrium. Each of the phases extends in a direction of constant M:X ratio along which AlO and SiN are mutually replaceable; homogeneity ranges in other directions are small. The similarity in bond lengths between Si-N (1.75Å) and Al-O (1.75Å) and the difference between Al-N (1.87) and Si-O (1.62) offer some explanation for this.

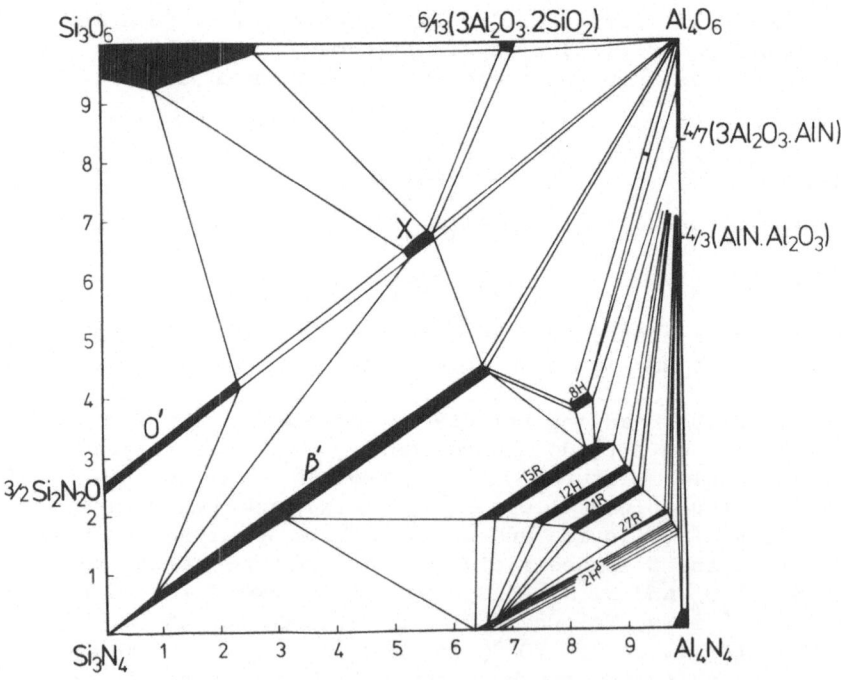

Fig. 3. The Si_3N_4-AlN-Al_2O_3-SiO_2 system.

2.1 The β'-sialon phase

The β' silicon-nitride-type sialon with composition $Si_{6-z}Al_zO_zN_{8-z}$ is the only one so-far examined in sufficient detail for technological evaluation, and most measurements have been on compositions containing about equal concentrations of silicon and aluminium, i.e. with z = 3. Because of its structure, its physical and mechanical properties are similar to those of β-silicon nitride, but chemically it is closer to aluminium oxide. Thus (see Fig. 4) its thermal expansion coefficient (2.7×10^{-6}°C^{-1}) is less than that of β-Si_3N_4 (3.5×10^{-6}°C^{-1}) and so its thermal shock properties are at least as good. Oxidation resistance is better, probably because a coherent and protective layer of mullite is formed on the surface. Compatibility with molten metals, e.g. aluminium copper and iron, is excellent and the use of β'-sialon for holding and conveying molten metals, including steel, is perhaps a more important application than for ceramic gas turbine components.

One potential advantage of β'-sialon over silicon nitride is in fabrication. The usual ceramic techniques of extrusion, pressing and slip-casting can be used to produce shapes of the mixed components and then these can be fired to near-theoretical density in an inert atmosphere at about 1600°C. As will be shown later, densifying agents which promote liquid-phase sintering can subsequently be incorporated in the sialon structure.

Now that there is a reliable behaviour diagram for the Si-Al-O-N system (Fig. 3) it is possible to produce β'-phase or (β'+15R) mixed-phase sialons free from glass and because the creep of nitrogen ceramics is essentially related to their glass content, the glass-free materials are highly creep-resistant.

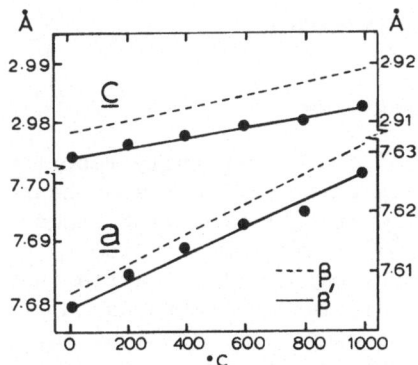

Fig. 4. Thermal expansion of β-Si_3N_4 and β'-sialon (z = 3).

2.2. The O'-sialon phase

Silicon oxynitride reacts with alumina to extend along the 2M:3X join giving sialons with the Si_2N_2O-type structure and slightly larger unit-cell dimensions. As with β', it can be prepared from a variety of oxide-nitride mixtures.

2.3. X-phase

The phase "X" - so-called because it could not at first be characterized - is identical with phase II reported by Oyama & Kamigaito (5,9) who gave it the composition $3Al_2O_3.2Si_3N_4$. It is usually the minor crystalline constituent in the production of β' from mixtures of silicon nitride and alumina and has also been variously designated as "Oyama-phase" and "J-phase". Its X-ray and electron diffraction patterns have been interpreted in terms of several different unit cells. For example Drew & Lewis (10) claim a triclinic structure with cell dimensions

$$\underline{a} = 9.9, \underline{b} = 9.7, \underline{c} = 9.5\text{Å}, \alpha = 109°, \beta = 95°, \gamma = 95°$$

while Gugel et al. (11) propose an orthorhombic lattice with

$$\underline{a} = 7.85, \underline{b} = 9.12, \underline{c} = 7.965\text{Å}$$

Neither these nor other suggestions are compatible with the X-ray data obtained from specimens of at least 95% purity prepared at Newcastle. The composition is more Al_2O_3-rich than the originally proposed $SiAlO_2N$ (12), and the unit cell has been determined (13; see 8) unequivocally as monoclinic with dimensions:

$$\underline{a} = 9.728, \underline{b} = 8.404, \underline{c} = 9.572\text{Å}, \beta = 108.96°$$

A complete structure determination is proceeding and although not yet complete it suggests that the initial description of X-phase as a "nitrogen-mullite" is not too misleading.

2.4. The tetrahedral AlN-polytype structures

In addition to the sialon phases β', O' and X originally reported (2,6), Gauckler et al. (7) prepared six unidentified phases nearer to the AlN corner of the Si_3N_4-AlN-Al_2O_3-SiO_2 system with ranges of homogeneity extending along lines of constant M:X ratio. These have been fully characterized by Thompson at Newcastle (see 8) and he has interpreted their X-ray diffraction patterns in terms of AlN-type polymorphs containing excess non-metal atoms, the structures of which are directly related to their compositions M_mX_{m+1}. Details of these will be discussed by Dr. Thompson at the present meeting. As shown in Fig. 3 the phases are designated by their Ramsdell symbols 8H, 15R, 12H, 21R and 27R. Along the $\underline{c}$-dimension, the structure consists of "n" layers where "n" is the numeral of the Ramsdell symbol. The nR polytypes consist of three rhombohedrally-related blocks each of m = n/3 layers, while the hexagonal nH polytypes have two blocks each containing m = n/2 layers.

In the above five polytypes the number (m) of MX layers per block is therefore 4, 5, 6, 7 and 9 respectively but one layer in each block has an additional non-metal atom, MX_2. The 2H AlN structure is the end member of this series but a similar 2H structure with an expanded c-dimension is obtained at M:X ratios greater than 9:10. The concentration of additional non-metal atoms is too small for an ordered arrangement and so the phase is designated $2H^\delta$ where δ indicates disorder in the sequence of MX_2 layers.

These and other polytypes occur in the Mg-Si-Al-O-N and Li-Si-Al-O-N systems. They are also structurally related, except that the roles of the metal and non-metal atoms are reversed, to the unidentified phases reported by Huseby et al. (14) in the Be-Si-O-N system and recently characterized by Thompson (15) as $M_{m+1}X_m$ polytypes.

2.5. The technological significance of AlN-polytype sialons

Komeya et al. (16) suggest that the desirable fibrous micro-structures formed in $AlN-Y_2O_3-SiO_2$ ceramics are due to an "Al-Si-O-N" phase which we have identified as a 9M:10X AlN-polytype. In general, the morphologies of the polytypes, like those of α-SiC, are acicular and it seems possible that the fracture toughness of a β' + 15R composite is better than that of a pure homogeneous β'-sialon. In β'-sialons produced by nitriding mixtures of volcanic ash and aluminium (17), the major unknown phase reported is a 15R polytype and this appears not to affect the properties adversely.

3. METAL-SIALON SYSTEMS

Mg, Li and other metal-silicon nitrides and oxynitrides all have structures built up of MX_4 tetrahedra and can be regarded as orthorhombic superlattices of the hexagonal AlN-type (see Fig. 5). It seemed likely that these metals could be incorporated into sialons. The representation of "quaternary systems of the third kind" (18) is by Jänecke's triangular prism (19) outlined in Fig. 6 for the Mg-Si-Al-O-N system and based on the standard $Si_3N_4-Al_4N_4-Al_4O_6-Si_3O_6$ square with Mg in equivalent units along a third dimension. Each of the phases of the basal square plane extend into the prism volume, generally along planes of constant M:X ratio which are not parallel with any of the prism faces and which cut other triangular planes representing pseudo-ternary systems e.g. $6MgO-Si_3N_4-4AlN$.

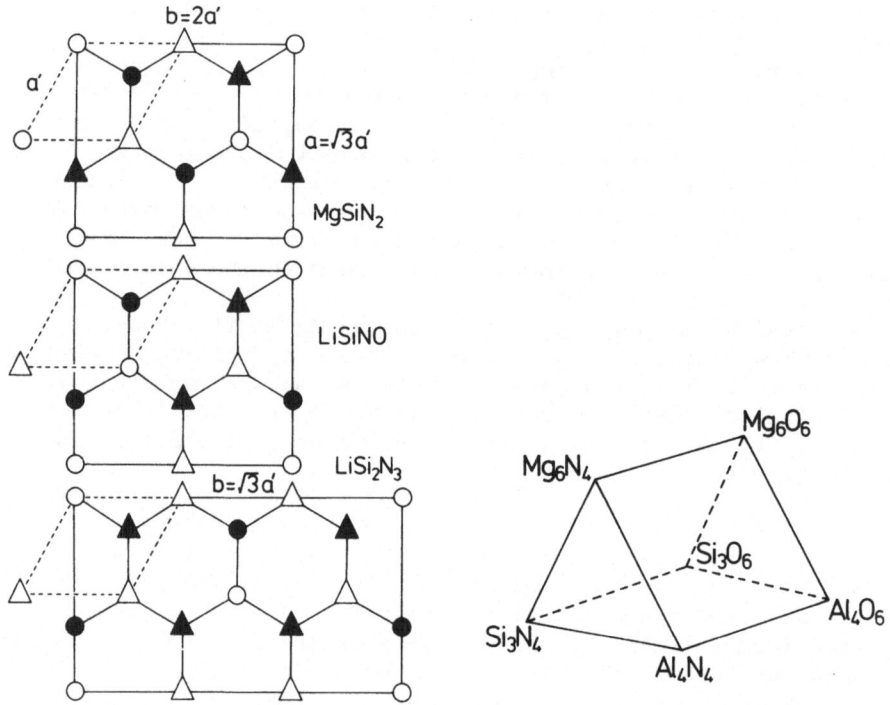

Fig. 5. Metal-silicon nitride and oxynitride structures.

Fig. 6. Outline representation of the Mg-sialon system.

Only the lithium, magnesium and yttrium sialon systems have been investigated in any detail and even these are far from complete. The systems were chosen because of their technological potential, magnesia and yttria being the most useful additives in hot-pressing silicon nitride and lithium sialons offering possibilities in their electrical properties; also, lithium, magnesium and yttrium are respectively uni-, di- and trivalent so that successive differences in phase relationships seemed likely. A few of the more significant features in the three systems are described.

3.1. The Mg-Si-Al-O-N system

The MgO-Si_3N_4-Al_2O_3 section represented by an equilateral triangle in Fig. 7 cuts across planes of constant M:X ratio and so shows few single-phase regions. In addition to β' occurring along the

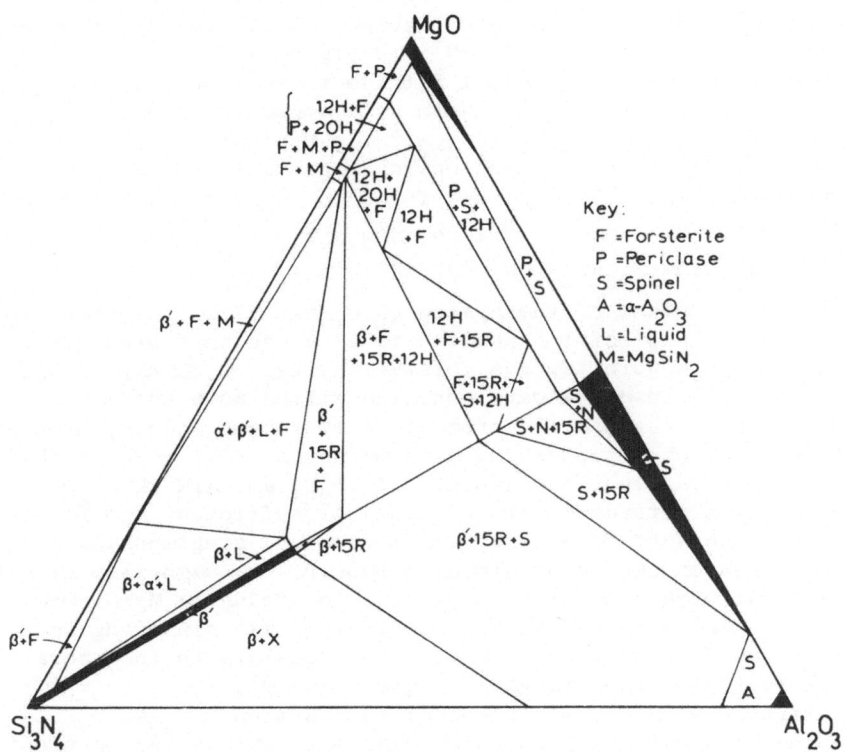

Fig. 7. The MgO-Si$_3$N$_4$-Al$_2$O$_3$ section of the Mg-Si-Al-O-N system at 1800°C.

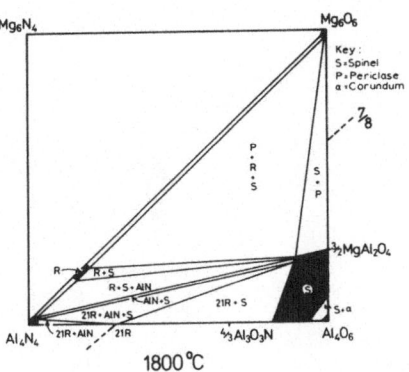

Fig. 8. The Mg-Al-O-N prism face.

Si_3N_4-$MgAl_2O_4$ join, phases α' (isostructural with α-silicon nitride), x, 12H, 15R and a nitrogen-spinel are observed. The homogeneity range of the nitrogen-spinel is fairly extensive and is better shown by the Mg-Al-O-N prism-face of Fig. 8.

The first quaternary metal nitride to be reported is observed along the $MgSiN_2$-AlN join at a composition $MgSiAlN_3$; it is isostructural with $LiSi_2N_3$ with Mg replacing Li in four-fold sites and with Si + Al occupying eight-fold sites in a random manner. New polytypes not found in the Si-Al-O-N system are also observed: 2OH at $Mg_4Si_{2.7}Al_{3.3}O_4N_7$ (10M:11X); and 12R at about $Mg_{4.25}Si_{0.25}Al_{0.5}ON_3$ (5M:4X).

The 3M:4X plane of the system (see Fig. 9) is important in showing that the β'-sialon phase extends along this plane from the Si_3N_4-Al_3O_3N join towards forsterite, $Mg_2SiO_4 \equiv 2MgO.SiO_2$. Silicon nitride powder always contains silica as a surface layer and this prevents the production of a homogeneous, single-phase material by hot-pressing with magnesia. Fig. 9 shows that by addition of appropriate amounts of Al_2O_3 and AlN ($\equiv Al_3O_3N$) to the silicon nitride together with just sufficient MgO to react with the surface silica, a homogeneous, single-phase β'-magnesium sialon can be obtained. Moreover, the magnesium silicate reacts first with some silicon nitride to produce a Mg-Si-O-N liquid which aids densification by liquid-phase sintering and then, by suitable heat-treatment, it is possible to incorporate this in solid solution to give single-phase β'.

Starting with any β'-sialon of composition $Si_{6-z}Al_zO_zN_{8-z}$ along the Si_3N_4-Al_3O_3N join, the z-value is maintained constant

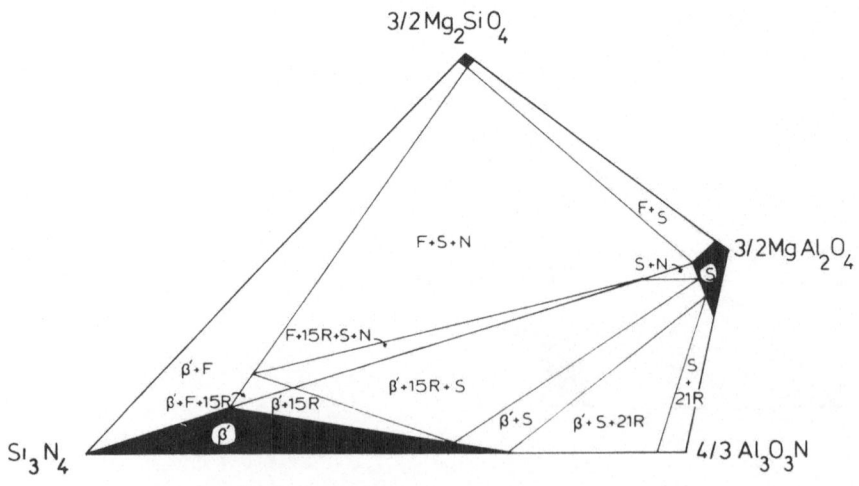

Fig. 9. The 3M:4X plane of the Mg-Si-Al-O-N system

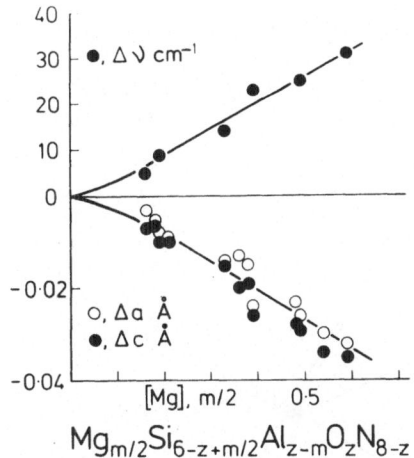

Fig. 10. Unit-cell dimension changes and infra-red absorption shifts as MgSi replaces 2Al in β'-sialons.

if each Al is replaced by $Mg_{0.5}Si_{0.5}$ along a direction parallel with the $MgAl_2O_4$-Mg_2SiO_4 join to give a composition

$$Mg_{0.5m} Si_{6-z + 0.5m} Al_{z-m} O_z N_{8-z}$$

Fig. 10 shows that irrespective of its z-value, the unit-cell dimensions of β'-sialon decrease smoothly as $Mg_{0.5}Si_{0.5}$ replaces Al. Infra-red absorption measurements made by Wild (20) show that this same replacement is accompanied by an increase in the frequency of a band characteristic of the β'-structure ($\nu = 580 cm^{-1}$ for $\beta-Si_3N_4$) even though it decreases with increasing z; see Fig. 10. The phase relationships shown by Fig. 9 offer the only reasonable explanations for these unit-cell dimensional and infra-red absorption changes.

3.2. The Li-Si-Al-O-N system.

As with the magnesium sialons, completely new phases occur in the Li-Si-Al-O-N system as well as those which extend from the basic Si-Al-O-N behaviour diagram of Fig. 3. Reaction of Si_3N_4 with the spinel $LiAl_5O_8$ gives a β'-lithium sialon, and with $LiAlO_2$ a sialon (α') with the structure of α-silicon nitride. Other nitrogen-containing phases have structures based on β-eucryptite (Eu'), spinel (S), silicon oxynitride, tetragonal cristobalite (γ) and the polytype 15R. The section Li_2O-Si_3N_4-Al_2O_3 of Fig. 11 shows these and although there are again no extensive single-phase regions, the variations of unit-cell dimensions observed

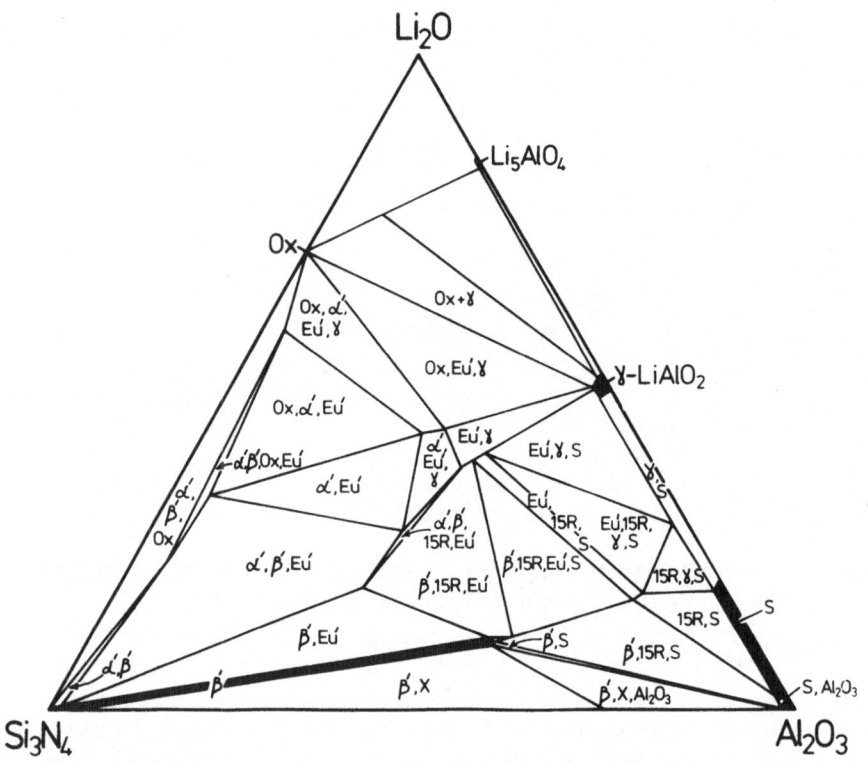

Fig. 11. The $Li_2O-Si_3N_4-Al_2O_3$ section of the Li-Si-Al-O-N system at 1550°C.

in the multi-phase products show that each phase has an appreciable range of homogeneity.

4. YTTRIUM SIALONS

Gazza (21) showed that improved properties are obtained in hot-pressed silicon nitride by using yttria additions instead of magnesia and suggested that these produced a more refractory grain-boundary phase identified at Newcastle (22) as a yttrium-silicon oxynitride $Y_2Si[Si_2O_3N_4]$. The latter has the same tetragonal structure as akermanite $Ca_2Mg[Si_2O_7]$ and gehlenite $Ca_2Al[SiAlO_7]$, members of the melilite series of silicates. Sheets of $Si(O_{0.4}N_{0.6})_4$ tetrahedra are stacked one on top of the other and are held together by yttrium ions between them; see Fig. 12. The oxynitride forms a continuous series of solid solutions with both akermanite and gehlenite and even the 50:50

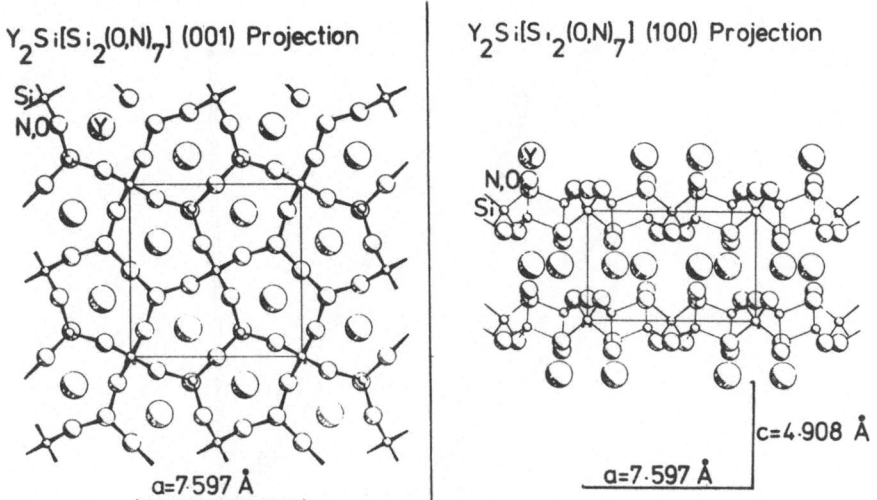

$Y_2Si[Si_2(O,N)_7]$ (001) Projection

$Y_2Si[Si_2(O,N)_7]$ (100) Projection

a=7·597 Å

a=7·597 Å

c=4·908 Å

Fig. 12. Projections of $Y_2Si[Si_2O_3N_4]$ on (001) and (100).

intermediates listed as (2) and (4) in Table 1 have melting points above 1600°C. Thus, appreciable amounts of Ca, Mg, Al and probably other silicon nitride impurities which would otherwise form a low-softening-temperature glass are accommodated in the structure without any loss of refractoriness. Two additional phases designated in the early investigation (22) as "H" and "J" have been subsequently identified in a number of laboratories (23, 8, 24, 25, 26) although there is not yet complete agreement about compositions and crystal structures or about the character-ization of a fourth quaternary phase originally called "D" at Newcastle (8) and labelled "K" by Lange et al. (26). It is now designated in the present paper as "H'".

4.1. The Y-Si-O-N system

The behaviour diagram of Fig. 13 deduced from hot-pressings at 1600°C and 2600 psi for 1h summarises the most recent Newcastle work. At 1800°C there are large liquid or liquid + solid phase-regions; see Fig. 14.

N-apatite has an appreciable homogeneity range near (Y_4Si) $(Si_3O_{11}N)N$ and can not be described solely by the composition $Y_5(Si_3O_{12})N$ initially proposed (23).   Wills et al. (27) report an oxidation product $Y_{4.67}(Si_3O_{12})O$ but we believe that this is the oxygen-rich end of the solid solution series. Other nitrogen-apatites have been reported (28) with the general formula

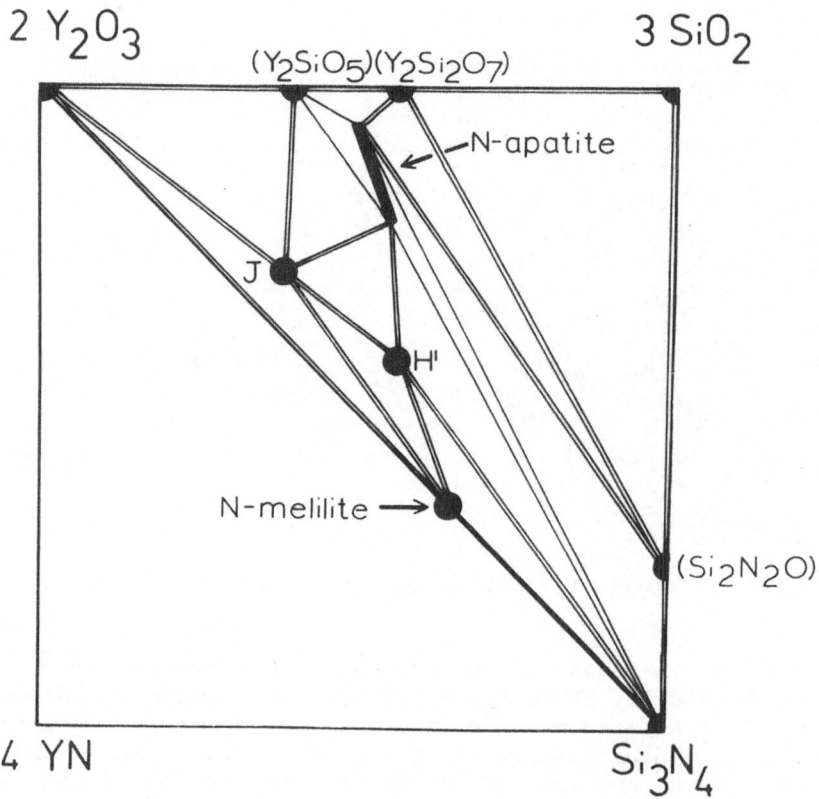

Fig. 13. The Y_2O_3-YN-Si_3N_4-SiO_2 section of the Y-Si-Al-O-N system at 1600°C.

Y-N melilites

(1) Åkermanite $Ca_2Mg[Si_2O_7]$
     1454 °C

(2) (1) + (3) $(Y,Ca)_2(Mg,Si)[Si_2O_5N_2]$

(3) "Ysion" $Y_2Si[Si_2O_3N_4]$
     $Y_2O_3 \cdot Si_3N_4$

(4) (3) + (5) $(Y,Ca)_2(Si,Al)[(Si,Al)_2O_5N_2]$

(5) Gehlenite $Ca_2Al[SiAlO_7]$
     1590 °C

Table 1. Solid solution series in the N-melilites

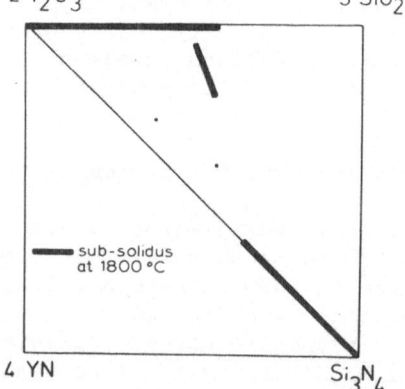

Fig. 14. The Y_2O_3-YN-Si_3N_4-SiO_2 section at 1800°C.

$(Ln,M)_5Si_3(O,N)_{13}$ and it is clear from this work at Rennes that a wide variety of N-apatites of the lanthanide elements exists.

J-phase was separately reported as $2Y_2O_3.Si_3N_4$ (29) and $3Y_2O_3.Si_3N_4$ (30) but it lies off the $Y_2O_3-Si_3N_4$ join with a composition $2Y_2O_3.Si_2N_2O$ (24, 25, 8). Although Wills et al. suggest (25) that it is isostructural with cuspidine $Ca_4Si_2O_7F_2$, we believe from similarities in their X-ray patterns and the existence of solid solutions that it is isostructural with the monoclinic yttrium-aluminate $2Y_2O_3.Al_2O_3$ ("YAM"). In the latter, substitution of 2(AlO) by 2(SiN) leads to the proposed composition. Unit-cell dimensions are:

	$\underline{a}$	$\underline{b}$	$\underline{c}$	β
$2Y_2O_3.Si_2N_2O$	7.558	10.446	10.812	111.0
$2Y_2O_3.Al_2O_3$	7.378	10.466	11.115	108.6

The H'-phase with composition $YSiO_2N$ is the same as the K-phase reported by Lange et al. (26) but whereas the latter give monoclinic cell dimensions:

$\underline{a}$, 9.208; $\underline{b}$, 13.157; $\underline{c}$, 8.675Å; β, 97.471°

we have shown that its structure is based on α-wollastonite $(CaSiO_3)$ with a monoclinic (pseudo-hexagonal) cell of dimensions:

$\underline{a}$, 7.012; $\underline{b}$, 12.168; $\underline{c}$, 18.202; β, 90.76°

The atomic arrangement consists of three-membered tetrahedron rings $[Si_3O_6N_3]$ sandwiched between sheets of yttrium cations.

4.2 SiN⇌AlO replacement in yttrium-sialons

The H' and J phases lie on the $Y_2O_3-Si_2N_2O$ join in the $Y_2O_3-SiO_2-Si_3N_4$ system (see Fig. 13) and so are also on the 2M:3X plane in the $Y_2O_3-Al_2O_3-Si_2N_2O$ pseudo-ternary system; Fig. 15. If, as suggested, the phases H' and J are respectively isostructural with the yttrium aluminates $Y_3Al_5O_9$ and $Y_4Al_2O_9$ solid solutions are expected. Figs. 16 and 17 show that the unit-cell dimensions of intermediate compositions lie with those of the end numbers on smooth curves. In each case a continuous series of solid solutions occurs by gradual replacement of AlO by SiN, e.g.

$$Y_4Al_2O_9 \rightarrow Y_4SiAlO_8N \rightarrow Y_4Si_2O_7N_2$$

Valency requirements are maintained by the replacement and because the Al-O and Si-N bond lengths are almost identical there is no structural change. Many more substitutional series of the same type are expected in sialon systems.

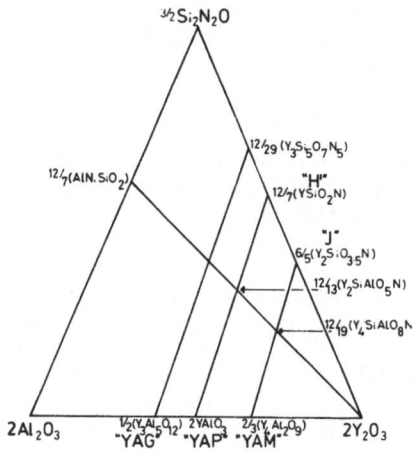

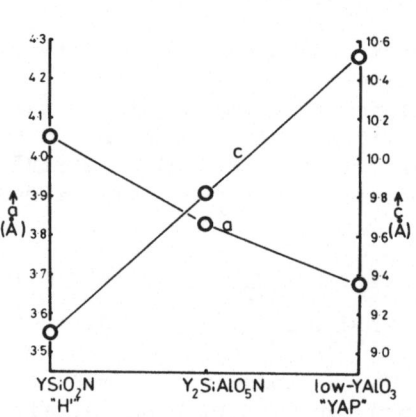

Fig. 15. The Y_2O_3-Al_2O_3-Si_2N_2O
pseudo-ternary system.

Fig. 16. Unit-cell dimensions of
the α-wollastonite solid-solution
series.

4.3 Hot-pressing Si_3N_4 with Y_2O_3

Yttria seems an ideal hot-pressing additive because it reacts
initially with surface silica and silicon nitride to give a
liquid which allows densification by liquid-phase sintering. As
reaction proceeds the liquid combines with more silicon nitride
to give refractory bonding phases. The four quaternary phases
N-melilite, N-α-wollastonite, N-YAM and N-apatite are analogues
of calcium silicates or aluminosilicates and can all accommodate
in refractory solid solution impurities which, with magnesia as
a hot-pressing additive, would give a glass. The final grain-boundary
phases in the silicon nitride will depend upon its initial silica
content and the amount removed by the reducing environment during
pressing.

5. CONCLUSIONS

The sialons are essentially silicates or aluminosilicates in
which oxygen is partly replaced by nitrogen. So-far, structure
types have been characterized in which the $(Si,Al)(O,N)_4$ tetrahedra occur as isolated units, as rings, as sheets and as three-
dimensional net-works. Whether or not the field is as large as
that of the mineral silicates, it is certainly extensive. It
includes vitreous materials as well as crystalline phases, and
the mutual replacement of oxygen and nitrogen gives an additional

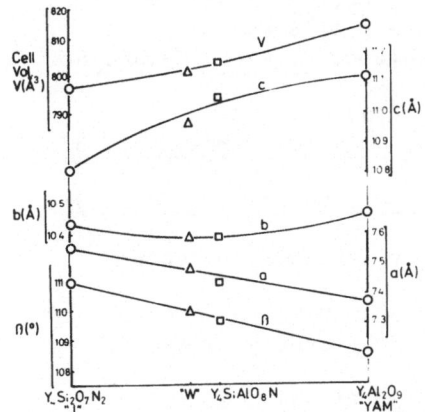

Fig. 17. Unit-cell dimensions of the YAM solid-solution series.

degree of freedom which is worth scientific and technological exploration. Until recently this exploration has been superficial and mistakes have been made in the interpretation of experimental observations. Sufficient of the principles are now known, however, to establish reliable preparative methods and to characterize the products more precisely.

REFERENCES

1. S. Wild, P. Grieveson and K.H. Jack, "The Crystal Chemistry of Ceramic Phases in the Silicon-nitrogen-oxygen and Related Systems", 1968 Progress Report No. 1, Ministry of Defence Contract N/CP.61/9411/67/4B/MP 387. See "Special Ceramics, 5", edited by P. Popper, Brit. Ceram. R.A., Stoke-on-Trent, 1972 p. 289.
2. K.H. Jack and W,I. Wilson, Nature Phys. Sci. (London) 238 (1972) 28.
3. Y. Oyama and O. Kamigaito, Japan J. Appl. Phys. 10 (1971) 1637.
4. A. Tsuge, H. Inoue and K. Komeya, Tenth Symposium on Basic Ceramics, Osaka, (1972).
5. Y. Oyama, Yogyo-Kyokai-Shi 82 (1974) 351.
6. R.J. Lumby, B. North and A.J. Taylor, "Special Ceramics 6" Brit. Ceram. R.A., Stoke-on-Trent, (1975) p. 283.
7. L.J. Gauckler, H.L. Lukas and G. Petzow, presented at the Second Powder Metallurgical Seminar at the Max-Planck-Institut, Stuttgart, (1974). See J. Amer. Ceram. Soc. 58 (1975) 346.
8. K.H. Jack, J. Mater. Sci. 11 (1976) 1135.
9. Y. Oyama and O. Kamigaito, Yogyo-Kyokai-Shi 80 (1972) 327.
10. P. Drew and M.H. Lewis, J. Mater. Sci. 9 (1974) 1833.

11. E. Gugel, I. Petzenhauser and A. Fickel, Powder Met. Int. 7 (1975) 66.

12. K.H. Jack, Trans. J. Brit. Ceram. Soc. 72 (1973) 376.

13. D.P. Thompson and K.H. Jack, to be published.

14. I.C. Huseby, H.L. Lukas and G. Petzow, J. Amer. Ceram. Soc. 58 (1975) 377.

15. D.P. Thompson, J. Mater. Sci. 11 (1976) 1377.

16. K. Komeya, H. Inoue and A. Tsuge, J. Amer. Ceram. Soc. 57 (1974) 411.

17. S. Umebayashi and K. Kobayashi, Amer. Ceram. Soc. Bull 54 (1975) 534.

18. J. Zernike, "Chemical Phase Theory" (Kluwer, Deventer, Netherlands, 1955)

19. E. Jänecke, Z. anorg. Chem. 53 (1907) 319.

20. S. Wild, Science Department, The Polytechnic of Wales, private communication.

21. G.E. Gazza, J. Amer. Ceram. Soc. 56 (1973) 662.

22. A.W.J.M. Rae, D.P. Thompson, N.J. Pipkin and K.H. Jack "Special Ceramics 6" (Brit. Ceram. R.A., Stoke-on-Trent, 1975) p. 347.

23. D.P. Thompson, contribution to the discussion of reference (22); "Special Ceramics 6" (Brit. Ceram. R.A., Stoke-on-Trent, 1975) p. 358.

24. P.E.D. Morgan, J. Amer. Ceram. Soc. 59 (1976) 86.

25. R.R. Wills, S. Holmquist, J.M. Wimmer and J.A. Cunningham, J. Mater. Sci. 11 (1976) 1305.

26. F.F. Lange, S.C. Singhal and R.C. Kuznicki, "Phase relations and stability studies in the Si_3N_4-SiO_2-Y_2O_3 pseudo-ternary system", Westinghouse Research Report 76-9D4-POWDR-R1, April 1976; private communication.

27. R.R. Wills, J.A. Cunningham, J.M. Wimmer and R.W. Stewart, J. Amer. Ceram. Soc. 59 (1976) 269.

28. C. Hamon, R. Marchand, M. Maunaye, J. Gaude and J. Guyader, Rev. Chimie minerale 12 (1975) 259.

29. A. Tsuge, H. Kudo and K. Komeya, J. Amer. Ceram. Soc. 57 (1974) 269.

30. R.R. Wills, J. Amer. Ceram. Soc. 57 (1974) 459.

DISCUSSION (Leake, Morgan)

Gauckler: (1) What is the temperature for which the $Si_3N_4-4(AlN)-2(Al_2O_3)-3(SiO_2)$ diagram is valid?
(2) The $\beta-Si_3N_4$ solid solution in the system Si, Al, Mg, N, O has a larger portion on your diagram in the case of the subsystem $Si_3N_4-6MgO-2Mg_3N_2-SiO_2$ than in the subsystem $SiO_2-MgO-Si_3N_4-Al_2O_3$. Can you comment on this?
(3) Why do you use for one system Si, Al, Mg, N, O, and for the subsystems, different sorts of units?
(4) What evidence do you have for your subsystem $Mg_2SiO_4-Al_2O_3$. MgO, in the system Si, Al, Mg, N, O? It is in contrast with much earlier data for the system $Mg_2SiO_4-MgO-Al_2O_3$.
(5) In your subsystem $Si_3N_4-AlN.Al_2O_3-Al_2O_3-MgO-MgSiO_4$ you showed no liquid phase regions. I believe there are at least two liquid regions at temperatures 1760 to 1800°C.

Jack: (1) It is based on results of reactions mainly at 1750°C but occasionally at lower temperatures (where there would otherwise be liquid phases) and sometimes at higher temperatures. Minimum temperature (in X phase region) 1550°C, and maximum 2000°C.
(2) You are wrong in saying that the β'-Mg sialon phase has a larger composition range in the subsystem $MgO-Si_3N_4-Al_2O_3$ than in the figure and model for the Mg-Si-Al-O-N system. The latter presents concentrations in equivalents and the former in gram moles. The corners of each of the two different diagrams are labelled appropriately and correctly, e.g. as 6MgO in one and as MgO in the other.
(3) As stated in (2), the units are different because in the pseudo-ternary system $MgO-Al_2O_3-Si_3N_4$ ceramists are used to the molar representation. I recommend the Jänecke prism representation but it is certainly not always convenient and not always recognisable by those familiar with conventional representation of ceramic systems.
(4) By all means correct whatever mistakes we have made in representing well-known subsystems such as $MgO-Al_2O_3-Mg_2SiO_4$. You will find it different in the present paper from that in J. Mat. Sci. 11, 1135.
(5) We do not show the liquid-phase regions in the subsystem $Si_3N_4-Al_3O_3N-MgAl_2O_4-Mg_2SiO_4$ but of course they exist and in another paper I shall show that one region at least in this system is liquid and cools to give a glass. In general we do not pretend that our Si-Al-O-N behaviour diagram or those derived from it for the Mg and Li sialons require no revision or modification. Only time, and much more careful experimental work, will show whether or not the results we present are valid. The same is true of the Gauckler, Lukas and Petzow results (J. Amer. Ceram. Soc. 1975, 58, 346).

Cannon: How wide are the ranges of homogeneity of the phases in the Al–Si–O–N reciprocal salt system?

Jack: We do not know exactly how wide these ranges are at right angles to the line of constant M:X ratio. Expressed in terms of this M/X value, e.g. 0.75 for β'-sialon, it is probably not more than about $\pm$ 2%, i.e. 0.73–0.77.

Räuchle: You expected good thermal shock properties of the sialons, because of their low thermal expansion. But even if the thermal expansion of sialon is lower than that of Si_3N_4, thermal shock behaviour can be worse. There are other factors influencing thermal shock behaviour, for instance elastic modulus and thermal conductivity; and especially concerning the last point, it is to be expected that alloying lowers thermal conductivity.

Jack: I agree that low coefficient of expansion does not necessarily mean good thermal shock resistance but good thermal shock resistance is impossible without it. The high modulus and high fracture stress seem to compensate for any reduction in thermal conductivity because actual measurements by Lucas Research Centre show that the thermal shock resistance of β'-sialon up to about z=3 is at least as good as silicon nitride; this confirms our more limited laboratory tests.

Jelacic: What is the nature of your starting materials?

Jack: Lucas 6 µm and Starck Berlin 8 µm silicon nitride; Alcoa XA16 and XA17 5 µm alumina; Starck Berlin < 50 µm AlN; SiO_2 (precipitated) was submicron size; crushed fused quartz or crushed fused crystal were both 50 µm. These are all milled during mixing by vibro mill for 2 hours.

Venables: Do the single phase β'-sialons exhibit any optical transparency in the visible region? One might expect that they should since CVD/Si_3N_4 is transparent.

Jack: No; they are black. I had thought the black colour was due to impurities in the silicon nitride starting materials. β'-sialons prepared from SiO_2/Al_2O_3 gels are nearly white. However, the observation made by Greskovich that very pure 99.99% Si_3N_4 is black when doped with Mgo refutes this.

THE CRYSTAL STRUCTURES OF 8H AND 15R SIALON POLYTYPES

D.P. Thompson

The Wolfson Research Group for High-Strength Materials,
Crystallography Laboratory, The University,
Newcastle upon Tyne, England.

ABSTRACT. The crystal structures of the two most silica-rich
aluminium nitride polytypes, 8H and 15R, have been determined by X-
ray powder methods, and consist of a layer of AlO_6 octahedra sandwiched in between wurtzite-type stacking sequences of metal and non-
metal atoms. The additional non-metal atom required by the formula
is situated approximately half way between each octahedral layer.
The geometry of the octahedron accounts satisfactorily for the large
increase in inter-layer spacing compared with aluminium nitride.

1. THE SIALON POLYTYPES

Gauckler et al (1) reported six uncharacterized phases in the Si-
Al-O-N system at compositions intermediate between aluminium nitride
and the high-z region of the β' phase field. Each phase occurs with
a range of homogeneity extended along a line of constant metal:non-
metal ratio of the type $M_m X_{m+1}$. Jack (2) reported the indexing of
the X-ray diffraction patterns in terms of rhombohedral or hexagonal
unit cells with similar a dimensions but large varying c dimensions
(Table 1) which could be simply related to the c dimension of
aluminium nitride (Figure 1) by the equations:

$$c_R = (n-3)\frac{c}{2}_{AlN} + 3(4.014); \quad c_H = (n-2)\frac{c}{2}_{AlN} + 2(4.014),$$

where c_R and c_H refer to the c dimensions of rhombohedral and
hexagonal structures respectively and n is an integer representing
the number of Al-N type layers stacked on top of one another in
the c repeat distance. These relationships are characteristic of the
unit cell dimension variation observed in a polytypic series.

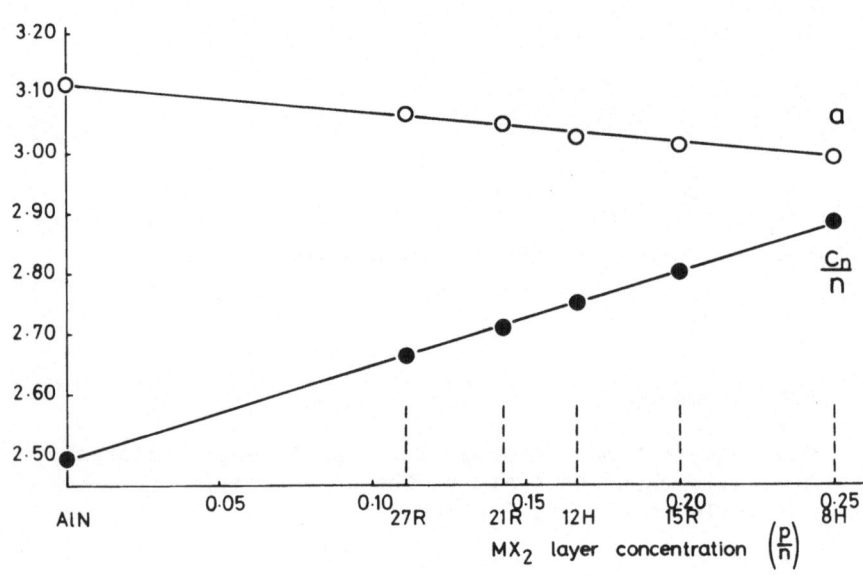

Fig. 1. Unit cell dimensions of the AlN-polytype sialons.

2. THE BASIC STRUCTURES

Several structural details can be deduced from the systematically absent reflexions in the X-ray diffraction pattern.   In the rhombohedral polytypes (designated nR), n must be divisible by 3 and each n/3-layered block of structure must be identical and displaced from the one below by the translation (2/3,1/3,1/3). The hexagonal polytypes (designated nH) all show a c-glide plane indicating that the upper n/2 layers of structure must be identical to the lower n/2 layers and related to them by c-glide symmetry.

Table 1. AlN-polytypes in the Si-Al-O-N system

M/X	type	a	c	c/n
4/5	8H	2.988	23.02	2.88
5/6	15R	3.010	41.81	2.79
6/7	12H	3.029	32.91	2.74
7/8	21R	3.048	57.19	2.72
9/10	27R	3.059	71.98	2.67
>9/10	2H$^\delta$	3.079	5.30	2.65
1/1	2H	3.114	4.986	2.49

From these considerations two possible metal stacking sequences
are possible for the 8H and 15R polytypes; approximate intensity
calculations were sufficient to confirm that the arrangements shown
in Figure 2(a) were correct. The (110) section has been used for
all the diagrams in this paper since all the atoms in the structure
occur on this plane. The sequence departs from hexagonal close-
packing by the insertion of one group of three metal atoms in
cubic close-packing in each symmetry related block of structure.
By placing non-metal atoms statistically in tetrahedral sites
immediately below each metal atom it can be seen (Figure 2(b))
that there is only one vacant tetrahedral site in each unique
block of structure which can be occupied by the extra non-metal
atom without requiring base sharing of tetrahedra. These models
were therefore used as trial structures. The range of homogeneity
for each polytype along a line of constant metal:non-metal ratio
shows that aluminium must replace silicon and simultaneously oxygen
replace nitrogen in some if not all of the atomic sites.

3. EXPERIMENTAL

Intensity data were collected photographically using a 19 cm
Unicam powder camera, and the films were measured on a direct

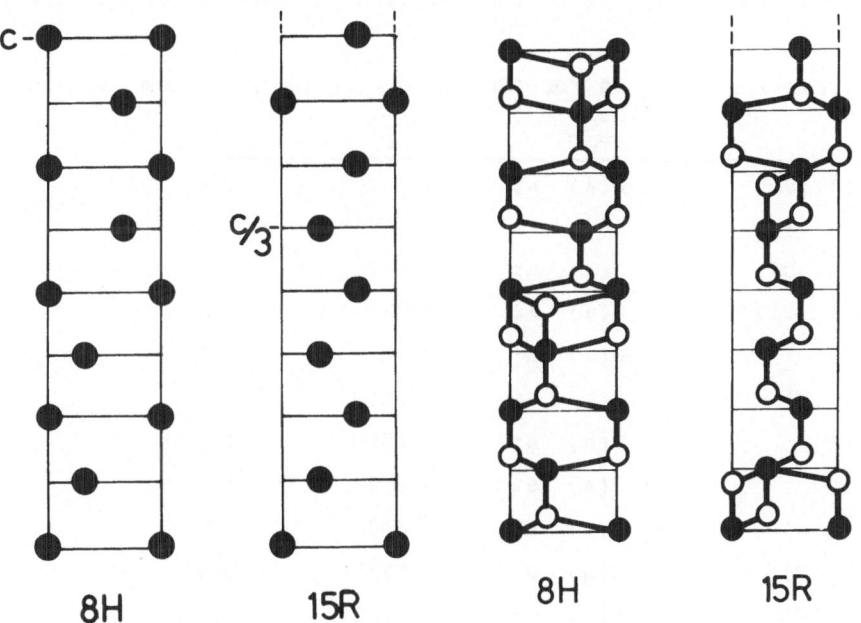

Fig. 2. (a) Metal stacking sequence, and (b) Trial structures,
of 8H and 15R sialon polytypes.

Table **2**. Final parameters for 8H sialon (Space Group $P6_3/mmc$, No. 194)

				z	B
2 Al	in	2(a)	at	0	2.1
4 (Si,Al)	in	4(f)	at	0.1219	2.2
2 (Si,Al)	in	4(e)	at	0.2294	1.8
4 O	in	4(f)	at	0.0443	3.7
4 (O,N)	in	4(f)	at	0.1423	3.7
2 (O,N)	in	2(c)	at	$\frac{1}{4}$	3.5

central metal atom in a fairly regular octahedral environment. In the 8H structure, the metal atom half way between stacking breaks showed a clear tendency to split into two positions on either side of the mirror plane at $z = \frac{1}{4}$, and in the 15R structure, the two metals equidistant from the stacking break showed low occupation numbers and an irregular electron density distribution. The final parameters for the two structures are given in Tables 2 and 3, and Figure 3 shows the bondlengths. R-factors at this stage of refinement were 9.0 and 11.4 respectively.

Table 3. Final parameters for 15R sialon (Space Group R3m, No. 160)

				z	B
3 (Al,Si)	in	3(a)	at	0.0	5.2
3 (Al,Si)	in	3(a)	at	0.1264	2.4
3 (Al,Si)	in	3(a)	at	0.5922	2.9
3 (Al,Si)	in	3(a)	at	0.7331	2.1
3 Al	in	3(a)	at	0.8577	4.4
3 (O,N)	in	3(a)	at	0.0500	3.8
3 (O)	in	3(a)	at	0.1690	1.1
3 (O)	in	3(a)	at	0.5492	1.5
3 (O,N)	in	3(a)	at	0.6590	0.2
3 (O,N)	in	3(a)	at	0.7798	0.7
3 (O,N)	in	3(a)	at	0.9332	1.4

reading microdensitometer (3); the usual corrections for
absorption, polarisation and Lorentz factors were applied.
Structure refinements were carried out on an IBM 370/168 computer
using the SHEL-X program written by Dr. G.M. Sheldrick.

4. STRUCTURE REFINEMENT

In the trial structures (Figure 2(b)) two of the metal atoms in
the cubic close-packed metal sequence have non-metal coordinations
of seven and five. In the preliminary stages of refinement these
non-metals shifted appreciably in z-coordinate to leave the

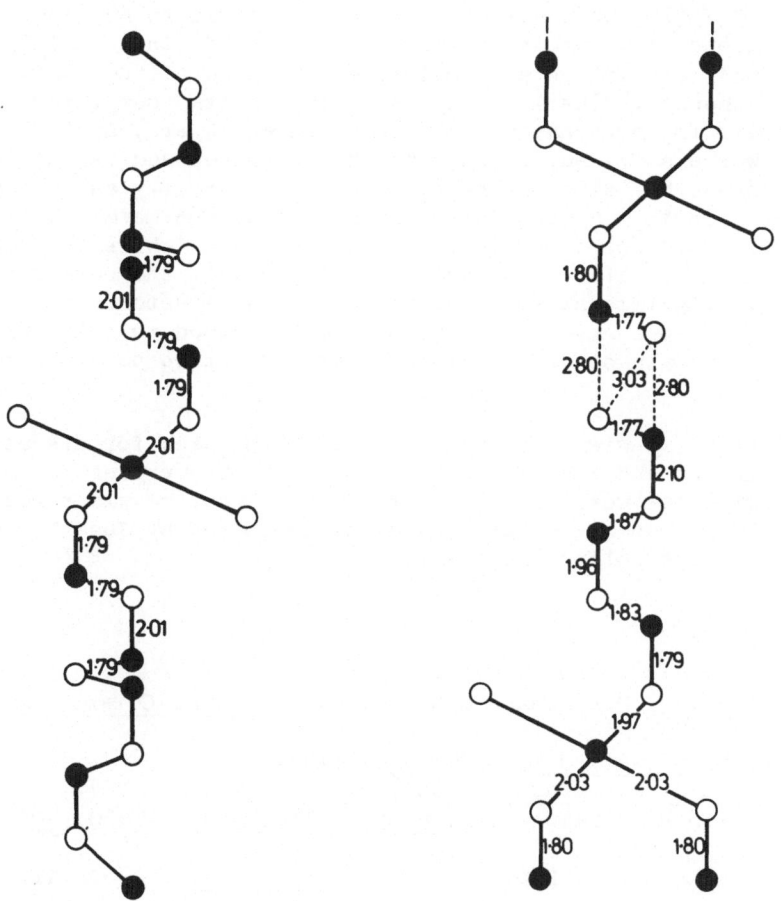

Fig. 3. Bondlengths in 8H and 15R sialons.

5. DISCUSSION

Both structures show that the prominent structural feature which distinguishes the polytypes from aluminium nitride itself is an aluminium-centred octahedron in the centre of the cubic close-packed metal sequence. From Figure 2(b) it can be seen that the six non-metal atoms coordinating this metal site form an octahedron which is considerably "squashed" by compression along the z-axis. The transformation to a regular octahedron involves a large expansion in the z-direction and a contraction in the x-y plane. In fact the amount of expansion is approximately 1.1Å per octahedron which accounts for most of the increase in inter-layer distance ($\sim$1.5Å) observed in the c parameter. The rest is taken up by the insertion of the additional non-metal atom.

The 8H structure can be compared directly with β-Be$_3$N$_2$ (4) and also Al$_5$C$_3$N (5). Both structures are anti-typic to 8H, the latter being one of a series of polytypic compounds in the Al$_4$C$_3$-AlN system (6). β-Be$_3$N$_2$ is a 4H polytype with ABAC nitrogen stacking. Originally Rabenau & Eckerlin (4) placed the "extra" beryllium atoms in (5)-fold coordinated sites in nitrogen layers, but subsequent workers (7) suggested that they occupied equivalent (4)-fold sites on either side of the nitrogen layer exactly as in the 8H structure. In Al$_5$C$_3$N, one carbon atom in four non-metal layers occurs in (6)-fold coordination as in the 8H sialon whilst the other non-metals occur in distorted tetrahedra in between the octahedral layers. It is significant that neither in these structures nor in the sialon polytypes have the R factors been reduced to below 9%. Clearly there remain other structural features yet to be explained.

ACKNOWLEDGEMENTS. I wish to thank Professor K.H. Jack for discussion and encouragement and also Mr. P.H.A. Roebuck for carrying out much of the experimental work. The present study is part of wider research on nitrogen ceramics at Newcastle and was supported by Joseph Lucas Limited and by the Wolfson Foundation.

REFERENCES

1. L.J. Gauckler, H.L. Lukas & G. Petzow, J. Amer. Ceram. Soc., 1975, 58, 346.
2. K.H. Jack, J. Mat. Sci., 1976, 11, 1135.
3. A. Taylor, J. Sci. Instr., 1951, 28, 200.
4. P. Eckerlin & A. Rabenau, Z. anorg. allg. Chem., 1960, 304, 218.
5. M. v. Stackelberg & K.F. Spiess, Z. Phys. Chem., 1935, A173, 140.
6. G.A. Jeffrey & V.Y. Wu, Acta Cryst., 1966, 20, 538.
7. D. Hall, G.E. Gurr & G.A. Jeffrey, Z. anorg. allg. Chem., 1969, 369, 108.

DISCUSSION (Leake, Morgan)

Popper: The term "polytype" has been used for SiC and ZnS where the cubic and hexagonal structures have almost the same energy and regular stacking sequences are very common. The type of structures where a basic coarse layer is repeated with the regular interposition of other atoms is well known in other cases, e.g. in the Si-Ti-O system where one or several layers of perovskite are repeated depending on the composition. I believe it would be confusing to use the term "AlN-polytypes" when there are compositional differences. A term "repeated layer structure" or "crystallographic shear structures" as proposed by J.S. Anderson for the partially reduced oxides (e.g. TiO_{2-x}) would be preferable.

Thompson: I agree that the term "polytype" is at present used for both constant composition and systematically varying composition series of layer-type structures. For the past two years I have been referring to these compounds as "compositional polytypes" but the unfavourable reaction of scientists from a variety of disciplines is good evidence of the need for a better title. The AlN-polytypes differ in two structural respects from aluminium nitride. Firstly, a layer of octahedra is incorporated at regular intervals along Z, and secondly, an additional non-metal atom is inserted into the layers of tetrahedra. Whereas the octahedral layer does not seem to give rise to structural irregularities, the additional non-metal atom could well cause defects leading to mis-stacking of lattice planes.

Lee: Could you give us some physical property data on your 90% 15R samples?

Thompson: Very little information is available. It oxidises more rapidly than β' sialon, and the thermal expansion is similar to AlN but less anisotropic (7×10^{-6}//a; 4×10^{-6}//c). As far as I know, no mechanical property tests have been carried out on 15R or β' + 15R materials, but the improved properties of β + N-melilite mixtures in the Y-Si-O-N system illustrates that thermal mismatch may not be critical.

CRYSTAL CHEMISTRY OF NITRIDED COMPOUNDS CONTAINING LANTHANIDES.
NEW POSSIBILITIES OF CATIONIC SUBSTITUTIONS RELATED TO NITROGEN
INTRODUCTION IN ANIONIC FRAMEWORK.

J. Gaudé, R. Marchand, Y. Laurent & J. Lang

Laboratoire de Chimie Minérale C – Université de Rennes
35031 Rennes-Cedex – France.

A systematic study of the possible substitution of oxygen by
nitrogen in many fundamental types of structures has been undertaken.

The introduction of nitrogen into the anionic framework
allows many various substitutions in the cationic framework, particularly of trivalent or higher oxidation state elements. The
lanthanides are preferred because they give strong bonds with
nitrogen, whereas some transition elements do not occur in their
maximum oxidation state in nitrides.

We describe here the obtained results for the apatite, K_2NiF_4
and $\beta-K_2SO_4$ type structures.

APATITE STRUCTURE

The $Ln_{10}Si_6N_2O_{24}$ (Ln = La, Nd, Sm, Gd) compounds are prepared
by heating appropriate mixtures at 1250°C of $Ln_2O_3-Si_3N_4-SiO_2$ or
$LnN-Ln_2O_3-SiO_2$ in sealed nickel tubes under nitrogen pressure.

Single crystals of $Sm_{10}Si_6N_2O_{24}$ have been obtained allowing
the structural determination. The unit cell has hexagonal symmetry. The space group is $P6_3$ instead of the usual $P6_3/m$ for the
apatite structure. Several types of environment are found for the
cations according to their sites.
- type A : a 6c position is filled by heptacoordinated samarium
atoms which are at the centre of a distorted pentagonal bipyramid.
- type B : two 2b positions. Four samarium atoms (2B + 2B') are
placed at the centre of a triangular prism which is distorted by
a relative rotation of the two bases.
- type C : a 6c position is filled by tetrahedrally surrounded
silicon atoms.

The structure can be described as an assembly of Si-4X tetrahedra connecting the Ln-7X polyhedra ; this involves the creation of holes which constitute the type B positions.

Some lanthanide atoms can be substituted by a trivalent element. With Cr(III) *e.g.*, the phase exists when the number of Cr atoms is < 2 (Ln = La, Nd, Sm, Gd, Dy). The two chromium atoms may be placed *a priori* in one of the two special $2b$ positions (B or B' position) according to the difference between the value of the Cr^{3+} ionic radii and that of Ln^{3+} ions. The $Sm_8Cr_2Si_6N_2O_{24}$ single crystal structural study shows that the two Cr atoms and two Sm atoms are randomly distributed among the B and B' positions. The ratio Cr/Sm is close to unity in the two positions. It is not the filling of one or the other position which explains the limitation of the exchanges, but more probably the property of the apatite structure to accept deviations therein due to substitutions.

To determine the true composition of the prepared apatites, it has been necessary to carry out density measurements together with X-ray analysis. The confirmation of the nitrogen content has been done by thermogravimetry under oxygen.

Replacing the trivalent lanthanide by another element of a greater charge involves more than two nitrogen atoms in the cell, according to the increase in the cationic charge. So with Ti(IV), we have obtained the $Ln_8Ti_2Si_6N_4O_{22}$ compounds (Ln = La, Nd, Sm, Gd). Other elements like Ge(IV), Mn(IV) and V(V) are also suitable for substitutions.

In these structures, the 26 anions are spread on one $2a$ position and four $6c$ positions. Our X-ray structural studies do not differentiate nitrogen and oxygen. In spite of this difficulty, we have placed the nitrogen with the oxygen atoms in a disordered way in a $6c$ position rather than in the $2a$ position for the following reasons :
- when the number of nitrogen atoms is greater than 2, some of them necessarily must fill partly a $6c$ position.
- the silicon and lanthanide type A atoms are surrounded by two of the four anionic positions which do not take a part in the coordination polyhedron of the B and B' positions ; so, nitrogen has no link with the subtituting element. This is corroborated by the examination of the interatomic distances which shows particularly that, in the tetrahedron surrounding silicon, three distances are homogeneous and equal to the Si-O distance, while the fourth is longer and intermediate between Si-O and Si-N distances.
- anions in $6c$ positions are sp^3 hybridized while those in the $2a$ position have a sp^2 hybridization which is more uncommon for nitrogen than for oxygen.

However, only studies using other methods would confirm the exact position of the nitrogen atoms.

K_2NiF_4 STRUCTURE

The cross-substitution mechanism is :

$$Ln^{3+} + N^{3-} = M^{2+} + O^{2-}$$

The Ln_2AlO_3N compounds (Ln = La, Nd, Sm) are prepared by heating at 1350°C Ln_2O_3-AlN mixtures in nickel sealed tubes. They have a K_2NiF_4 type structure which is derived from perovskite. The crystal symmetry is tetragonal (space group : I_4/mmm). The lanthanide atoms occupy 9-coordinated positions and aluminium atoms occupy octahedral holes.

Each unit cell contains 6 oxygen and 2 nitrogen atoms which are distributed on *4c* and *4e* positions of the space group.

It is not possible by X-ray diffraction to determine if an anionic order does exist.

β-K_2SO_4 STRUCTURE

When aluminium is replaced in the above compounds by silicon the latter atom always has a tetrahedral environment.

The cross-substitution :

$$Ln^{3+} + Al^{3+} = Eu^{2+} + Si^{4+}$$

gives the $Ln^{III}Eu^{II}SiO_3N$ compounds (Ln^{III} = La, Nd, Sm, Gd). They belong to the orthorhombic system with β-K_2SO_4 type structure (space group : Pnma).

Europium II is produced *in situ* during the reaction :

$$3\ Ln_2O_3 + 3\ Eu_2O_3 + 2\ Si_3N_4 \xrightarrow{1350°C} 6\ LnEuSiO_3N + N_2\uparrow$$

CONCLUSION

The introduction of nitrogen into the anionic framework increases the number of structures in which lanthanide atoms can be placed.

Several compounds can be prepared such as $Ln_4Si_2O_7N_2$: cuspidine type structure or $Ln_2Si_3O_3N_4$: melilite type structure, the study of which is nearing completion.

The data presented here form a part of a wider project and were obtained in this laboratory by a team including J. Guyader, P. L'Haridon, M. Maunaye, P. Verdier and S.A.A. Jayaweera°.

°On sabbatical leave from School of Environmental Sciences, Plymouth Polytechnic, Plymouth PL4 8AA.

THE α/β SILICON NITRIDE PHASE TRANSFORMATION

D.R. Messier[*] and F.L. Riley[**]

[*]AMMRC, Watertown, Mass., U.S.A.
[**]University of Leeds, Leeds, U.K.

INTRODUCTION

Early work on silicon nitride showed that the α crystallographic form can be converted to the β modification by heat treatment at temperatures exceeding 1500°C (1,2). Thompson and Pratt (3) confirmed these findings and noted that prolonged heating of β silicon nitride in argon or nitrogen at temperatures from 1100 to 1350°C failed to bring about the reverse transformation.

Although the importance of the transformation to the fabrication of high strength, hot-pressed silicon nitride has been recognised (4), its exact role has remained obscure. Coe et al. (4) reported that α silicon nitride powder could be densified without transformation; they added, however, that high strength was only achieved through the phase transformation. Lumby and Coe (5) further emphasised the importance of the transformation to hot pressing. It was suggested by others that strength enhancement was a result of transformation induced grain refinement (6,7), but detailed mechanisms were not considered.

Certain features that may be transformation related have often been observed in hot-pressed silicon nitride. Among such features are elongated grains and preferred orientation (8,9). Gazza (10), reporting on the hot pressing of silicon nitride with yttria additions, observed an abrupt change in microstructure that appeared to be associated with the transformation.

The importance of the α/β phase transformation to silicon nitride technology is clear. The purposes of this paper are to

review the literature on the α/β transformation, to present a
few relevant new results, and to evaluate this information in
view of existing knowledge on the general subject of phase transformations.

CRYSTAL STRUCTURES

Although it is not intended to discuss the crystal
structures of α and β silicon nitride in detail, disagreement in
the literature on this subject makes it necessary to briefly
review structural aspects that may relate to the transformation
mechanism.

After publication of preliminary structure data (11,12),
Ruddlesden and Popper (13) provided the first complete determinations of the α and β silicon nitride structures. Both forms
were found to be hexagonal with formulae of $Si_{12}N_{16}$ for α, and
Si_6N_{12} for β.

It was later proposed that α silicon nitride was necessarily
an oxynitride of composition $Si_{11.5}N_{15}O_{0.5}$ (14), and a subsequent
structure determination (15) was consistent with this theory.
Additional supporting evidence was provided by a thermodynamic
study (16), and by work reported by Feld et al. (17).

Recent results, however, do not support the oxynitride
theory. Priest et al. (18) reported an oxygen content of only
0.30 w/o for α silicon nitride prepared by CVD, an amount substantially less than the 0.90 to 1.48 w/o that is required for
the oxynitride. It was also reported that this material was untransformed to β after heating for several hours at 1800°C in
nitrogen at a pressure of 2 MPa. Edwards et al. (19) found oxygen
contents in reaction bonded silicon nitride that were at most half
as large as required for the oxynitride formula to be correct.
More recently, Kijima et al. (20) measured oxygen contents of
0.05 and 0.09 w/o on α silicon nitride single crystals.

Redeterminations of the α crystal structure by Kohatsu and
McCauley (21), and by Kato et al. (22), agree with the determination of Marchand et al. (23), but not with the work of Wild et
al. (15). In none of the former determinations was it necessary
to assume that α comprised anything but Si_3N_4. In addition, the
determination of Kato et al. (22) was done on the same crystal
known to have a low oxygen content (0.05 w/o) from analytical
work (20).

LITERATURE ON TRANSFORMATION

The literature includes no results of systematic, quantitative
studies of the kinetics of the α/β transformation. The data

included here are therefore generally from work in which the study of the transformation was of secondary interest.

Coe et al. (4) reported a linear relationship between % β and log time for silicon nitride powder hot pressed with 1% MgO at 1740°C. Naruse et al. (24) gave data on transformation vs temperature for three hour heat treatments of pressed pellets at temperatures from 1300-1800°C. Some isothermal data were given at 1600°C, but the results appear erratic. A few additional data were published by Weston and Carruthers (25) for hot pressed material, and by Inomata (26) for pressed pellets. The results of Feld et al. (17) were obtained on specimens that were apparently contaminated with silicon carbide and silicon oxynitride. In none of these studies was the transformation mechanism considered.

Wild et al. (6) proposed that, since they considered α to be an oxynitride, the transformation involved a chemical reaction between α and enstatite (formed by reaction of the MgO additive with SiO_2) as follows:

$$4Si_{11.5}N_{15}O_{0.5} + Mg_2SiO_4 = 15Si_3N_4 + 2MgSiO_3$$
$$(\alpha) \qquad\qquad\qquad (\beta)$$

A detailed interpretation of the mechanism, however, was not given.

In view of the probable effect of the solubility of silicon nitride in oxide melts on the transformation, it is pertinent to review evidence for such solubility. Jack (27) has demonstrated solubility of silicon nitride in molten magnesium silicates. Nuttall and Thompson (9), in a TEM study of hot pressed silicon nitride, observed a hexagonal grain that appeared to have grown from a melt. Kossowsky (28), in a wetting study, also reported that silicon nitride was soluble in molten magnesium silicate. Also consistent with this evidence are proposals that sintering in the silicon nitride-magnesium oxide system occurs via a liquid phase mechanism (28-30).

The TEM observations of Drew and Lewis (31) strongly support their conclusion that the transformation occurs via a solid/liquid/solid mechanism. The same authors (32) also presented convincing microstructural evidence for the formation of β' silicon aluminium oxynitride compounds by a similar mechanism.

RECENT RESULTS

In order to observe transformation kinetics without the complications of sintering, a series of experiments was done on commercial purity (85% α) silicon nitride powder specimens loaded loosely into graphite crucibles and heated in flowing nitrogen at 1600°C. The powders were heated for various times as received, and with additions of magnesium oxide and yttrium oxide. Figure 1 shows the results of these experiments. Although the rate of transformation was greatly accelerated by magnesium oxide, particularly at the 10 w/o level, it was unaffected by the yttrium oxide addition at 1600°C.

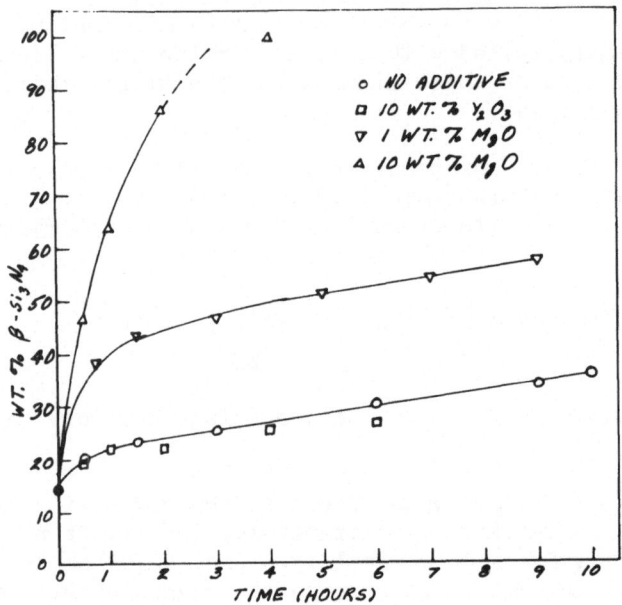

FIGURE 1. EFFECTS OF ADDITIVES ON THE RATE OF TRANSFORMATION OF COMMERCIAL PURITY 85% α POWDER TO β AT 1600°C IN FLOWING NITROGEN.

Further runs were then done on high purity, 100% α silicon nitride powder obtained by grinding pieces made by CVD. This powder was untransformed after six hours of heating at 1600°C. The same powder with 10 w/o magnesium oxide added showed a significant amount (3.5 w/o) of β phase after heating for ½ hour at the same temperature.

Figure 2 shows the appearance of the commercial purity powder before and after transformation. The "log" shaped morphology of the β product is apparent.

Figure 2. S.E.M. photomicrographs of commercial purity silicon nitride powder (A) as-received and (B) after conversion to β silicon nitride by heating with 10 w/o magnesium oxide at 1600°C. (bar = 4 μm)

Recent quantitative measurements of the transformation rate under pressure sintering conditions are discussed elsewhere in this volume (33).

DISCUSSION

Existing experimental evidence strongly supports the conclusion of Drew and Lewis (31) that the α/β transformation occurs by a solution-precipitation mechanism. An early, tentative suggestion that a solid state mechanism is involved (7) can therefore be discounted. In view of the crystal structure and transformation data, a mechanism involving "deoxidation" of an oxynitride (α) to form the "pure" nitride (β) (6) is also unlikely.

As shown in the Results section, there is ample evidence in the literature for the solubility of silicon nitride in molten silicates and there are microstructural observations supporting the solution-precipitation mechanism. The recent experimental results given here are entirely consistent with such a mechanism. The strong dependence of the transformation rate on the content of magnesium oxide suggests that the rate is proportional to the amount of magnesium silicate liquid that is present. It is also noteworthy that the transformation rate at 1600°C was unaffected by the addition of yttria, another commonly used hot pressing aid (10). These results are readily explained by considering that the lowest eutectics occur at 1543°C in the magnesia-silica system (34) and at 1660°C in the yttria-silica system (35). One would therefore not expect an enhanced rate of transformation in the yttria system until the temperature was above the eutectic and liquid was formed.

That transformation occurred in the commercial powder without additives must be attributed to the presence of impurities in the powder that combine with the silica to form liquid phases. As already mentioned, Priest et al. (18) observed no transformation in high purity CVD silicon nitride at 1800°C. Similarly, in this study, transformation only occurred at 1600°C when magnesium oxide was added to the CVD powder. The observation of Jack (27) that transformation occurs without magnesium oxide may be attributable to the relatively low purity of the powders that were used in his work.

Whether or not the α/β transformation produces grain refinement (6,7) is still unresolved. The observations reported here (Fig. 2) are consistent with the development of a material that is relatively fine-grained (1-4 µm), but, in general, with grains somewhat larger than those in the starting powder. It may be significant, however, that extremely large grains are seldom observed, and the high strength of the hot-pressed material may be due to the development of a very uniform, fine-grained structure.

It has been suggested from structural considerations that the α/β transformation should be reconstructive (3,37), and it does conform to Buerger's (38) classification of a reconstructive transformation of secondary co-ordination, i.e. one which involves disruption of the structure, but in which the first co-ordination is the same in both crystallographic forms. Such transformations are sluggish, and often require the presence of a solvent in order to form the new structure (39). Among other examples of this type of transformation are those between any of the pairs of the set quartz-tridymite-cristobalite (SiO_2), and the wurtzite-sphalerite transformation in ZnS (39).

FUTURE WORK

Although the qualitative features of the α/β transformation are now reasonably well understood, further work is needed on the transformation kinetics. Unanswered questions exist with respect to the rate-controlling step, e.g. solution, precipitation, or diffusion, and with regard to the details of nucleation and growth of the β phase. Better characterisation of the morphology of the β phase particles could lead to the fabrication of better materials. It may be possible to control that morphology by suitable heat treatment, and by the use of additives that promote growth in desirable crystallographic directions. The demonstrated solubility of silicon nitride in molten silicate fluxes suggests that it may be possible to grow β silicon nitride single crystals from such fluxes.

REFERENCES

1. Turkdogan, E.T., Bills, P.M. and Tippet, V.A., J. Appl. Chem., 8 (1958), 296.
2. Forgeng, W.A. and Decker, B.F., Trans. AIME, 212 (1958), 343.
3. Thompson, D.S. and Pratt, P.L. Science of Ceramics, Vol. 3. New York: Academic Press (1967), 33.
4. Coe, R.F., Lumby, R.J. and Pawson, M.F. Special Ceramics, Vol. 5. Manchester: BCRA (1972), 361.
5. Lumby, R.J. and Coe, R.F. Proc. Brit. Ceram. Soc. No. 15 (1970), 91.
6. Wild, S., Grieveson, P., Jack, K.H. and Latimer, M.J. Special Ceramics, Vol. 5. Manchester: BCRA (1972), 377.
7. Evans, A.G. and Sharp, J.V., J. Mater. Sci., 6 (1971), 1292.
8. Terwilliger, G.R. and Lange, F.F., J. Amer. Ceram. Soc., 57 (1974), 25.
9. Nuttall, K. and Thompson, D.P., J. Mater. Sci. 9 (1974), 850.
10. Gazza, G.E., to be published in Proc. U.S. Army Sci. Conf., (1976).
11. Popper, P. and Ruddlesden, S.N., Nature, 179 (1957), 1129.
12. Hardie, D. and Jack, K.H., Nature, 180 (1957), 332.
13. Ruddlesden, S.N. and Popper, P., Acta Cryst., 11 (1958), 465.
14. Grieveson, P., Jack, K.H. and Wild, S., Special Ceramics, Vol. 4. Manchester: BCRA (1968), 237.
15. Wild, S., Grieveson, P. and Jack, K.H., Special Ceramics, Vol. 5. Manchester: BCRA (1972), 385.
16. Colquohoun, S., Wild, S., Grieveson, P. and Jack, K.H., Proc. Brit. Ceram. Soc., No. 22 (1973), 207.
17. Feld, H., Ettmayer, P. and Petzenhauser, I., Ber. Deut. Keram. Ges., 51 (1974), 127.
18. Priest, H.F., Burns, F.C., Priest, G.L. and Skaar, E.C., J. Amer. Ceram. Soc., 56 (1973), 395.
19. Edwards, A.J., Elias, D.P., Lindley, M.W., Atkinson, A. and Moulson, A.J., J. Mater. Sci., 9 (1974), 516.
20. Kijima, K., Kato, K., Inoue, Z. and Tanaka, H., J. Mater. Sci., 10 (1975), 362.
21. Kohatsu, I. and McCauley, J.W., Mater. Res. Bull., 9 (1974), 917.
22. Kato, K., Inoue, Z., Kijima, K., Kawada, J., Tanaka, H. and Yamane, T., J. Amer. Ceram. Soc., 58 (1975), 90.
23. Marchand, R., Laurent, Y. and Lang, J., Acta Cryst., B25 (1969), 2157.
24. Naruse, W., Nojiri, M. and Tada, M., Nippon Kinzoku Gakkaishi, 35 (1971), 731.
25. Weston, R.J. and Carruthers, T.G., Proc. Brit. Ceram. Soc., No. 22 (1973), 197.
26. Inomata, Y., Yogyo Kyokai Shi, 82 (1974), 522.

148

27. Jack, K.H., _Ceramics for High Performance Applications._ Chestnut Hill, Mass.: Brook Hill Pub. Co. (1974), 265.

28. Kossowsky, R., J. Mater. Sci., _9_ (1974), 2025.

29. Terwilliger, G.R. and Lange, F.F., J. Mater. Sci., _10_ (1975), 1169.

30. Bowen, L.J., Weston, R.J., Carruthers, T.G. and Brook, R.J., Ceramurgia International, In Press.

31. Drew, P. and Lewis, M.H., J. Mater. Sci., _9_ (1974), 261.

32. Drew, P. and Lewis, M.H., J. Mater. Sci., _9_ (1974), 1833.

33. Brook, R.J., Carruthers, T.G., Bowen, L.J. and Weston, R.J., Proceedings of this Institute.

34. Bowen, N.L. and Andersen, O., Amer. J. Sci., _37_ (1914), 488.

35. Toropov, N.A. and Bondar, I.A., Izv. Akad. Nauk SSR, Otd. Khim. Nauk (4) (1961), 547.

36. Kossowsky, R., J. Mater. Sci., _8_ (1973), 1603.

37. Henderson, C.M.B. and Taylor, D., Trans. and J. Brit. Ceram. Soc., _74_ (1975), 49.

38. Buerger, M.J., _Phase Transformations in Solids._ New York: John Wiley and Sons (1951), 183.

39. Verma, A.R. and Krishna, P. _Polymorphism and Polytypism in Crystals._ New York: John Wiley and Sons (1966).

ACKNOWLEDGMENT

This work was supported by a Fellowship awarded to D.R. Messier by the Secretary of the U.S. Army whose support is gratefully acknowledged.

DISCUSSION (Leake, Morgan)

Lange: We have obtained similar results, viz. the simultaneous fibrous growth of β grains and the α–β transformation. But further, we hypothesised that if it is assumed that the fibre diameter does not change during growth, the aspect ratio (length/diameter) of the β-fibres will be $1 + \alpha/\beta$. This result is obtained by allowing the _solution_ of α in the liquid and the precipitation of β onto pre-existing β grains in the powder. That is, the growing β grains will consume the α particles. Thus, control of the microstructure is obtained by the particle size of pre-existing β particles and the α/β ratio which, in turn, controls the aspect ratio of the fibrous grains.

Messier: The aspect ratios of the β-silicon nitride grains that we have observed from the transformation of 85% α powder are similar to the 1.6:1 predicted by Iskoe and Lange. It should be noted, however, that we observed some transformation when phase pure (by X-ray diffraction analysis) α powder was used. In that case, pre-existing β grains were absent, and the β phase must have nucleated from the melt.

Morgan: I believe the α-β Be_3N_2 conversion is also irreversible and this is not a polytypic conversion.

Godfrey: With respect to the influence of impurities on the α-β transformation, I reported that even small additions of Al_2O_3 increased β contents on nitriding Si. As commercial purity Si often has >0.5^W/o Al, the influence of Al_2O_3 content on the transformation is worth considering since the effect apparent in Si_3N_4 formation may operate in the transformation.

Messier: The only additives used in this study were MgO and Y_2O_3 but it would be interesting to determine the effect of Al_2O_3 as well. It should be remembered, however, that the transformation proceeds by a liquid phase, and the additive must be able to form such a phase at the transformation temperature.

THERMODYNAMICS AND
GENERAL KINETICS

SOME CONSIDERATIONS OF THE THERMODYNAMICS OF GAS-SOLID REACTIONS.

Paul Grieveson

Department of Metallurgy, University of Strathclyde,
Glasgow.

ABSTRACT.   Some consideration is given to the thermodynamic properties of Gibbs free energy, chemical potential and equilibrium
constant.   The integral free energies for some gas reactions are
presented.   The equilibria involved in nitriding with nitrogen
gas and ammonia-hydrogen mixtures are assessed and the data for
the metastable equilibria with iron nitrides are considered
together with information on the equivalent nitrogen gas fugacities.
In addition, the effect of pressure on the chemical potential of
gases is explained.
    A presentation is made of data on the standard free energy of
formation of some nitride phases and the stabilities with temperature are discussed.   Bonding energies of nitride phases are compared with those of metals, carbides and silicides and are suggested
as useful in the estimation of data.
    Finally, some demonstrations of the use of thermodynamic cal-
culations in assessing the value of nitride phases as high temperature phases are given.   Of particular interest are the relative
stabilities of nitrides compared to oxides, the interaction of
nitrides with liquid and solid metals and the free evaporation of
a nitride phase in vacuo.   Some consideration is also given to
the storage and retrieval of thermodynamic data.

1.   INTRODUCTION

Heterogeneous reactions make up the most significant group of
interactions of interest in ceramics and metallurgy.   In particular, those reactions involving gases and solids are of prime
importance, e.g. reduction of solid oxides, preparation of ceramic
phases and the corrosion of metals.   The extent of such reactions

depends upon the position of equilibrium between the reacting substances and their products.   At the high temperatures often
encountered in ceramic and metallurgical processes, reaction rates
are usually so high that equilibrium is generally attained.   In
these cases, thermodynamic considerations allow the computation of
the extent of reaction and even when kinetics are slow, these considerations govern the force driving the reaction towards equilibrium.
Some knowledge of thermodynamics is, therefore, essential for a complete understanding of gas-solid reactions.
    This communication attempts to present various aspects of
thermodynamics which are of special interest in the field of nitrogen ceramics.   The limitation of space does not allow the development of a full treatise on thermodynamics of gas-solid reactions.
In this article, attention is focussed on a limited number of
aspects related to the factors which are important in the production of nitrides and their environmental interactions.

2.   FREE ENERGY, CHEMICAL POTENTIAL AND EQUILIBRIUM CONSTANT

For many decades, scientists searched to find a function which
represented the 'driving force' for a reaction.   From the work of
Gibbs and Helmholtz in particular, it became apparent that a thermodynamic function, now called free energy, conformed to the con-
ception of driving force.   In particular, Gibbs free energy, G,
which expresses the first and second laws of thermodynamics
explicitly in terms of temperature and pressure is the function of
widest utility in our considerations of gas-solid reactions. Gibbs
free energy is defined by the relation

$$G = H - TS \tag{1}$$

where H is enthalpy, S is entropy and T is absolute temperature.
Thus, any process which takes place is accompanied by a change in
free energy of the system and this change is equal to the net reversible work done by or absorbed during the process.   This work
is represented by the expression:

$$\Delta G = G_{products} - G_{reactants} \tag{2}$$

and is the maximum mechanical work that can be carried out by any
reaction which takes place reversibly at constant pressure.   It
should be remembered that free energy in common with all other
forms of energy cannot be measured in absolute terms but only
relative to some predetermined reference or standard state.   Thus,
it is usual to consider free energy change for any process relative
to the reference state.   It can be stated that the free energy
change for any system which undergoes spontaneous reaction at constant temperature and pressure is negative.   The other criterion
is that the free energy change of a system undergoing a reversible

process at constant temperature and pressure and doing work only against pressure is zero.

Although any process for a system of constant pressure, temperature and mass for which $\Delta G < 0$ is thermodynamically possible, it may not proceed at a perceptible rate if the reaction rate is slow.

Generally, many reactions require some knowledge of the free energy of one component in a solution or compound and difficulty arises. This problem was not solved directly but avoided by the invention of partial molar quantities, such that

$$\Delta \overline{G}_1 \equiv \left(\frac{\partial G'}{\partial n_1} \right)_{P,T,n_2, \ldots} \tag{3}$$

where this partial molar quantity $\overline{G}_1$ is the increment of the total free energy with composition.

From the combined expression of the first and second laws of thermodynamics at constant temperature, it is found that for a perfect gas:-

$$dG = RT \, d \ln P \tag{4}$$

Similarly for each component of the system, the partial molar free energy equation is:-

$$d\overline{\overline{G}}_i = RT \, d \ln p_i \tag{5}$$

Upon integration

$$\overline{G}_i = G_i^0 + RT \ln p_i \tag{6}$$

where p_i is the partial pressure of component, i, in an ideal solution and G_i^0 is the free energy of component, i, in its reference or standard state. The partial molar free energy, $\overline{G}_i$, is also called the chemical potential of the ith component.

In real gases, particularly at low temperatures and high pressures, the ideal gas law is not obeyed. In order to maintain the generality of the simple relationships (4-6), the concept of fugacity, f, was introduced such that:-

$$RT \, d \ln f = dG \tag{7}$$

and for a species in a mixture

$$RT \, d \ln f_i = d\overline{G}_i \tag{8}$$

Thus for pure species,

$$RT \ln f_i = G_i + I \tag{9}$$

where the integration constant I is chosen so that the fugacity
approaches the pressure as the pressure approaches zero.   At
atmospheric pressures and moderate to high temperatures, the fugacity and the partial pressure are almost equal and, therefore,
subsequently partial pressures are used with the assumption that
the ideal gas law is obeyed.

Now, let us consider a solution in equilibrium with its vapour;
the vapour pressure of the ith component in the solution is equal
to the partial pressure, p_i, of the component in the gas phase.
If the partial pressure of the pure i^{th} component is p_i^o, then the
ratio $pi/p_i^o = a_i$ was defined as the activity of the ith component
in the solution.   Although in most cases the pure component is
chosen as the standard state, any other state can be the reference
state, the choice of standard state and the inter-relationships
between standard states is considered by Darken and Gurry[1].
However, it is obvious that the activity of the component in its
standard state is of course unity.

From this definition of activity, the chemical potential derived for gases can be applied to solid or liquid solutions by the
use of activity, a_i, in place of partial pressure or fugacity,
thus at constant pressure and temperature and at complete internal
equilibrium,

$$\overline{G}_i = G_i^o + RT \ln a_i \tag{10}$$

or $$\Delta \overline{G}_i = \overline{G}_i - G_i^o = RT \ln a_i \tag{11}$$

The value determined experimentally is the difference $\overline{G}_i - G_i^o = \Delta \overline{G}_i$,
i.e. the difference between the free energy value for a given
solution and that in its standard state.   The difference $\Delta \overline{G}_i$ is
called the relative partial molar free energy, similarly $\Delta \overline{H}_i$ and $\Delta \overline{S}_i$
are the relative partial molar heat and entropy of component i.

The relative partial molar quantities of a solution are related to the integral functions by the equation

$$\Delta G = N_1 \Delta \overline{G}_1 + N_2 \Delta \overline{G}_2 \tag{12}$$

For any equilibrium occurring at constant pressure and temperature

$$xX + yY = uU + vV \tag{13}$$

it is found that

$$\Delta G^o - \Delta G = RT \ln \frac{(a_U)^u . (a_V)^v}{(a_X)^x . (a_Y)^y} \tag{14}$$

If we consider the equilibrium state, the activity product is
defined as K, the thermodynamic equilibrium constant

$$K = \left[\frac{(a_U)^u (a_V)^v}{(a_X)^x (a_Y)^y} \right]_{equilibrium} \tag{15}$$

Upon substituting this indentity into equation (14) together with the fact that at equilibrium $\Delta G = 0$, we obtain

$$\Delta G^o = - RT \ln K \tag{16}$$

If ΔG^o is presented in calories,

$$\Delta G^o = - 4.575T \log K \tag{17}$$

It is important to note that in a solid or liquid solution, the activity can be considered to represent the effective concentration of a component to participate in a reaction. Furthermore, it follows from our definitions that in reactions involving gases and vapours, the equilibrium constant is given in terms of the partial pressures of reacting gases.

3. THERMODYNAMICS OF FORMATION OF NITRIDE PHASES

3.1 Integral free energy information for some gas reactions

A compilation of free energy equations for several gas reactions of interest in many experimental investigations is presented in Table 1.

TABLE 1. Free energy equations for gas reactions

Reaction	ΔG^o calories	Estimated Uncertainty $\pm$ K. cals
$C + O_2 = CO_2$	$- 94,200 - 0.20T$	0.1
$C + \frac{1}{2}O_2 = CO$	$- 27,250 - 20.40T$	0.5
$C + 2H_2 = CH_4$	$- 21,940 + 26.56T$	0.1
$H_2 + \frac{1}{2}O_2 = H_2O$	$- 59,680 + 13.64T$	0.05
$\frac{1}{2}N_2 + 3/2H_2 = NH_3$	$- 13,000 + 28.10T$	0.5
$\frac{1}{2}S_2 + O_2 = SO_2$	$- 86,410 + 17.34T$	0.1
$\frac{1}{2}S_2 + 3/2O_2 = SO_3$	$-109,350 + 39.09T$	0.5

The thermodynamic terms are those defined above such that for the gas reaction:-

$$N_2(g) + 3H_2(g) = 2NH_3(g) \tag{18}$$

The equilibrium constant is

$$K = \frac{(p_{NH_3})^2}{(p_{N_2}) \cdot (p_{H_2})^3} \tag{19}$$

and $\Delta G^o_{18} = 2 \times \Delta G^o_{NH_3} - (\Delta G^o_{N_2} + 3 \times \Delta G^o_{H_2})$ (20)

Within the limits of uncertainty of the data, the temperature dependence of the change in free energy can be represented satisfactorily by a linear equation. The free energies are referred to the elements in their standard states, which are 1 atmosphere for all gases and graphite for carbon, at 1 atmosphere pressure. In all cases, the values are taken from the comprehensive compilation given in JANAF Thermochemical Tables[2].

It is important to realise that the free energies for complex reactions may be derived by combining values for simpler systems, such as those presented in Table 1. However, the free energy change of a reaction derived in this way may not agree exactly with directly measured values. The discrepancies between the experimental values of the equilibrium constant and those derived from selected thermodynamic data are very disconcerting. Whenever possible, the equilibrium constant of a reaction studied should be determined by direct experiment rather than placing too much reliance on values computed from compilations of thermodynamic data.

3.2 Nitriding equilibria and the iron-nitrogen system

Nitriding may be carried out in the simplest way by the use of molecular nitrogen, where the nitriding reaction is

$$\tfrac{1}{2}N_2(g) = \underline{N} \tag{21}$$

for which the equilibrium constant is

$$K_{21} = \frac{a_N}{(p_{N_2})^{\frac{1}{2}}} \tag{22}$$

where a_N is the activity of nitrogen in the nitrided phase.

The nitriding potential is determined by the partial pressure of nitrogen and this can be adjusted by the use of diluents of which hydrogen is the most popular.

The nitriding reaction involving ammonia is of particular
interest, as it is of commercial importance and serves as an outstanding example of a metastable reaction.   The nitriding
potential of an ammonia-hydrogen gas mixture is determined by the
ratio of the concentrations of ammonia and hydrogen,

$$NH_3(g) = \underline{N} + 3/2\ H_2(g)\ ;\ K_{23} = \frac{a_N \cdot (p_{H_2})^{3/2}}{(p_{NH_3})} \tag{23}$$

Other possible complicating reactions are:-

$$NH_3(g) = \tfrac{1}{2}N_2(g) + 3/2\ H_2(g) \tag{24}$$

$$NH_3(g) = \tfrac{1}{2}N_2(g) + 3\underline{H} \tag{25}$$

$$\underline{N} = \tfrac{1}{2}N_2(g) \tag{26}$$

Without experimental information, it would be impossible to
decide which of these reactions prevailed.   It is found experimentally that reaction (25) takes place to a negligibly small
extent and at low temperatures and high flowrates, the extent of
reaction (24) is small.   Generally, the thermal decomposition of
the nitride phase, reaction (26), is virtually non-reversible
at one atmosphere pressure.
    If the catalytic effect of the metal sample and the reaction
tube upon the cracking of ammonia is constant, then the degree to
which the gas approaches the equilibrium $NH_3/H_2/N_2$ mixture is
determined by the gas flow.   The flowrate must therefore be controlled as well as the gas composition if the required nitrogen
potential is to be maintained.   Clearly, the nitriding action of
ammonia on metals can be very complex since it depends upon the
balance of the above reactions.   These reactions are affected by
temperature, by gas composition, by the catalytic effect of surfaces to which the gas is exposed, by poisoning and side effects
of gas contaminants, by gas flow, and by previous treatment and
impurities in the metal itself.
    The metastable equilibria involving iron nitrides have been
extensively investigated[3-7].   The univariant equilibrium curves
as found by Lehrer are shown in Figure 1.   By combination of
these results with those of the homogeneous gas reaction (24), it
is possible to calculate the fugacity of nitrogen for each equilibrium at each temperature.   The values of $\log f_N$, thus calculated are plotted against temperature in Figure 2.   The meta-
stability of the ammonia reaction is clearly demonstrated by these
figures.

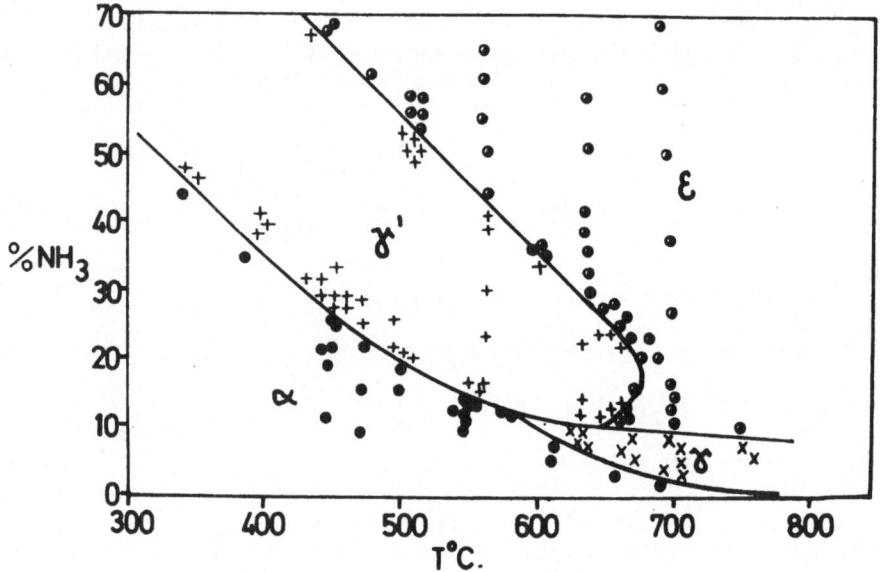

Figure 1.   Equilibrium between NH3-H2 mixtures and iron nitrides
(from Lehrer[3]).

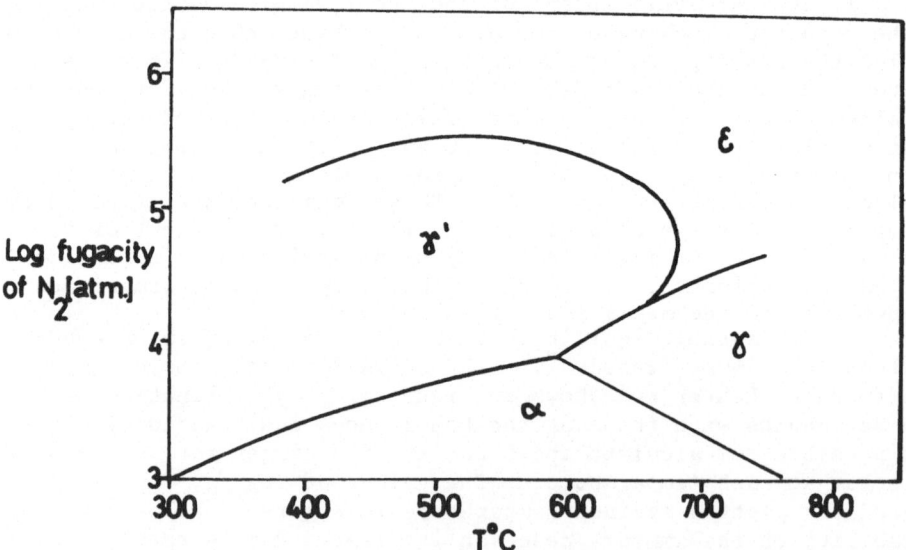

Figure 2.   Fugacity-temperature diagram for the Fe-N system
(from Lehrer[3]).

3.3 Effect of pressure on the chemical potential of gases

In recent years high pressure research has been undertaken to
synthesize new materials which cannot be manufactured under ordinary conditions. Advantages can be achieved in the production of
nitrogen ceramics by the use of such techniques with high pressure
molecular nitrogen at high temperatures.

However, gases do not obey the ideal gas laws at high press-
ures and the chemical potential is given by $\overline{G}_i = G_i^o + RT \ln f_i$.
The fugacity coefficient, $\eta_i = f_i/P_i$ is a measure of the deviation
from ideality, such that

$$\ln \eta_i = \int_0^P (V_i/RT - 1/P) \, dP \tag{27}$$

where V_i is the molar volume. The fugacity coefficients, η_i^o,
of pure gases are usually obtained by an integration of a volume-
pressure curve, according to equation (27). Theoretically, η_i
for a component, i, in a mixture can be determined in a similar
manner if the partial molar volume is known as a function of
pressure. Such information is not available for most gas mixtures
and it is necessary to determine the fugacities experimentally.
For most gases the fugacity coefficients can be correlated if the
reduced temperatures T/T_c and pressures P/P_c (index c refers to
the critical point) are used as variables[8], as shown in Figure 3
for the high temperature range ($T_R > 3.5$).

3.4 The free energy of formation of nitride phases

In order to assess the relative stabilities of nitride phases,
consideration must be given to the free energy changes accompanying these reactions at various temperatures and pressures. The
advantages of compiling free energy data in graphical form were
first demonstrated by Ellingham[9] and later by Richardson and
Jeffes[10]. Such a graph showing the standard free energy of formation per mole of nitrogen consumed as a function of temperature
for a number of nitrides is presented in Figure 4. The more
negative values of ΔG^o indicate a greater stability of the nitride
with respect to the elements. It can be seen for example that
zirconium and titanium form nitrides of high stabilties, while iron
forms nitrides of very low stabilities and silicon,magnesium and
calcium are intermediate with respect to their stability of their
nitrides.

It is often useful to express these stabilities in terms of
the nitrogen partial pressures (P_{N_2}) of the gas phase with which
the metal and its lowest nitride co-exist in equilibrium. The
equilibrium nitrogen pressure is related to the standard free
energy of formation by the equation

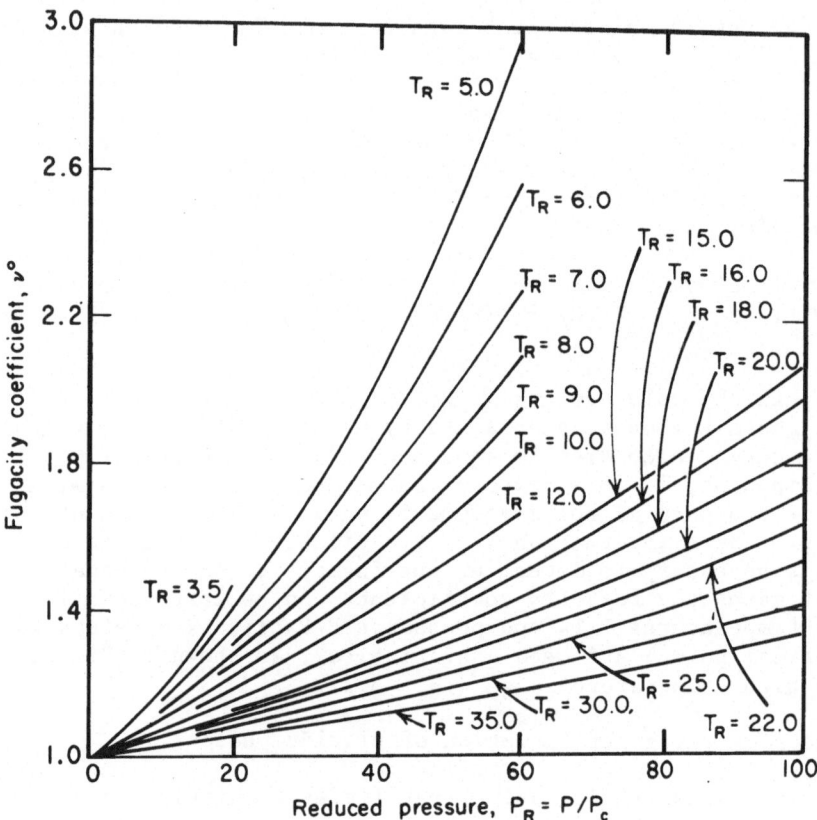

Figure 3. Fugacity coefficient of gases at high temperatures.

$$\Delta G^{O} = 4.575T \log p_{N_2} \tag{28}$$

and Figure 4 can be referred to as a nitrogen potential diagram.

It follows from equation (28) that for a given partial pressure of nitrogen, ΔG^{O} varies directly with temperature; at its
standard state, $p_{N_2} = 1$ atm; ΔG^{O} is zero at all temperatures and
at the absolute zero, $\Delta G^{O} = 0$ for all nitrogen pressures. Therefore, lines of equal $\log p_{N_2}$ drawn from a point representing
$\Delta G^{O} = 0$ at 0°K, on the ordinate are shown as broken curves in
Figure 4. The intersection of such a curve with a solid line
defines the nitrogen pressure prevailing when metal and its lowest
nitride coexist in univariant equilibrium. If the solid line
representing equilibrium coexistence of a metal-metal nitride pair
is located below a chosen dashed curve at any particular temperature,

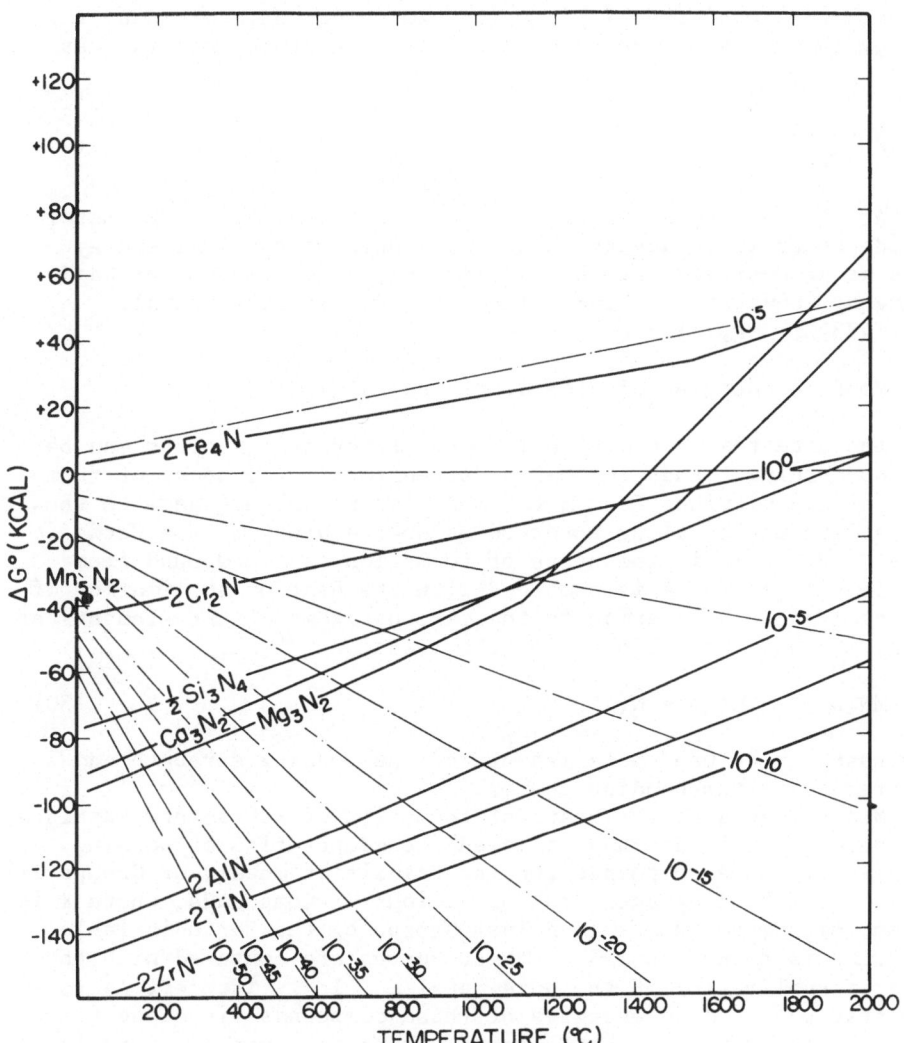

Figure 4: Standard free energies of formation of nitrides from the elements as a function of temperature.

the nitride is the stable phase at the chosen nitrogen pressure.
Conversely, if the solid line is located above a dashed line, the
metal is the stable phase.

One feature of particular interest is that the free energy
curves have a positive slope. The slopes are determined by the
entropy change associated with the formation of the nitride, as
shown by

$$d \frac{(\Delta G^O)}{dT} = - \Delta S^O \tag{29}$$

At the melting and boiling points of the metal or the metal
nitride there is an abrupt change of slopes of the free energy
lines as demonstrated by the particularly steep slopes for Ca_3N_2
and Mg_3N_2 above the boiling points of the respective metals
(Ca, 1483°C; Mg, 1103°C).

3.5 Bonding energies of nitride phases

From the data that are available, some interesting points can be
made concerning trends in bonding strengths. The heats of formation of the nitrides from metal and nitrogen do not seem to show
any pattern either along a Period or down a Group of the Periodic
Table. However, if the heats of formation of a compound from
the elements at 298°K is combined with the heat of sublimation of
the metal and dissociation to the gaseous atoms of the elements at
298°K, one obtains the heat for the reaction

$$MN(s) \rightarrow M(g) + N(g) \tag{30}$$

This heat can be used as a measure of the bonding strength and is
referred to as the bonding energy.

Measurements of the heats of formation of refractory carbides,
nitrides and silicides have allowed the computation of bonding
energies of these compounds for the transition metals of Groups III
to VI. The bonding energies for various MX compounds, where X is
carbon, nitrogen or silicon against Groups of the Periodic Table
are summarised in Figure 5. These curves are compared with the
heats of sublimation of the pure metals. It is interesting to
note that the bonding energies of these compounds lie close to
those for the metals. Furthermore, there is a general correlation
with the melting points of the elements and compounds. Whereas
it is difficult to use these trends to say anything significant
about the bonding in these compounds, the similarities in the
curves allow this type of correlation to be used in the estimation
of thermodynamic data for other nitride, carbide and silicide phases.

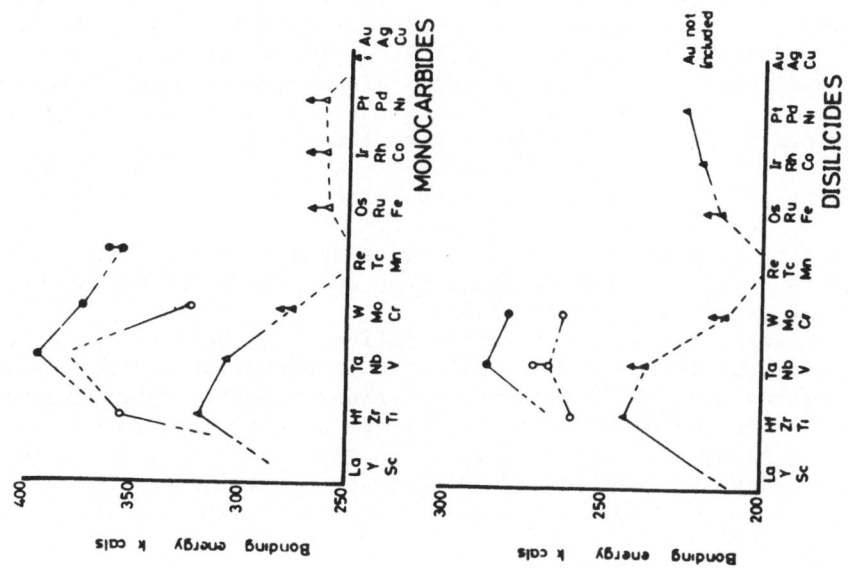

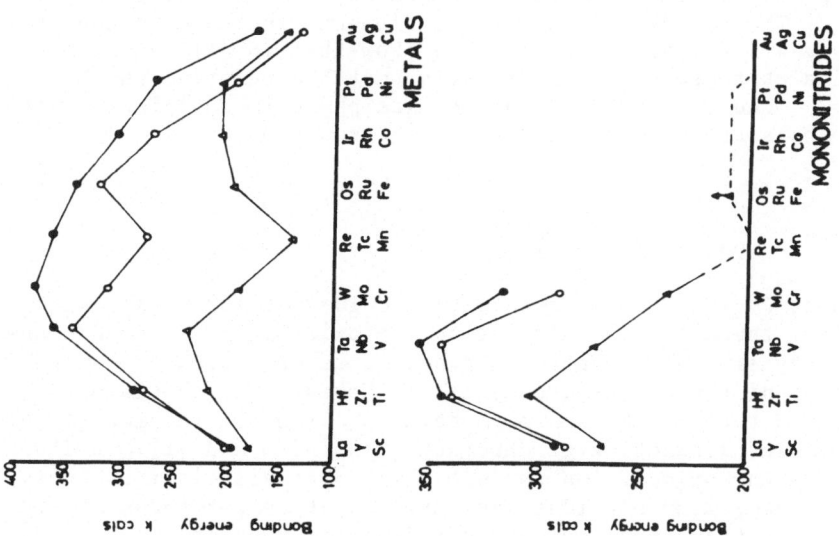

Figure 5: Bonding energies for mononitrides, transition metals, monocarbides and disilicides.

4. RELATIVE STABILITIES OF NITRIDE PHASES

The assessment of the value of nitride materials as useful ceramic
materials particularly at high temperature and in hostile environments requires some knowledge of their relative stabilities in
those environments.   A few examples are presented which indicate
how thermodynamic calculation can be used to assess the suitability
of various materials for different applications.

4.1 Nitride stabilties relative to oxides

The ability of nitrides to withstand oxidation in oxidising atmospheres is related to the relative free energies of formation of
the metal oxide and nitride phases.
     The stabilities of nitrides relative to oxides are illustrated
in Figure 6 by curves showing differences $\Delta(\Delta G^o)$ between the standard free energies of formation of nitrides and the corresponding
oxides.   The use of this diagram can be demonstrated by the
following examples.
     Consider the reaction of an oxide to a nitride according to
the equation

$$2M_xO + yN_2 = 2M_xN_y + O_2 \tag{31}$$

where one mole of oxygen exchanges with y moles of nitrogen (see
Figure 6.
     The less positive the $\Delta(\Delta G^o)$ value for this reaction is, the
greater the tendency for the reaction to proceed from left to
right.   The $\Delta(\Delta G^o)$ values for the elements included in the diagram are all strongly positive and show that in all cases, the
nitrides of these elements are very unstable relative to the
oxides.   The ratio P_{N_2}/P_{O_2} of the gas phase in equilibrium with
oxide and nitride is determined by the equation

$$\log \frac{(p_{N_2})^y}{(p_{O_2})} = \frac{\Delta(\Delta G^o)}{4.575T} \tag{32}$$

Ratios of P_{N_2}/P_{O_2} would have to exceed 10^{10} at 1500°C in order to
stabilise the nitrides relative to the oxides.   Such conditions
are hardly likely to happen in practice except, perhaps, where
very high nitrogen pressures exist.   However, it should be
remembered that kinetics are important.   Many of the metals which
form stable nitrides, e.g. chromium, aluminium and silicon also
form coherent oxide films through which diffusion is relatively
slow.   Thus, many nitrides have utility at temperatures considerably higher than those which thermodynamics predict for the decomposition of the nitride phase.
     In addition to the formation of stable solid oxides, the
formation of volatile sub-oxides is of some importance.

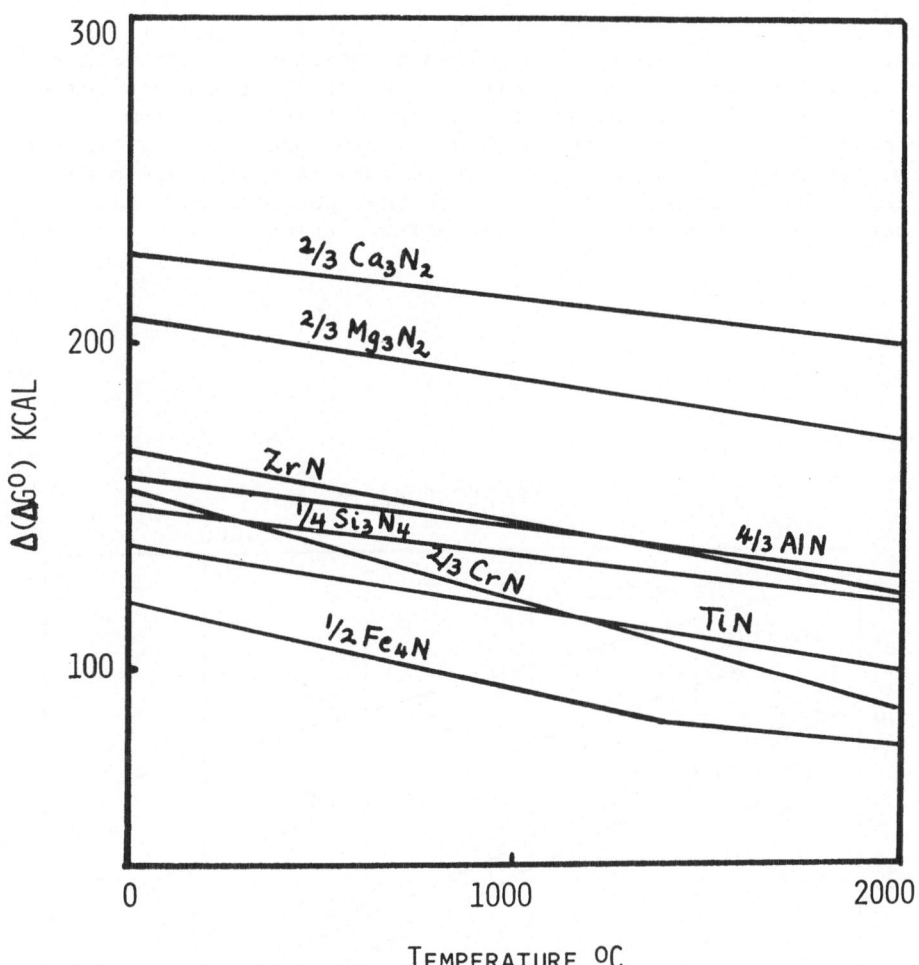

Figure 6.   Difference in free energies of formation between
            nitrides and corresponding oxides.

Volatile suboxides can play an effective role in the formation of
nitride materials from the vapour phase.   In addition, the possible production of volatile species when nitride ceramics are held
in atmospheres of low oxygen potential at high temperature can

168

lead to the disintegration of the material. It is useful, therefore, to have some knowledge of the thermodynamics of these volatile suboxides. Let us consider the reaction

$$MO_2(s, \ell) + H_2(g) = H_2O(g) + MO(g) \tag{33}$$

where M represents Si, Ti, Zr. The equilibrium constant for
reaction (32) is given in Figure 7 as a function of temperature
for several systems. It is self evident that for a given temperature and P_{H_2O}/P_{H_2} ratio, aluminium monoxide and zirconium monoxide
are considerably less volatile than silicon monoxide. It is for
this reason that silicon nitride refractory materials are more
susceptible to decomposition in low oxygen potentials at high
temperatures than aluminium or zirconium nitride.

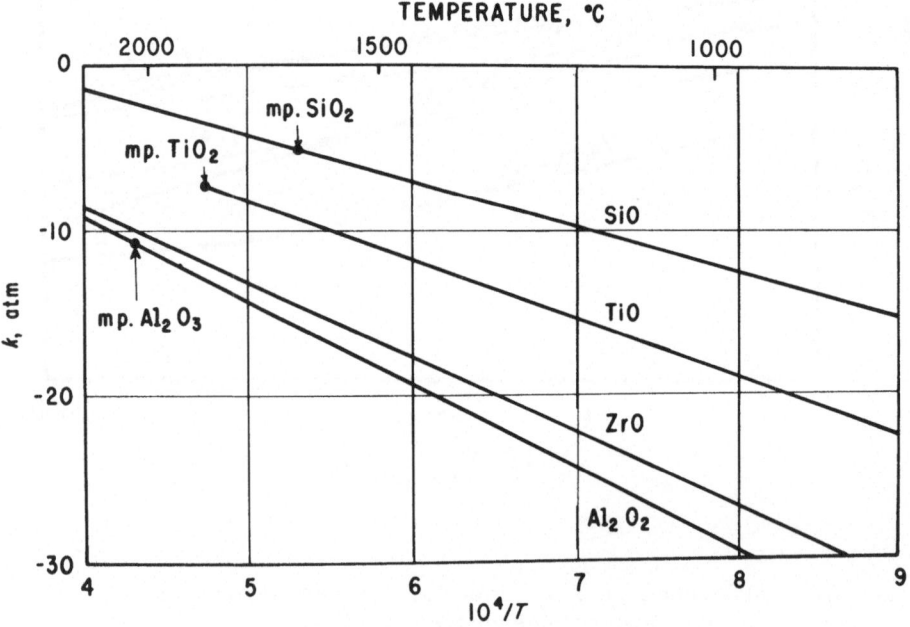

Figure 7.   Temperature dependence of the equilibrium constant
for reaction (32).

4.2 Liquid metal – nitride refractory interactions

One of the principle uses of high temperature ceramic materials
is to contain liquid metals. The extent of reaction of liquid
metals and nitride is of great importance if these materials are
to be of value for the melting of metals and alloys. At the
high temperatures required to melt metals, kinetics are very rapid
and the thermodynamic equilibria assume considerable importance.

Let us consider as an example, the amount of nitride dissolved
by pure liquid iron at 1600°C when held in a titanium nitride
crucible. The appropriate chemical reaction is

$$TiN(s) \;=\; \underline{Ti} + \underline{N} \tag{34}$$

In order to obtain the free energy change corresponding to
equation (33), the following known thermodynamic data will be used:

$$TiN(s) \;=\; Ti(s) + \tfrac{1}{2}N_2(g) \tag{35}$$

$$\Delta G^o_{35} \;=\; 81,000 - 22.9T$$

$$Ti(s) \;=\; \underline{Ti}\ (1\%\ \text{solution in Fe}) \tag{36}$$

$$\Delta G^o_{36} \;=\; -13,000 - 10.7T$$

$$\tfrac{1}{2}N_2(g) \;=\; \underline{N}\ (1\%\ \text{solution in Fe}) \tag{37}$$

$$\Delta G^o_{37} \;=\; 900 + 5.7T$$

By combination of these equations, it is possible to evaluate
$\Delta G^o_{33} = 68,900 - 27.9T$ and from this it is possible to compute the
equilibrium constant for reaction (34) at 1873°K as 1.14×10^{-2}.

Since using the 1% solution as a standard state

$$K_{34} \;=\; (\%Ti).(\%N) \tag{38}$$

and TiN must decompose in atomic proportions such that, one atom
of titanium is produced for each atom of nitrogen, it is possible
to evaluate the corresponding quantities of titanium and nitrogen
dissolved in liquid iron, i.e. %Ti as 0.18t.% and %N as 0.06wt.%.
Thus, it can be seen that nitride refractories can be of some
utility for holding liquid metals without a considerable extent
of dissolution.

4.3 Evaporation of nitride ceramics in vacuo

It is quite often necessary to use high temperature ceramic materials in vacuum furnaces, where it is important to consider the
loss in weight and thickness of material which will occur due to
vaporisation. If it is assumed that the equilibrium vapour

pressure is exhibited under conditions of free evaporation, the calculation of the rate of weight loss can easily be made.   As an example of this type of computation, the estimated weight loss of titanium nitride at a temperature of 2000°K will be demonstrated.

The mechanism for evaporation of titanium nitride is assumed to be:

$$TiN(s) = Ti(g) + \tfrac{1}{2}N_2(g) \qquad (39)$$

and the free energy change for this process may be obtained from data on the free energy of formation of TiN and the free energy of vaporisation of titanium.   This yields a simplified equation

$$\Delta G^0 = 189,000 - 53.1T \text{ cal/mole of nitride} \qquad (40)$$

from which

$$K = \frac{p_{Ti}\, p^{\frac{1}{2}}_{N_2}}{a_{TiN}} \qquad (41)$$

Since we are concerned only with pure titanium nitride ($a_{TiN} = 1$), there must be a relation between p_{Ti} and p_{N_2} resulting from the stoichiometry of equation (39), i.e. two atoms of titanium are produced for each mole of nitrogen gas to maintain the stoichiometric ratio of the solid phase.

Thus the weights of the two elements leaving the surface must be in proportion to their atomic and molecular weights.

$$\frac{n_{Ti}}{n_{N_2}} = \frac{wt.Ti}{M_{Ti}} \cdot \frac{M_{N_2}}{wt.\%N_2} = \frac{2}{1} \qquad (42)$$

The weight of each constituent leaving unit area of the solid in unit time is given by the Knudsen equation:

$$wt.Ti = 44.4\, p_{Ti} \sqrt{\frac{M_{Ti}}{T}} \qquad (43)$$

Hence,
$$\frac{n_{Ti}}{n_{N_2}} = \frac{p_{Ti}}{p_{N_2}} \sqrt{\frac{M_{N_2}}{M_{Ti}}} = \frac{2}{1} \;;\; p_{N_2} = \frac{p_{Ti}}{2} \sqrt{\frac{M_{N_2}}{M_{Ti}}} \qquad (44)$$

Substituting in the equilibrium constant, then

$$K = 0.62\, p_{Ti}^{3/2} = 8.9 \times 10^{-10}$$

$$p_{Ti} = 1.3 \times 10^{-6} \text{ atm.}$$

The weight loss per unit area of the solid is then:

$$\Sigma wt = wt_{Ti} + wt_{N_2} = 44.4 \frac{P_{Ti}}{\sqrt{T}} \ (\sqrt{\overline{M}_{Ti}} + 0.62 \ \sqrt{\overline{M}_{N_2}})$$

$$= 1.32 \times 10^{-5} \ gm.cm^{-2}.sec^{-1}.$$

From this calculation, it can be seen that titanium nitride would make a good choice as a ceramic material for operation in vacuo at relatively high temperatures.

4.4   Interaction of nitrides with solid metals

The use of nitrides as fibre-reinforcing and dispersion strengthen-ing materials in metals is of some importance.   The performance of such materials at high temperature is related to the reactivity of the nitride with its metallic matrix.   Once again, some know-ledge of the free energies of formation of the respective phases is most useful.

Let us consider the possible use of silicon nitride in this context.   It can be seen from a perusal of figure 4 that silicon nitride is stable with respect to some of the probable matrix materials, i.e. iron and nickel over the whole temperature range with respect to decomposition to the metal nitride.   However, titanium, zirconium and aluminium nitrides are much more **stable** than silicon nitride and considerable reaction would be anticipated with these metals at high temperatures.

With respect to decomposition to form silicides, it is useful to refer once again to an Ellingham-type diagram as shown in Figure 8.   As before, the larger the negative value of ΔG^O is, the greater is the stability of a silicide with respect to its elements.   Hence the further down the graph a curve is located, the more stable is the silicide represented by that curve.   It can be seen that carbon and iron form silicides of low stability while, titanium, zirconium and calcium form the most stable silicides.

The free energy of formation line for silicon nitride for nitrogen in its standard state of 1 atmosphere is shown as a solid line with a significant positive slope.   The interaction of this line with that for any silicide represents the equilibrium between that silicide and silicon nitride.   Thus, for example, the sili-con nitride curve intersects that for FeSi at 1250°C which tells us that below this temperature silicon nitride is stable with respect to iron silicide and that FeSi will not form.   However, at high temperatures, FeSi will be the stable phase and it is expected that silicon nitride will decompose.

The stability of silicon nitride is governed by its free energy

$$\Delta G^O = N_{Si} \ RT \ \ell n \ a_{Si} + N_N \ RT \ \ell n \ a_N \tag{45}$$

172

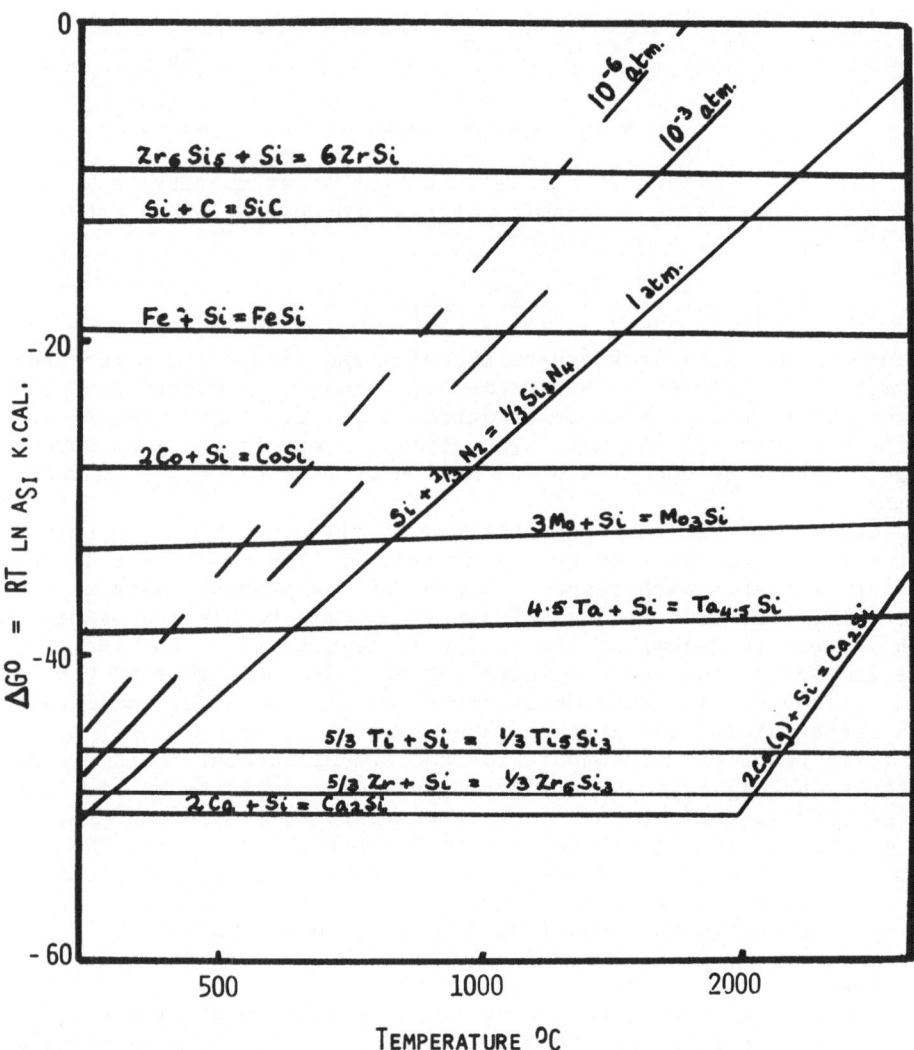

Figure 8: Standard free energies of formation of metal silicides compared with silicon nitride at various nitrogen pressures.

From this relationship, it is obvious that at any temperature, if the nitrogen potential of the system is lowered, the silicon activity and hence its reactivity to form a metal-silicide must increase correspondingly. So, if silicon nitride and a metal are reacted in an environment which has some fixed nitrogen potential which is less than one atmosphere, the formation of a metal silicide will occur more easily. A demonstration of this effect is made in Figure 8 where the broken lines represent the curve for silicon nitride in equilibrium with nitrogen gas at 10^{-3} and 10^{-6} atmospheres. From this we see that as the nitrogen pressures drop to 10^{-3} and 10^{-6} atmospheres, the temperature at which FeSi becomes stable drops from 1250°C to 1100°C and then to 900°C, indicating the substantial increase in the silicon activity in silicon nitride as predicted by equation (45).

5. CONCLUSION

An attempt has been made to demonstrate the utility of thermodynamic understanding of gas-solid reactions and the reactivity of solid phases. However, to make use of this ability, it is essential thermodynamic data are available. Those of us who actively use thermodynamics know how seemingly little data are available and how much we would like. However, at the same time we recognise that an enormous quantity of thermodynamic information is scattered throughout the literature. The major problem for most scientists is to know what information exists and how these values can be assimilated for easy and efficient use. Fortunately, there are several compilations of thermodynamic data for solids, liquids and gases[2,11,12,13,14].

In addition to absolute values, a need exists for the evaluation of data which brings us to the major problem of science today – the storage and retrieval of quantitative scientific information. Of such data, thermodynamic values are among the most important. Fortunately, the Chemical Standards Division of the National Physical Laboratory are actively involved in the computer storage of thermodynamic data. Data on thermodynamic properties of metallurgical and ceramic materials are becoming increasingly important to modern science and technology as is the availability of these values.

REFERENCES

1. L.S. Darken and R.W. Gurry, Physical Chemistry of Metals, McGraw Hill, New York, 1953.
2. JANAF Thermochemical Tables, U.S. Department Commerce, Washington, D.C., 1965, PB 168370; and 1966, PB 168370-1.
3. E. Lehrer, Z. Elektrochem., 1930, 36, 383.
4. K.H. Jack, Proc. Roy. Soc., 1948, 195A, 41, 56.

174

5. L.S. Darken, R.P. Smith and E.W. Filer, Trans. AIME, 1951, 191, 1174.
6. L.J. Dijkstra, Trans. AIME, 1949, 185, 252.
7. V.G. Paranjpe, M. Cohen, M.B. Bever and C.F. Floe, Trans.AIME, 1950, 188, 261.
8. R.H. Newton, Ind. Eng. Chem., 1935, 27, 305.
9. H.J.T. Ellingham, J. Soc. Chem. Ind., 1944, 63, 125.
10. F.D. Richardson and J.H.E. Jeffes, J. Iron Steel Inst., 1948, 160, 261.
11. Selected Values of Chemical Thermodynamic Properties, U.S. National Bureau of Standards, Washington, D.C., 1947, Series III.
12. D.R. Stull and G.C. Sinke, Advances in Chem. Ser., 1956, 18.
13. K.K. Kelley, Bureau of Mines Bulletin, 1960, 584; 1961, 592.
14. O. Kubaschewski, E.Ll. Evans and C.B. Alcock, Metallurgical Thermochemistry, Pergammon, London, 1967.

DISCUSSION (Schlichting, Singhal)

Desmaison: I would like to emphasise the importance of the kinetic factors on assessing the oxidation resistance of nitrides. Although all the nitrides are thermodynamically unstable in oxygen they exhibit different behaviour. This is primarily due to the fact that the reaction kinetics are very dependent on the crystal structure and morphology of the oxide formed. This is particularly true in the case of the nitrides of Ti and Zr which exhibit the same physicochemical properties, but show markedly different oxidation behaviour.

Popper: Professor Jack has indicated that molten iron could be contained in a Si-Al-O-N crucible. In practical ferrous metallurgy the containers have to be inert with regard to the slag. My colleagues from the Refractory Division of the B.C.R.A. have shown that a particular sialon composition was very readily attacked by the slag.

REACTIONS IN THE SiO_2-C-N_2 SYSTEM

J. G. Lee and Ivan B. Cutler

University of Utah
Salt Lake City, Utah   84112

ABSTRACT. Although thermodynamic calculations predict that silica, carbon and nitrogen will yield silicon nitride at temperatures below about 1550°C and silicon carbide above that temperature, kinetic considerations lower this boundary to about 1450°C. In the interval between 1400°C and 1550°C, silicon carbide forms more rapidly than silicon nitride but reverts to silicon nitride when the carbon to silica molar ratio is less than three. Under these conditions, silicon nitride and silicon monoxide form through the following two-step reactions even at temperatures higher than the boundary temperature:

$$SiO_2 + 3C \rightarrow SiC + 2CO \qquad \text{followed by}$$

$$2SiC + SiO_2 + 2N_2 \rightarrow Si_3N_4 + 2CO \qquad \text{or}$$

$$SiC + 2SiO_2 \rightarrow 3\ SiO + CO$$

1.   INTRODUCTION

The interest in silicon nitride and its related systems is increasing among many investigators (1,2,3) who recognize its superior properties for high temperature engineering applications. Among many methods of preparation of silicon nitride described in the literature (2), the carbothermal reduction of silica followed by nitridization provides an inexpensive route for silicon nitride production.

It is well known that there is a boundary temperature between the formation of SiC and Si_3N_4 in the SiO_2-C-N_2 system (4,5,6).

Several reports agree that SiC forms above this boundary temperature and Si_3N_4 forms below it as predicted by thermodynamic calculations. This boundary temperature will vary depending on the nature of the reactants as well as products. The porosity and size of the sample will also affect it due to the CO concentration produced by the reaction. Thus, different boundary temperatures were reported from different workers ranging from 1400 to 1500°C (4,5,6).

At temperatures ranging from 1300 to 1600°C, four major products are known to form with evolution of carbon monoxide in the SiO_2-C-N_2 system; they are SiC, Si_3N_4, Si_2ON_2 and SiO formed by the overall reactions:

$$SiO_2 + 3C \rightarrow SiC + 2CO \qquad (1)$$

$$3SiO_2 + 6C + 2N_2 \rightarrow Si_3N_4 + 6CO \qquad (2)$$

$$2SiO_2 + N_2 + 3C \rightarrow Si_2N_2O + 3CO \qquad (3)$$

$$SiO_2 + C \rightarrow SiO + CO \qquad (4)$$

It is noted that nitrogen gas is a reactant in Reaction (2) and (3) while it is inert in Reaction (1) and (4).

The formation of silicon nitride and silicon oxynitride depend on the partial pressures of oxygen and nitrogen as reported by Wild et.al. (7) and Blegen (8). A partial pressure of oxygen less than ~10^{-20} atm. is known to be sufficient to suppress the formation of silicon oxynitride in 1 atm. of N_2 at 1800°K. This means that the formation of silicon oxynitride can be eliminated as long as carbon is present because carbon maintains the partial pressure of oxygen much lower than 10^{-20} atm. in the system.

One should remember that this boundary temperature can exist in the SiO_2-C-N_2 system only when the molar ratio between carbon and silica is higher than three. If carbon is deficient for Reaction (1), which means that the carbon to silica ratio is less than three, SiC may form first and may react with excess silica producing either Si_3N_4 or SiO even above the boundary temperature through the following reactions:

$$2SiC + SiO_2 + 2N_2 \rightarrow Si_3N_4 + 2CO \qquad (5)$$

$$SiC + 2SiO_2 \rightarrow 3SiO + CO \qquad (6)$$

Once SiC is formed, all three reactions, Reaction (1), (5) and (6), can take place simultaneously or step by step such as Reaction (5) after Reaction (1) or Reaction (6) after Reaction (1). A recent report by Hendry and Jack (9) may represent the above case since the carbon content in their sample was stoichiometric for Reaction (2) but was deficient for Reaction (1). They observed the formation of Si_3N_4 up to 1650°C at 1 atm. N_2. They also encountered excessive weight losses for Reaction (2) possibly due to SiO formation by Reaction (6).

Thus it is the purpose of this paper to give a better understanding of the important reactions taking place in the SiO_2-C-N_2 system. This could lead to efficient and economic production of silicon nitride by reacting silica with carbon under a nitrogen atmosphere.

2. EXPERIMENTAL PROCEDURE

The raw materials and chemicals used in this study are listed in Table 1. Silica and carbon were wet-mixed using water. After drying the slurries, the samples were pressed uniaxially at 2000 psi and were cut into cubes about 0.5 cm long on each side.

TABLE 1
Raw Materials and Chemicals

MATERIALS	PURITY	PARTICLE SIZE	SUPPLIER
Silica (Cabosil M-5)	99.0%	∿ 0.012 μm	Cabot Corp.
Carbon Black (Raven 450)	99.7%	∿ 6.0 μm	Cities Service
Silicon Carbide	Commercial αSiC	1000 Grit	Carborundum
Fe	99.0%	∿10.0 μm	Baker Chemical
N_2 Gas	O_2 < 8 ppm	---	Liquid Air

As shown in Table 2, four different samples were prepared by the method described above.

TABLE 2
Samples Prepared for this Study

SAMPLE NO.	MOLAR RATIO	Fe ADDITION
1	C/SiO_2=4	None
2	C/SiO_2=2	None
3	C/SiO_2=2	5 w/o
4	SiC/SiO_2=2.5	5 w/o

A thermogravimetric appartus, which is similar to the one used in previous work (6) was used for kinetic studies of carbothermal reduction of silica under a N_2 atmosphere. The flow was kept at 2 ℓ/min (∿ 170 cm/min) throughout the study.

The mineral composition of the samples was determined qualitatively by comparing the intensities of each phase shown in the x-ray diffraction patterns. This is only approximate at best but experience has made this a fast and reliable technique for sorting out reactions.

3.   RESULTS AND DISCUSSION

3.1 Determination of Boundary Temperature

Sample #1, which has excess carbon for SiC formation was reacted at 1400, 1450 and 1500°C for 2 hours. X-ray diffraction analysis of each sample showed that SiC started to form at temperatures above 1450°C. Only silicon nitride forms at 1400°C and below in agreement with previous work. This indicates that the boundary temperatures is near 1450°C under the present experimental conditions.

X-ray diffraction also revealed that SiC formed in the β-modification while Si_3N_4 formed largely in the α-modification.

3.2 Formation of Si_3N_4 Above the Boundary Temperature

Sample #2 and #3, which have deficient carbon contents for SiC formation, but stoichiometric amounts of carbon for Si_3N_4 formation, were reacted at 1500°C, which is higher than the boundary temperature. Figure 1 shows the weight changes along with the reaction time for both samples. Sample #3 which has 5 wt.% Fe reacted much faster than sample #2 initially. This may be due to the catalytic effect of iron on the carbothermal reduction of silica as previously reported (5,6). Weight losses greater than that of complete conversion to Si_3N_4 may be due to SiO formation by Reaction (6).

The progressive composition changes on the sample #2 during the reaction were estimated by x-ray diffraction and are tabulated in Table 3. It shows that the presence of SiC as well as Si_3N_4 which supports the assumption mentioned in the Introduction, i.e., the formation of Si_3N_4 by Reaction (1) followed by Reaction (5).

Silicon oxynitride started to appear at the later stage of the reaction where the reaction time was longer than three hours. This may be due to the absence of carbon which increases the partial pressure of oxygen in the system high enough to form Si_2ON_2 instead of Si_3N_4.

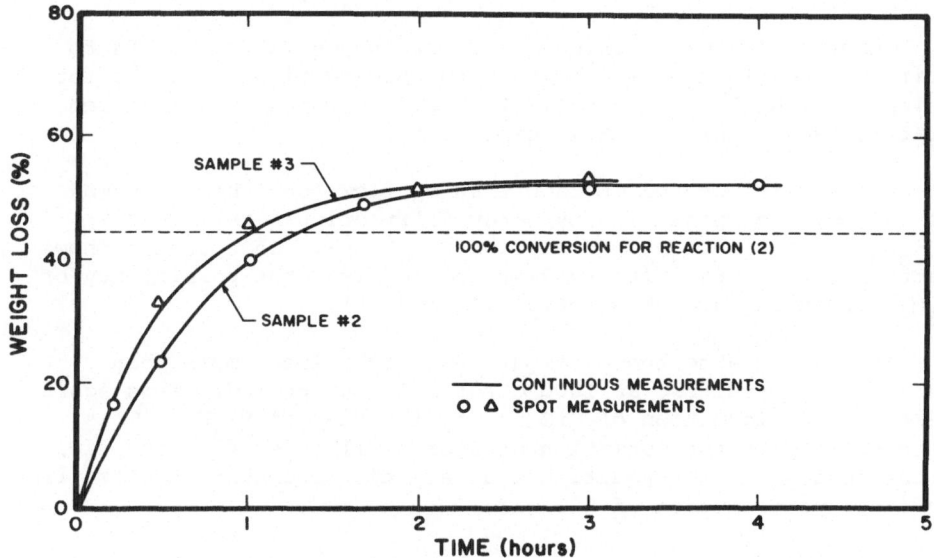

Figure 1. Weight Changes in the Sample #2 and #3 at 1500°C (Data
for Sample #3 was adjusted to exclude the weight of Fe).

TABLE 3
Composition Changes During the Reaction in Sample #2 at 1500°C

| | TIME (hr) | | | |
PHASE	0.25	0.5	1.0	3.0
SiO_2 (Cristobalite)	∿90%	∿85%	∿75%	∿30%
$\beta-SiC$	∿10%	∿10%	∿15%	∿20%
$\alpha-Si_3N_4$	--	∿5%	∿10%	∿25%
$\beta-Si_3N_4$	--	--	--	∿10%
Si_2ON_2	--	--	--	∿15%

After a 3 hour reaction, both sample #2 and #3 showed about
the same weight loss. However, this does not necessarily imply
the same degree of reaction since Reaction (5) occurs without any
weight loss. X-ray diffraction analysis on the sample #3 after
the 3 hour experiment showed the following phases: $\beta-Si_3N_4$, 55%;
$\alpha-Si_3N_4$, 20%; Si_2ON_2, 25%; SiC, trace. This result gives the
following information about the role of iron.

a) Iron enhances the rate of Reaction (5) as well as that
of Reaction (1) and (2).

b) Iron assists the formation of silicon nitride in the
β-modification.

3.3 Formation of Si_3N_4 from the SiO_2-SiC-N_2 System

Reaction (5) was carried out separately using sample #4 to verify the conclusion made on the previous section; i.e. the formation of Si_3N_4 by Reaction (1) followed by Reaction (5) at temperatures above the boundary temperature.

The sample reacted at $1500^\circ C$ in N_2 flow for 3 hours showed the following compositions by x-ray diffraction: α-SiC $\sim$ 40%; β-Si_3N_4 $\sim$ 30%; α-Si_3N_4 $\sim$ trace; Si_2ON_2 $\sim$ 30%. This result shows Reaction (5) occurs quite readily and confirms the possibility of Si_3N_4 formation through Reaction (1) and (5).

The sample also showed about 10% weight loss, more than equation (5) predicts, at $1500^\circ C$ for a 3 hour period. This again indicates the formation of SiO along with that of Si_3N_4. This was confirmed by the formation of fibrous mixture of Si and SiO_2 in the cooler part of reaction chamber which probably resulted from the disproportionation of SiO gas.

4. CONCLUSIONS

4.1 In the SiO_2-C-N_2 system, a temperature boundary exists at about $1450^\circ C$ above which silicon carbide forms and below which silicon nitride forms only when the carbon to silica molar ratio exceeds three.

4.2 Silicon nitride and silicon monoxide can form above this boundary temperature when the carbon to silica molar ratio is less than three through the following two-step reactions:

$$2(SiO_2 + 3C \rightarrow SiC + 2CO)$$
$$2\ SiC + SiO_2 + 2N_2 \rightarrow Si_3N_4 + 2CO$$
$$\overline{3SiO_2 + 6C + 2N_2 \rightarrow Si_3N_4 + 6CO} \qquad \text{and}$$

$$SiO_2 + 3C \rightarrow SiC + 2CO$$
$$SiC + 2SiO_2 \rightarrow 3\ SiO + CO$$
$$\overline{3(SiO_2 + C \rightarrow SiO + CO)}$$

4.3 β-Silicon nitride prefers to form in the presence of impurities such as iron.

REFERENCES

1. K. H. Jack and W. I. Wilson, "Ceramics Based in the Si-Al-O-N and Related Systems," Nature Physical Science, __238__ (80), 28-29 (1972).

2.   I. B. Cutler and W. J. Croft, "Silicon Nitride," Powder Met. International, 6 (2), 92-94 (3), 143-146 (1974).

3.   J. W. Edington, D. J. Rowcliffe and J. L. Henshall, "The Mechanical Properties of Silicon Nitride and Silicon Carbide," Powder Met. International, 7 (2), 82-86, (3), 136-147 (1975).

4.   K. Komeya and H. Inoue, "Synthesis of the α Form of Silicon Nitride from Silica," J. Mat. Sci., 10, 1243-1246 (1975).

5.   J. G. Lee, R. Casarini and I. B. Cutler, "SIALON Derived from Clay to Provide an Enconomical Refractory Material," Industrial Heating, 43 (4), 50-53 (1976).

6.   J. G. Lee and Ivan B. Cutler, "Formation of Silicon Carbide From Rice Hulls," Ceram. Bull., 54 (2), 195-198 (1975).

7.   S. Wild, P. Grieveson and K. H. Jack, "The Thermodynamics and Kinetics of Formation of Phases in the Ge-N-O and Si-N-O Systems," Special Ceramics 5, Brit. Ceram. R. A., Stoke-on-Trent (1972), 217-287.

8.   K. Blegen, "Equilibria and Kinetics in the Systems Si-N and Si-N-O," Special Ceramics 6, Brit. Ceram. R. A., Stoke-on-Trent (1975), 223-244.

9.   A. Hendry and K. H. Jack, "The Preparation of Silicon Nitride from Silica," Special Ceramics 6, British Ceram. R. A., Stoke on-Trent, (1975) 199-208.

DISCUSSION (Schlichting, Singhal)

Billy: Our own studies on the stability of silicon nitride in the presence of carbon (M. Billy and F. Colombeau, C.R.A.S. Paris, 264, 1967, série C, 392) have shown that α-Si$_3$N$_4$ is more reactive than β and reacts with carbon above 1400°C giving β-SiC.

Discussion contribution
THERMODYNAMICS OF SILICON NITRIDE AND OXYNITRIDE

A. Hendry (contributed by K.H. Jack)

The Wolfson Research Group for High-Strength Materials,
Crystallography Laboratory, The University,
Newcastle upon Tyne, England.

Considerable disagreement exists in the published values of
the free energy of formation of α- and β-silicon nitride and silicon oxynitride (1-5). The present recalculation by A. Hendry shows
that the disagreement is a result of the use of different sources
of reference data, and that the results can be reconciled if a
single set of values for the activity of silicon in liquid iron,
and the same value of the free energy of formation of silica are
used. Reviews of the thermodynamic properties of Fe-Si alloys by
Chart (6) and of the free energy of formation of silica by Ancey-
Moret et al.(7) have been used in this recalculation.

The results of the recalculation of the free energy of formation of β-silicon nitride are shown in Fig. 1 and excellent agreement is obtained between the results (3-5). The apparent disagreement reported by Blegen (5) is resolved by use of common values of
the activity of silicon in iron for the three sets of results.

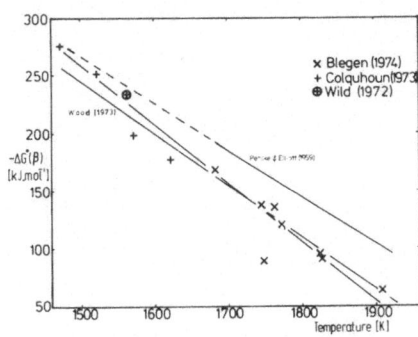

Figure 1

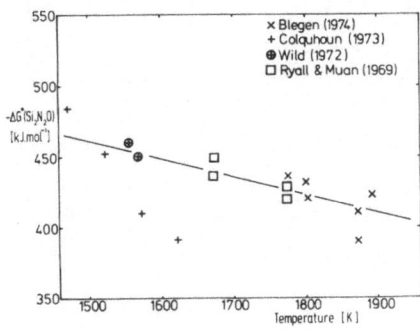

Figure 2

184

The equation of the best straight line is:

$$\Delta G^\circ \; (\beta) \;\; = \;\; -987.3 + 0.490T \;\; kJ \; mol^{-1} \qquad \ldots \; (1)$$

The room temperature heat of formation of β (8) has been combined with specific heat data (9,10) and the heat of fusion of silicon (10) to calculate the line shown as Wood (1973) in Fig. 1. The equation of this line is:

$$\Delta G^\circ \; (\beta) \;\; = \;\; -925.2 + 0.450T \;\; kJ \; mol^{-1} \qquad \ldots \; (2)$$

Equation (2) is preferred to equation (1) as it is based on an experimentally determined heat of fusion and not simply on a least squares analysis of the ΔG° against T points.

Fig. 2 shows the results of the recalculated free energies of formation of Si_2N_2O (2-5). The values of $\Delta G^\circ(Si_2N_2O)$ at 1300°C and 1350°C (4) are apparently in error with respect to the other results and this was found to be true for all of the free energy values calculated from these two results. The equation of the best line is:

$$\Delta G^\circ \; (Si_2N_2O) \;\; = \;\; -658.3 + 0.131T \;\; kJ \; mol^{-1} \qquad \ldots \; (3)$$

The free energy of formation of silica has been calculated from the results of Wild et al. (3) and Colquhoun et al.(4) as a test of their experimental method. Good agreement is obtained with the value given by Ancey-Moret et al.(7), Fig. 3.

$$\Delta G^\circ \; (SiO_2) \;\; = \;\; -940.0 + 0.20T \;\; kJ \; mol^{-1} \qquad \ldots \; (4)$$

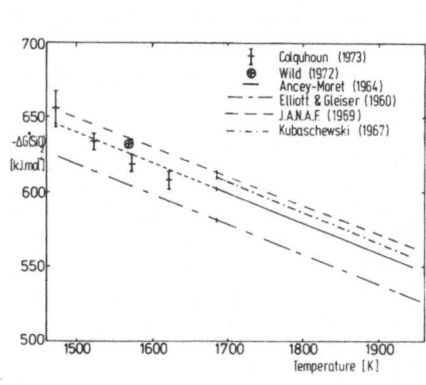

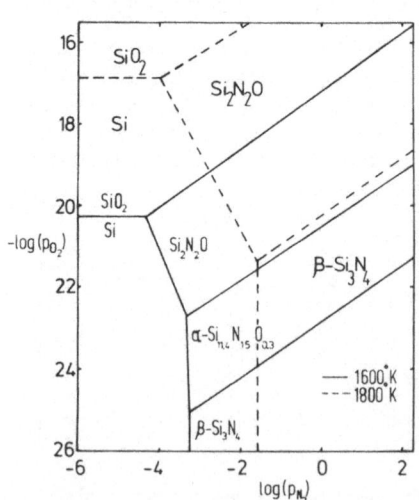

Figure 3   Figure 4

The difference between the values of $\Delta G°$ (SiO_2) at 1300°C and 1350°C calculated from Colquhoun et al. (4) and those given by equation (4) above show that a loss of $2^W/o$ Si from the Fe-Si alloy used ($20^W/o$ Si) - by volatilisation as SiO, for example - accounts for the systematic error noted above for these two results.

The free energy of formation of α-silicon nitride has been recalculated from the results of Wild et al. (3) and of Colquhoun et al. (4) (corrected for the error in their results at 1300°C and 1350°C) using the accepted values of the activity of silicon in iron (6). The relationship obtained is:

$$\Delta G° (\alpha) = -1167.3 + 0.594T \quad kJ\ mol^{-1} \qquad \ldots (5)$$

Equations (2)-(5) have been used to calculate phase stability diagrams for this system, Fig. 4. At a partial pressure of oxygen $p_{O_2} = 10^{-20}$ atm, α becomes unstable with respect to β and oxynitride at T > 1380°C; adopting the stoichiometry for α of $Si_{11.4}N_{15}O_{0.3}$ (4).

ACKNOWLEDGMENTS

The use of results contained in a private communication from Mrs. K. Blegen, Institute of Silicate Science, Trondheim, Norway, is gratefully acknowledged.

REFERENCES

1. R.D. Pehlke & J.F. Elliott. Tr. AIME, 215, 1959, 781.
2. W.R. Ryal & A. Muan. Science, 165, 1969, 1363.
3. S. Wild, P. Grieveson & K.H. Jack. "Special Ceramics 5", B. Ceram. R.A. (Stoke-on-Trent), 1972, p.271.
4. I. Colquhoun, S. Wild, P. Grieveson & K.H. Jack. Proc. Brit. Ceram. Soc., 22, 1973, 207.
5. K. Blegen. "Special Ceramics 6", B. Ceram. R.A., 1975, 223.
6. T.G. Chart. High Temp.-High Pres., 2, 1970, 461.
7. M.F. Ancey-Moret, M. Olette & M. Richardson. Mem. Sci. Rev. Met., 61, 1964, 169.
8. J.L. Wood, G.P. Adams, J. Mukerji & J.L. Margrave. "Third Int. Conf. on Chemical Thermodynamics", Vol. 1 (Baden, Austria), 1973.
9. JANAF Thermochemical Tables, 2nd Edition (Dow Chem. Co.), 1971.
10. J.F. Elliott & M. Glieser. "Thermochemistry for Steelmaking", Addison-Wesley (London), 1963.
11. O. Kubaschewski, E.Ll. Evans & C.B. Alcock. "Metallurgical Thermochemistry", Pergamon (London), 1967.

DEFECT EQUILIBRIA IN THE SOLID STATE

R.J. Brook

Department of Ceramics, University of Leeds, U.K.

Introduction

The assumption is commonly made in the discussion of nitrogen ceramics that the departures from stoichiometry are negligible and that the various atomic species are present in their correct ratio. Since this assumption is necessarily an approximation, and since the consequences of small concentrations of point defects can be pronounced, it is important that the ability of point defects to affect the processing and properties of this group of ceramics should not be overlooked.

In the following, the use of defect equilibria in the characterisation of ceramic systems (1) will be reviewed in terms of examples drawn from oxide systems. Generally, it will be seen that such equilibria can be of value in the interpretation of rate processes (sintering, creep, solid reactions) although the complexities of ceramic compositions and microstructures rarely allow an unambiguous interpretation of the rate controlling step. In the closing section, data for nitrogen ceramics are reviewed and discussed in the light of the oxide experience.

Importance of point defects

In addition to their role in effecting nonstoichiometry, point defects are the active elements in many transport processes of concern to the ceramist. These processes include diffusion, sintering, grain growth, creep, electrical conductivity, permeability, and reactivity, and the wish to control these processes more deliberately has been a strong driving force for the study of point defects in oxides. As an example, there are now some ten systems which can be sintered to theoretical density by way of additives or sintering aids (2); an understanding of defect

equilibria offers the promise of predicting suitable additives for
other systems that may be of interest in the future. The extension
of such an approach to nitrogen ceramics is of obvious interest.

Classification

Calculations of equilibrium concentrations of point defects
and the dependence of these concentrations on such experimental
variables as temperature and the partial pressures of the constituents can be performed using the techniques of defect chemistry
(3,4). These methods are most readily applied when the defects
can be treated ideally, that is, as isolated non-interacting species.
As defect concentrations increase, interactions can result in a
variety of more complicated structures, the particular type depending on the material under consideration. A classification of such
structures is given in Figure 1; in general, the techniques of
defect chemistry are only applicable in the boxed region, where
defect interactions are negligible or readily accounted for as in
the case of associates.

This figure may be compared with the classification of oxides
proposed by O'Keeffe (5) and given in Table 1. Since the oxides
are grouped according to the concentration of defects which they
can accommodate, one can readily identify the systems to which
calculations of defect equilibria can most readily be applied.
Thus the calculations will become progressively less valid as one
moves from Group A down the Table; in this respect it is encouraging that the belief in a small range of nonstoichiometry for the
nitrogen ceramics would place them in the category where defect
chemistry is most valid.

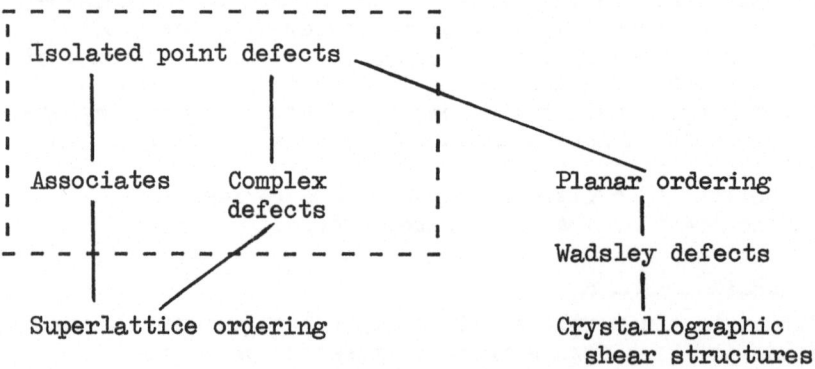

Figure 1.   A classification of point defects showing (inside
            the box) the categories to which the techniques
            of defect chemistry may be applied.

Class	Examples	Defect Concentration (%)
A	MgO, Al_2O_3	$< .01$
B	NiO, ZnO	$.01 - 1$
C	FeO, UO_2	> 1

Table 1. Classification of oxides (O'Keeffe)

Procedure for defect calculations

Defect chemistry treats defects as equilibrium species which may participate in chemical reactions and for which one may use the mass action law. Using Si_3N_4 as an example and ensuring (6) that the equations preserve mass balance, charge balance and site ratio, one may write as possible defect reactions

$$0 \longrightarrow e' + h^{\bullet} \qquad ; \text{ creation of electronic defects.}$$

$$0 \longrightarrow 3V_{Si}'''' + 4V_N^{\bullet\bullet\bullet} \qquad ; \text{ creation of ionic defects.}$$

$$2N_2(g) \longrightarrow 3V_{Si}'''' + 4N_N^x + 12h^{\bullet} \quad ; \quad \begin{array}{l} \text{interaction with the} \\ \text{gas phase.} \end{array}$$

The notation is that proposed by Kröger and Vink (3) (Table 2). The equations relate to four unknowns, namely the equilibrium concentrations of the defects, e', $h^{\bullet}$, V_{Si}'''' and $V_N^{\bullet\bullet\bullet}$; by writing mass action expressions for each of the equations and making an assumption concerning the defects predominantly responsible for the achievement of charge neutrality in the sample, one finds four equations for the four unknowns. The dependence of the concentrations on the nitrogen pressure, for example, may then be calculated and the result presented in diagrammatic form. Figure 2 is an example for Si_3N_4 in which it has also been assumed that a certain fixed oxygen partial pressure exists in the system and that aluminium is present as an impurity.

Although the diagram is hypothetical, it can act as a basis for interpreting the nitrogen pressure dependence of such processes as diffusion, sintering, and conductivity in silicon nitride. Such a procedure is in its earliest stages for the nitrogen ceramics, but it has made substantial advance in the case of oxides; it is consequently reasonable to review the oxide findings to see what progress has been made and to indicate the difficulties that can be encountered.

$$\text{Main symbol}^{\text{Superscript}}_{\text{Subscript}}$$

Main symbol = chemical nature of the defect (V = vacancy)

Subscript = lattice site of the defect (i = interstitial)

Superscript = electrical charge of the defect relative to that present in the perfect crystal (' negative; $\cdot$ positive; x neutral).

V_{Si}'''' $Si_i^{\cdot\cdot\cdot\cdot}$ Mg_{Si}'' Al_{Si}'

$V_N^{\cdot\cdot\cdot}$ N_i''' $O_N^{\cdot}$ N_N^{x}

Table 2. Notation for Point Defects (Kroger-Vink); Si_3N_4

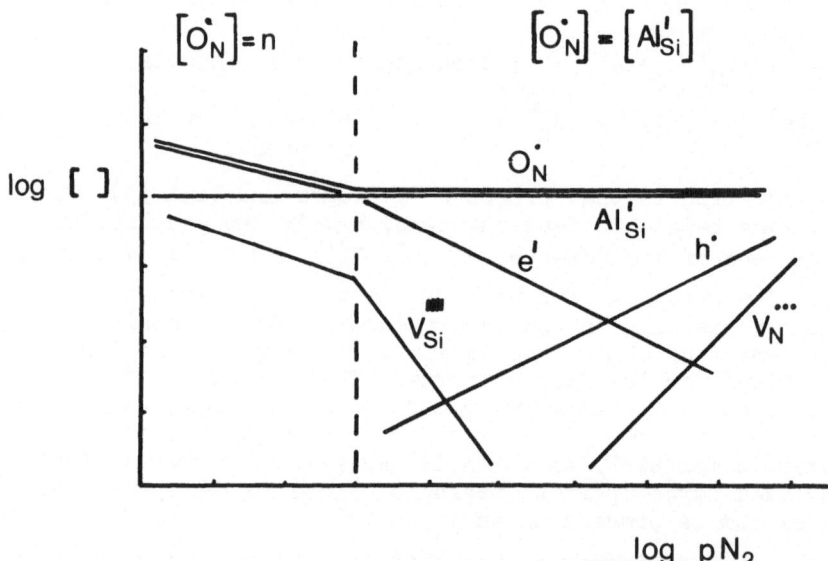

Figure 2. Equilibrium concentrations of defects in Si_3N_4 shown as functions of the nitrogen partial pressure. It is assumed that the defects are fully ionised, that a small constant oxygen pressure is maintained, and that aluminium is present as an impurity.

Defect equilibria in Oxides

The greatest progress has been made in treating oxides in Group B; this can be readily understood in that this group represents an optimum where concentrations are low enough for the methods of defect chemistry to be applied and high enough for the experimental measurement of these concentrations to be made with some precision.

As a result, a substantial body of evidence has appeared relating to diffusion, nonstoichiometry, and conductivity in oxides such as CoO, MnO, NiO and ZnO, and the understanding has reached the point where useful compendia (7) of data relating to defect processes in these oxides can be assembled.

For the type C oxides, progress has been less clear because of the important interactions between the large concentrations of defects and because of the alternative ways in which nonstoichiometry may be incorporated into the lattice. The existence of shear faulting in oxides, such as is found in TiO_2 (8), has called into question some of the earlier interpretations made on the basis of point defect models, and it will now be necessary to identify for each system the relative contributions made by point defects and by other forms of lattice disorder.

The type A oxides, which, as noted, seem the most likely analogue to the nitrogen ceramics, are systems in which the requirement for low concentrations is best met, but in which the experimental and interpretative difficulties have been found considerable. However, this is an area in which recent progress has been substantial, and it will be helpful to review both the problems and the approach which has had most success. Alumina is a suitable example for discussion.

Some difficulties

Three areas of difficulty can be identified in the study of defect processes in alumina; these can be briefly treated in turn.

(i) The role of impurities.

Since the native defect concentrations in oxides such as alumina are low owing to the large energies of formation, relatively low concentrations of impurities can have a dominating effect. For example, in the case of measurements (9) of the electrical conductivity of alumina containing iron at the 100 ppm level, it was originally assumed that the impurity would have negligible effect; defect models on that basis were proposed. More recently, it has been suggested (10) that the observations can be more consistently explained on the basis of impurity dominated behaviour. Thus to explain dependence of conduction (ionic) on the oxygen partial pressure in the form

$$\sigma = f(pO_2^{-3/16}),$$

it is suggested that oxidation of the impurity be written as

$$2Al_i^{\cdots} + 6Fe_{Al}' + \frac{3}{2} O_2(g) \longrightarrow 6Fe_{Al}^x + 2Al_{Al}^x + 3O_o^x + V_i^x,$$

whence $\left[Fe_{Al}'\right]^6 \left[Al_i^{\cdots}\right]^2 pO_2^{3/2} = K\left[Fe_{Al}^x\right]^6 \left[Al_{Al}^x\right]^2 \left[O_o^x\right]^3 \left[V_i^x\right].$

Since $\left[V_i^x\right]$, $\left[O_o^x\right]$ and $\left[Al_{Al}^x\right]$ are all constant, the required dependence can be derived if it is assumed that

$$\left[Fe_{Al}'\right] << \left[Fe_{Al}^x\right] \approx \left[Fe\right]_{Total} = \text{constant},$$

and that overall neutrality is achieved by $\left[Fe_{Al}'\right] = 3\left[Al_i^{\cdots}\right].$

As the example shows, it is necessary when treating materials of this type to take explicit account of the impurities present, and to include them in the analysis. Another clear example (11) of this need is in the analysis of sintering rates in TiO_2-doped alumina, where traces of a divalent impurity are believed to exert a pronounced influence.

(ii) The existence of alternative mechanisms

A common procedure for investigating the defects present in a system is to measure the dependence of some rate process such as sintering or electrical conductivity on temperature or on partial pressure, and then to relate the experimentally determined trends to predicted trends in defect concentrations. This procedure requires that the observed process be correctly analysed in terms of its rate controlling mechanism; the diversity of possible mechanisms has proved a source of major difficulty.

A clear example concerns electrical conductivity measurements on insulators such as alumina; here the process, that of charge transport, can occur by alternative mechanisms (bulk, surface, or gas phase conduction) and particular care must be taken to ensure that the intended mechanism is the actual one. It seems likely that many conduction measurements on alumina have in fact been made on the surrounding gas phase (12).

A second example relates to the existence of alternative paths in diffusional creep studies where the bulk grains and the grain boundaries can both act as diffusion channels. For a binary compound, four diffusion coefficients can therefore be recognised (Figure 3), and since the overall process will be controlled by the high flux pathway for the slower of the two species, there are 4 possible controlling mechanisms. Further, since the boundary fluxes are expected to be inversely proportional to the grain size (or directly proportional to the boundary area), the controlling

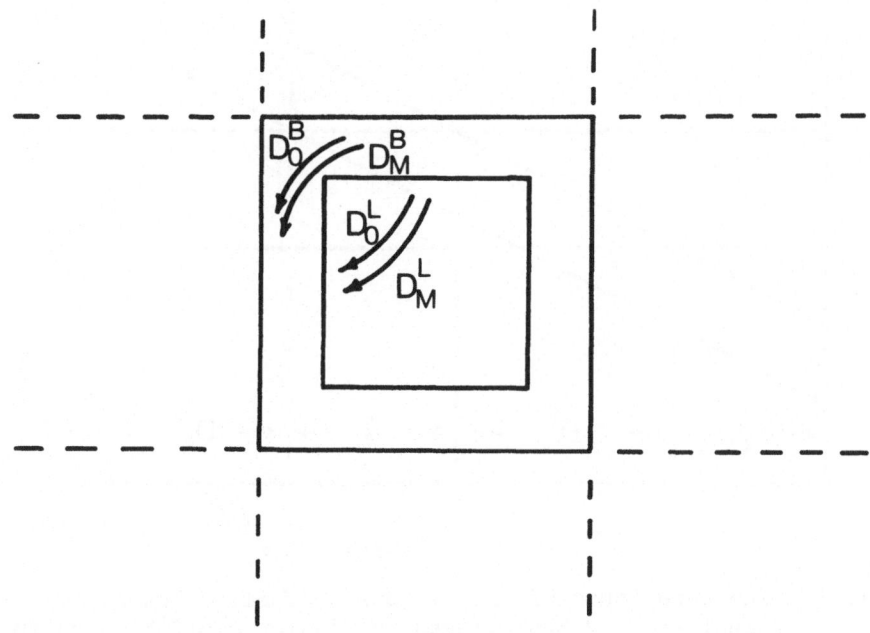

Figure 3. A schematic diagram of a grain in a polycrystalline ceramic showing diffusion paths for the anion (0) and cation (M) in the grain boundary region (B) and in the bulk lattice (L).

mechanism is expected to change with grain size. A possible scheme for alumina is given in Figure 4; experimental verification for such changes has been found for magnesium oxide (13). This choice of at least four controlling steps is common in ceramics; ambiguity of interpretation is a natural consequence.

Sintering and grain growth are examples of processes where an even wider choice of mechanisms is possible. Ashby (14) identifies six possible paths for mass transport in sintering which, for a binary compound, allows twelve possible controlling steps. Similarly, eleven possible paths can be identified (2) for grain growth.

The existence of alternative mechanisms makes it important that the true controlling mechanism be identified and that naive interpretations in terms of defect concentrations be treated with some caution. It is an advantage, where possible, to work with an experimental system or with a microstructure where circumstances particularly incline to a specific mechanism. Measurements on systems containing grain boundary liquid phases are an example.

194

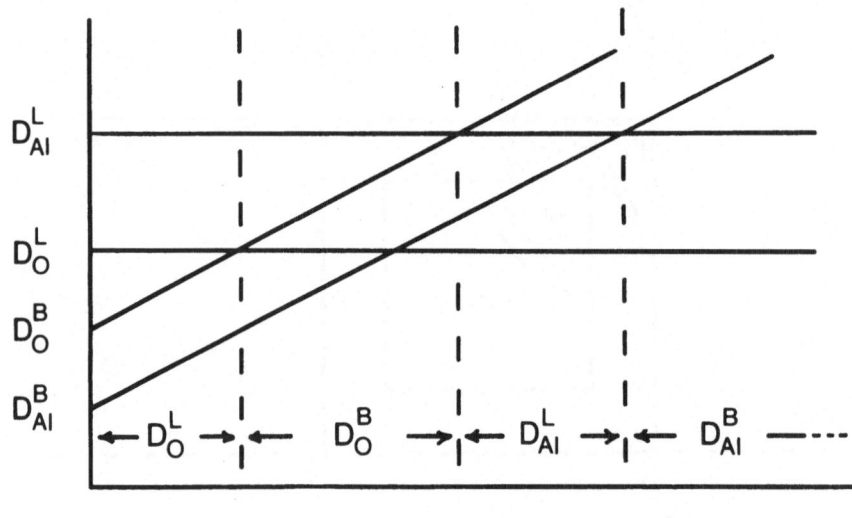

Figure 4. The dependence of the various diffusion fluxes
(indicated by the relevant diffusion coefficients) on
the reciprocal of the grain size for polycrystalline
Al_2O_3. The controlling diffusion mechanism is
indicated for each of the regions of grain size.

(iii) The interpretation of defect data.

The most commonly employed technique for studying defect
equilibria, namely, the determination of the partial pressure
dependence of some rate process, is itself open to a number of
difficulties.

In the first instance, the gradients of the lines on the
Brouwer diagram (e.g. Figure 2) are often similar for different
interpretations. An example is the sintering of FeO-doped alumina
(15); if the defect Fe'_{Al} is compensated by $Al_i^{\cdots}$, the rate of
sintering depends on $pO_2^{3/4}$, whereas for compensation by $V_O^{\cdots}$, the
rate depends on $pO_2^{2/3}$. Distinction between these two cases can
be difficult on the basis of sintering data.

A second problem lies in the fact that intermediate gradients
are expected in those partial pressure regions which lie between
the ranges where the different approximations for overall charge
neutrality apply. This problem also occurs under conditions
where more than one defect makes a substantial contribution to
the rate process.

A final problem lies in the inherent ambiguity of the defect calculations. As an example, the dependence of conductivity in alumina on $pO_2^{-3/16}$ described earlier can also be explained on the basis of the following reaction:

$$\tfrac{3}{2}\,O_2(g) + 2Al_i^{\cdots} + 6e' \longrightarrow 3O_o^x + 2Al_{Al}^x \; ;$$

whence $\left[O_o^x\right]^3 \left[Al_{Al}^x\right]^2 = Kn^6\left[Al_i^{\cdots}\right]^2 pO_2^{3/2}.$

The required dependence then follows from the assumption, $n = 3\left[Al_i^{\cdots}\right]$, as the approximation for neutrality. Other models (12) can also lead to this same dependence.

These problems of interpretation make it unusual for any single set of measurements to allow the drawing of convincing conclusions with respect to defect equilibria in the type A oxides. As a consequence, it has been found necessary to make measurements of different rate processes under a variety of conditions, and to look for interpretations that are consistent for the different processes. This procedure has had some success in the case of alumina.

An approach to the problem

A first requirement for the understanding of defect controlled rate processes is the availability of diffusion coefficient data; in the case of alumina, these are available both for the aluminium ion (16) (polycrystal) and for the oxygen ion (17) (single crystal and polycrystal). An immediate, and much used, clarification of the situation then arises from the fact that the coefficients are widely different in magnitude; consequently, even with approximately valid rate equations for processes such as creep and sintering, an indication of the controlling ion type is given.

Against this background, a number of measurements of sintering, creep, and conductivity have been made in alumina doped with deliberate additions of alivalent impurity (commonly FeO and TiO_2) chosen to enhance the concentration of one or other of the dominating defect types. The consistent pattern that can emerge from this approach when the different measurements reinforce one another has been found the surest way of building an understanding of defect equilibria in the material.

For alumina, in the grain size range commonly encountered, comparison of diffusion coefficients shows that the aluminium ion controls rates of sintering and creep; the necessary transport of oxygen is believed to be accomplished by way of high flux grain boundary paths (18). Measurements of sintering with TiO_2 additives show $V_{Al}^{\prime\prime\prime}$ as one dominant defect species (11), and measurements of sintering (15) and creep (19), with FeO additives, show $Al_i^{\cdots}$ as

another. Measurements also suggest that in typically 'pure'
materials, charge compensation is achieved by equivalent quantities of higher and lower valence cation impurities (5,11). The
present intimation is therefore that the defect structure in a
genuinely pure material will most probably be that of Frenkel dis-
order on the aluminium sublattice, $[Al_i^{\cdot\cdot\cdot}] = [V_{Al}^{'''}]$; experimental
observation of such intrinsic behaviour has however not yet been
achieved.

The conclusion that the aluminium ion controls mass transport
processes and that disorder is probably of the Frenkel type suggests
that substantial additions of any alivalent impurity should accelerate rate processes, and evidence to support this view exists.
Thus sintering rates appear to be accelerated both by TiO_2 addition
(20) and by MgO addition (21) under conditions where the additive
is in solid solution. Conversely, the argument suggests that where
processes have been suppressed, the additive has been added in
excess of the solid solution limit, and other rate controlling
mechanisms have been brought into play.

In making comparisons between the mass transport processes
and electrical conductivity data, a first requirement is to
identify the nature of the charge carrier or rather to separate
out the ionic contribution to the conductivity. This contribution
can then be compared with measurements of diffusion or other mass
transport processes. In this respect, it must be remembered that
conduction is attributable to the high flux ion in contrast to
the situation in sintering and creep.

For alumina, the ionic contribution has been identified (22)
and since the work relates to single crystals with no grain
boundary pathways, the carrier is as expected found to be the
aluminium ion. Good consistency is then found with other measurements in that data for samples containing FeO are, as noted earlier,
best interpreted (10) in terms of $Al_i^{\cdot\cdot\cdot}$ as the particular ionic
carrier. Thus broad agreement has now been achieved across the
full range of mass transport processes in this material. The picture is also consistent with defect related measurements such as
density variation (23) or absorption spectroscopy (22), which do
not involve atom migration but which are nonetheless guides to
the nature of the defects present.

In summary, it should be said that the progress of work on
alumina, while slow and not without difficulty, has had its rewards
in terms of improvements in processing and properties. It will
probably be necessary to follow a similar path for the nitrogen
ceramics particularly in view of the stringent requirements placed
on them by the intended applications.

Defects in nitrogen ceramics

If one attempts to follow the procedure outlined above for alumina by applying it to even the most studied of the nitrogen ceramics, Si_3N_4, it is clear that very few of the necessary measurements have been made.

Thus, for the undoped material, there have as yet been no reported measurements of diffusion coefficients. Electrical measurements on crystalline Si_3N_4 have been few (24) and have for the most part not attempted to identify the charge carriers. An exception lies in the work of Gorbatov (25) who has studied nitrogen concentration cells in the temperature range 300-700°C and found a large (50%) ionic contribution to the conductivity. This is a most promising line of enquiry, but questions remain concerning the nature of the sample and particularly concerning the nature of any grain boundary phases or oxide coatings present on the nitride grains.

One area where defect models can be employed is in the interpretation of oxygen solubility data for silicon nitride. Although originally seen (26) as evidence for a second phase of fixed oxygen content, the data can also be considered in terms of solution of oxygen as point defects. In accord with the latter viewpoint, defect concentrations can be calculated as a function of oxygen partial pressure over the Si_3N_4 sample; it is then a useful simplification that the several alternative models each based on a different assumption for the dominating defects in the low oxygen pressure state, such as n = p, or $3 \left[V_N^{\bullet\bullet\bullet} \right] = 4 \left[V_{Si}^{''''} \right]$ or $\left[O_N^{\bullet} \right] = \left[Al_{Si}^{'} \right]$, all show the same range $\left[O_N^{\bullet} \right] = 4 \left[V_{Si}^{''''} \right]$ as the oxygen content is raised. The diagram for the third alternative (assumptions as for Figure 2) is shown in Figure 5. On the basis of such a diagram, one can calculate the stability of the material under different oxygen and nitrogen pressures, if the dependence on these variables is known for one composition. A comparison of such a calculated line with the experimental data is shown in Figure 6; it may be seen that defect models are indeed capable of providing one explanation for the findings.

In other processes, such as creep (27) and hot pressing (28), the measurements have generally been made on heavily doped materials and rate controlling steps are consequently likely to involve the grain boundary phase. Under these conditions, defect equilibria within the silicon nitride grains are not effective in determining rates.

In sum, our knowledge of point defects in the nitrogen ceramics must be seen as virtually negligible at this stage. For the hot pressed materials, whose properties are most probably dominated by boundary processes, this level of ignorance is perhaps understandable. However, for reaction-bonded material, and for the single-phase sialon materials, it must be expected that

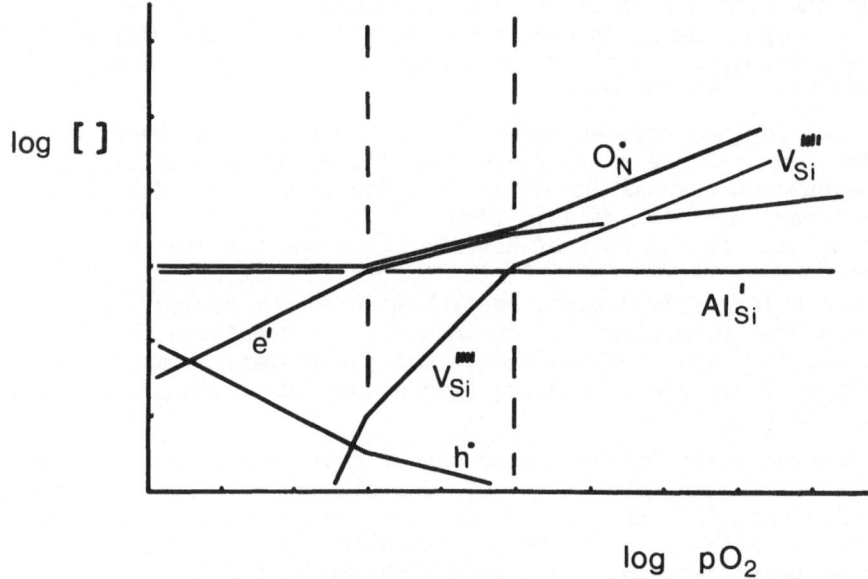

Figure 5. Equilibrium concentrations of defects in Si_3N_4 shown as functions of the oxygen partial pressure. It is assumed that the defects are fully ionised, and that a constant nitrogen partial pressure is maintained, equal to a value in the right-hand region of Figure 2.

properties will be found to depend on defect behaviour, particularly as the materials become more refined, and it is important that some basis for the understanding of this behaviour be prepared.

<u>Conclusions</u>

Their refractory nature, the absence of transition metals, and the indications of narrow ranges of nonstoichiometry in the sialons, all suggest that in the consideration of defect equilibria in the nitrogen ceramics, a situation similar to that of alumina will be encountered. This means that impurities will be important, that rate limiting processes will require careful identification, and that consistent measurements of a number of processes will be required on doped samples before a convincing picture can be assembled.

There is no doubt, however, that this picture will be required if confidence is to be placed in the performance of these materials in severe environments; for processing and high temperature properties will, on the basis of experience with oxides, depend upon defect parameters.

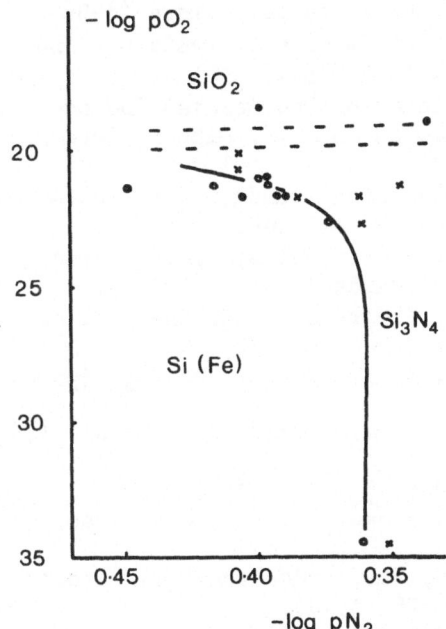

Figure 6. Phase stability data (26) for the N-O-Si system.
Crosses indicate Si_3N_4 (α or β), open circles Si, and
solid circles SiO_2. The solid line is the phase
boundary Si/Si_3N_4 calculated on the basis of the defect
model of Figure 5. It is assumed that the experimental
oxygen partial pressures are high enough to correspond
to those of the right-hand range of Figure 5.

It is therefore important that, alongside the exploration of
new systems of nitrogen ceramics, intensive study of selected
single phase materials be attempted. In this category, the need
for diffusion data seems paramount. The rewards for such study,
both in terms of scientific understanding and in terms of technological performance, are, on the basis of experience with oxides,
potentially high.

References

1. A historical account of the development of the subject is
   given in: F.A. Kroger, pp. 3-24 'Defects and Transport in
   Oxides', ed. M.S. Seltzer and R.I. Jaffee, Plenum Press (1974).
2. R.J. Brook, pp. 331-364, Treatise on Materials Science and
   Technology 9, 'Ceramic Fabrication Processes',
   ed. F.F.Y. Wang, Academic Press (1976).

3. F.A. Kroger and V.J. Vink, pp. 307-435, Solid State Physics 3, ed. F. Seitz and D. Turnbull, Academic Press (1956).

4. F.A. Kroger, 'The Chemistry of Imperfect Crystals', 2nd revised edition, North-Holland (1973).

5. M. O'Keeffe, pp. 57-96, 'Sintering and Related Phenomena', ed. G.C. Kuczynski, N.A. Hooton and C.F. Gibbon, Gordon and Breach (1967).

6. R.J. Brook, pp. 179-267, Electrical Conductivity in Ceramics and Glass A, ed. N.M. Tallan, Marcel Dekker (1974).

7. P. Kofstad, 'Nonstoichiometry, Diffusion, and Electrical Conductivity in Binary Metal Oxides', Wiley (1972).

8. L.A. Bursill and B.G. Hyde, Progr. Sol. St. Chem. 7, ed. H. Reiss and J.O. McCaldin, 177-253 (1972).

9. J. Pappis and W.D. Kingery, J. Am. Ceram. Soc. 44, 459-64 (1961).

10. B.V. Dutt and F.A. Kroger, J. Am. Ceram. Soc. 58, 474-476 (1975).

11. R.J. Brook, J. Am. Ceram. Soc. 55, 114-115 (1972).

12. R.J. Brook, J. Yee and F.A. Kroger, J. Am. Ceram. Soc. 54, 444-451 (1971).

13. R.T. Tremper, R.A. Giddings, J.D. Hodge and R.S. Gordon, J. Am. Ceram. Soc. 57, 421-428 (1974).

14. M.F. Ashby, Acta Met. 22, 275-289 (1974).

15. W. Raja Rao and I.B. Cutler, J. Am. Ceram. Soc. 56, 588-593 (1973).

16. A.E. Paladino and W.D. Kingery, J. Chem. Phys. 37, 957-962 (1962).

17. Y. Oishi and W.D. Kingery, J. Chem. Phys. 33, 480-486 (1960).

18. R.E. Mistler and R.L. Coble, J. Am. Ceram. Soc. 54, 60-61 (1971).

19. G.W. Hollenberg and R.S. Gordon, J. Am. Ceram. Soc. 56, 140-147 (1973).

20. R.D. Bagley, I.B. Cutler and D.L. Johnson, J. Am. Ceram. Soc. 53, 136-141 (1970).

21. J.G.J. Peelen, Fourth International Conference on Sintering and Related Phenomena, Notre Dame, May 25-27 (1975), U.S.A.

22. B.V. Dutt, J.P. Hurrell and F.A. Kroger, J. Am. Ceram. Soc. 58, 420-427 (1975).

23. J.J. Rasmussen and W.D. Kingery, J. Am. Ceram. Soc. 53, 436-440 (1970).

24. J.S. Thorp and R.I. Sharif, J. Mat. Sci. 11, 1494-1500 (1976).

25. A.G. Gorbatov and V.M. Kamyshov, pp. 53-74, 'Geterogennye Protsessy Uchastiem Tverd. Faz.', ed. V.M. Kamyshov, Sverdlovsk, USSR (1970).

26. I. Colquhoun, S. Wild, P. Grieveson and K.H. Jack, Proc. Brit. Ceram. Soc. 22, 207-227 (1973).

27. R. Kossowsky, D.G. Miller and E.S. Diaz, J. Mat. Sci. 10, 983-997 (1975).

28. L.J. Bowen, R.J. Weston, T.G. Carruthers and R.J. Brook, Ceramurgia Int. 2 (1976).

DISCUSSION (Schlichting, Singhal)

Vasilos: In work recently completed at our laboratory an estimated diffusion coefficient for Si in Si_3N_4 was obtained by stable isotope diffusion of Si^{29} through a dense polycrystalline sample of Si_3N_4. The diffusion coefficient obtained was 1.5×10^{-13} cm^2/sec at 1400°C. This number is based on analyses of but a few sections in a diffusion sample profile. Also the assumption is made that grain boundary diffusion is not important.
Reference: B.J. Wuensch & T. Vasilos "Self-Diffusion in Silicon Nitride". Final Report Contract N00014-73-C-0212, No. NR 032-537/ 8-18-72(471) (1975).

Brook: The value is close to those found in the hot pressing of MgO-doped Si_3N_4 which are described in another paper at this meeting. Since the hot pressing data are for diffusion in the grain boundary phase, the assumption made here, that the boundary is not important in Si diffusion in hot pressed samples, may perhaps be unwarranted. At any event, the appearance of diffusion data for the Si_3N_4 system is certainly welcome.

Clarke: Is it at all likely that there are crystallographic shear structures in covalent compounds such as Si_3N_4 and the sialons where (a) the bond angles and lengths are exactly determined, and (b) where M-M bonds are rare in comparison with M-X bonds (because of their much higher energy)?

Brook: The factors that lead a particular system to sustain non-stoichiometry by shear structure formation rather than by point defect formation have not been unequivocally identified; some thoughts are given by O'Keeffe. In general, the immediate co-ordination shell, e.g. the oxygen co-ordination of the titanium ion in reduced rutile, is unchanged on the shear structure, so point (a) may not be too much of a disadvantage. The Ti-Ti co-ordination is locally changed so that point (b) would apply; I would however have thought this interaction equally important in an ionically bonded system in view of the Coulombic repulsion term.

Lange: What are the difficulties in relating defect kinetics (for chemistry) directly to densification phenomena? Why is liquid phase sintering neglected when additives are used to sinter oxides and defect chemistry imposed?

Brook: As noted, the difficulties are three, namely (i) the interference of impurities, (ii) the need for correct identification of the rate controlling process, and (iii) ambiguities in the interpretation of defect data. The problem with liquid phase sintering is included in the second of these: if it is rate controlling, then the techniques of defect chemistry are not appropriate; if it is not, then they may be depending on whether the controlling process is solid state diffusion controlled or not.

KINETICS OF GAS-SOLID REACTIONS

M. Billy

Department of Ceramics[*], University of Limoges, France.

The purpose of this lecture is to review the main mechanisms occuring in any gas-solid reaction, in order to provide information about nitrogen systems. So we will consider the attack of a metal M by a gas G

$$M + G \rightarrow MG,$$

with reference to nitridation of metals by nitrogen, then the reaction of the type

$$M + GG' \rightarrow MG + G'$$

related to nitridation with ammonia, whilst oxidation of nitrides, oxynitrides or related compounds corresponds to a reaction

$$MG + G' \rightarrow MG' + G.$$

A reverse reaction of the first type is given by thermal decompositions of nitrogen compounds.

1. THE BASIC STAGES OF A GAS-SOLID REACTION

We will describe separately the early stages of formation of the reaction product, and then the development of the coating.
The initial step is gas <u>adsorption</u>. The adsorbed species will

* Supported in part by the C.N.R.S., ERA n°539 : "Nitrogen Ceramics".

dissolve into the solid and form a two-dimensionnal structure on the surface called <u>nucleus</u>.

For transition metals that may dissolve appreciable amounts of nitrogen, no detectable nitride formation will occur at this stage (1). But for other metals or oxidation of nitrides, the surface nuclei grow to cover the whole surface. This is the stage of <u>nucleation and growth</u>.

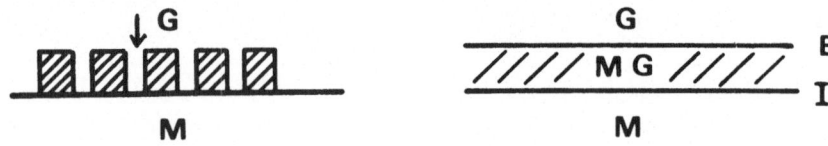

Fig. 1. The development of the coating
         a) Non protective scale ; b) Protective scale.

The second step is concerned with the growing of the film separating the gas and the original solid. Now two cases must be considered where the surface scale may be protective or not.

1.1 Porous coatings

If a porous scale is formed, the reacting gas can reach the initial solid through pores or cracks, so that the reaction is to be located at the inner phase boundary.

It is also the case, of course, when the solid product continuously evaporates (SiO for instance). But then, the gaseous diffusion through the reactive atmosphere or through the pores may become the lowest step of the overall kinetics. Such a case is illustrated by the thermal decomposition of silicon oxynitride and silicon nitride in the presence of silica (2).

1.2 Compact coatings

When a compact scale is formed, the reacting species are separated from each other ; further reaction is only possible if one of the reactants can diffuse through the surface layer of the scale. Other processes, however, are taking place at the same time.

First, the gas supply to the outer surface. It is in fact a gaseous diffusion that proceeds in order to balance the gaz consumption during the reaction. This kind of diffusion is normally faster than the other steps, provided a sufficient gas supply is available.

Secondly, we have the chemisorption step on the surface with splitting of the molecules.

Then the outer phase-boundary reaction E, that is a dissolution of the chemisorbed species with the formation of anions and
simultaneous electron exchange.

At the opposite interface I, we have the inner phase boundary reaction that corresponds to the transfer of the metal from
the metallic phase in the form of cations and electrons.

It must be kept in mind, at last, that all these processes
take place simultaneously, and so, it becomes obvious that the
slowest process determines the overall kinetics.

2. SORPTION KINETICS AND EQUILIBRIA

The gas molecules are fixed in certain positions at the surface
called adsorption sites. It is easy to predict that the adsorption rate v is both proportional to the number S of vacant sites
at a given time and to the number of molecules that impinge on
the surface, this number being directly related to the gaz pressure P :

$$v = kPS.$$

At the same time, we have also to consider the reverse phenomenon of desorption that implies a simple evaporation, the rate
of which is proportional to the number n of adsorbed molecules :

$$v' = k'n.$$

The effective rate of sorption can be expressed as the difference between v and v' :

$$\frac{dn}{dt} = kPS - k'n. \tag{2.1}$$

Now, each molecule, when attached to the surface, does not
cover just one site but a certain number that depends on the
cross-section b of the molecule and on the number S_o of adsorption sites per unit area. Thus, the disappearance of the sites
will be given by the expression :

$$- \frac{ds}{dn} = bS_o \text{ , or : } S = S_o(1-bn).$$

By inserting this value for S into the rate expression (2.1)
we get :

$$\frac{dn}{dt} = kPS_o(1-bn) - k'n. \tag{2.2}$$

Suppose the adsorption step is rate-determining. The mole-cules that impinge on the surface will be instantaneously taken away by the other reaction steps which are supposed to be faster. Thus, the number n of adsorbed species is always small and can be neglected in the rate expression (2.2). It follows

$$\frac{dn}{dt} = kPS_o,$$

so that the reaction rate remains constant and we have both a linear law with time and a linear dependence with pressure. Such a case seems to have been observed for the nitridation of silicon by nitrogen, according to Atkinson's data [3].

Suppose now the adsorption step is not rate-determining. An equilibrium will be then rapidly established and by setting dn/dt equal to zero in the rate equation (2.2) we get an expression of the form

$$n = \frac{\alpha}{1 + \beta P},$$

which is the well known expression of the Langmuir's isotherm. Pressure dependence, if any, is given by this expression, provi-ded of course the Langmuir's concept is appropriate.

3. NUCLEATION AND GROWTH PHENOMENA

This step is very fast in most cases. However, there are some in-teresting examples among nitrides where the nucleation and growth step is rate-determining ; for example the germanium nitride for-mation by reaction of Ge [4] or GeO_2 [5] with ammonia, and the hot corrosion of HfN by oxygen [6].

In such cases, the kinetic curves, like those represented on fig. 2 for the oxidation of HfN (degrees of reaction α have been plotted versus time at each temperature), have a sigmoïd shape which is always obtained when nucleation and growth is rate-deter-mining.

On the contrary, it is worth noting that a sigmoïd curve does not necessarily imply a nucleation rate-determining step. Oxida-tion of Zr carbonitride [7] or of Zr nitride [7] [8] are good examples of sigmoïd curves which have nothing to do with the nu-cleation and growth phenomena.

3.1 Rate of nucleation and growth

Suppose, at the very beginning of the reaction, that an atomic cluster or embryo has appeared on the surface of the solid. It follows from general nucleation theory that there is a critical

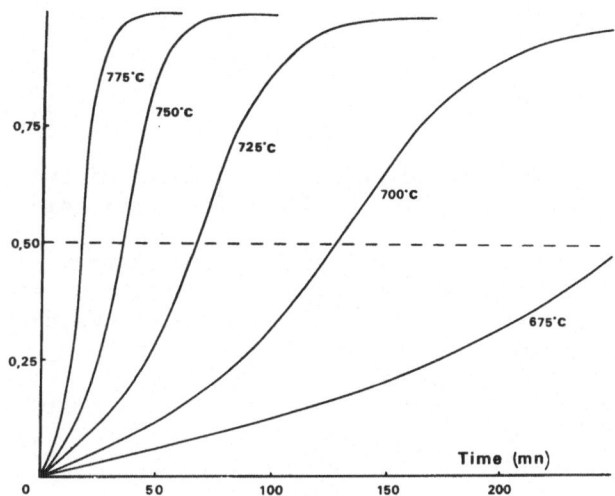

Fig. 2. Hot corrosion of HfN powders by oxygen (p = 32 torr).

size that must be exceeded for the cluster to become stable.
Below this critical volume, clusters can lower their free energy
by shrinking in size, whilst clusters above this critical volume
will normally grow and act as nuclei for further growth. Thus
there are two steps to be considered : the so-called nucleation,
that is the formation of nuclei from potential nucleation sites,
and the growth of these nuclei after they have been formed.

Now, suppose that a nucleus has been formed at a given time
t_0 ; its volume v at time t depends of course on both t_0 and t.
But, there is a certain number of nuclei that have been formed at
t_0 given by the nucleation rate $(dN/dt)_{t=t_0}$, where N denotes the
nuclei able to grow. And if we consider all the nuclei created to
the present time, their total volume V will be given by the following expression

$$V(t) = \int_0^t v(t,t_o) \left(\frac{dN}{dt}\right)_{t=t_o} dt_o \qquad (3.1)$$

which implies that the nucleation and growth steps cannot be
separated from each other.

The growth term can be generally expressed in a simple way
since the growth of the nuclei obeys a linear law of time in most
cases. For a given direction in space, a dimension L of a nucleus
will be of the form

$$L(t_1, t_0) = k \ (t-t_0),$$

and the volume :

$$v(t_1, t_0) = FL^\lambda = Fk^\lambda (t-t_0)^\lambda, \tag{3.2}$$

where F and λ are coefficients depending on the geometry of the nucleus ; λ takes values of 1, 2, 3 respectively for mono-, di- or tridimensional nuclei *.

By inserting the volume of each nucleus into the general expression (3.1) and in noting that the degree of reaction α is proportional to the total volume of the reaction product, we can obtain a rate law of the form

$$\alpha(t) = a \int_0^t (t, t_0)^\lambda \ (\frac{dN}{dt})_{t=t_0} \ dt_0 \tag{3.3}$$

which can be solved if we know the nucleation term, that is the nucleation mechanism involved.

3.2 Nucleation mechanisms

Among the many concepts that have been proposed up till now (9) to explain experimental results, we may distinguish two main groups relating to a step nucleation or to a chain nucleation.

3.2.1 <u>Step nucleation</u>. Suppose at the surface of the solid a number N_0 of potential nucleation sites, N of which are covered with growing nuclei. In a one-step process, the nucleation rate is proportional to the remaining potential sites N_0-N, that is :

$$\frac{dN}{dt} = k(N_0-N) \tag{3.4}$$

But the nucleation process generally implies a certain number of steps and not necessarily just one. The growing nucleus depends on a series of surface reactions giving step by step aggregates of binuclear, trinuclear and, finally, a p-nuclear species which is the only stable. In this model, the nucleation rate at each step is proportional to the concentration of the reacting species, as shown above. And if we assume, as a first approxima-

---

* For spherical nuclei, where $v = 4\pi R^3/3$, we have of course

$F = 4\pi/3$, $L = R$ and $\lambda = 3$.

tion, that each species remains in large excess with regard to the product, we obtain for the p-nuclear species a number n_p of the form

$$n_p = N = Kt^{p-1}. \tag{3.5}$$

On substitution of (3.5) into the general kinetic expression (3.3) of the nucleation and growth process, this results in a power law of the form

$$\alpha = At^n, \tag{3.6}$$

where A is a constant term and $n = \lambda+p-1$. This implies that the reaction becomes faster and faster.

This expression however is only valid for values of α up to about 0.2, because it has been established on the grounds of nuclei growing independently of one another. This is not true, of course, when nuclei come into contact as the reaction proceeds and also since a nucleus, when growing, covers a number of potential nucleation sites around it. Hence, the effective rate becomes lower than the theoretical one as the reaction proceeds. This may be expressed in writing that the experimental rate is of the form

$$\frac{d\alpha}{dt} = (1-\alpha) \frac{d\alpha'}{dt},$$

where α' denotes the theoretical degree of reaction. After integration, one gets a final expression

$$-Log(1-\alpha) = \alpha' = At^n,$$

or :

$$\alpha = 1 - \exp(-At^n), \tag{3.7}$$

which agrees with the observed sigmoid curves.

This expression corresponds to the theoretical power law (3.6) at the beginning of the reaction, for small values of α or t. Such an expression has been verified in the case of the thermal decomposition of calcium and barium azides [10]. Here the reaction obeys a 6 power of time and as the nuclei have been shown to be spherical ($\lambda = 3$), this results in a p-value of 4. In other words, nuclei have to accumulate 4 Ba atoms before being stable and growing.

3.2.2 <u>Chain nucleation</u>. Another concept of nucleation introduced by Garner and Hailes [11] many years ago is related to a chain process.

In this model, each growing nucleus leads to a certain number of new ones and so on. The nucleation rate may then be expressed by a formula of the type

$$\frac{dN}{dt} = k(N_O-N) + k_cN,$$

where the first term denotes a one-step initiating nucleation according to (3.4) and the second the chain process. The integral expression is exponential in time :

$$N = \frac{kN_O}{K} \left(\exp(Kt)-1\right)$$

where $K = k_c-k$. Finally, the nucleation and growth kinetics will follow a general expression of the form

$$\alpha = A \left(\exp(Kt)-1\right) + Bt^\lambda - Ct^{\lambda-1} + Dt^{\lambda-2},$$

in which the exponential term becomes predominant as the reaction proceeds (increasing time), so that :

$$\alpha = A \exp(Kt). \tag{3.8}$$

Here again, this law is appropriate to describe phenomena for α values up to 0.2 since the overlapping of the nuclei and the desappearance of potential sites during nucleation and growth are not taken into account. In order to describe the experimental results, we could always introduce a correction factor like the one mentioned above for the step mechanism, but a more classical way is to use the well known Prout and Tompkins law

$$\text{Log} \frac{\alpha}{1 - \frac{\alpha}{2\alpha_i}} = kt + ct, \tag{3.9}$$

where α_i corresponds to the α value at the inflection point of the experimental curve. This law results, in fact, from an empirical expression for the rate

$$\frac{d\alpha}{dt} = k\alpha(1 - \frac{\alpha}{2\alpha_i}),$$

which matches any sigmoïd curve, and not necessary a chain mechanism. Thus, in order to prove such a mechanism, we need further information ; the splitting open of the reacting solid is one of them.

In the case of germanium nitridation (4) or Hf nitride oxidation (6), for example, a swelling of the solid is observed as its specific area increases by a factor of 10 to 20. Besides, an activation energy of 90 kcal/mol has been found for Ge nitridation which can be compared with the energy of germanium vaporisation (87 kcal) and, thus, with the cohesion energy of the germanium network. This implies that nucleation results in a splitting of the solid with the formation of new surfaces, new nucleation sites, and so on.

4. FORMATION OF NON-PROTECTIVE COATINGS

In most cases, the nucleation process is so fast that a monolayer of the solid product has been formed as soon as the reaction starts. The basic stage is that of film growth.

Suppose a planar specimen and a porous coating to be formed. If we assume a linear growth rate for the nuclei, the scale formation in a direction normal to the initial surface will obey a linear law of time

$$\alpha = Kt. \tag{4.1}$$

Let us consider now the case of powder specimens of spherical grains with an initial radius R_0 which becomes R at time t. The degree of reaction α can be expressed in comparing the remaining volume V with the initial one

$$\alpha = \frac{V_0 - V}{V_0} = 1 - \frac{R^3}{R_0^3} .$$

α then depends on a cubic law of the radius, and if this decreases linearly with time

$$R = R_0 - Kt,$$

the kinetic law takes the form :

$$G(\alpha) = 1-(1-\alpha)^{1/3} = \frac{K}{R_0} t, \tag{4.2}$$

where the first term is directly proportional to time.

Such a law has been found for the oxidation of γ aluminium oxynitride powders (12), where the overall kinetics are governed by a rate-determining step located at the inner phase boundary. Fig. 3 precisely shows that the variations of the first member of the previous equation (4.2) are linear with time up to $G(\alpha)$ values of about 0.8 corresponding to α values of 0.9. It follows that the coating is never protective during the whole course of the reaction.

In most cases, however, the coating may become protective, due to crystal growth phenomenon or a pore closure mechanism as

212

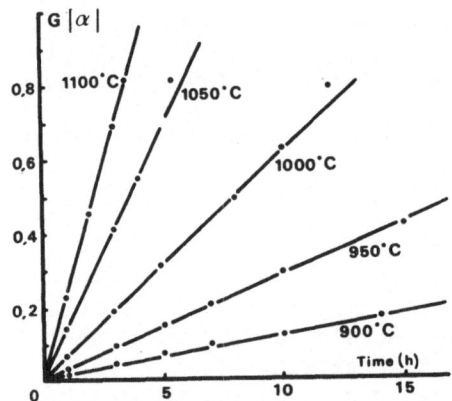

Fig. 3. Kinetic behaviour of γ aluminium oxynitride powders
in oxygen (p = 32 torr).

the reaction proceeds. Examples of this have been found for the
hafnium nitridation (1) and for oxidations of Ge nitride (13),
Si nitride and oxynitride (14). Here, the so-called "linear
step" is only observed at the beginning of the reaction (during
half an hour or less). That is the reason why it has escaped
notice in many kinetic studies. This is a pity because interesting information can be drawn from this early stage of the reaction. About the oxidation of Si nitride (14), for example, we
could conclude that the α phase is more reactive than β and that
the initial process is related to the breaking down of Si-N bonds.
        Let us now consider the different case of a compact coating
that becomes porous with time as a result of crystallisation or
internal stresses. Here, a linear law with time will be obtained
as a limit and the overall kinetics will accelerate. The oxidation
of ZrN (8) is an example though, in this case, the true mechanism
is slightly different because the rate-determining step is located
at the outer phase boundary.
        In connection with this subsequent formation of a porous
coating, we have to consider <u>whiskers</u> or other shapes of crystals,
such as those of $\alpha-Si_3N_4$ or Al_2O_3 observed during oxidations of
AlN (15) and sialons (16) respectively. What role do they play on
the course of the reaction ? Though it is really difficult to
answer this question, it can be said that :
        - whiskers are never in contact with the reacting solid ;
        - their growth never occurs during the initial stages of the
reaction ; so they are probably a result of effects and changes in
the coating, namely surface evaporation/condensation processes or
stresses through the film ;
        - their volume is always of minor importance with regard to
the total volume of the coating.

For these reasons the course of the reaction must not be really modified by their formation.

5. FORMATION OF PROTECTIVE COATINGS

According to the old criterium of Pilling and Bedworth, a layer might be adherent and pore free if the ratio of the equivalent volumes of the product and the reacting solid were greater than one.

This ratio is probably important but not decisive, as there are many cases of porous layers where the expansion coefficient is greater than one. On the contrary, we know cases where, despite a coefficient less than unity, there is a thin protective coating close to the reacting solid. A good example of this was given by Streiff (17) for the Ca/N system. Thus, we cannot predict whether a given product will be a protective coating or not.

Furthermore, a porous scale can become protective as mentioned above. In the pore closure model proposed by Evans (18), the reaction product is supposed to produce a blocking of gas paths along the pores. Then the reaction rate, which is proportional to the pores still open, will decrease until a compact scale is formed. If both reactants are unable to diffuse through this coating, the reaction may nearly stop and the overall kinetics will be either asymptotic or logarithmic. Such a case seems to be observed for silicon nitridation below 1300°C (19) (3). Another example has been recently observed for the oxidation of sintered specimens of silicon oxynitride using BeO as additive (20) in the temperature range of 1400-1500°C.

5.1 The parabolic law

Let us consider now a compact scale through which the reactants can diffuse. Diffusion is given by the well known Fick's first law

$$\frac{1}{S} \frac{dQ}{dt} = -D \frac{\partial c}{\partial x} \text{ ,}$$

where the flow rate of the diffusing species across a plane surface S is proportional to the concentration gradient normal to the plane ; D is the diffusion coefficient and x the thickness.

$$\frac{dQ}{dt} = D(C_E - C_I) \frac{S}{x} = k \frac{S}{x} \text{ ,}$$

If one keeps in mind the fact that the degree of reaction α is both proportional to the thickness x and to the gas uptake Q, the reaction rate will be of the form

$$\frac{d\alpha}{dt} = k' \frac{S}{\alpha} ,$$

which implies the classical parabolic law :

$$\alpha^2 = Kt. \tag{5.1}$$

Many nitridations have been found to match this parabolic law. Among them, let us quote nitridation of titanium and zirconium $(21)(22)$, the nitridation of tantalum (23) and uranium (24). The formation of vanadium mononitride from the lower nitride V_2N still obeys a parabolic law (25). Other examples are oxidations of silicon nitride $(14)(26)(27)$, silicon oxynitride (14) and also titanium mononitride (28) for which, however, results disagree with the linear dependence with time obtained at Limoges (29).

Now, it is important to notice that a rate-determining diffusion does not imply necessary a parabolic law. In the Ca/N system (17) for instance, there is a thin protective layer near the metal and a porous outer one. As the reaction proceeds, the thickness of the compact nitride remains constant, as does the flow rate of the diffusion species according to the basic equation $dQ/dt = kS/x$. The overall kinetics then obey a linear process whilst diffusion is rate-determining.

5.2 The case of powder specimens

Here again, the parabolic law is not verified as the first Fick's law must take into account the sphere surface of each grain. The concentration gradient then depends on R inside the diffusion space

$$\frac{dQ}{dt} = -D \ (4\pi R^2) \ \frac{dc}{dR} .$$

By separating variables for a given time and using the boundary limits of the coating

$$(\frac{dQ}{dt}) \int_{R_I}^{R_E} \frac{dR}{R} = -4\pi D \int_{C_I}^{C_E} dc ,$$

and by integrating, one finds the expression of the rate flow

$$\frac{dQ}{dt} = 4\pi k \ (\frac{1}{R_I} - \frac{1}{R_E}) , \tag{5.2}$$

where the radii R_I and R_E depend on time according to the following expressions :

$$R_I = R_o(1-\alpha)^{1/3}$$

$$R_E = R_o\left(1+(\Delta+1)\alpha\right)^{1/3}$$

where R_o denotes the initial mean radius of each grain and Δ the expansion coefficient of Pilling and Bedworth. Inserting these expressions into the rate equation (5.2), then integrating, one obtains the following kinetic law :

$$\frac{\Delta}{\Delta-1} - (1-\alpha)^{2/3} - \frac{1}{\Delta-1}\left(1+(\Delta-1)\alpha\right)^{2/3} = \frac{2V_o}{R_o^2}\,kt, \qquad (5.3)$$

where V_o means the molar volume of the reacting solid.

This expression, often called Carter's equation (30) was first proposed in fact by Valensi (31) as early as 1936. It has been reported to describe nitridation of silicon powders by Mrs Blegen (32), though her results disagree with Atkinsons's (this author found a linear dependence on time excluding diffusion as a rate-determining step). Another application is concerned with the oxidation of the δ sialon phase (16), that is the 15R phase of Jack and Thompson (33). Oxidation kinetics of aluminium nitride (15) powders could be matched to Valensi's equation provided that some correction factors were introduced to describe the powder changes in specific area during the oxidation.

6. CONCLUSIONS

Kinetics laws, as well as pressure and temperature dependences, are heavily dependent on the microstructure of the coating. Its role is always decisive, the actual mechanism resulting from the extent of the protective character of the coating, as well as from the alteration in this coating.

As far as nitrogen systems are concerned, it can be concluded that there is still much to do, particularly in comparison with oxides. It is always surprising to note, for instance, the lack of kinetic studies about nitridation by ammonia or about thermal decomposition. But there are many other problems to be solved :
- the mechanism of silicon nitride formation which is still rather uncertain ;
- the role of SiO in any reaction involving the subsequent formation of this volatile compound ;
- the influence of nitrogen releasing from nitrides or nitrogen ceramics during oxidation when a compact coating of oxide is formed.

216

It would be admirable, if even one of these problems could be partly solved during this meeting.

REFERENCES

1. M. Billy and B. Teyssedre, Bull. Soc. chim., 1973 (5), 1537.
2. P. Lortholary and M. Billy, C.R.A.S. Paris, 278C, 1974, 1342.
3. A. Atkinson, in Trans. of the Brit. Ceramic Soc., 74, 1975, XV.
4. J.C. Labbe, F. Duchez and M. Billy, C.R.A.S. Paris, 273C, 1971, 1750.
5. J.C. Labbe and M. Billy, C.R.A.S. Paris, 277C, 1973, 1137.
6. M. Billy and B. Teyssedre, C.R.A.S. Paris, 276C, 1973, 421.
7. H.F. Ayedi, Thesis,Grenoble, June 1976.
8. J. Desmaison, M. Billy and W.W. Smeltzer, Proc. of the VIIIth Int. Symp. on React. of Solids, Göttenburg, June 1976.
9. B. Delmon, Introduction à la cinétique hétérogène, Technip, Paris, 1969 ; W.E. Garner, Chemistry of the Solid State, Butterworths, London, 1955.
10. A. Wischin, Proc. Roy. Soc., A172, 1939, 314.
11. W.E. Garner and H.R. Hailes, Proc. Roy. Soc., A139, 1933, 576.
12. P. Goursat, P. Gœuriot and M. Billy, Materials chemistry, 1, 1976, 131.
13. J.C. Labbe, Limoges (un.published results).
14. P. Goursat, P. Lortholary, D. Tétard and M. Billy, Proc. of the 7th Int. Symp. on React. of Solids, Chapman and Hall, Bristol, 1972, 315.
15. D. Tétard and M. Billy, Proc. of the 25th Int. Meeting of the Soc. chim.-phys., Dijon, 1975, 512.
16. D. Brachet, Thesis, Limoges,June 1976.
17. R. Streiff, Thesis, Nancy, April 1967.
18. U.R. Evans, the Corrosion and Oxidation of Metals, Edward Arnolds Ltd, London, 1960.
19. J. Lamure and M. Billy, C.R.A.S. Paris, 245, 1957, 1951.
20. M.H. Négrier, Thesis, Limoges, June 1976.
21. J.P. Bars, E. Etchessahar and J. Debuigne, Proc. of the 25th Int. Meeting of the Soc. chim.-phys., Dijon, Elsevier, 1975, 124.
22. E.A. Gulbransen and K.F. Andrew, Metals Trans., 185, 1949, 741.
23. K. Östhagen and P. Kofstad, J. Less-Common Metals, 5, 1963, 7.
24. M.W. Mallett and A.F. Gerds, J. electrochem. Soc., 102, 1955, 292.
25. S.K. Iyer and W.L. Worrel, Proc. of the 7th Int. Symp. on React. of Solids, Chapman and Hall, Bristol, 1972, 294.
26. I. Fränz and W. Langheinrich, Proc. of the 7th Int. Symp. on React. of Solids, Chapman and Hall, Bristol, 1972, 303.
27. M. Mitomo and J.H. Sharp, Yogyo-kyokaï-shi, 84, 1976, 33.
28. A. Münster and G. Schlamp, XVIe Cong. chim. pure et appl., Paris, 1957, 691 ; Z. für Phys. chem., 13, 1957, 59.
29. J. Desmaison and P. Lefort, Limoges (to be published).
30. R.E. Carter, J. chem. Phys., 34, 1961, 2010.

31. G. Valensi, C.R.A.S. Paris, 202, 1936, 309.
32. K. Blegen, in J. of the Brit. ceram. Soc., 74, 1975, XV.
33. D.P. Thompson, Crystal structures of the 8H and 15R "AlN-polytypes" in the Si-Al-O-N system (this meeting).

DISCUSSION (Schlichting, Singhal)

Jelacic: Have you observed in any solid-gas reaction an induction period? I observed this effect in the nitridation of aluminium.

Billy: An induction period is observed of course when the nucleation and growth step is rate determining. This can be found also in cases of surface contamination and, in this respect, the nitriding of aluminium must be quite dependent on the presence of surface alumina.

Singhal: What is the stability of aluminium oxynitride? This compound has been observed to form in many sialons at temperatures as high as 1750°C, as is shown in the Si_3N_4-SiO_2-AlN-Al_2O_3 phase diagram.

Billy: Υ-aluminium oxynitride has a good stability in inert and N_2 atmospheres, but a poor resistance to oxidation above 900°C.

Lange: The conditions causing active and passive oxidation must be included in considering the nitriding kinetics of Si. Gas phase reactions, e.g. $3SiO + 2N_2 \rightarrow Si_3N_4 + 3/2\ O_2$, would thus be important during nitridation.

Billy: Gas phase reactions must not be neglected during nitridation of silicon when the nitrogen contains small amounts of oxygen. This explains the formation of some silicon oxynitride, via SiO_2, according to a reaction of the type:

$$3Si + SiO_2 + N_2 \rightarrow Si_2N_2O.$$

SOLID STATE DIFFUSION IN BINARY CERAMIC COMPOUNDS

W.W. Smeltzer and J.G. Desmaison*

Department of Metallurgy and Materials Science,
McMaster University, Hamilton, Ontario, Canada.
L85-4M1.
*Laboratoire de Chimie Minérale et Cinétique
Hétérogène, Université de Limoges, 87100 Limoges, France.

ABSTRACT. A brief survey is presented of experimental and theoretical work pertinent to diffusion in carbides, oxides and nitrides. Phenomenological theory is utilized to relate the self, tracer and chemical diffusion coefficients. Diffusional properties of oxides can be correlated in several instances by models involving Frenkel and Schottky point defect disorder. Although metal carbides exhibit broad ranges of nonstoichiometry, point defect models have proven useful for elucidating diffusion mechanisms. Diffusion studies on nitrides are to the present a subject of limited interest; a tabulation is given of measured diffusivities for these solids.

1. INTRODUCTION

Mass transport in carbides, nitrides and oxides is an important parameter in assessing the feasibility of those refractory materials for high temperature technology because the rate determining step in sintering, grain growth, creep and the formation of reaction product layers invariably involves boundary and volume diffusion. Our purpose is to review methods and results which are appropriate for elucidating diffusion mechanisms. Most work has been completed on oxides with the consequence that volume diffusional properties have been shown to correspond to predictions from point defect models. These models have also been applied successfully to transition metal carbides. Present knowledge with respect to nitrides is qualitative and extremely limited but diffusion studies appear to offer a promising avenue of future work.

The volume diffusion coefficients most commonly used are the self (D), tracer D^T) and chemical ($\tilde{D}$) diffusivities. We present the phenomenological equations defining these coefficients. Since diffusion occurring under a chemical potential gradient (ambipolar diffusion) is most important in materials processing, equations relating $\tilde{D}$ to D and point defects are emphasized. Furthermore, investigations have often been completed on polycrystalline materials and, in these cases, one is obliged to consider the boundary diffusion coefficient (D^B) and an effective diffusion coefficient (D^E) to describe mass transport.

2. DIFFUSION EQUATIONS

2.1 Self and Tracer Diffusion

The self-diffusion coefficient according to random walk theory for an isotropic solid is

$$D = \frac{1}{6} \nu a^2 \qquad (2.1)$$

where ν is the atomic jump frequency and a is the jump distance. This diffusivity is commonly determined by a tracer isotope. Correlation effects, however, exist if a defect must be next to the tracer atom for it to execute a jump. Accordingly tracer and self-diffusion coefficients are related by

$$D^T = fD \qquad (2.2)$$

where f is the correlation factor determined by the crystallo-graphy of the solid and the diffusion mechanism. If migration of an ionic species occurs independently under the influence of an applied electric field, its mobility B_i is related to its self diffusivity by the Nernst-Einstein equation

$$B_i = \left| \frac{q_i}{RT} \right| D_i \qquad (2.3)$$

where $|q_i|$ is the species charge. Thus, the correlation factor may be determined by comparing the diffusivities obtained by tracer and electrical measurements.

D_i is related to the diffusivity of the appropriate defect upon knowing the diffusion mechanism. Interstitial and self diffusivities are identical. When migration occurs by vacancies, the concentrations and mobilities of atoms and vacancies are related by

$$c_i B_i = c_v B_v \qquad (2.4)$$

and from (2.3)

$$D_i = \frac{D_v c_v}{c_i} \simeq D_v c_v \qquad (2.5)$$

In a similar but more complicated manner, D_i can be related to the defect diffusivity for mechanisms involving divacancies, interstitialcies and clusters. Moreover, diffusivity terms in D are additive if more than one mechanism is operative. D values usually obey an Arrhenius equation

$$D = D^o \exp - E/RT \qquad (2.6)$$

where D^o and E are of the form,

$$D^o = a^2 \nu^o \exp(\Delta s_f + \Delta s_m)/R \; ; \quad E = \Delta H_f + \Delta H_m \qquad (2.7)$$

Here ν^o is the fundamental vibration, ΔS_f and ΔS_m are entropies, and $\Delta H_f + \Delta H_m$ are the enthalpies of defect formation and migration. The obeyance of D to different Arrhenius equations is attributed to changes in the diffusion mechanism.

2.2 Chemical Diffusion

D is defined by Fick's laws

$$J = - \tilde{D} \frac{dc}{dx} \; ; \quad \frac{d}{dx} \tilde{D} \frac{dc}{dx} = \frac{dc}{dt} \qquad (2.8)$$

where J is the mass flux and c is a component concentration. $\tilde{D}$ was shown to be a composite of the D's of both components by Wagner [1] for ionic and semiconducting compounds and by Darken [2] for metallic solids. Both formalisms, which invoke the phenomenological diffusion theory of Onsager [3] , are utilized to interpret diffusion mechanisms in ceramics.

According to this theory,

$$J_i = \Sigma_k -L_{ik} X_k \qquad (2.9)$$

where the proportionality constants L_{ik} directly relate a species flux to an electrochemical or chemical potential gradient X_i. Assuming that $L_{ii} \gg L_{ik}$ eq. (2.9) reduces to

$$J_i = -L_{ii} X_i = -c_i B_i X_i \qquad (2.10)$$

Wagner assumed that charged species in a binary compound migrate under the influence of the electro-chemical potential gradient

$$J_i = -c_i B_i \left\{ \frac{du_i}{dx} + q_i \frac{dV}{dx} \right\} \tag{2.11}$$

where u_i is its chemical potential and $q_i V$ is the electrical
potential. The flux was assumed to proceed by ambipolar diffusion
(mass transport under no net flow of electric current).When the
total flux is compared to that given by Fick's law (eq.2.8),the
expression of $\tilde{D}$ for a semi-conductor is

$$\tilde{D} = \left\{ \frac{Z_M}{|Z_X|} D_M + D_X \right\} \frac{c}{RT} \frac{du_x}{d\tilde{c}} \tag{2.12}$$

where the subscripts M and X denote the cation and anion,Z is a
valency,c is a total concentration of one of the above species and
$\tilde{c}$ is the excess concentration.

Darken considered the chemical potential gradient as the
driving force for diffusion in binary alloys.Defining an intrinsic
diffusion coefficient as follows

$$J_i = -D_i^I \frac{dc_i}{dx} = -B_i c_i \frac{du_i}{dx} \tag{2.13}$$

then $$D_i^I = D_i c_i \frac{d\ell n a_i}{dc_i} \tag{2.14}$$

From an analysis of the Kirkendall effect in a diffusion couple,
Darken derived

$$\tilde{D} = N_X D_M^I + N_M D_X^I = N_X D_M + N_M D_X \left\{ 1 + \frac{d\ell n \gamma_M}{d\ell n N_M} \right\} \tag{2.15}$$

where the concentration is expressed as a mole fraction N and γ
is an activity coefficient.

Expressions (2.12) and (2.15) relating $\tilde{D}$ to D_M and D_X are
representations of an experimentally determined inter-diffusion
coefficient used to interpret diffusion mechanisms.

2.3 Boundary and Short-Circuit Diffusion

Atom transport through grain boundaries is much more rapid than in
the lattice of a solid at temperatures less than ½ its melting
point. D^B is defined using Fick's laws to describe this preferential
diffusion. A combination of boundary and lattice diffusion leads

to short-circuit diffusion and uniform penetration of atoms when the distances between boundaries are small. Assuming random walk of atoms, Hart 4 derived the following expression for the effective diffusion coefficient

$$D^E = D(1-f) + D^B f \tag{2.16}$$

where f is the fraction of diffusion sites in boundaries.

3. EXPERIMENTAL METHODS

D^T is determined by constructing a diffusion couple with one of the identical sections containing an isotope in small concentration which is radioactive or activated by irradiation after the diffusion anneal for its detection. D^T for N or O in compounds is often most readily determined by measuring the decrement of the isotopic concentration in the gas phase. Appropriate equations satisfying these types of measurements are given by Crank [5], De Poorter and Wallace [6], and Carman and Haul [7]. One must be able to assign the proper correlation factor to obtain D from D^T. An isotope mass effect influences atom migration. Peterson |8| has critically reviewed these factors and the methods by which the diffusion mechanism can be correlated to defects from the inter-related values of the correlation factor and isotope effect. D can be determined by electrical measurements; Wagner [9] has shown that this method is applicable even when migration of charged species are influenced by interaction effects. Mossbauer spectra have been used to relate atomic jump frequencies to D [10].

Diffusion of one species usually predominates in binary compounds such that the sublattice of the second component can be regarded as fixed. Accordingly eq (2.12) and (2.15) for chemical diffusion of metal reduces to

$$\tilde{D} = \frac{Z_M}{|Z_X|} D_M \frac{c}{RT} \frac{du_X}{d\tilde{c}} \; ; \quad \tilde{D} = N_X D_M \left\{ 1 + \frac{d\ln\gamma_M}{d\ln N_M} \right\} \tag{3.1}$$

The classical method for determining D is based upon interpreting the concentration profile by the Matano method [6]. the profile being determined by chemical or physical means. When it is impractical to determine this profile, the compound can frequently be exposed to an atmosphere at constant composition and the kinetics of weight or electrical conductivity changes interpreted by the appropriate diffusion equations [11]. Two additional techniques have been used but the $\tilde{D}$ evaluations often are open to question due to the influence of boundary diffusion in the polycrystalline materials: growth of a compound layer from reaction of the metal with the X component and sintering kinetics [6,12].

Mathematical solutions are available for evaluating the magnitude of D^B relative to D[13,14]. Knowing the values of D^B and D,
D^E can be estimated knowing the grain size.

4. APPLICATION OF DIFFUSION EQUATIONS

At the stoichiometric composition,defects are present either as
equivalent concentrations of vacancies in each component sublattice
(Shottky disorder) or of interstitials and vacancies in the same
sublattice (Frenkel disorder). Various combinations of point
defects,however,describe compositions, $c = \delta$, ranging away from
the stoichiometric formula : hypostoichiometric compounds,$MX_{1-\delta}$,
result from metal interstitials and/or non-metal vacancies; conversely hyperstoichiometric compounds, $MX_{1+\delta}$ result from non-metal
interstitials and/or metal vacancies. Thus diffusion on either side
of the MX composition will correspond to different defect mechanisms.

The problem of relating mass transport to point defects has
been partially successful in several instances for transition
metal oxides, carbides and nitrides existing as MX type compounds
in the rock-salt (B_1) structure. Oxides often exhibit diffusion
predominantly by one species, the defect concentration and type
being characteristically related to the chemical activity of the
ambient atmosphere. This problem is aptly demonstrated by
reference to properties of NiO. The carbides are extremely stable
exhibiting defect structures established by their preparatory composition. We illustrate diffusional properties of these solids by
referring to investigations on TiC and UC.Diffusion and defects in
nitrides appear to correspond to those of oxides and carbides but
sufficiently extensive results are only available for interpreting
diffusion in UN.

4.1 Metal Oxides (NiO)

The defect and diffusional properties of oxides are treated in the
texts by Kröger [15] and Kofstad [12]. Interaction between defects
predominate when appreciable concentrations exist,and it is only
for very small ranges of nonstoichiometry,$\delta \sim 10^{-3}$ to 10^{-5},that
simple models are applicable. Defect concentrations,electrical and
diffusional properties of nickel oxide have been interpreted in
terms of cation vacancies obeying the following reaction :

$$\tfrac{1}{2} O_2 = O_O^X + V_{Ni}^m + mh^{\cdot} \tag{4.1}$$

where V_{Ni}^m is a cation vacancy of effective negative charge m', h$^{\cdot}$is
an electron hole,and O_O^X is an oxygen ion with no effective excess
charge. Upon applying the mass law,the concentrations of vacancies

and positive holes are related to the oxygen pressure by $a(P_{O_2})^{1/n}$ dependence where $n = 2(m+1)$. A non-integral value of n arises from a non-integral value of the effective ionization degree m resulting from the co-existence of more than one type of ionized vacancy which changes its charge many times during its lifetime. Investigations carried out on this subject show that $^1/4 < n < ^1/6$ signifying that the effective charge alternates between 1 and 2.[12].

This defect equilibrium leads to a simple expression for $\tilde{D}$ using eq (3.1) and (2.5). For $z_M = |z_O| = 2, \tilde{c} = v_{Ni}^m$,

$$D_{Ni} = D_v c_v \gg D_o, \quad du_O = \frac{RT}{2} d\ell np_{O_2}, \quad \frac{d\ell n\, P_{O_2}}{d\ell n\, c_v} = 2(m+1),$$

$$\tilde{D} = (m+1)\, D_v = \frac{(m+1)}{c_v}\, D_{Ni} = \frac{(m+1)\, D^T}{c_v\, f} \tag{4.2}$$

Thus, the various diffusivities are related by the concentration and the ionization degree of the nickel vacancies, and the correlation factor (0.782 for a B_1 compound). Plots are given in Fig.1 of D^T_{Ni} at 1245° and 1380°C which correspond to values of $n = 2\,(m+1)$ of 6.35 and 4.0, respectively [16]. Thus, diffusion involves both singly and double charged vacancies. By measuring D^T for ^{57}Ni and ^{66}Ni, Volpe et al [17] were able to conclude that migration of non-associated vacancies was indeed the predominant diffusion mechanism. NiO exhibits a very small degree of

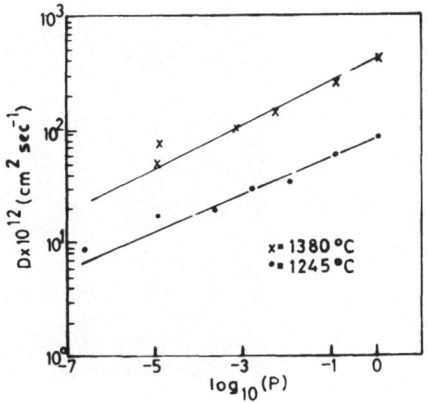

Fig.1. Self-diffusion coefficient of Ni in NiO [16].

226

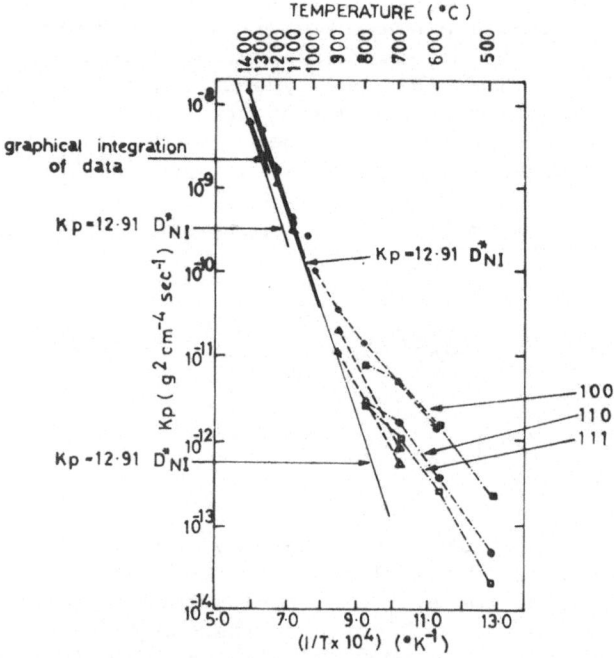

Fig. 2. The temperature coefficient of the parabolic oxidation
constant for NiO growth on polycrystalline and
crystal faces of Ni.

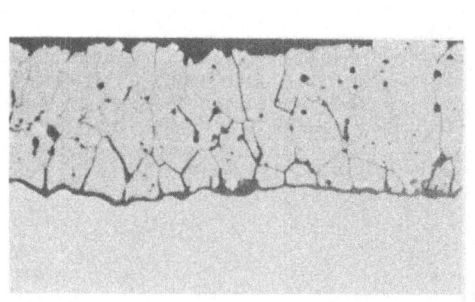 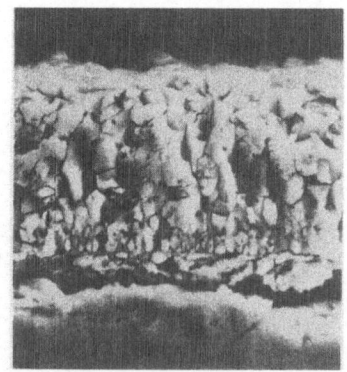

Fig. 3. Microstructures of NiO scales on Ni at 1100°C
(left) and 800°C.

nonstoichiometry, $\delta=0.001$. Accordingly, Price and Wagner[18] followed the kinetics of the electrical conductivity change upon instantaneously changing the oxygen pressure to determine values of $\tilde{D}$ which they used to calculate D_{Ni} and D_v by means of eqs.(3.1) and (4.2). The values of D_{Ni} were in close agreement to those presented in Fig.1.

A method often applied to evaluate $\tilde{D}$ and hence D_i for a ceramic compound is to determine the parabolic rate constant for its growth as a reaction product layer. NiO growth on nickel aptly demonstrates the application of this principle. Integrating the species flux across the layer from its inner(i) to outer(o) inter-face using eq.(2.8) and the expression (3.1) for $\tilde{D}$, the parabolic rate constant k_p is obtained

$$k_p = \bar{c} \int_{a_o(i)}^{a_o(o)} \frac{z_{Ni}}{|z_o|} D_{Ni} \, d\ln a_o \qquad (4.3)$$

from which D_{Ni} can be evaluated. We illustrate an inherent limita-tion of this method by doing the converse of using values of D_{Ni} from tracer experiments to evaluate k_p and then compare these values to those obtained experimentally. The comparison is shown in Fig.2 [19]. Both values correspond at temperatures higher than $900^{\circ}C$ but at lower temperatures the experimental values are larger, the deviation from the high temperature Arrhenius relation increasing with decreasing temperature.The microstructure of the NiO layers formed at $1100^{\circ}C$ and $800^{\circ}C$ are shown in Fig.3.Apparently the grain size is sufficiently large only at temperatures exceeding $900^{\circ}C$ for diffusion to occur by point defects.At lower temperatures one can only determine an effective diffusion coefficient for nickel migration by volume and boundary diffusion eq.(2.16).This is a common behavior in layer formation experiments which negates many evaluations of $\tilde{D}$ from parabolic rate constants determined at temp-eratures < ½ the melting point of the compound [19].

4.2 Metal Carbides (TiC,UC)

Diffusional properties of metal carbides are reviewed by De Poorter and Wallace[6],and Williams[20], the former authors tabulating diffusivities known up to 1970. In general diffusion in monocarbides is interpreted in terms of vacancies in the octahedral sites of each sublattice with metal and carbon migrating independentally by vacancy mechanisms involving either direct octahedral jumps or jumps via intermediate tetrahedral sites.Vacancies at a relatively

small concentration in the metal sublattice result from thermal
energy whilst the much larger concentration of vacancies in the
carbon sublattice is established by the carbon/metal ratio in the
hypostoichiometric carbides. As a consequence D_C is several orders
of magnitude larger than D_M. In the only hyperstoichiometric
carbide examined, $UC_{1+\delta}$, the prevalent view is that carbon diffusion
proceeds by excess carbon as C_2 complex units on octahedral sites.
Self-diffusion coefficients calculated from growth of carbide reaction layers are often at variance with and of larger magnitude
than tracer diffusivity determinations using single carbide crystals
due to the presence of short-circuit diffusion[6] as previously
discussed for the growth of nickel oxide on nickel.

In Fig.4, Arrhenius plots are shown for D_C^T and D_M^T of single
crystal TiC[21,22] and UC[23,24]. Carbon diffusivities are several
orders of magnitude larger than those of the metal. Furthermore, the
activation energies for carbon diffusion are approximately one-half
the values for metal diffusion. This charactersitic feature has
given rise to the concept that carbon diffusion only involves
migration of vacancies whose concentration is determined by the
carbide composition; on the other hand, metal diffusion requires an
activation energy both for formation and migration of vacancies.

Despite these relationships, carbon diffusion in the hypostoichiometric range is not simply determined by random migration
of non-interacting vacancies. If this condition did apply, eq.(2.7)

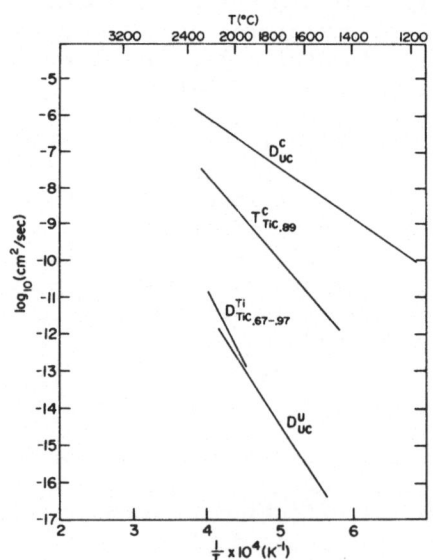

Fig.4. D_C^T and D_M^T vs. $1/T$ for TiC and UC.

giving the dependence of the carbon diffusivity on composition becomes

$$D_C = a^2 \nu \delta \exp \Delta S_m / R \exp -\Delta H_m / RT \qquad (4.4)$$

where the degree of nonstoichiometry δ is a measure of the concentration of constitutional vacancies. Thus, the value of the pre-exponential factor should be directly dependent upon the carbon vacancy concentration. This direct relationship is approximately obeyed for $UC_{1-\delta}$ at values close to the stoichiometric formula, $\propto \delta < 0.03$ [25]. It is not a good approximation, however, for the influence of composition on carbon diffusion in $TiC_{1-\delta}$ where the degree of nonstoichiometry is much larger, $\propto \delta < 0.4$ [6]. Apparently interactions between vacancies influence the diffusion of carbon in a manner not presently understood.

Attempts have been made to obtain additional information from the theory coupling chemical and self-diffusion. For predominant carbon diffusion, eqs. (2.15) and (3.1) lead to

$$\tilde{D} = N_M D_C^I = N_M D_C \left\{ 1 + \frac{d \ln \gamma_M}{d \ln N_M} \right\} = N_M D_C \frac{d \ln a_C}{d \ln N_C} \qquad (4.5)$$

upon relating the chemical potentials of carbon and metal by the Gibbs Duhem relation. Two methods have been used to obtain information on D_C using eq. (4.5) and values of $\tilde{D}$ found upon butting graphite to single crystals of the carbides and determining the carbon gradient resulting from the diffusion anneal.

For $TiC_{1-\delta}$, Kohlstedt et al [26] expressed the activation energy for $\tilde{D}$ as a sum of the activation energy for D_C and a thermodynamic contribution resulting from the carbon gradient:

$$E(\text{chemical diff}) = E(\text{self-diff}) - R \frac{d \ln}{d 1/T} \left\{ \frac{d \ln a_C}{d \ln N_C} \right\} \qquad (4.6)$$

where the thermodynamic term is evaluated from a defect model. The experimental and evaluated activation energies for $\tilde{D}$ were in good agreement, ~ 100 kcal/mole, but the predicted value of the pre-exponential term D^o was independent of composition in contradiction to the experimental results. Wallace [27] completed the opposite calculation for $UC_{1+\delta}$ calculating D_C from the measured $\tilde{D}$ and available thermodynamic data at several temperatures. In this case the measured and calculated activation energies for D_C^o were also in excellent agreement, $\sim 60 - 100$ kcal/mole, but the corresponding values for D_C^o showed poor agreement.

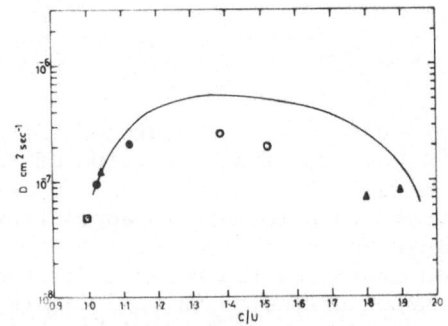

Fig. 5. D^C in $UC_{1+\delta}$ at $1827^\circ C$ [28].

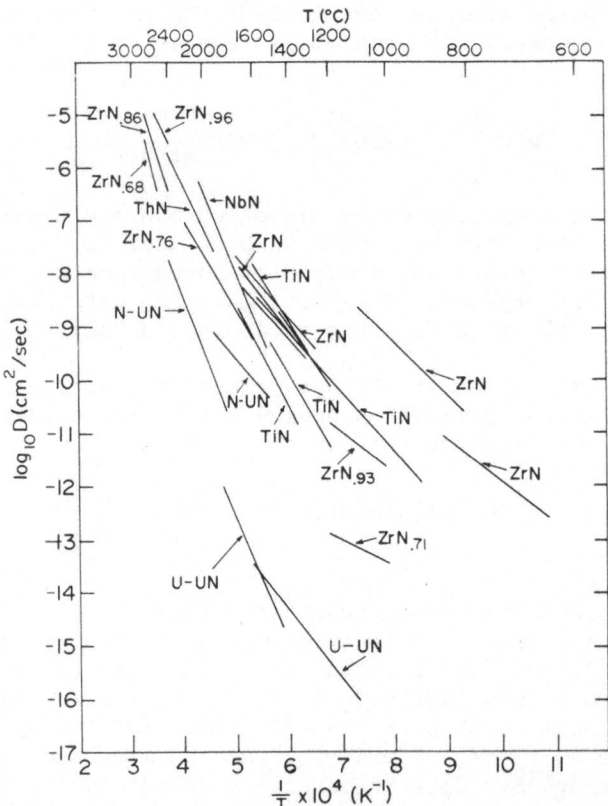

Fig. 6.   The temperature coefficient of the diffusivities for metal mononitrides.

Despite the good agreement when activation energies are determined by the relationships between D and D_C, the simple defect models and experimental results for the thermodynamic contributions are inadequate for precisely predicting the dependence of either the chemical or self diffusion coefficients on composition 6 . Murch [28] has attempted to obtain a more precise evaluation of the dependence of the carbon diffusion rate on carbon content in $UC_{1+\delta}$ based upon the mechanism advanced by Sarian [25] involving carbon atoms migrating from doubly occupied octahedral C_2 sites to singly occupied C_1 sites. The equation for the diffusion D_C^T was refined to include in addition to the jump frequency and the fraction of C_2 defects, N_{c-c}/N_C, a correlation factor f dependent upon carbon concentration, and a path probability equal to the fraction of available diffusion paths given by the ratio of singly occupied to the total number of sites, N_C-2N_{c-c}/N_U. The resultant expression for the diffusivity is

$$D_C = \nu a^2 \frac{N_{c-c}}{N_C} \frac{N_C - 2N_{c-c}}{N_U} \exp \frac{\Delta S_m}{R} \exp - \frac{\Delta H_m}{RT} \qquad (4.7)$$

and a plot of D_C vs. the C/U ratio is shown in Fig.5. The calculated curve gives a reasonable representation of the compositional variation, the deviations from the experimental values probably resulting from the simple assumption of an ideal solution of defects. There is at present insufficient knowledge on defect interactions and on the degree to which diffusion of each species can be considered as migrating independently.

4.3 Metal Nitrides ($TiN, Ti_2N, ZrN, VN, NbN, Cr_2N, Fe_4N, ThN, UN$)

It is apparent from an appraisal of the literature on the measured diffusion coefficients that these studies on the nitrides remain at the infancy stage. Investigations have only been carried out on polycrystalline materials often of unknown structures and impurity content. The diffusion coefficients are given in Table 1. In Fig.6, Arrhenius plots are given of the temperature dependence of the diffusion coefficients for the rock-salt structured mononitrides. These transition metal nitrides behave similarly to the corresponding carbides in that both types of solids behave electrically as metals and exhibit large concentrations of vacancies in both sublattices.

Tracer diffusion of nitrogen and uranium into large-grained UN demonstrate that the diffusivities increase with increasing nitrogen pressure 50,52 . Thus diffusional properties correspond to those of oxides when defect concentrations are determined by an atmospheric activity. Also $D_N >> D_U$ which is characteristic of $D_C >> D_M$ found for the carbides. The results available for UN at N/U ratios from 0.995 to 0.997 at 1946°C give best fit to $D_N \propto (P_{N_2})^{\frac{1}{2} \cdot 8}$ [50].

Table 1. Diffusion coefficients for binary nitrides.

Nitride	Diffusivity	Temp. ($^{\circ}$C)	$D^{\circ}(cm^2/sec)$	E(Kcal/mole)	Ref.
TiN	$\tilde{D}$,N	900–1570	$5.4x10^{-3}$	52.0	29
TiN	$\tilde{D}$,N	1300–1600	4.4	72.8	30
TiN	$\tilde{D}$,N	1350–1700	$2.0x10$	90.0	31
TiN	$\tilde{D}$,N	1200–1450	$5.8x10^{-1}$	66.7	32
TiN	$\tilde{D}$,N	1300–1670	$2.3x10^{-3}$	50.2	33
TiN	$\tilde{D}$,N	950–1500	$2.1x10^{-3}$	47.6	55
ZrN	$\tilde{D}$,N	650–850	$7.9x10^{-5}$	35.9	35
ZrN	$\tilde{D}$,N	1350–1700	$6.0x10^{-2}$	60.0	31
ZrN	$\tilde{D}$,N	1260–1720	$1.7x10^{-2}$	53.6	34
ZrN	$\tilde{D}$,N	800–1100	$2.5x10^{-2}$	44.2	36
ZrN	$\tilde{D}$,N	1200–1500	3.1	79.4	32
ZrN(.76)	$\tilde{D}$,N	1600–2200	$7.5x10^{-1}$	78.3	37
ZrN(.68)	D,N	2550–2700	$3.5x10^{12}$	245.0	38
ZrN(.86)	D,N	2400–2750	$7.9x10^{5}$	150.0	38
ZrN(.96)	D,N	2400–2600	$2.6x10^{2}$	97.0	38
ZrN(.95)	D^T,C^{14}	2030–2690	$1.6x10^{-3}$	92.1	39
ZrN(.93)	D^T,N^{15}	1000–1200	$4.1x10^{-6}$	36.6	40
ZrN(.75)	D^T,N^{15}	1000–1200	$3.0x10^{-10}$	23.0	41
VN	$\tilde{D}$,N	1500	$D=2.5x10^{-8}$		32,42
NbN	$\tilde{D}$,N	1510–2035	$2.1x10^{4}$	112.0	43
Cr_2N	$\tilde{D}$,N	1200	$D=4.2x10^{-8}$		44
Fe_4N	$\tilde{D}$,**N**	504	$D=3.2x10^{-12}$		45
εFe_4N	$\tilde{D}$,N	580–730	$4.4x10^{-3}$	27.1	46
ThN	$\tilde{D}$,N	1900–2400	$2.5x10^{2}$	99.4	47
UN	D^T,N^{15}	1500–1900	$2.6x10^{-4}$	55.0	48
UN	D^T,N^{15}	1792	$D=8.1x10^{-11}$		49
UN	D^T,N^{15}	1700–2000		$\sim$56.0	50
UN(1-δ)	D^T,U^{233}	1100–1600	$3.2x10^{-7}$	60.0	51,52
UN(1+δ)	D^T,U^{233}	1420–1830	$7.5x10^{-2}$	105.0	52
UN	$\tilde{D}$,N	1800–2400	$1.2x10$	120.0	53

Reimann et al [52] suggest that nitrogen transfer involves the disorder equations

$$\tfrac{1}{2}N_2 + V_N^x = N_N^x \ ; \quad 0 = V_U^x + V_N^x \ ; \quad N_N^x = N_i^x + V_N^x \tag{4.8}$$

for neutral nitrogen interstitials and vacancies. These equations give N_i^x and V_U^x α $(P_{N_2})^{\frac{1}{2}}$ and hence the diffusivities $\alpha (P_{N_2})^{\frac{1}{2}}$. The observed nitrogen dependence on D_N is qualitatively explained. The relationship on D_U was not obeyed, however, since this diffusivity correlated directly to the pressure. The authors therefore suggest that uranium diffusion occurs by divacancies which will proportionally increase with pressure.

Activation energies reported for nitrogen diffusion show a tendency of increasing with temperature (Table 1 and Fig.6). Most values were obtained from results of reaction layer formation on the metals assuming that $\tilde{D}$ corresponds to D_N. For carbon diffusion in TiC, Kohlstedt et al [26] found that the contribution to the activation energy from the activity term in eq. (4.6) is minor. They therefore suggest that the unexpectedly low activation energies reported for diffusion in this polycrystalline solid result from short-circuit diffusion. Such considerations probably apply to nitrogen diffusion in several of the nitrides. The activation energies, < 50 kcal/mole, at temperatures < 1200°C may represent nitrogen diffusion largely through low resistance paths such as ingrown dislocations and boundaries in the polycrystalline reaction layers and materials. The transition of values to ⱱ 100 kcal/mole as the temperature is elevated then represents the decreasing influence of short-circuit diffusion as the activation energy becomes that for a point defect mechanism. The activation energies >> 100 kcal/mole and the high pre-exponential factors above 2000°C obtained from sintering experiments subsequently indicate participation of metal diffusion in the mass transport process [38].

(Postscript) Kijima and Shirasaki [54] recently measured the self-diffusion of nitrogen in polycrystalline α and β-Si_3N_4 at temperatures in the range 1200° - 1410°C. The diffusion coefficients for single-crystal grains $D_N (cm^2/sec) = 1.2 \times 10^{-12} \exp - 55,700/RT$ for α -Si_3N_4 and $D_N = 5.8 \times 10^6 \exp - 185,700/RT$ for β-Si_3N_4. Diffusion of nitrogen along grain boundaries in the reaction-sintered materials was considerably faster than diffusion through the bulk.

REFERENCES

1. C. Wagner, Z.Physik Chem. B21, 25, 1933.
2. L. Darken, Trans.AIME, 174, 184, 1948.
3. L. Onsager, Ann.N.Y.Acad.Sci. 46, 241, 1945-46.

234

4. E.W. Hart, Acta Metall. 5, 597, 1957.
5. J. Crank. Mathematics of Diffusion, Clarendon Press, Oxford, 1956.
6. G.L. De Poorter and T.C. Wallace, Advances in High Temperature Chemistry, ed. by L. Eyring, Academic Press, New York, 4, 107, 1971.
7. P.C. Carman and R.A.W. Haul, Proc.Roy.Soc.London, A22, 109, 1954.
8. N.L. Peterson, Diffusion in Solids, Recent Developments, ed. by A.S. Nowick and J.J. Burton, Academic Press, New York, 1975.
9. C. Wagner, Progress in Solid State Chemistry, ed. by J.McCaldin and G.Somorjai, Pergamon Press, New York, 10, 3, 1975.
10. N.N. Greenwood and A.T. Howe, Proc.7th Int.Symp.Reactivity of Solids, 1972.
11. J.B. Wagner,Jr., Mass Transport in Oxides, ed by J.B. Wachtman, Jr. and A.D. Franklin, Nat. Bur. Standards Special Publ. 296, Washington.
12. P. Kofstad, Nonstoichiometry, Diffusion and Electrical Conductivity in Binary Metal Oxides, John Wiley & Sons, New York, 1972.
13. R.T. Whipple, Phil.Mag. 45, 1225, 1954.
14. T. Suzuoka, Trans.Jap.Inst.Metals, 2, 25, 1961.
15. F.A. Kröger, Chemistry of Imperfect Crystals, John Wiley & Sons, New York, 1964.
16. M.L. Volpe and J. Reddy, J.Chem.Phys. 53, 1117, 1970.
17. M.L. Volpe, N.L. Peterson and J. Reddy, Phys.Rev. B3, 1417, 1971.
18. J.B. Price and J.B. Wagner, Jr., Z.Physik Chem. (N.F.) 49, 257, 1966.
19. W.W. Smeltzer and D.J. Young, Progress in Solid State Chemistry, ed. by J.McCaldin and G. Somorjai, Pergamon Press, New York, 10, 17, 1975.
20. W.S. Williams, Progress in Solid State Chemistry, ed. by H.Reiss and J.McCaldin, Pergamon Press, New York, 6, 57 (1971).
21. S. Sarian, J.Appl.Phys. 39, 3305, 1968.
22. S. Sarian, J.Appl.Phys. 40, 3515, 1969.
23. G.G. Bentle and G. Ervin, Jr., Atomics International Report, Al-AEC-12726, 1968.
24. H. Matzke, J.L. Roubort and H.A. Tasman, J.Appl.Phys. 45, 5187, 1974.
25. S. Sarian, J.Nucl.Mater. 49, 291, 1973-74.
26. D.L. Kohlstedt, W.S. Williams, and J.B. Woodhouse, J.Appl.Phys. 41, 4476, 1970.
27. T.C. Wallace, Fundamentals of Refractory Compounds, ed. by H.H. Hausner and M.G. Bowman, Plenum Press, New York, 133, 1969.

28. G.E. Murch, J.Nucl.Mater. 58, 244, 1975.
29. R.J. Wasilewski and G.L. Kehl, J.Inst.Metals, 83, 94, 1954-55.
30. Y.V. Levinskii, Y.D. Strogonov, S.E. Salibekov, M.K. Levinskaya and S.A. Prokofev, Izv.Akad.Nauk SSSR, Neorg.Mater. 4, 2068, 1968.
31. V.S. Eremeyev, Y.M. Ivanov and A.S. Panov. Izv.Akad.Nauk SSSR, Metal, 4, 262, 1969.
32. S.K. Iyer, Ph.D. Thesis, University of Pennsylvania, Philadelphia, 1971.
33. V.D. Repkin, G.V. Kurtukov, A.A. Kornilov and V.V. Bespalov, Metalloterm, Protsessy Khim. Met., 320, 1971.
34. Y.V. Levinskii, S.S. Kiparisov and Y.D. Strogonov, Izv. Vyssh, Ucheb. Zaved, Tsvet. Met., 13, 91, 1970.
35. C.J. Rosa and W.C. Hagel, J.Electrochem.Soc., 115, 467, 1968.
36. J. Paidassi and R. Le Delliou, C.R.Acad.Sc.Paris, Ser.C., 272, 249, 1971.
37. I.I. Spivak, Izv.Akad.Nauk SSSR, Neorg.Mater., 5, 1138, 1969.
38. R.A. Andrievskii, I.I. Spivak and K.L. Chevasheva, Porosh. Met., 8, 65, 1968.
39. Y.F. Khromov, V.P. Yanchur and V.S. Eremeev, Fiz.Metal. Metalloved, 33, 642, 1972.
40. J.G. Desmaison and W.W. Smeltzer, J.Electrochem.Soc., 122, 354, 1975.
41. J.G. Desmaison and W.W. Smeltzer, J.Electrochem.Soc., (in press).
42. S.K. Iyer and W.L. Worrell, Proc.7th Int.Symp.Reactivity of Solids, 1972.
43. Y.V. Levinskii, S.S. Kiparisov and Y.D. Strogonov, Izv.Akad. Nauk SSSR , Metal, 1, 70, 1973.
44. K. Schwerdtfeger, Trans.AIME, 239, 1432, 1967.
45. K. Schwerdtfeger, P. Grieveson and E.T. Turkdogan, Trans. AIME, 245, 2461, 1969.
46. B. Prenosil, Kovove Mater. 3, 69, 1965.
47. R. Benz, J.Electrochem.Soc. 119, 1596, 1972.
48. T.J. Sturiale and M.A. Decrescente, PWAC-477, 1965.
49. R.H. Condit, J.B. Holt and L. Himmel, J.Electrochem.Soc., 114, 1100, 1967.
50. J.B. Holt and M.Y. Almassy, J.Amer.Ceram.Soc., 52, 631, 1969.
51. D.K. Reimann and T.S. Lundy, J.Met., 20, 54A, 1968.
52. D.K. Reimann, D.M. Kroeger and T.S. Lundy, J.Nucl.Mater., 38, 191, 1971.
53. R. Benz and W.B. Hutchinson, J.Nucl.Mater., 36, 135, 1970.
54. K. Kijima and S-I. Shirasaki, J.Chem.Phys., (to be published Oct. 1976).
55. F.W. Wood, P.A. Romans, R.A. McCune and O.G. Pasche, United States Bureau Mines Report, RI7943, 1974.

DISCUSSION (Schlichting, Singhal)

Stuijts: In the carbides the carbon is the fastest diffusing species. In covalently bonded materials, like silicon carbide and silicon nitride, however, one would expect that the carbon or nitrogen would be the slowest diffusing species. Apart from a large difference in diffusional behaviour in different crystallographic phases, one also has to consider the large difference in self-diffusion coefficients for p- and n- type materials. This was shown by Ghoshtagore to be the case for C-diffusion in p- and n- type silicon carbide. The diffusion coefficient of Si and SiC lies orders of magnitude higher.

Cannon: Although metallic carbides appear to behave in interdiffusion similar to binary metal pairs these compounds are not only electronic conductors but exhibit high deviations from stoichiometry. For the nitrides like Si_3N_4 the deviation from stoichiometry is apparently quite small. Even if they are electronic conductors the requirement to maintain site balance (i.e. M:X ratio) will result in interdiffusion behaviour like that expected for ionics rather than for metals. This means that for interdiffusion experiments between compounds Nernst-Planck types of interdiffusion relations will be more appropriate than Darken's relation.

THE FEASIBILITY OF STUDYING NITROGEN TRANSPORT IN CERAMIC
MATERIALS BY PROTON ACTIVATION ANALYSIS

Leslie D. Major, Jr., and Alfred R. Cooper, Jr.

Department of Metallurgy and Materials Science,
Case Western Reserve University,
Cleveland, Ohio 44106, U.S.A.

ABSTRACT

A technique utilizing the $^{15}N(p,\alpha_o)^{12}C$ nuclear reaction is
described for studying the transport of nitrogen in ceramic
materials. The reaction has a high cross section and is virtually
background free, which allows the precise measurement of ^{15}N
concentrations as low as the natural abundance, 0.37%, with submicron resolution. This technique is being used to study nitrogen
diffusion in silicon nitride. By combining the resonance technique with the single spectrum technique, measurement of D_t down
to 10^{-14} cm^2 or diffusivities as low as 10^{-20} cm^{2}s^{-1} is anticipated.

DISCUSSION (Messier, Siebels)

Singhal: Because of the presence of a glass and phases other
than Si_3N_4 in hot-pressed silicon nitride material, it seems
doubtful that it would be possible to get diffusion coefficient
of nitrogen in Si_3N_4 by using hot-pressed material. Have you seen
any interference peaks in your experiments due to the presence of
other phases in the hot-pressed material?

Major: The diffusion coefficient that would be measured for
nitrogen in hot-pressed silicon nitride would as you say not be
for nitrogen in pure silicon nitride, but for nitrogen in the hot-
pressed material. As far as interference reactions from other
phases in the hot-pressed material are concerned, the technique
is sensitive only to isotopes (regardless of chemical or physical
state) that exhibit (p,α) reactions. In the hot-pressed sample I
looked at, I saw no interfering nuclear reactions. We are planning to use CVD silicon nitride in future work so we can eliminate
the problem of other phases.

Desmaison: With this proton activation technique, would it be possible to differentiate grain boundary diffusion from volume diffusion?

Major: The proton activation technique allows the determination of ^{15}N concentration versus depth profiles in samples with sub-micron resolution. By varying the diffusion time, diffusion temperature, and grain size in a controlled manner, it should be possible to separate grain boundary diffusion from bulk diffusion by deriving various concentration versus depth profiles based on models (Fisher: J. App. Phys. 22 74 (1951); Whipple: Phil. Mag. 45 1225 (1954); Levine and MacCallum: J. App. Phys. 31 595 (1960); Suzuoka: J. Phys. Soc. Japan 19 839 (1964)), involving more than one diffusion mechanism.

Section D

THE VITREOUS STATE

THE STRUCTURE OF GLASS, AND ITS RELATION TO THE SOLUBILITY
AND MOBILITY OF GASES

Helmut A. Schaeffer

Institut für Werkstoffwissenschaften (Glas und Keramik),
Universität Erlangen-Nürnberg, Erlangen, Germany

ABSTRACT. Concepts of the vitreous state and the rôle of network
formers and network modifiers in silicate glasses are reviewed.
Data concerning the molecular incorporation of gases in glass
(physical solubility) and the chemical reaction of gases (H_2O, H_2,
N_2) with the silicate network (chemical solubility) are discussed.
Emphasis is placed on the diffusivities of gases and network formers and their influence on diffusion-controlled processes. The
effect of OH groups on the mobility of sodium impurities in silica
glass, as well as the correlation between molecular and lattice
oxygen diffusion are investigated.

1. INTRODUCTION

The enhanced interest in the vitreous state originates not only
from recent technological developments in the field of glasses
such as optical wave-guides, laser rods, glass ceramics etc., but
also from processes which cause the formation of amorphous layers.
Of particular interest in this connection is the oxidation process
to non-crystalline silica either from silicon or from high temperature ceramics such as silicon carbide and silicon nitride. The
oxidation of silicon is extensively exploited in the semiconductor
industry for passivation purposes and for metal-oxide semiconductor
devices. On the other hand, silica scales formed on surfaces of
SiC and Si_3N_4 act as protective layers against further attack by
oxidizing agents.

2. GLASS STRUCTURE

The progress in developing a theory of liquids and glasses has been
rather slow compared to the advancements in the field of crystalline solids and gases. The idea of a periodic crystal lattice for
the solid state and the idea of molecular chaos for the gaseous
state served as a convenient basis for a theoretical treatment.
Unfortunately, in the case of liquids and glasses a generally accepted basis has not been found, thus leading to a variety of concepts in characterizing the vitreous state. Owing to these different
approaches only a few general aspects of the vitreous state will
be reviewed in terms of thermodynamic, structural, and bond nature
considerations.

2.1. Thermodynamic aspects

In the phenomenological treatment glasses are conceived as frozen-
in supercooled liquids. The freezing process depends on the cooling
rate, i.e. the immobilization of a structural configuration in the
metastable supercooled melt is characterized by the freezing-in
temperature (glass transformation temperature T_g). Therefore, the
glass structure and glass properties depend on the thermal history.
The ease with which silicate melts can be supercooled is reflected
in the extent of attainable supercooling; this is in the order of
$0.1^{\circ}C$ for metallic melts, $10^{\circ}C$ for ionic, but is several hundreds of
degrees for silicate melts; i.e. a low crystallization rate favors
glass formation. On the other hand almost any material can be obtained in a non-crystalline form provided that high enough cooling
rates are applied (splat-cooling or condensation from the vapor
phase). To summarize, glasses are in a thermodynamic non-equilibrium compared with the crystallized phase and in a thermal structural non-equilibrium compared with the supercooled liquid.

2.2. Structural aspects

The term "glassy" or "vitreous" designates a non-crystalline state
(glasses are "amorphous" in the sense of X-ray and electron beam
diffraction), but does not specify the degree of deviation from
crystallinity. Therefore, the question of the nature of glass structure raises the problem of order parameters. From the very beginning of glass science two models of glass structure have been developed which originally appeared to be contradictory: the "crys-
tallite model" and the "random-network model". Nowadays it is well
recognized that glasses may possess a highly complex microstructure
ranging in an order scale somewhere between complete randomness
and polycrystalline configurations, i.e. a glass structure can be
envisaged as a deviation from crystallinity or as a pre-ordering
of a statistically disordered ensemble.

Restricting the following considerations to silicate systems the glass structure is generally characterized by a short-range order, consisting in the simplest case of SiO_4 tetrahedra, and by a certain degree of long-range order depending on the way in which different SiO_4 units are linked together.

In silica glass the strict short-range order is characterized by a mean silicon-oxygen bond distance of 1.6 Å; the long-range order can be characterized by a mean Si-O-Si bond angle of about 140°. The introduction of different cations (network modifiers) into the silicate network, i.e. formation of non-bridging oxygens (Si-O-X), changes predominantly the short-range order by altering the Si-O bond distance. Table 1 shows schematically an attempt to define defects in the silicate network as deviations from a mean bond distance and bond angle. Since glasses are not in thermodynamic nor structural equilibrium they are not subject to the treatment of the law-of-mass-action formalism in determining defect concentrations. In view of defect-dependent properties, such as diffusion phenomena, it is believed that the relatively open network of silica glass (27% free volume) provides the necessary steric conditions for interstitial defects, i.e. network modifiers and soluble gases will be incorporated into "interstitial sites". The nature of the defects of the network formers (oxygen and silicon) will be more complex. It was suggested that the process of glass formation causes a frozen-in oxygen defect structure which is temperature independent below a certain temperature [1].

Another aspect of concern is the elucidation of the influence of the oxygen partial pressure on the defect concentration of oxygen in glasses. The first theoretical studies have been undertaken applying the principles of defect equilibria to vitreous silica and dilute silicates [2].

Table 1. Defect structure in the silicate network.

Silicate network

type of linkage of SiO_4 tetrahedra (long-range order)	type of Si-O bond (short-range order)

"defects"

perturbation of tetrahedra linkage (deviation from mean bond angle)	perturbation of short-range order (deviation from mean bond length)
correlated (microcrystallinity) uncorrelated (microheterogeneity)	intrinsic (Si^{III}, $-O^-$) extrinsic (Si-O-X)
	vacancies interstitials
≙ dislocations	≙ point defects

2.3. Bond nature aspects

Traditionally for silicate glasses ionic bond interaction has been
assumed. However, it has been realized that the Si-O bond is to a
large extent covalent. A substantial contribution to the covalent
character arises from the overlap of the lone-pair 2p electrons of
oxygen with the 3d orbitals of silicon[3]. This mixed bond character is the reason for the poor knowledge of bond energies, because
the calculation of the covalent contribution from "first principles"
has proven practically impossible for atoms more complex than hydrogen.
A variety of empirical approaches has been developed correlating
significant properties of ions or atoms with their effect on the
glass structure. Emphasizing the importance of Coulombic interaction
Dietzel[4] has defined "field strength" values, which are proportional to the attractive force between the cation under consideration
and the oxygen anion. This assignment of field strength values to
cations leads to a classification which determines the rôle of cations in the glass structure, namely as network formers or modifiers,
and has been proven to be successful in a heuristic sense, especially in predicting phenomena of glass formation and phase separation.
In order to estimate the extent of the ionic or covalent bond character of compounds, electronegativity data have been utilized. The
fractional ionic character of the Si-O bond is about 40% on the
Pauling scale[5]. The dielectric description of ionicity developed
by Phillips[6] was successfully employed to complex crystal structures[7]; an ionicity of $f_i = 0.57$ is assigned to quartz.
A correlation between the Si-O distance and the bond nature in the
presence of other elements was pointed out by Noll[8], demonstrating
that covalent bonding of Si-O in a Si-O($\rightarrow$X) configuration is higher

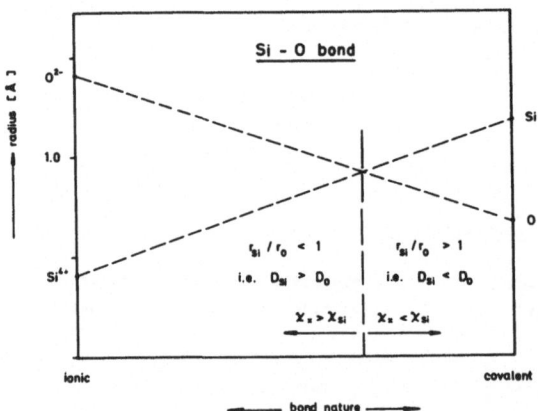

Fig. 1. Schematic diagram of the correlation between network
former radii and extent of covalency in the Si-O bond.

than in a Si-O(→Si) configuration, provided that the electronegativity of element X is smaller than that of silicon. In view of
predicting relative mobilities of oxygen and silicon this concept
was further developed in relating the bond nature with the effective
size and charge of oxygen and silicon [9] . It was shown that increasing covalency favors oxygen diffusivity compared with silicon, $D_O>$
D_{Si}, because of the radius ratio $r_O/r_{Si}<1$, cf. Fig. 1. This concept
is supported by studies of the electron densities of various compounds which have indicated that the radii are significantly larger
for the metal atoms and smaller for the non-metal atoms than the
usual "ionic radii". Moreover, experiments using positron annihilation [10] have shown the expected decrease of effective charge with
increasing covalency.

3. PHYSICAL SOLUBILITY OF GASES

The solubility of gases above glass melts or glasses is considered
to be governed by Henry's law

$$c_{gas}^l = k \cdot p_{gas} \tag{1}$$

where c_{gas}^l is the concentration of gas dissolved in the liquid phase,
and p_{gas} is the pressure of gas in the vapor phase.
Applying the ideal gas law, the pressure can be expressed in terms
of concentration and the constant k in eq. (1) becomes

$$k = c_{gas}^l/p_{gas} = (c_{gas}^l/c_{gas}^v) (1/RT) = k_{Ostw.}/RT \tag{2}$$

The ratio of the gas concentrations in the liquid phase and in the
vapor phase is called the coefficient of solubility after Ostwald.
A different way of defining gas solubilities consists in determining
the ratio of the gas volumes in the vapor phase (STP) to the volume
of solvent (glass) at experimental temperatures and 1 atm gas pressure (coefficient of solubility after Bunsen), i.e. eq. (1) can be
written as

$$n/v^l = k(nRT/v^v) \tag{3}$$

with eq. (2) this changes into

$$v^v(STP)/v^l = k \cdot R \cdot 273 = k_{Ostw.} \cdot (273/T) = k_{Bunsen} \tag{4}$$

Coefficients of solubility in glasses are determined either by the
"saturation-outgassing" or by the "membrane-permeation" technique.
The first technique provides data in the form after Bunsen (cm^3(STP)
gas/cm^3 glass·atm). The second technique is the more common one
and makes use of the relation

$$k_{Bunsen} = P_{gas}/D_{gas} \tag{5}$$

where P_{gas} is the permeability (permeation rate) obtained for the case of steady-state flow rate of a gas through a membrane, and D_{gas} is the diffusion coefficient obtained by the time-lag method (case of non-stationary permeation). The calculation of solubility acc. to eq. (5) requires that P_{gas} is measured in the units of $(cm^3(STP) \; gas/cm^2 \cdot s)/(atm/cm)$ and D_{gas} in cm^2/s. Permeability data are often measured in the units of $cm^3(STP) \; gas/cm^2 \cdot s$ for a 1 mm thick wall with 10 torr gas pressure difference; eq. (5) changes than into $k_{Bunsen} = 7.6 \; P_{gas}/D_{gas}$.

In Table 2 solubility data are collected which demonstrate that the solubility decreases with increasing diameter of the gas molecules and with increasing content of network modifiers in the glass structure; a temperature dependence of solubility is either not existent or very small.
The He solubility indicates a maximum free volume of about 3%, whereas a free volume of about 27% in silica glass was estimated from structural considerations. The discrepancy can be solved by assuming a high portion of "porosity" which is not accessible to gases, because the pores are too small in diameter (<2Å) or because they are isolated in the network.

4. CHEMICAL SOLUBILITY OF GASES

In contrast to the physical solubility of gases a chemical solubility involves chemical reaction of the gas molecules with the glass network, thereby achieving incorporation of gases to a much higher extent. In the following the chemical interaction of water, hydrogen, and nitrogen will be treated (see also [15]).

4.1. Water

The chemical reaction of water molecules with glasses or glass melts is well recognized, and neglecting the detailed reaction mechanism the overall reaction can be characterized by

$$H_2O + O^{2-} \rightleftharpoons 2 \; OH^-$$

The chemical solubility is, therefore, dependent on the concentration of reactive oxygen in the silicate network, i.e. dependent on the basicity of the melt, which increases with increasing portions of network modifiers such as alkalis and alkaline earths.

Concerning the partial pressure dependence of water vapor on the formation of OH groups it follows that the concentration of OH groups increases with the square root of the partial pressure of water vapor, a correlation which has been found valid for a wide variety of glass compositions.

Table 2. Physical solubilities of gases.

gas $\varnothing(\text{Å})$	T ($^{\circ}$C)	$k_{Ostw.}$ in 10^{-3}	glass oxides (mol%)			reference
He(2.0)	20-900	20-34	100 Si			[11]
	1480	21	75 Si,	25 Li		[12]
	1480	23	75 Si,	25 Na		"
	1480	28	75 Si,	25 K		"
	1200-1480	18-23	74 Si,	16 Na,	10 Ca	"
Ne(2.3)	100-1000	20	100 Si			[11]
	1480	11	75 Si,	25 Li		[12]
	1480	18	75 Si,	25 K		"
	1200-1480	7-10	74 Si,	16 Na,	10 Ca	"
H_2(2.4)	20-600	40-25	100 Si			[12]
Ar(3.0)	650-900	10	100 Si			[12]
O_2(3.0)	950-1080	10	100 Si			[13]
N_2(3.2)	400	0.1-0.01	96 Si (Vycor)			[14]
	1480	7.5	75 Si,	25 Li		[12]
	1300-1480	2-4	74 Si,	16 Na,	10 Ca	"

Table 3. Chemical solubilities of H_2O, H_2, and N_2 at 1 atm.

gas	T ($^{\circ}$C)	$k_{Bun.}$	$k_{Ostw.}$	ppm	glass oxides (mol%)			reference
H_2O(OH)	600	4.9	15.7	1700	100 Si			[18]
	1750	1.9	14.1	660	100 Si			[19]
	1250	7.6	42.4	2470	72 Si,	28 Li		"
	1250	8.7	48.5	2700	72 Si,	28 Na		"
	1250	11.7	65.3	3640	72 Si,	28 K		"
	1250	6.5	36.3	2000	74 Si,	16 Na,	10 Ca	"
H_2(OH)	1000	0.024	0.11	8	100 Si			[20]
N_2	1400[1]	0.014	0.086	7	74 Si,	16 Na,	10 Ca	[21]
	1400[2]	13.9	85.3	7000	74 Si,	16 Na,	10 Ca	"

[1] melt bubbled with H_2/N_2 (80/20) for 5h

[2] melt bubbled with NH_3 for 5h

The detailed mechanism of water incorporation into the silicate
network is still a matter of debate [16]; however, certain basic
mechanism have been outlined:

(a) $\equiv$Si-O-Si$\equiv$ + $H_2O \rightleftharpoons \equiv$Si-OH + HO-Si$\equiv$

i.e. depolymerization of the network.
This mechanism is assumed to be operative in SiO_2 glass and SiO_2
rich glass; to which extent it occurs in alkali rich glasses depends on the type of network modifiers.

(b) $\equiv$Si-O-X   + $H_2O \rightleftharpoons \equiv$Si-OH + XOH

i.e. ion exchange between a single-bonded network modifier X and
a hydrogen ion.

Both reactions (a) and (b) lead to the formation of Si-OH groups,
and three basic types have been detected in the silicate network
depending on the degree of interaction of the OH group with its
environment [17]: unassociated OH groups (type 1), OH groups interacting via hydrogen with neighboring single bonded oxygens (type 2),
and OH groups interacting via hydrogen with isolated SiO_4 tetrahedra (type 3).
It has been shown that the extent of type 2 and 3 increases with
increasing alkali content; however, the portion of free OH groups
is not necessarily decreasing at the same time. The latter depends
on the type of alkali, e.g. in Li_2O-SiO_2 glasses the portion of
free OH groups increases, in K_2O-SiO_2 glasses this portion decreases
with increasing alkali content. This behavior is believed to reflect
the tendency of increased phase separation in the Li_2O-SiO_2 system
[16]. The total amounts of chemically dissolved water in the form
of OH groups are listed in Table 3.

The effect of chemically dissolved water on electrical and mechanical properties, viscosity, crystallization etc. has been reviewed
elsewhere [22].

4.2. Hydrogen

The formation of OH groups can also be achieved by a reaction
with hydrogen

$\equiv$Si-O-Si$\equiv$ + $1/2 \ H_2 \rightleftharpoons \equiv$Si-OH + Si$\equiv^{(3+)}$ or

$\equiv$Si-O-Si$\equiv$ + $H_2 \rightleftharpoons \equiv$Si-OH + $\equiv$Si-H

The H_2 solubility is proportional to the square root of the partial
pressure of H_2; in silica glass the chemical solubility exceeds the
physical one by a factor of 4 at $1000^{\circ}C$, cf. Table 2 and 3.
The formation of OH groups via H_2 occurs much faster than via
water molecules or ion exchange with hydrogen ions, since H_2 possesses a higher mobility (see section 5).

4.3. Nitrogen

A chemical solubility of nitrogen can be achieved only under conditions when network-breaking events take place in connection with
a simultaneous removal of oxygen, i.e. reducing conditions of the
atmosphere. Basically a Si-O bond is replaced by a Si-N bond.

Reactions with nitrogen have been investigated as follows [21] ; as
long as reducing conditions are preserved the reaction takes place
either in the presence or absence of water vapor.
Reactions in the absence of water:
(a)   reactions with N_2 and reducing agents (e.g. carbon)
(b)   reactions with N_2 and H_2
(c)   reactions with atomic nitrogen (nitrogen plasma torch)
(d)   reactions with NH_3

The reactions listed can be displayed schematically

(a) $3 (\equiv Si-O-Si\equiv) \;+\; N_2 \;+\; 3C \;\rightleftharpoons\; 2 (\equiv Si-N-Si\equiv) \;+\; 3CO$,
       with $Si\equiv$ below the N

(b) $\equiv Si-O-Si\equiv \;+\; N_2 \;+\; 3H_2 \rightleftharpoons 2 (\equiv Si-NH_2) \;+\; H_2O$,

(c) $6 (\equiv Si-O-Si\equiv) \;+\; 4N \;\rightleftharpoons\; 4 (\equiv Si-N-Si\equiv) \;+\; 3O_2$
       with $Si\equiv$ below the N

(d) $\equiv Si-O-Si\equiv \;+\; NH_3 \;\rightleftharpoons\; \equiv Si-NH_2 \;+\; HO-Si\equiv$

       $\equiv Si-N-Si\equiv \;+\; H_2O$
       with H below the N

       $+$

$\equiv Si-N-Si\equiv \;+\; H_2O \;\rightleftharpoons\; Si-OH$
with $Si\equiv$ below the N

The reactions (a) to (d) clearly demonstrate that their equilibrium
depends on the partial pressure of the atmospheric gases; e.g.
under oxidizing conditions nitrogen may be released from a glass,
which has been dissolved under reducing conditions (see eq. (c));
the influence of water vapor is shown by eq. (d): decreasing water
vapor in the atmosphere increases the amount of dissolved nitrogen
(as nitrides), which also means that water can be removed from
glass melts very effectivly by bubbling with dry N_2 or NH_3 [21] .
The changes of glass properties owing to the incorporation of nitrogen have been investigated for 96% silica glass [23] .

5. TRANSPORT PHENOMENA

5.1. Diffusion of gases

Diffusivity data of gases can be obtained by applying the "membrane-permeation" (cf. section 3) or the "bubble-contraction" technique in which the decrease of bubble radius is correlated with molecular gas diffusion [24].
In Table 4 diffusivities of different gases are listed, demonstrating that the diffusion coefficients decrease with increasing diameter of the gas molecules and by replacement of network formers by modifiers.
Two different approaches have been discussed in explaining the mechanism of gas diffusion in glasses - either treating the gas diffusion as a movement of a sphere in a viscous liquid (Stokes-Einstein equation), or assuming a molecular streaming process through voids and/or channels of molecular dimensions (Knudsen diffusion). The first assumption would predict a strong temperature dependence of the gas diffusivity, since the viscosity of a glass is strongly temperature dependent; however, this is not the case. Moreover, the

Table 4. Diffusivities of gases.

gas	T ($^{\circ}$C)	D_o (cm^2/s)	Q (kcal/mol)	D at 1100°C (cm^2/s)	glass oxides (mol%)	ref.
He	24-1038	3×10^{-7}*	4.8*	7.1×10^{-5}	100 Si	[25]
	1200-1400	3×10^{-3}	10	8.4×10^{-5}	74 Si, 16 Na, 10 Ca	[26]
Ne	194- 457	1×10^{-7}*	9.5*	4.2×10^{-6}	100 Si	[27]
	948-1196	5×10^{-4}	13.7	3.3×10^{-6}	73 Si, 12.5 Na +K, 13.5 Ca+Mg	[24]
H$_2$	700-1000	9×10^{-4}	15.8	2.9×10^{-6}	100 Si	[20]
Ar	644- 905	1×10^{-4}	28.7	4.2×10^{-9}	100 Si (film)	[28]
O$_2$	950-1080	3×10^{-4}	27	1.4×10^{-8}	100 Si	[13]
	1100-1300	4×10^{1}	53	1.5×10^{-7}	Si, Na, Ca	[29]
N$_2$	1100	-	-	$\leqslant 6.5 \times 10^{-8}$	73 Si, 12.5 Na +K, 13.5 Ca+Mg	[30]
H$_2$O	600-1200	1×10^{-6}	18.3	1.2×10^{-9}	100 Si	[18]
	1000-1400	2×10^{-2}	27.5	8.4×10^{-7}	74 Si, 16 Na, 10 Ca	[31]

*$D = D_o \cdot T \cdot \exp(-Q/RT)$, D_o (cm^2/s deg)

application of the Stokes-Einstein equation leads to diffusion
coefficients which are orders of magnitude too small, i.e. the
glass is too viscous to account for the observed high gas diffusivities.

The second approach relates the gas diffusivity to the accessible
free volume in the glass structure. The activation energy of diffusion would then be governed by the strain energy which is required to deform the openings in the glass network in such a way
that a channel or interstitial diffusion mechanism is achieved.
Indications for this mechanism are provided by the fact that the
activation energy of diffusion Q depends on the radius r of the
molecules, i.e. $\sqrt{Q} \propto r$. This type of relation follows also from
theoretical considerations, assuming an elastic dilation of ori-
fices of certain size; for the unstrained orifice a diameter of
about 1.1 Å has been obtained [11] .
A combined mechanism of molecular streaming through voids or channels and migration through a viscous liquid has also been visualized [24] .

5.2. Diffusion of glass components

Glass components (formers and modifiers) differ strongly in their
mobilities, giving cause to network-modifier-controlled and network-
former-controlled processes. The influence of network modifiers in
transport phenomena such as ionic conductivity, ion exchange, electrical and mechanical relaxation is well recognized and the identification of rate-controlling species is achieved. In the case of
network-former-controlled processes there is still some doubt about
the rate-controlling species owing to the lack of diffusion coefficients (especially oxygen and silicon). Fig. 2 shows the different
magnitudes of diffusion coefficients in a soda-lime-silica glass
system; with the exception of silicon, the diffusivities of all
species have been determined, including water diffusion. Oxygen
diffusion is slowest and seems to be rate-determining for processes
such as viscous flow, multicomponent diffusion, and phase separation. From rate constants of these phenomena diffusion coefficients
were calculated which can be compared with the oxygen self-diffusion
data. In the case of viscous flow the Stokes-Einstein equation holds
for oxygen diffusion induced viscosity [32] , at least at high temperatures (for lower temperatures deviations occur as polymerization
of the network increases the effective size of flow units). The
data of the interdiffusion experiment indicate a combined transport
of calcium and oxygen, the latter being rate-controlling [33]. The
calculated diffusion coefficient from phase separation data D_{ph} [34]
is even lower than the oxygen diffusion coefficient. Since silicon
diffusion data are not available the question of the rate-controlling species remains open.

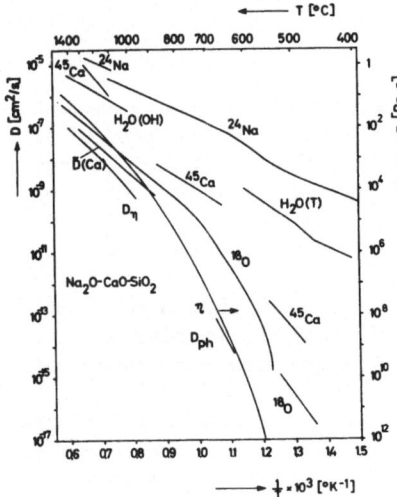

Fig. 2. Diffusion coefficients and network-former-controlled
processes in the Na$_2$O-CaO-SiO$_2$ system.

Oxygen diffusion coefficients for differently composed glasses are
collected elsewhere [9]. For silica glass only oxygen diffusion
coefficients have been determined thus far; data from the literature are depicted in Fig. 3.

5.3. Transport phenomena governed by the glass/gas interaction

In view of the importance of transport processes in non-crystalline
silica two examples will be discussed in more detail which will
demonstrate the influence of an interaction glass/gas (atmospheric
conditions) on transport processes.

(a) Network modifier/solute gas interaction

SiO$_2$ glasses or films contain different amounts of impurities depending on their method of fabrication. Especially the OH content
of chemically dissolved water can vary orders of magnitude ranging
from about 1200 ppm in Suprasil I (flame hydrolysis of SiCl$_4$) to
about 0.1 ppm in Suprasil W (vapor phase oxidation of SiCl$_4$), whereas the sodium impurity content varies much less (0.1-10 ppm). The
interaction which has been observed [39] exists between the Na mobility and the OH content, and is of such a nature that increasing OH
content decreases the sodium mobility drastically. Since the electrical conductivity in SiO$_2$ is caused by the charge transport of
alkali impurities, the effect of immobilisation in the presence of
OH groups is expected to reduce it. An interaction between differ-

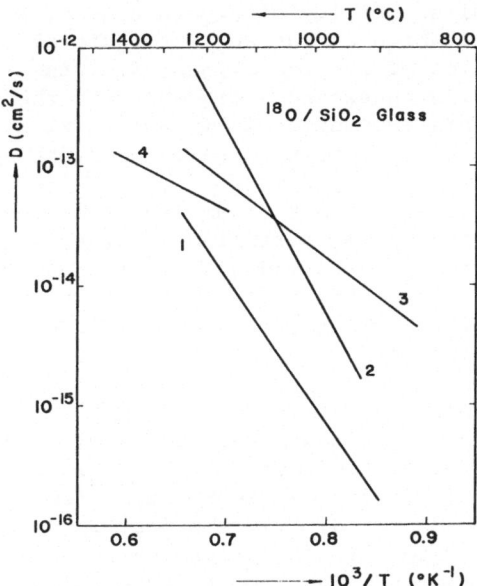

Fig. 3. Lattice oxygen diffusion coefficients in silica glass
Q = 56(1), 71(2), 29(3), 20(4) kcal/mol [35-38].

ent alkali ions with the effect of reducing the electrical conductivity is well known in glasses (mixed-alkali effect). It therefore
appears conceivable that a mixed-alkali effect also occurs between
sodium and hydrogen ions [40].
The detrimental effect of sodium impurities in SiO_2 layers oxidized
on silicon semiconductors may be diminished by the presence of
dissolved water in the form of OH groups.

(b) Network former/solute gas interaction

Studies of the lattice oxygen self-diffusion in SiO_2 glass has
raised the question of the diffusion mechanism and a possible relation to the diffusion of molecular oxygen, which is 5 to 6 orders
of magnitude larger, cf. Table 4 and Fig. 3. Moreover, it has been
suggested that oxidation kinetics of silicon are correlated to the
permeation of molecular oxygen through the SiO_2 layer to the Si/
SiO_2 interface [41]. A correlation between all these phenomena was
presented as follows [42].

(1) $D_O \cdot c_O$ = $D_{O_2} \cdot k_{O_2}$, according to [35]

(2) P_{O_2} = $(D_{O_2} \cdot k_{O_2})/7.6$, definition (see section 3)

(3) K_{SiO_2} = $2D_{O_2} \cdot k_{O_2}/c_O$, according to [43]

Eq. (1) implies that the physical solubility of oxygen causes interstitial molecular oxygen sites which act as an "interstitial defect structure" for the migration of lattice oxygen, i.e. the lattice oxygen can migrate only via an exchange process with the interstitial oxygen molecules. This assumption is in accord with the fact that the lattice oxygen diffusion is linearly proportional to the oxygen partial pressure.

Eq. (3) describes the parabolic rate constant of an reaction process under the assumptions that the rate-controlling species is gaseous and that the concentration of dissolved gas (K_{O_2}) is small compared to the concentration of gas bound to the solid (lattice oxygen) c_O, i.e. $K_{O_2} \ll c_O$, a condition which is easily fulfilled in the case of SiO_2: $K_{O_2}/c_O \approx 2 \times 10^{-6}$ [42].

It follows from eq. (1) to (3) that $K_{SiO_2} = 2D_O = 7.6 \, (2P_{O_2})/c_O$, showing that the parabolic rate constant can either be calculated from the molecular oxygen diffusivity, the lattice oxygen diffusivity, or the permeation rate. Since hardly any temperature dependence of oxygen solubility can be detected, i.e. the heat of solution is negligible, identical energies of activation are to be expected for all oxygen related phenomena. The reported data for K_{SiO_2}, P_{O_2}, and D_{O_2} range between 20-30 kcal/mol; therefore, from the four different oxygen self-diffusion experiments reported in the literature, those with the lower activation energies (20 [38] and 29 [37] kcal/mol) appear to be more reliable.

Increasing the oxidation rate of silicon by exposure to wet atmosphere is a well-established practice in the planar technique for fabricating Si/SiO_2 devices. It is suggested that the oxidation occurs now via the permeation of water or OH, i.e. $K_{SiO_2} \propto P_{H_2O} \propto D_{H_2O} \cdot k_{H_2O}$ ($k_{H_2O} = 2 \, k_{OH}$). Although the diffusivity of water is smaller by a factor of 10 at $1100^\circ C$ compared with the diffusivity of O_2 (cf. Table 4), the solubility of water exceeds that of oxygen by a factor of 10^3, i.e. the permeation rate and thus the parabolic rate constant is expected to increase by a factor of 10^2, very similar to what has been observed [44]. Moreover, the measured activation energy of the wet oxidation (16 kcal/mol) compares rather favorably with the activation energy of H_2O diffusion (cf. Table 4).

REFERENCES

1. H.A. Schaeffer and H.J. Oel, Glastechn. Ber. 42, 493 (1969).
2. R.A. Huggins, J. solid state Chem. 2, 385 (1970).
3. A.G. Revesz, J. Non-Crystalline Solids 4, 347 (1970).
4. A. Dietzel, Glastechn. Ber. 22, 41 (1948).
5. L. Pauling, J. Phys. Chem. 56, 361 (1952).
6. J.C. Phillips, Rev. Mod. Phys. 12, 317 (1970).
7. B.F. Levine, J. Chem. Phys. 59, 1463 (1973).
8. W. Noll, Angew. Chem. 75, 123 (1963).

9. H.A. Schaeffer, Silicon and oxygen diffusion in oxide glasses, in: Mass Transport Phenomena in Ceramics, ed. by A.R. Cooper and A.H. Heuer, Plenum Press, New York, 1975.

10. G.M. Bartenev, S.M. Brekhovskikh, A.Z. Varisov, L.M. Landa, and A.D. Tsyganov, Inorg. Mat. 6, 1371 (1970).

11. R.H. Doremus, Glass Science, John Wiley&Sons, New York, 1973.

12. H.O. Mulfinger, A. Dietzel, and J.M.F. Navarro, Glastechn. Ber. 45, 389 (1972).

13. F.J. Norton, Nature 191, 701 (1961).

14. C.C. Leiby and C.L. Chen, J. Appl. Phys. 31, 268 (1960).

15. H. Scholze, Gases in Glass, in: Proc. VIII International Congress on Glass, London 1969.

16. H. Franz and T. Kelen, Glastechn. Ber. 40, 141 (1967).

17. H. Scholze, Glastechn. Ber. 32, 81, 142, 278 (1959).

18. A.J. Moulson and J.P. Roberts, Trans. Farad. Soc. 57, 1208 (1961).

19. H. Franz and H. Scholze, Glastechn. Ber. 36, 347 (1963).

20. T. Bell, G. Hetherington, and K.H. Jack, Phys. Chem. Glasses 3, 141 (1962).

21. H.O. Mulfinger, J. Amer. Ceram. Soc. 49, 462 (1966).

22. E.N. Boulos and N.J. Kreidl, J. Can. Ceram. Soc. 41, 83 (1972).

23. T.H. Elmer and M.E. Nordberg, Nitrided Glasses, in: Proc. VII International Congress on Glass, Brussels, 1965.

24. G.H. Frischat and H.J. Oel, Phys. Chem. Glasses 8, 92 (1967).

25. J.E. Shelby, J. Amer. Ceram. Soc. 54, 125 (1971).

26. H.O. Mulfinger and H. Scholze, Glastechn. Ber. 35, 495 (1962).

27. J.E. Shelby, Phys. Chem. Glasses 13, 167 (1972).

28. W.G. Perkins and D.R. Begeal, J. Chem. Phys. 54, 1683 (1971).

29. R.H. Doremus, J. Amer. Ceram. Soc. 43, 655 (1960).

30. G.H. Frischat and H.J. Oel, Glastechn. Ber. 40, 311 (1967).

31. H. Scholze and H.O. Mulfinger, Glastechn. Ber. 32, 381 (1959).

32. Y. Oishi, H. Ueda, and R. Terai, Oxygen diffusion in liquid silicates and relation to their viscosity, in: Mass Transport Phenomena in Ceramics, ed. by A.R. Cooper and A.H. Heuer, Plenum Press, New York, 1975.

33. E.W. Sucov and R.R. Gorman, J. Amer. Ceram. Soc. 48, 426 (1965).

34. J.J. Hammel, J. Mickey, and H.R. Golob, J. Colloid and Interface Science 27, 329 (1968).

35. R. Haul and G. Dümbgen, Z. Elektrochem. 66, 636 (1962).

36. E.W. Sucov, J. Amer. Ceram. Soc. 46, 14 (1963).

37. E.L. Williams, J. Amer. Ceram. Soc. 48, 190 (1965).

38. K. Muehlenbachs and H.A. Schaeffer, submitted to Can. Mineral.

39. G.H. Frischat, Glastechn. Ber. 45, 309 (1972).

40. H.A. Schaeffer, J. Steinmann, and J. Mecha, to be published.

41. K. Motzfeldt, Acta Chem. Scand. 18, 1596 (1964).

42. K. Muehlenbachs and H.A. Schaeffer, submitted to J. Non-Crystalline Solids.

43. P.V. Dankwerts, Trans. Farad. Soc. 46, 701 (1950).

44. B.E. Deal and A.S. Grove, J. Appl. Phys. 36, 3770 (1965).

DISCUSSION (Clarke, Tien)

Brook: The expression $D_O = (\frac{k_{O_2}}{c_O})D_{O_2}$ looks surprisingly simple.
Does it not assume that the O and O_2 species migrate with identical enthalpies and jump distances?

Is it reasonable to assume that size is the dominating factor affecting diffusion, as has been done in relating the ratio $\frac{D_{Si}}{D_O}$ to the extent of covalency in the glass bonding?

Schaeffer: The expression for D_O was first proposed by Haul and Dümbgen (Z. Electrochem. 66, 636, 1962) to explain the linear oxygen partial pressure dependence of the oxygen tracer diffusion coefficient in silica glass; and later Meek (J. Amer. Ceram. Soc. 56, 341, 1973) gave an interpretation of the relation in terms of the probability that a dissolved oxygen molecule is found adjacent to a specific lattice oxygen. It is assumed that the diffusion mechanism for lattice oxygen occurs via the dissolved molecular oxygen, i.e. similar to an interstitialcy mechanism. The lattice oxygen diffusion is therefore expected to be proportional to the concentration of interstitially dissolved oxygen molecules and their mobility.

The assumption is that the diffusivity of network formers is governed by a cooperative diffusion process, since the Si-O bond has to be activated or broken. It is then conceivable that steric factors are dominant for the relative mobilities of silicon and oxygen.

Evans: I was intrigued to notice that the reaction between hydrogen and silica occurs very rapidly. Is it possible to infer from this that the reaction between water and silica may involve a reaction with the hydrogen ion? I ask the question because there has been a recent debate about the relative roles of the hydroxyl and hydrogen ions in the stress corrosion of amorphous materials. It is important to resolve this issue, because the theories of hydroxyl and hydrogen in stress corrosion lead to quite different stress corrosion relationships.

Schaeffer: In the case of alkali-containing glasses, the dominant process for the incorporation of OH groups is ion exchange between the alkali and the hydrogen ion. However, for pure SiO_2 glass the only way for OH formation is via a bond breaking process with H_2O (aqueous or gaseous) or H_2. The chemical solubility at 1 atm is about 100 times greater for water than for hydrogen, but the formation rate of OH via H_2 exceeds that via H_2O by 3-4 orders of magnitude. Thus, the effect of hydrogen in stress corrosion phenomena may be of importance.

SIALON GLASSES

K.H. Jack

The Wolfson Research Group for High-Strength Materials,
Crystallography Laboratory, The University,
Newcastle upon Tyne, England.

ABSTRACT. Vitreous phases in the Mg-Si-Al-O-N and Y-Si-Al-O-N
systems contain up to about 10 a/o nitrogen. In addition to their
role in the densification of silicon nitride and sialons, nitrogen
glasses and glass-ceramics offer prospects for technological
exploitation.

1. Mg-Si-O-N GLASSES

In the hot-pressing of silicon nitride with magnesia a phase which
is liquid at high temperature cools to give a glass. Fig. 1 from
Nuttall & Thompson (1) shows a well-formed hexagonal β-Si_3N_4
crystal which has nucleated and grown from the liquid. If the
glass is merely a magnesium silicate, it would be expected from
the MgO-SiO_2 phase diagram (see Fig. 2) to devitrify on suitable
heat-treatment to give enstatite and cristobalite. Instead, it
gives (2) enstatite and silicon oxynitride, Si_2N_2O, after 24h at
1350°C; see Fig. 3

$$\text{(i)} \qquad 45\,MgO.55SiO_2 \xrightarrow{\;1350°C\;} MgSiO_3 \;+\; SiO_2$$
$$\text{enstatite} \quad \text{cristobalite}$$

$$\text{(ii)} \qquad +\text{ dissolved }Si_3N_4 \longrightarrow MgSiO_3 \;+\; Si_2N_2O$$

Clearly, the original high-temperature liquid and the glass that
forms on cooling must contain nitrogen.

Small samples of this nitrogen-glass were obtained by both
hot-pressing and by pressureless heat-treatment of mixtures of
MgO, SiO_2 and Si_3N_4 at 1700°C in a boron nitride crucible. Table 1

258

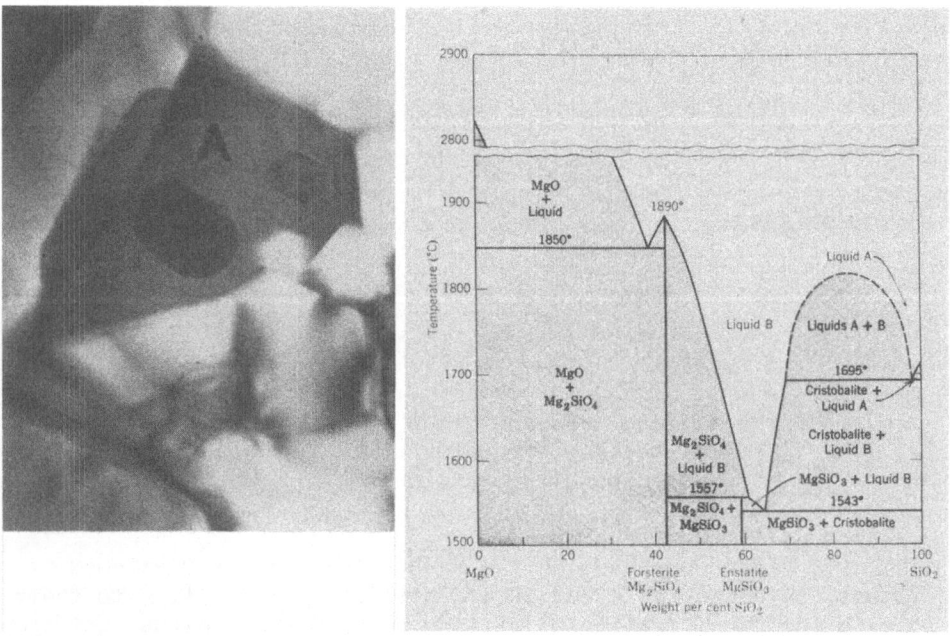

Fig. 1. Transmission electron
micrograph of hot-pressed
silicon nitride-magnesia
showing an amorphous
(liquid) region "A" from which
has grown a well-defined
hexagonal β-Si_3N_4 crystal,
X 20,000.

Fig. 2. The MgO-SiO_2 phase diagram.

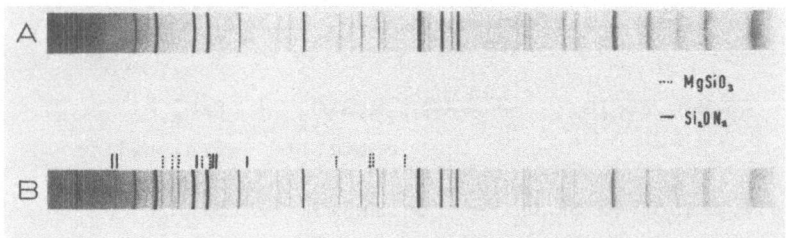

Fig. 3. X-ray photographs showing the devitrification of the
glassy phase in silicon nitride hot-pressed with 10 w/o MgO at
1700°C.   A = as hot pressed
            B = aged at 1350°C for 24h

Table 1. Mg-Si-O-N glass formation

run	reactants	conditions	products
A	20 w/o MgO 80 w/o SiO_2	(1)	80% cristobalite 20% proto-enstatite
B	4 w/o Si_3N_4	(1)	80% glass 20% {cristobalite + proto-enstatite
C	10 w/o Si_3N_4 90 w/o A	(1)	100% glass
D	15 w/o Si_3N_4 85 w/o A	(1)	85% glass 10% Si_2N_2O 5% cristobalite
E	10 w/o Si_3N_4 24 w/o MgO 66 w/o SiO_2	(1)	100% glass
F	as in E	(2)	glass trace of Si_2N_2O
G	as in C	(3)	100% glass

(1) = hot-pressed in boron nitride at 1700°C for 10 min.

(2) = pressureless heat-treatment in nitrogen at 1600°C for 30 min.

(3) = pressureless heat-treatment at 1700°C for 10 min.

lists the products obtained from different starting mixtures. Without silicon nitride, mixtures of 20 w/o MgO:80 w/o SiO_2 gave a product showing only diffraction patterns of enstatite and cristobalite but with addition of 10 w/o Si_3N_4 to this mix no diffraction pattern at all was observed. The X-ray photograph showed a diffuse band, typical of glass, at about d = 4Å. When heated with or without pressure, runs E and F gave glass containing about 6 a/o N i.e. $Mg_{12}Si_{26}O_{56}N_6$, with a density 2.58 g cm^{-3}. Devitrification of these glasses at 1500°C gave mixtures of cristobalite, enstatite and silicon oxynitride.

2. Mg-SIALON GLASSES

Addition of Al_2O_3 to Mg-Si-O-N compositions extends the vitreous region and mixtures of MgO, SiO_2 and AlN cold-compacted, heated to 1650°C for 30 min and then cooled, give glasses containing up to 10 a/o N; a typical composition is $Mg_9Si_{21}Al_{10}O_{50}N_{10}$.

All attempts so-far to prepare nitrogen glasses have been made without rapid quenching from the melt and even with 10 g

specimens the cooling rate is quite slow - about 30°C per minute
between 1650° and 1000°C. There is little doubt that larger
vitreous samples with wider limits of homogeneity could be obtained
by accelerated cooling.

3. Y-SIALON GLASSES

Other nitrogen glasses have been shown to occur in the lithium-
sialon and yttrium-sialon systems. For example, by melting a
mixture of molar composition

$$14 \ Y_2O_3 : 59SiO_2 : 27AlN \ (\text{i.e. } Y_9Si_{20}Al_9O_{53}N_9)$$

in a graphite crucible lined with boron nitride at 1650°C under
nitrogen, and then cooling to room temperature, a completely
vitreous product was obtained. Electron probe microanalysis
(kindly provided by Dr. A. Mocellin using Camebax equipment at
the Ecole des Mines, Paris) showed that the glass was homogeneous
with a composition close to that of the starting mix. In thin
sections it was transparent with a refractive index 1.76.
Devitrification by heat-treatment for 16h at 1200°C gave a
crystalline product showing diffraction patterns of β-Y_2O_3.$2SiO_2$,
$3Y_2O_3$.$5Al_2O_3$ (yttrium-aluminium garnet, YAG) and Si_2N_2O.
In exploring glass-forming regions in the Y_2O_3-SiO_2-AlN
system the maximum nitrogen content so-far found in a completely
vitreous product is 12 a/o in a composition $Y_{12}Si_{15}Al_{12}O_{49}N_{12}$.

4. PROPERTIES OF NITROGEN GLASSES

Even small concentrations of nitrogen not exceeding 1 a/o in
oxide glasses are reported (3, 4, 5) to increase the softening
temperature, viscosity, and the resistance to devitrification.
Glasses with 10 a/oN or more might be expected to have more unusual
properties. If in the tetrahedral net-work the nitrogen is
co-ordinated by three ligands the structure should be more rigid
and hence have a higher viscosity than the silicate glasses. The
prospect of producing glasses more refractory and more resistant
to devitrification than vitreous silica seems worth exploration.

The ease of shaping glasses, and the possibility of producing
glass-ceramics in which the crystalline phases are refractory
nitrides and oxynitrides suggest that nitrogen-containing glasses
might eventually be just as important, from a technological viewpoint, as the crystalline sialons.

REFERENCES

1. K. Nuttall & D.P. Thompson, J. Mater. Sci. 9 (1974) 850.
2. S. Wild, P. Grieveson, K.H. Jack & M.J. Latimer, "Special Ceramics 5" Brit. Ceram. R.A., Stoke-on-Trent, (1972) p. 377.
3. F.L. Harding & R.J. Ryder, Glass Tech. 11 (1970) 54.
4. T.H. Elmer & M.E. Nordberg, J. Amer. Ceram. Soc. 50 (1967) 275.
5. H.O. Mulfinger, J. Amer. Ceram. Soc. 43 (1966) 462.

DISCUSSION (Clarke, Tien)

Popper: With regard to hot pressing, does this mean that a carefully chosen heating/cooling schedule is very important? One might hold at one temperature to get a certain amount of nitrogen dissolved in the glass and then hold at another temperature to devitrify the glass. Have any effects of the schedule on the crystalline content and/or mechanical properties (e.g. high temperature creep) been observed?

Jack: This is exactly the procedure I suggested in an earlier paper for producing a liquid which would allow liquid phase densification and which then could be cooled and devitrified in the appropriate temperature range to give a refractory grain-boundary phase. The effects of an appropriate heat-treatment schedule on creep properties have certainly been observed and the principle has been applied to produce more creep-resistant materials both by hot-pressing and pressureless sintering.

Katz: Is it necessary to have a reducing atmosphere (as mentioned by Schaeffer) to incorporate the N into the glass?

Jack: No - because the solubility is so high it is incorporated into the glass structure.

Cannon: How does the solubility of nitrogen gas change with temperature in these glasses?

Jack: We do not know since we only prepared the glasses at 1600 to 1700°C.

Bowen: In hot-pressing silicon nitride with magnesia, a magnesium silicate grain boundary phase of eutectic composition is believed to form at approximately 1540°C. Your work suggests that nitrogen can dissolve in large quantities in glasses of this composition. Would you expect a significant lowering of the temperature at which liquid forms in hot-pressing by the presence of silicon nitride in the grain boundary phase?

Jack: Up to at least 10% N dissolves in this phase but we do not know whether this lowers the liquidus temperature.

Brook: It would be interesting to analyse the linkages in the nitrogen glass networks using the methods applied by Stevels to oxygen glasses. Working from the composition and assuming that the ions take up modifier or former positions as in the oxides, it would be possible to assess the average number of triply, doubly or singly linked corners for the tetrahedra present in the network. Such an analysis might suggest a structural reason for the upper limit of nitrogen concentrations that can be introduced into the glass.

Section E

FORMATION OF NITRIDES

NITRIDATION AND REACTION BONDING

F.L. Riley

Department of Ceramics,
The University of Leeds, U.K.

1. INTRODUCTION

The reactions of the metallic and semi-metallic elements with nitrogen have not, in general, been as extensively studied as the corresponding reactions with oxygen. Where detailed studies have been made many have been more concerned, for metallurgical reasons, with solid solution or dispersion formation than with generation of the nitride in bulk. Of those nitrides having ceramic applications, silicon nitride has received by far the most attention, and in reflecting the interest shown in these materials during the past 20 years this review will be concerned to a large extent with the silicon-nitrogen reaction and associated processes.

'Reaction bonding' is taken to mean the composite process of a formation reaction, coupled with the apparently simultaneous development of strength in the bulk product. This term is in practice applied almost exclusively to the case of silicon nitride, though it is equally applicable to certain methods for the production of silicon carbide and sialon ceramics, for example.

Because pure silicon nitride powder appears not to be sinterable without the use of additives the reaction bonding process is at present an important fabrication route for this material. There is an additional bonus in that conversion of the green silicon body to the nitride ceramic takes place with almost negligible dimensional change ($< 0.1\%$), so that complex engineering shapes can be produced with relative ease. A thorough understanding of the silicon nitridation reaction is therefore of crucial importance from the point of view of understanding the reaction bonding process, and in turn for the development of reaction bonded silicon nitride as a high quality engineering material with fully controlled sets of properties.

2. NITRIDATION REACTIONS

Most metallic and semi-metallic elements will react directly with nitrogen. Gulbranson and Jennsen (1) noted in their early studies on selected transition metals that reactions with nitrogen tended to be slower than those with oxygen. Because the oxide is always thermodynamically more stable than the nitride an oxide film tends to be formed on an element in air, and may then interfere with subsequent attempts to study the nitridation. Moderate temperature (600-1000°C) is usually required, and nitrogen pressures higher than 1 atm, or ammonia, may be used to facilitate reaction. An important feature from the mechanistic point of view is that with many elements considerable volume change occurs (Table 1), so that adherent protective scales would not be expected.

Element	Nitride	$\Delta V\%$
Magnesium	Mg_3N_2	−11.3
Calcium	Ca_3N_2	−28.2
Titanium	TiN	8.1
Zirconium	ZrN	2.5
Hafnium	HfN	3.7
Vanadium	VN	27.0
Niobium	NbN	18.3
Tantalum	TaN	25.2
Boron	BN	151.7
Aluminium	AlN	25.7
Silicon	Si_3N_4	21.7
Germanium	Ge_3N_4	27.5

Table 1: Examples of volume change occurring on nitridation.

Nitridation reactions may be broadly classified according to the nature of the combining element as follows:

(a) Group IA and IIA metals

Not all these metals form nitrides by direct reaction, but examples of the group which have been studied in detail are Mg (2) and Ca (3). Partially protective nitride films form readily but tend to be mechanically unreliable. Breakaway reactions are commonly observed, sometimes involving the metal vapour. When the thin films do remain intact, outwards ionic diffusion of the metal has been assumed to be the rate controlling process.

(b) Transition metals

Attention has been paid to the nitridation of many of these elements, including Ti, Zr, Hf, V, Nb, Ta, Cr and Fe (4-9). Nitrogen tends to be appreciably soluble in these metals (Table 2) and solid-solution formation is often the dominating process with parabolic kinetics. Thin surface films of the nitrides are formed at slower rates and through which nitrogen appears to be able to pass relatively easily. In the case of Ti the overall nitridation enthalpy of activation has been associated with the mobility of nitrogen in the metal, and not in the nitride (5). Mechanisms of nitrogen transport in these largely covalent nitrides have in general not been clearly established.

	Temperature/°C	Solubility/ ^{a}/o
Aluminium	1200–1500	< 0.002
Silicon	1410	0.02
Titanium	1400	20
Zirconium	1400	2.85
Hafnium	1700	29
Niobium	1100	0.01
Tantalum	1000	4
Chromium	1600	14
Iron	590	0.4

Table 2: Examples of nitrogen solubility (10).

(c) Other First Row and B Group Elements

For practical purposes this group consists of the nitrides of B, Al, Si and Ge. No single mechanism appears to be applicable to their formation reactions. Boron shows parabolic behaviour in the range 600-1300°C, which becomes linear and then logarithmic above this point (11). Aluminium presents difficulties because of a protective oxide film and the reaction, once started, is difficult to control (12,13). Germanium nitride is thermodynamically un-stable at reaction temperatures under 1 atm nitrogen but can readily be formed using ammonia, following a logarithmic law associated with the disruption of germanium grains by growing nitride nuclei (14). Silicon nitridation has received increasing attention because the nitridation reaction is the basis for the reaction bonding process and because α-silicon nitride is the starting point for hot-pressed silicon nitride. This reaction will therefore be examined in some detail.

3. SILICON NITRIDATION

3.1 The System Si-N

Silicon has a melting point of 1410°C (15) and becomes appreciably volatile towards this temperature (Figure 1). It is protected against extensive low temperature oxidation in air by a film of silica of the order of 3 nm thick (17). Thus a pure powder of median particle size 10 μm would be expected to contain 0.16 a/o oxygen; oxygen solubility in silicon is small (3.5×10^{-3} a/o at 1410°C (10)). In the presence of 1 atm nitrogen, silicon yields the gaseous subnitrides, SiN and Si_2N, and two crystalline phases, α- and β-Si_3N_4. The solid nitrides are also volatile at high temperature (Figure 2), although the nitrogen dissociation pressure does not reach 1 atm until approximately 1900°C (18). The densities of the α and β phases are 3.169 and 3.192 Mg m^{-3} (19) respectively. Silicon nitridation, like its oxidation, is strongly exothermic. Heats of reaction with nitrogen and oxygen at 1410°C are (16) -724 kJ mol^{-1} Si_3N_4 and -900 kJ mol^{-1} SiO_2 respectively. A rapid nitriding reaction involving fine powders may lead to melting of the silicon.

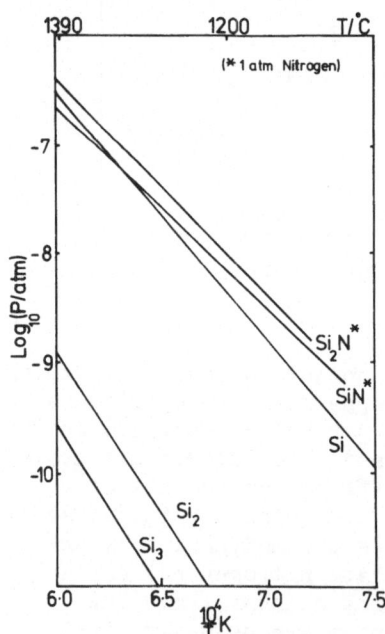

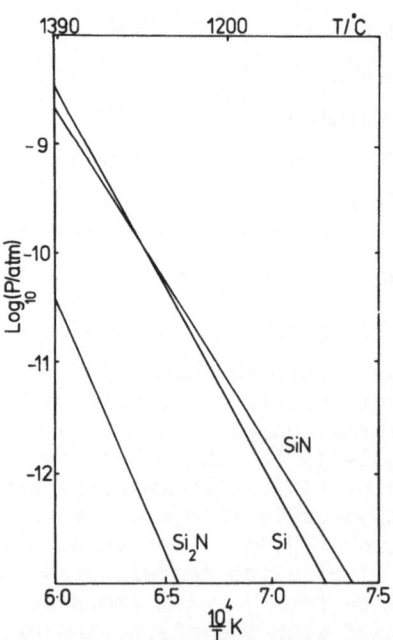

Figure 1: Theoretical vapour pressures over Si(c) as a function of temperature (16).

Figure 2: Theoretical vapour pressures over Si_3N_4 as a function of temperature (16).

Faster reactions have been reported with dissociated ammonia (20) and nitrogen/hydrogen mixtures (21) have replaced nitrogen. Most kinetic studies of silicon nitridation have made use of nitrogen alone, however, and the details of reactions in other nitriding atmospheres have not received the same attention.

3.2 The commercial process

Commercial grades of silicon are typically about $98^W/o$ pure with respect to metals, with Fe $(0.5-1.0^W/o)$, Al $(0-0.5^W/o)$ and Ca $(0-0.3^W/o)$ commonly being significant impurities. Additionally oxygen may be present at the 0.3 to $1.5^W/o$ level. The constitution of the impurity phases has been examined; many are complex silicates originating in the slag of the original silica reduction process (22). Free iron from the grinding and classification process can often be detected magnetically. Powders normally have a mean particle size in the region of 10 μm, but silicon fragments of 100 μm or more may also be present. In the production of the reaction bonded material silicon packing efficiencies of 55-75%, giving green densities in the range 1.3 to 1.7 Mg m^{-3}, can be achieved.

Although reaction can be observed at 1125°C, with 1 atm nitrogen (22,23), silicon proves to be difficult to nitride completely at temperatures below its melting point and a two-stage process is standard practice (20,21,24). A typical schedule might be 24-50h at 1350°C followed by 10-24h at 1450°C (23). Because there is no overall volume change in the compacted powder on nitriding (in spite of the 21.7% molar increase in volume) the final density depends only on the green density. A green density of 1.913 Mg m^{-3} should theoretically nitride to give fully dense silicon nitride, but in practice the conversion obtainable falls off at green densities above about 1.6 Mg m^{-3} (21) and maximum densities are in the region of 2.7 Mg m^{-3}. Evidence for the volatility of the solid phases in this system is provided by the fact that, even after an apparently complete reaction, the theoretical weight gain (66.49%) is often not observed, and loss of material corresponding to up to 5% of the original silicon is not uncommon (21).

The products of nitridation consist of a mixture of the α and β-phases with proportions in the region of 75:25 as long as the nitridation temperature is held below 1410°C, and richer in the β-phase is higher temperatures are used (25,26). That formation of the β-phase is favoured by higher temperatures has long been recognised (20,27), the usual assumption being that the α-phase is formed through vapour phase reactions involving $Si_{(g)}$ or SiO, and the β-phase from reaction between nitrogen and liquid silicon. A degree of control over the proportions of the two phases, as well as the microstructure of the reaction-bonded material, is thus obtainable through the time-temperature programme (24,26).

3.3 Kinetic studies of the silicon-nitrogen reaction

Because of the interest in reaction bonded silicon nitride as a ceramic material, the majority of the many studies carried out over the past 20 years have been on compacts of commercial grades of silicon powder. While this approach might have been justifiable from the point of view of obtaining rapidly a viable production process, it is not so well suited to the development of an understanding of the nitridation mechanism and fine details of the interface processes. Powders are in any case far from ideal as the basis for precise work of this type because of the assumptions necessary about particle size, shape and size distribution. It is also difficult to avoid contamination, particularly by iron, during the preparation procedures, and the nature and distribution of secondary phases may be hard to determine. The important, if not dominating, role of trace impurities in the nitridation reaction was recognised in the earliest comprehensive investigations (21,28). Nonetheless it has proved to be surprisingly difficult to define the exact functions of these impurities, and the notorious irreproducibility of silicon nitridation data, even from one furnace to another (29) must be ascribed in part to imperfect control over their presence in the system used. Recent improvements in understanding of the reaction mechanism have been obtained through the use of ultra-high purity silicon in carefully controlled gaseous environments (17,30), and it seems that this approach is the one most likely to provide the degree of understanding required if full control over the nitridation process is to be achieved.

3.3.1 Physical processes

With rapid reactions absorbing possibly several $cm^3 min^{-1}$ of nitrogen, it has been recognised that the permeability of silicon powder compacts to gas may become rate determining. For 20 μm particles 80% of the voids will in theory have diameters between 0.1 and 1.0 μm. The mean free path of nitrogen at 1400°C is 0.6μm, and the effect of pore size on gas diffusion should become noticeable. This may well be the case for cylindrical compacts greater than 55 mm diameter and high green densities (31). For smaller compact sizes, however, it has been assumed that the supply of nitrogen is adequate, and that processes occurring at the particle-gas interface are rate-determining (31,32). The picture may not be quite clear-cut however because of the demonstrated rate-controlling action of small amounts of oxygen (33), and it is conceivable that it is the outwards flow of silicon monoxide from a large compact which determines the overall nitridation rate rather than the inwards flow of nitrogen. For this reason the gas flow rates at the outer surface of a compact could be an important factor in controlling the silicon monoxide pressure gradient and flux. In this context the remarkable effects on nitridation rate and product quality of gas flow rate through the nitriding furnace are of great interest (34). Even a very slow flow (10 $cm^3 min^{-1}$)

appears to be significant when comparisons are made with a 'static' system, and almost 50% loss of strength under such conditions has been observed.

A further and so far unexplained aspect of the nitridation of silicon compacts is the accelerating action of pre-formed silicon nitride 'filler', attributed variously to its effect on the pore structure of the silicon compact (35), and on the sintering rate of the silicon powder itself (36). An additional possible explanation is that the nitride provides a high density of nucleation sites for new growth, but there is as yet no conclusive evidence for this effect.

Because of the strongly exothermic nature of the $Si-N_2$ reaction, a temperature gradient will always exist from the centre to the outside of a nitriding compact, and which under rapid heating conditions, particularly with a large individual compact (or high density of furnace loading), can lead to an uncontrolled reaction, with unwanted melting of the silicon in the centre of the compact. Under normal conditions, with small experimental compacts, such gradients seem to be relatively small. Experimental and calculated values for temperatures within a 50 mm diameter cylindrical compact have been obtained (37) and found to be in good agreement. The maximum difference at the centre ranged from 17° to 38°C above ambient for 6-10% reaction, independent of ambient temperatures between 1295 and 1330°C, resulting in a 3% difference in nitrided density at this stage. In this particular instance a constant α/β-phase ratio was obtained across the compact, but in other cases variations in phase composition from the outside to the centre (richer in β phase) have been observed (38). The explanation for this effect may lie as much with accessibility of oxygen, as with reaction temperature, but it must be noted that different silicon powders show varying sensitivities to temperature so far as the α-phase and β-phase yields are concerned (39).

3.3.2 Processes occurring at the gas-solid interface

Most kinetic studies of the silicon-nitrogen reaction have been made on loose or compacted powders, with the experimentally simpler measurement of weight gain being preferred to that of gas uptake. The conventional metallurgical approach to corrosion studies using single or polycrystalline bars has generally failed because of the problem of silicon evaporation in nitrogen containing traces of oxygen (28). Although two (α- and β-nitride) phases are invariably formed, in almost all cases rate laws have been derived on the basis of the overall extent of reaction, and the implications of the possibility that two distinct reactions may be proceeding simultaneously have tended to be ignored. Because the α-phase is normally the main product (at the 70-90% level) the omission may for practical purposes be not too serious; however, this is an aspect which requires further examination.

Where analyses of the product phase compositions have been made, useful data may often have been ignored by expressing these in terms of "α/β ratio", rather than absolute phase yield. Much attention has been directed at following nitridation kinetics to a high degree of conversion, often using relatively impure material. Most of these studies have been at temperatures below 1410°C, the assumed liquidus temperature, and these will therefore be reviewed first.

For the reaction with nitrogen (normally at 1 atm pressure) a range of rate laws has been recorded with different materials (Table 3). This diversity strongly suggests that an uncontrolled variable, such as an impurity in the silicon, or oxygen contamination of the nitrogen, is having a decisive influence on the reaction. The balance of opinion is that the rate-controlling process for the larger part of the reaction is the permeation of a gas (which has commonly been assumed to be nitrogen, although which equally could be a silicon-containing species) through gradually closing pores or fissures in an adherent, but only partially protective nitride scale (28), and that solid-state diffusion of both silicon and nitrogen does not make a significant contribution to the reaction (40,46,47). For this reason once channels do become completely blocked either by recrystallisation of material (40), or by fresh additions (47,41), nitridation is effectively terminated. The factors affecting the length of time over which material transport is possible before channel closure occurs have not been completely evaluated although nitride nucleation density has been shown to be of importance in ultra-high purity systems (47) and may well be so in others. Raising the temperature at the reaction plateau stage may then produce a further period of reaction (32), presumably as a result of thermal expansion disruption of the protective scale.

Almost equally numerous, and on the whole remarkably high, enthalpy of activation values for the bulk process have been reported, again for the most part ignoring the possible dual nature of the reaction (Table 4). In some studies the value has been found to depend on reaction extent, and temperature. For certain commercial powders a transition, with acceleration, in the nature of the rate-determining process may be seen at temperatures around 1370°C (39), which suggests that there may be a tendency for enthalpy values based on a temperature range extending above this point to be exaggerated. A change in mechanism would also be expected on reaching the melting point of silicon at 1410°C. If this is so, then the lower value in the region of 420 kJ mol^{-1} may be the more meaningful, which could indicate that silicon evaporation (ΔH_{443} kJ mol^{-1}) is an essential rate-determining element of the nitridation process (Table 5). Very little comparative data is available, but in the case of titanium nitridation, fast inwards diffusion of nitrogen through a porous nitride scale gives rise to linear scale formation kinetics and a much

Silicon stated purity/%	Temp./°C	Special Conditions	Law	Ref.
99.95 (cast bar)	1200–1410	0.2% N_2/argon	Logarithmic	(28)
99.5	1315–1385 1385–1410	– –	Logarithmic) Linear)	(40)
–	1240–1420	–	'Approximately' parabolic	(21)
98	1050–1350	–	Logarithmic	(41)
>99	1200–1500	–	Linear	(42)
'Semi-conductor' disc	1250–1380	–	Linear (oxide product	(43)
98.5–99.75	1250–1450	–	Parabolic	(44)
98.5–99.5	1350–1450	To 25% 25–100%	"Complex") Jander)	(32)
99.999	1290–1410	–	Linear (corrected for particle size)	(45)
'Semi-conductor'	1370	Low pressure (first stage) Low pressure (second stage)	Linear) Asymptotic)	(30)

Table 3: Rate laws for reactions of silicon with nitrogen.

Purity/%	Conditions	ΔH/kJ mol^{-1}	Ref.
99	1314–1415°C	652	(42)
98.5–99.5	1250–1450°C	547	(44)
98	0–30% conversion 40–70% conversion	419 837	(31)
99.999	1290–1410°C	661	(45)

Table 4: Enthalpy of activation values for the reaction of silicon powders with nitrogen.

Process	$\Delta H / \text{kJ mol}^{-1}$
$2\text{Si}(c) + \frac{1}{2}\text{N}_2(g) \longrightarrow \text{Si}_2\text{N}(g)$	385
$\text{Si}(c) \longrightarrow \text{Si}(g)$	443
$\text{Si}(c) + \frac{1}{2}\text{N}_2(g) \longrightarrow \text{SiN}(g)$	364
$\frac{1}{2}\text{Si}(c) + \text{SiO}_2(c) \longrightarrow \text{SiO}(g)$	339
$\text{SiO}_2(c) \longrightarrow \text{SiO}(g) + \frac{1}{2}\text{O}_2$	789

Table 5: Reaction enthalpies in the temperature range 1300–1400°C [16]

lower enthalpy of activation (92 kJ mol^{-1}) tentatively associated with nitride nucleation at the α-Ti interface (5). More work on high purity silicon is clearly required before activation enthalpy data can be interpreted with any degree of confidence.

The role of traces of oxygen in silicon nitridation is without doubt a very important one, not least in the initial stages of reaction (17,21,23). Concentrations of oxygen at the part per million level and higher have a marked overall retarding action which has been assumed to be due to the formation of silica (21). Even in apparently oxygen-free systems significant amounts may still be present in the silicon, both in solid solution and as surface silica films, the latter source increasing in importance with decreasing particle size. It has been demonstrated that nitridation can occur at a silicon interface underneath thin oxide films (48,49,50), but the reaction is slow and the nitride film itself acts as a further diffusion barrier. An induction period in the high temperature nitridation of ultra-high purity silicon has been attributed to the silica film, which can be removed as silicon monoxide by heating under vacuum or suitably pure inert gas (17). High initial nitridation rates are then observed. Conversely even a very low partial pressure of oxygen (10^{-8} atm) in the nitridation atmosphere will suppress the evaporation of silica (51) and so can have an effect on the long-term reaction rate out of all proportion to its concentration. The difficulties of achieving satisfactory control in this respect over the nitriding atmosphere are considerable. Thus, the type of alumina furnace tubing used (33,52), the position or order of the sample in the reaction zone (22), the type and location of oxygen getter (32,53), water vapour levels (29), and gas flow rate (34), have all been shown to be capable of affecting reaction rate and product, an effect attributable in one way or another essentially to oxygen. For this reason the general lack of agreement with regard both to rate law and to activation enthalpy is not surprising and complete understanding of the reaction must require more stringent control of the nitriding atmosphere.

A number of careful studies of the details of the nitride growth mechanism have been attempted, using different experimental approaches, and focussing attention on the initial stages of the reaction (17,40,30,47,54). This has been treated as a nuclei formation and growth process, with nuclei density being a critical factor in determining the extent of reaction at constant temperature, since the impingement of developing nuclei on each other eventually leads to termination of the reaction. With ultra-high purity silicon and prior removal of the native silica film, a kinetically linear initial stage can be observed, with a rate constant proportional to p_{N_2}. This stage has been associated with the formation of nitride nuclei, and their growth involving surface diffusion of nitrogen (30). Similar pressure/reaction rate dependencies are seen in the initial stage oxidation of metals such as Cu (55) and Fe (56), where it has been assumed that the surface diffusion of oxygen to the nucleus results in a depletion zone within which no new nuclei can form. Nucleation density is therefore determined by oxygen pressure, surface migration rate and temperature. An alternative explanation for the nucleation density is the Ostwald ripening effect, in which incipient nuclei evaporate on to larger nuclei, nuclei growth in this case being dependent on surface diffusion of both metal and oxygen (57).

In the case of ultra-high purity silicon nitride, completion of the linear stage leads to a second stage which can be fitted to an asymptotic rate law (30), suggesting a self-blocking pore closure mechanism similar to the mutual pore closure (logarithmic) mechanism postulated in several studies of less pure material (28, 41). Onset of this second, decelerating, stage of nitridation is dependent on the initial nuclei density, however, and at sufficiently low nitrogen pressures (0.067 atm at 1370°C) may not be seen. Thus although the nitridation rate is relatively low, complete nitridation of large particles is possible at low pressures, and which cannot otherwise be achieved with ultra-high purity silicon (30). Probably for a similar reason, thick films (to 20 μm) of nitride have also been observed to form readily, following a logarithmic rate law, on high purity polycrystalline silicon bars when a low partial pressure (0.002 atm) of nitrogen in argon is used (but not under nitrogen at 1 atm (28)).

A second important factor, in addition to (though certainly in part related to) the oxygen in the nitridation reaction, is the presence of metallic elements such as iron in the silicon, either as impurities, or deliberate additions. Indeed it has been believed that high purity silicon was virtually un-nitridable (46), although thick films can be grown if careful attention is paid to oxygen gettering (58), and more effectively, as was pointed out above, by the use of low partial pressures of nitrogen. Several studies have been made of the effects of metals and fluorides in amounts ranging from parts per million to several percent (17,21, 28,59,60). Interestingly, in view of the important function of surface silica, the effect of these additives has often been seen most clearly in the earlier stages of reaction, and appears less pronounced when long term yields are compared (21). There is no doubt that one of the functions of these additives is an inter-action with silica films, and the effectiveness of iron at the part per million level in this respect has now been demonstrated (17), although the exact nature of the interaction has still to be clarified. Thermodynamically metallic iron should be without effect on silica under normal nitriding conditions, but it is not clear whether the element is present solely as the metal, or in part as FeO (or a silicate). Other functions for metallic elements have been proposed in the main stage of nitridation, following elimination, or disruption, of the primary silica films. These include the formation of lower melting point silicon alloys (60). Direct evidence has been provided for the presence of $FeSi_2$ in nitrided silicon powders (32). This phase melts at 1208°C and since the solubility of Fe in Si is low (7×10^{-7} $^{a}/o$ at 1300°C (10)), most of the iron present initially will ultimately appear in this form. One possible function of such liquids might be in assisting nitride formation through a VLS mechanism, and nitride whiskers have been shown to grow by this route (32,61). Liquid $FeSi_2$ would be expected to be an effective medium for silicon nitride nucleation and growth because the solubility of silicon in iron under 1 atm nitrogen at 1400°C is only $26.9^{w}/o$ (62), so that a phase corresponding initially to $FeSi_2$ should precipitate silicon nitride on exposure to nitrogen. Detailed attention needs to be paid to the kinetics of nitridation of compositions in the Si-Fe and similar systems if the mechanism of transition metal cat-alysis is to be clearly understood.

Studies of the microstructure of a growing metal corrosion film always make an important contribution towards an understanding of the processes involved, and this has been equally true in the case of silicon nitridation. It was recognised in the earliest work with compacted silicon powder that two types of material tended to be formed (24). A whiskery matt (predominantly α-phase) was observed to grow out into the void space in the silicon com-pact, and the residual silicon grains were slowly converted to more granular material (assumed to be largely β-phase) (63). The former process appeared to be favoured by temperatures below 1400°C; the latter, by higher temperatures and, perhaps surprisingly, by higher compaction pressures (20). Vapour phase reactions involving silicon vapour or SiO (23,24) have been assumed to be responsible for the growth of these whiskers although it has been shown more recently that the nitridation of silicon monoxide vapour, genera-ted from SiO_2/C mixtures, is also capable of yielding a product with a more equiaxed, dense, morphology (64,65). Whisker growth must therefore involve additional factors, the most likely of which is the presence of traces of impurity.

The availability of vapour phase processes readily accounts for the absence of overall volume change on reaction-bonding, the volume expansion on nitridation being easily accommodated by the ample void space, as well as for the absence of sintering shrinkage (21). However, while the broad pattern is well-established, and qualitatively fits kinetic data indicating vapour phase transport of material through non-protective nitride scales, a detailed picture has been slower to develop.

Improved understanding of the reaction mechanism has come from the more recent observation of extensive pore development at the silicon-silicon nitride interface. This was first observed in the case of ultra-high purity single crystal silicon (66), and it has become apparent that voids at the interface, and pores which appeared to migrate inwards into silicon particles (67), are also a common, if not universal, feature of nitridations of less pure grades of material, being most strikingly evident in nitridations at pressures of less than 1 atm (30,47) (Fig. 3). An assumption made is that the pore channels may move inwards until several meet at the centre of a large silicon particle, thereby possibly accounting for some of the larger regions of porosity clearly having as their original basis a silicon particle. An early

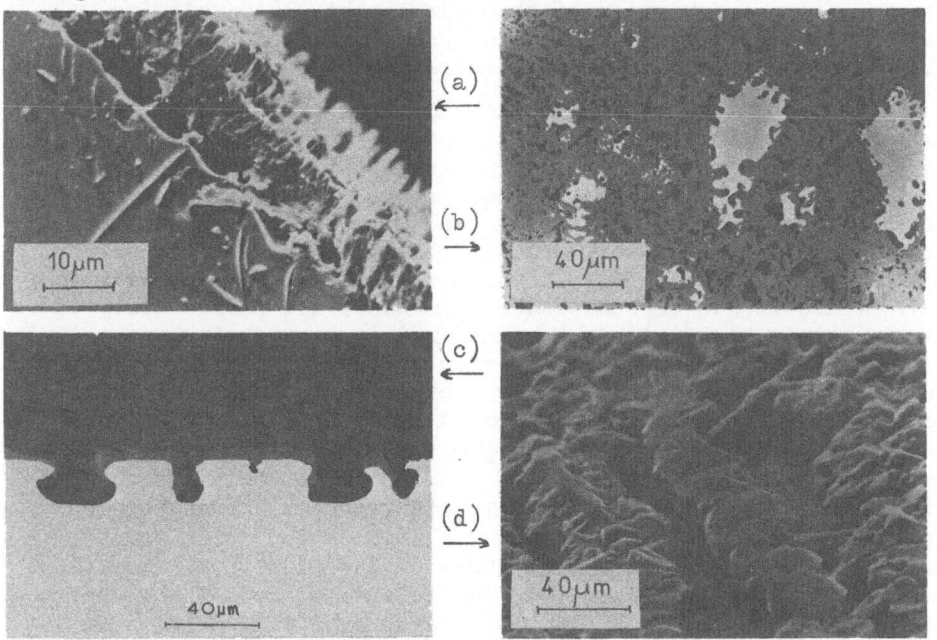

Figure 3: Pore development in silicon during nitridation
(a) Single crystal, 1 atm N_2, fracture face
(b) Commercial powder, 1 atm N_2, polished section (67)
(c) Single crystal, 0.067 atm N_2, polished section (30)
(d) Single crystal, 0.067 atm N_2 (30).

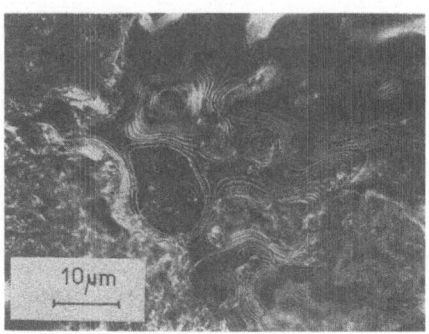

Figure 4: Dark field optical micrograph of terraced silicon (111) surfaces in pores at a single crystal-silicon nitride interface (courtesy E.W. Roberts).

suggestion was that a vacancy condensation mechanism similar to that assumed to operate in the fast oxidations of Fe and Cr was responsible for the generation of this porosity (67), but following detailed observations of slower growth patterns on single crystal silicon at low nitrogen pressure, and the presence of facets on the silicon surfaces of the pores (Fig. 4), this explanation was replaced by one involving the outwards evaporation, or surface diffusion, of silicon from exposed surfaces, with the progressive encapsulation of that surface by the developing silicon nitride (30). For inwards pore migration to continue therefore the nitride film is required to be permeable to nitrogen.

Microstructural features such as these are observed in metal-oxygen systems where significant vapour-phase transport of metal-containing species is possible by virtue of a high vapour pressure (10^{-8} to 10^{-9} atm) at reaction temperature. Chromium may show films and pore developments identical with those seen on single crystal silicon, and where chromium vapour is assumed to diffuse to the oxide surface when diffusion as the cation takes over (68, 69). While single crystal oxide film has a low permeability to oxygen, grain boundaries are considered to provide leakage paths and oxide formation may then be seen on the underlying metal. In the case of the less volatile iron extensive pore development is not seen, and where pores are formed reaction ceases (70). In the silicon-nitrogen case, it appears that evaporation of silicon, as $Si(g)$ or a sub-nitride (or as silicon monoxide in the presence of traces of oxygen), and its outwards transport through porosity and microcracks, is an important mechanism for nitride growth. Because of the high vapour pressures of these species theoretical evaporation rates are more than sufficient to account for observed nitridation rates. With chromium clean, facetted, metal-pore surfaces were taken as indicating an absence of oxygen. Silicon nitride growth into larger cavities certainly does appear to occur, however, and in some cases nitride deposits form on cavity floors. Since nitrogen is present in the pores, nitride nucleation on a clean, silicon surface must be a relatively improbable occurrence.

Several reports have appeared of direct observations on the early, nuclei formation, stages of the silicon-nitrogen reaction. Individual nuclei development on clean (111) and (311) silicon surfaces has been observed which suggest strongly that surface diffusion of silicon towards the growth point is occurring (54,71). This would then be substantiating evidence for an Ostwald ripening effect on the nucleation density dependence on gas pressure (57). Attempts to grow uniform thin films of nitride on single crystal silicon have, however, been frustrated by non-uniform formation of nuclei, each developing outwards at an appreciable rate (46). Thin scales of nitride on (111) silicon have been seen to show preferred orientation with respect to the crystallographic axes of the substrate, though the origin of this effect is not clear (58).

The growth mechanisms for the whisker form of silicon nitride require more detailed investigation, though there is no doubt that a VLS mechanism can operate and solidified droplets at whisker tips containing Fe and Al have been observed (61). Scanning electron micrographs also show that whisker and blade growth can apparently originate on underlying relatively dense material, and that the type of growth is apparently independent of the structure and nature of this underlying material (66). This suggests that whisker growth may in some instances be a process distinct from, and independent of, the primary nitridation reaction. Similar features have been observed in the oxidation of Fe (72) and Cu (73) where the whiskery material contains the metal in its highest oxidation state and appears not to affect growth of the base oxide film. It has been assumed that growth is controlled by local fluctuations in the M/O activity ratio. In the case of silicon nitride the assumption is generally made that the whiskers are almost always of α-phase material (23,25,63), and if oxygen is required for the growth of such material then similar local variations in oxygen activity within the reaction zone may be responsible for its development. Whether the β-phase can be obtained in whisker form is less clear.

3.3.3 Processes occurring at the gas-liquid interface

Commercial nitriding schedules involve a period above the nominal melting point of silicon in order to bring about completion of nitridation, and kinetic studies have been extended into the 1400-1500°C range (42). The fact that liquid phases may also be expected at lower temperatures in impure silicon has already been referred to. Very little work, however, seems to have been carried out specifically into the liquid silicon reaction.

An early study (74) estimated the solubility of nitrogen in silicon to be 2×10^{-2} a/o, and showed the growth into, and precipitation from, the melt of both α- and β-phase needles. This suggests that liquid silicon could be an effective transporting medium for nitrogen during normal nitridations at high temperature.

Polished sections of partially nitrided powder compacts known to
have contained molten grains show a distinctive growth of large,
well facetted, nitride crystallites within the grains (63), where
it would appear that nitrogen transport through the melt is
occurring. The volume expansion on nitridation is then accommodated by flow of the liquid phase. The nitride formed in this way
is usually assumed to be predominantly β-phase material (23).

3.4 Phase composition of the product

Earlier detailed studies conducted over a wide range of low
oxygen potentials suggested that the α-phase was formed under
"high" oxygen potentials and should be regarded as an oxynitride
of approximate composition $Si_{11.4}N_{15}O_{0.3}$. The β-phase appeared
to be formed under "low" oxygen potentials (75). The view that
the α-phase is an oxynitride has been questioned, however, on the
grounds that it can contain far less oxygen than that required by
the crystallographic and thermodynamic evidence (76,77,78). In
practice, with normal nitriding systems, conditions are such that
oxygen potentials higher than those required for the formation of
even silica (10^{-17} atm at 1650K) are likely to exist, and the
nitride phase formed therefore would depend on a mixture of kinetic factors, and the local availability of oxygen. The general
impression obtained is that below about 1410°C the α/β phase ratio
in the product is relatively insensitive to nitriding conditions
(including variations in oxygen potential over several orders of
magnitude) (29,79) and falls usually in the region of 40:60 to
90:10 for no very obvious reason. On the other hand almost pure
α-phase has been obtained under conditions of both low oxygen
potential and absolute quantity (58), suggesting, if oxygen is
required, an almost catalytic role.

Consideration of the separate rates of formation of the α-
and β-phases at temperatures below 1410°C from silicon powders
of varying purity and oxygen content shows that the nitridation
reaction can proceed in two stages (33). The first, slower,
stage provides all the α-phase and some β-phase; the second,
initially faster, stage generates the bulk of the β-phase (Fig.5).
Sigmoidal kinetic curves sometimes seen with impure powders may
be a reflection of this type of behaviour. Some of the α-phase
formed initially may subsequently disappear suggesting slow α-
phase to β-phase conversion, even at temperatures as low as 1370°C
(60). Deliberate additions of silicon monoxide to the nitriding
atmosphere have the effect of suppressing both α-phase and β-phase
forming reactions, but the β-phase reaction more effectively, so
that the overall effect is to give a product richer in α-phase
(33). At temperatures approaching the silicon melting point the
α-phase yield may diminish and can reach zero, while the β-phase
yield continues to increase as would be expected, so that 100% β-
phase can be obtained at a temperature as low as 1400°C (39). The
temperature at which α-phase formation becomes less important

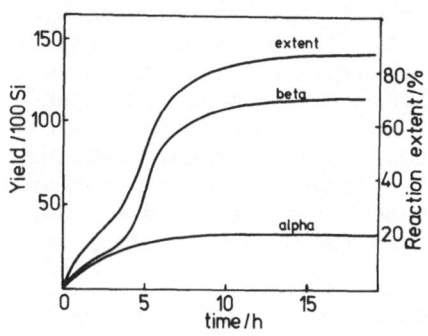

Figure 5: α-Si$_3$N$_4$ and β-Si$_3$N$_4$ yields, and overall extent of reaction, in the nitridation at 1365°C of commercial purity silicon powder (33).

appears to be related to the metal/oxygen content ratio of the powder. A high purity powder containing a small amount of oxygen may be less reactive, and provide a greater proportion of α-phase, than one containing a larger percentage of oxygen if this is balanced by a higher metal content.

It would appear on this evidence that the α-phase is formed while the silicon is contaminated by oxygen (specifically as silicon monoxide or silica) either from a SiO-N$_2$ reaction or at surfaces containing adsorbed oxygen, and that the β-phase is formed from silicon vapour (or a subnitride), or possibly in the presence of a liquid phase produced by interaction of silicon and reduced metallic contaminants. The nature of the surface at which reaction is occurring and the presence of adsorbed oxygen or halogen for example, might well be a factor determining the crystal structure of the product. Many such cases are documented, even where incorporation of adsorbed impurities into the crystal would seem to be extremely unlikely (80). The increased rate of the second stage process suggests either a reaction unhindered by the presence of silica, or possibly one taking place within a liquid alloy. The presence of a liquid is not in itself a sufficient condition for β-phase formation, however, because the α-phase can also be formed from liquid iron-silicon alloys under suitable nitrogen and oxygen potentials (81).

In the case of impure silicon powders already containing perhaps 1^W/o Fe, the addition of metals such as iron appears to have only a slight effect on the α-forming and β-forming reactions, but in the direction of increasing the β-phase yield (32,39). Additions of aluminium or alumina on the other hand appear to aid more strongly formation of the β-phase (21,35), presumably because Al and O now enter the nitride lattice forming β'-sialon compositions.

The reaction of GeO$_2$ with ammonia yields β-Ge$_3$N$_4$ (14), iso-structural with β-Si$_3$N$_4$ (75). This product is also favoured by high gas flow rates in the Ge/N$_2$ system, and it has been suggested that the β-phase might be stabilised by oxygen. It seems, however, that if germanium should parallel silicon in its behaviour then a close investigation of the effects of metallic contaminants would be required before definite conclusions could be drawn. α-Ge$_3$N$_4$ has been observed to form in the early stages of the reaction of germanium with nitrogen, and the β-phase later, features explained in terms of changing gas compositions at the Ge-Ge$_3$N$_4$ interface (82).

The question of the exact natures of the α- and β-phases of silicon nitride (and germanium nitride), and the possible stabilis-ing action of minute amounts of impurities such as oxygen and aluminium, is one which therefore still remains to be resolved. An unsatisfactory aspect of most of the experimental work to date is that it has been carried out using alumina, or mullite furnace tubing, and the future use of a different type of clean reaction environment is probably essential.

4. REACTION BONDING

The Si-N system is exceptional for its capacity to develop during reaction a nitride microstructure consisting of strong, interconnecting crystallites. The basis for this behaviour must be the volatility of the reactants, probably enhanced by traces of oxygen, and perhaps to a lesser extent of the product itself. These factors, together with the inherent low solid-state mobilit-ies of the elements, also account for the absence of sintering shrinkage. This system is representative therefore of the case of sintering by an evaporation-condensation mechanism, for which com-pact densification does not occur (83). Silicon nitride powder appears to be sinterable only if fairly high proportions of im-purity or secondary phases are present, presumably to assist a liquid phase sintering mechanism (84).

It seems a safe assumption that the bulk of the bonding micro-structure is developed during the primary crystal growth process, and that the nitride once formed is relatively involatile and effectively inert. In the case of processes involving the evapor-ation of molecular species not existing in the solid itself the sticking coefficients are generally small, and evaporation rates slow (85). The evaporation rate of silicon nitride has also been shown to be suppressed by the presence of free silicon (86). How-ever, as in nitridation, the significant involvement of trace amounts of oxygen must always be suspected unless proved to the contrary. The relatively high strength of the reaction-bonded material must depend on the formation of bridging nitride crystals linking nitride growth centres. There appears to be very little evidence for significant presintering of silicon grains prior to nitride formation unless deliberate efforts are made. Non-bridging nitride might not be expected to contribute to the over-all strength of a body and some strength improvement in reaction-bonded material might be obtained if control over the proportions of the two types of material could be achieved. Although much of the fine whiskery matt might have been expected to be of this non-bridging type, considerable strength is developed in material prepared at relatively low temperatures (24,87) when this form predominates. Moreover an increased proportion of fine-grained α-phase matt (resulting from the use of hydrogen-nitrogen mixtures) has been specifically associated with increased strength and de-creased creep rates (88). It is clear that understanding of the true basis of strength in reaction-bonded material is far from satisfactory. High temperature annealing has also improved strengths to some extent, though prolonged high temperature treat-ment appears to lead to extensive α-phase to β-phase conversion, and a fall-off in strength attributed to microstructural coarsen-ing (21).

It would be of great interest to be able to examine other pure systems in which there is clear evidence of sintering by an evaporation-condensation mechanism, so that the processes of micro-structure and strength development can be followed free from com-plications arising out of simultaneous chemical reactions. Unfort-unately well-documented cases where vapour phase transport of material is unequivocally the controlling process, and where information is available about the development of strength in the sintering compact, appear not to exist.

The original work (83) on sodium chloride, showing neck development without movement of particle centres, was on micro-spheres of 61-71 µm diameter. Compacts of finer (<53 µm) sodium chloride powder can be densified without difficulty (89), however, and plastic flow has been suggested as a possible contributing mechanism (90). Similarly, while microsphere ice (200-600 µm in diameter) appears to sinter predominantly by an evaporation-con-densation mechanism (91), fine powders (<1 µm) do densify, and again plastic flow is indicated (92). Particle size is clearly an important factor determining the sintering mechanism. The volatile chromium oxide system is complicated by the existence of several oxides of chromium, but shows negligible sintering shrinkage unless a critical (10^{-15} atm) oxygen pressure is main-tained (93). No data appears to be available however for sinter-ing rates and the strengths of sintered compacts.

It is clear that if a model system is to be developed for study, and this would be very desirable, much more work in these or similar systems is required. Silicon itself would be expected to be a good case, provided careful control over oxygen potentials could be maintained.

5. CONCLUSIONS

In spite of the large numbers of studies carried out, and the volume of information now amassed on the silicon-nitrogen system, understanding of almost every aspect of the nitridation and reaction-bonding processes is still rudimentary. After the first burst of progress some 15 or so years ago, further significant developments have been slow. Many of the difficulties of inter-pretation have in part undoubtedly centred around the multifunc-tional role of, and the difficulties of controlling, very small amounts of oxygen in a non-oxide system. A further difficulty has been that of obtaining silicon in powder form with a sufficiently high level of purity. Oxygen contamination is, of course, a feature also likely to be of importance in the cases of other nitride systems. There have, however, been encouraging recent developments associated with patient efforts to use very high purity materials and systems, and this approach must point the way for all further studies if more rapid progress is to be seen.

Attention was drawn some 10 years ago by Bénard (94), in a similar context, to the importance of purity in studies of the oxidation of metals. He then noted that "we began to make some progress when the two following principles were taken into account: First, to understand the mechanisms involved in these reactions it is necessary to make experiments on metals of high purity, and then to study, systematically, the effect of additional elements in increasing proportion. Second, a large number of factors initially assumed to be of little importance and neglected in theoretical treatments play, in fact, a decisive role in the kinetics of these reactions. As a consequence these factors should be studied separately". The "factors" referred to included oxide recrystallisation, impurity accumulation at interfaces, and pore formation. It was also noted that "if these principles had been followed earlier, a large number of wasted efforts would have been avoided, and ... we would have a much better understanding of what happens in these reactions". There is every reason for supposing that these principles are equally applicable in the field of metal nitridation, and certainly in that of silicon nitridation.

REFERENCES

1. E.A. Gulbransen and S.A. Jennsen. J. Electrochem. Soc., 1951, 98, 241.
2. R.J. Bickley and S.J. Gregg. J. Chem. Soc., 1966(A), 1849.
3. M.W. Roberts and F.C. Tompkins. Proc. Royal Soc., 1959, 251A, 369.
4. E.A. Gulbransen and K.F. Andrew. Trans. Met. Soc. AIME, 1950, 188, 586.
5. K.N. Strafford and J.M. Towell. Oxidation of Metals, 1976, 10, 41.

6. J.P. Bars, E. Etchessahar, J. Debuigne and A. Leroux. C.R. Acad. Sc. Paris, Serie C, 1974, 278, 581.
7. J. Paidassi and R. LeDelliou. C.R. Acad. Sc. Paris, Serie C, 1971, 272, 249.
8. M.D. Lyutaya and O.P. Kulik. Poroshk Metall., 1974, 21.
9. K. Schwerdtfeger, P. Grieveson and E.T. Turkdogan. Trans. AIME, 1969, 245, 2461.
10. Constitution of Binary Alloys, First Supplement. Ed. R.P.Elliott, McGraw-Hill, New York, 1965.
11. G.V. Samsonov and V.M. Slepstov. Kinetika I Kataliz, Akad. Nauk. SSSR Sb. Statei, 1960, 129.
12. G. Long and L.M. Foster. J. Amer. Ceram. Soc., 1959, 42, 53.
13. C.F. Cooper, C.M. George and L. Winter. Special Ceramics 4, Ed. P. Popper, B.Ceram.R.A., 1968, p.1.
14. J.C. Labbe, F. Duchez and M. Billy. C.R. Acad. Sc. Paris, Serie C, 1971, 273, 1750.
15. G.W.C. Kaye and T.H. Laby. Tables of Physical and Chemical Constants. 14th Edition, Longman, 1973, p.151.
16. JANAF Thermochemical Tables, 2nd Edition, NBS, 1971.
17. A. Atkinson and A.J. Moulson. Science of Ceramics 8, B.Ceram. Soc., 1976, p.111.
18. R.D. Pehlke and J.F. Elliott. Trans. A.I.M.E., 1959, 215, 781.
19. S. Wild, P. Grieveson and K.H. Jack. Special Ceramics 5, Ed. P. Popper, B.Ceram.R.A., 1972, p.385.
20. N.L. Parr, R. Sands, P.L. Pratt, E.R.W. May, C.R. Shakespeare and D.S. Thompson. Powder Met., 1961, 8, 152.
21. P. Popper and S.N. Ruddlesden. Trans. Brit. Ceram. Soc., 1960, 61, 603.
22. D.J. Godfrey. Soc. Automotive Engineers, Paper No. 740238, New York, 1974.
23. N.L. Parr and E.R.W. May. Proc. Brit. Ceram. Soc. 1967, 7, 81.
24. N.L. Parr, G.F. Martin and E.R.W. May. Special Ceramics, Ed. P. Popper, Heywood, 1960, p.102.
25. W.D. Forgeng and B.F. Decker. Trans. A.I.M.E., 1958, 212, 343.
26. A.G. Evans and R.W. Davidge. J. Mat. Sci., 1970, 5, 314.
27. E.T. Turkdogan, P.M. Bills and V.A. Tippett. J. Appl. Chem., 1958, 8, 296.
28. J.W. Evans and S.K. Chatterji. J. Phys. Chem., 1958, 62, 1064.
29. D.P. Elias and M.W. Lindley. J. Mat. Sci., 1976, 11, 1278.
30. A. Atkinson, A.J. Moulson and E.W. Roberts. J. Amer. Ceram. Soc., 1976, 59, 285.
31. A. Atkinson, P.J. Leatt and A.J. Moulson. Proc. Brit. Ceram. Soc., 1973, 22, 253.
32. D.R. Messier and P. Wong. J. Amer. Ceram. Soc., 1973, 56, 480.
33. D. Campos-Loriz and F.L. Riley. J. Mat. Sci., 1976, 11, 195.
34. D.P. Elias and M.W. Lindley. J. Mat. Sci., 1976, 11, 1288.
35. A. Atkinson, P.J. Leatt and A.J. Moulson. J. Mat. Sci., 1972, 7, 482.
36. R.J. Lumby. U.S. Pat. 3,591,337, 10 Apr. 1967.

37. A. Atkinson and A.D. Evans. Trans. Brit. Ceram. Soc., 1974, 73, 43.

38. W. Naruse, M.N. Ojiri and M. Tada. Japan J. Metals, 1971, 35, 731.

39. D. Campos-Loriz. Ph.D. Thesis, University of Leeds, 1976.

40. M. Billy. Ann. Chim., 1959, 4, 797.

41. D.S. Thompson and P.L. Pratt. Science of Ceramics 3, 1967, p.33.

42. K.J. Hüttinger. High Temperatures-High Pressures, 1969, 1, 221.

43. K.J. Hüttinger. High Temperatures-High Pressures, 1970, 2, 89.

44. R.F. Horsley. Ph.D. Thesis, University of Leeds, 1972.

45. Y. Inomata and Y. Uemura. Yogyo-Kyokai-Shi, 1975, 83, 42.

46. R.G. Frieser. J. Electrochem. Soc., 1968, 115, 1092.

47. A. Atkinson, A.J. Moulson and E.W. Roberts. J. Mat. Sci., 1975, 10, 1242.

48. S.J. Raider, R.A. Goula and J.R. Petrak. Appl. Phys. Letters, 1975, 27, 150.

49. B.H. Vromen. Appl. Phys. Letters, 1975, 27, 152.

50. E. Kooi, J.G. van Lierop and J.A. Appels. J. Electrochem. Soc., 1976, 123, 1117.

51. C. Wagner. J. App. Phys., 1958, 29, 1295.

52. R.B. Guthrie and F.L. Riley. J. Mat. Sci., 1974, 9, 1363.

53. D.R. Messier, P. Wong and A.E. Ingram. J. Amer. Ceram. Soc., 1973, 56, 171.

54. R. Heckingbottom and P.R. Wood. Surface Science, 1973, 36, 594.

55. J. Bénard. Z. Elektrochem., 1959, 63, 799.

56. W.E. Boggs, R.H. Kachik and G.E. Pellisier. J. Electrochem. Soc., 1965, 112, 539.

57. G.E. Rhead. Trans. Faraday Soc., 1965, 61, 799.

58. R.B. Guthrie and F.L. Riley. Proc. Brit. Ceram. Soc., 1973, 22, 275.

59. H. Suzuki. Bulletin of Tokyo Institute of Technology, 1963, 54, 163.

60. A. de S. Jayatilaka. D. Phil. Thesis, University of Cambridge, 1975.

61. V.N. Gribkov, V.A. Silaev, B.V. Schchetanov, E.L. Umantsev and A.S. Isaikin. Sov. Phys. Crystallogr., 1972, 16, 852.

62. P. Grieveson, E.T. Turkdogan and F.J. Beister. Trans. A.I.M.E. 1963, 227, 1258.

63. B.J. Dalgleish and P.L. Pratt. Proc. Brit. Ceram. Soc., 1973, 22, 323.

64. K. Kijima, N. Setaka, M. Ishii and H. Tanaka. J. Amer. Ceram. Soc., 1973, 56, 346.

65. A. Hendry and K.H. Jack. Special Ceramics 6, Ed. P. Popper, B.Ceram.R.A., 1975, p.199.

66. R.B. Guthrie. Ph.D. Thesis, University of Leeds (1974).

67. A. Atkinson, P.J. Leatt, A.J. Moulson and E.W. Roberts. J. Mat. Sci., 1974, 9, 981.

68. V.R. Howes. Nature, 1967, 216, 362.

69. J. Stringer, A.Z. Hed, G.R. Wallwork and B.A. Wilcox. Corrosion Science, 1972, 12, 625.

70. D. Caplan and G.I. Sproule. Oxidation of Metals, 1975, 9, 459.

71. R. Heckingbottom, S.S. Valleekanathan and P.R. Wood. Post Office Research Dept. Report No. 394, London, 1974.

72. I. Svedung, B. Hammar and N.G. Vannerberg. Oxidation of Metals, 1973, 6, 21.

73. E.A. Gulbransen, T.P. Copan and K.F. Andrews. J. Electro- chem. Soc., 1961, 108, 119.

74. W. Kaiser and C.D. Thurmond. J. Appl. Phys., 1959, 30, 427.

75. S. Wild, P. Grieveson and K.H. Jack. Special Ceramics 4, Ed. P. Popper, B.Ceram.R.A., 1968, p.237.

76. H.F. Priest, F.C. Burns, G.L. Priest and E.C. Skaar. J. Amer. Ceram. Soc., 1973, 56, 395.

77. A.J. Edwards, D.P. Elias, M.W. Lindley, A. Atkinson and A.J. Moulson. J. Mat. Sci., 1974, 9, 516.

78. K. Kijima, K. Kato, Z. Inoue and H. Tanaka. J. Mat. Sci., 1975, 10, 362.

79. M. Mitomo. J. Amer. Ceram. Soc., 1975, 58, 527.

80. H.E. Buckley. "Crystal Growth", Chapman & Hall, London, 1951, p.383.

81. I. Colquhoun, S. Wild, P. Grieveson and K.H. Jack. Proc. Brit. Ceram. Soc., 1973, 22, 207.

82. S. Wild, P. Grieveson and K.H. Jack. Special Ceramics 5, Ed. P. Popper, B.Ceram.R.A., 1972, p.271.

83. W.D. Kingery and M. Berg. J. Appl. Phys., 1955, 26, 1206.

84. G.R. Terwilliger and F.F. Lange. J. Mat. Sci., 1975, 10, 1169.

85. E.G. Wolff and C.B. Alcock. Trans. Brit. Ceram. Soc., 1962, 61, 667.

86. E.D. Whitney and H.D. Batha. J. Amer. Ceram. Soc., 1973, 56, 365.

87. B.F. Jones and M.W. Lindley. J. Mat. Sci., 1975, 10, 967.

88. J.R. Mangels. J. Amer. Ceram. Soc., 1975, 58, 354.

89. A.A. Ammar and D.W. Budworth. Proc. Brit. Ceram. Soc., 1965, 3, 185.

90. C.S. Morgan, L.L. Hall and C.S. Yust. J. Amer. Ceram. Soc., 1963, 46, 559.

91. W.D. Kingery. J. Appl. Phys., 1960, 31, 833.

92. H.H.G. Jellinek and S.H. Ibrahim. J. Colloid and Interface Sci., 1967, 25, 245.

93. P.D. Owenby and G.E. Jungquist. J. Amer. Ceram. Soc., 1972, 55, 433.

94. J. Bénard. J. Electrochem. Soc., 1967, 114, 139C.

DISCUSSION (Goursat, Krauth, Smeltzer)

Lange: Our views concerning the nitriding kinetics of Si are that active oxidation conditions are required, that Si_3N_4 is formed by the vapour phase reaction:

$$3SiO + 2N_2 \longrightarrow Si_3N_4 + 3/2 O_2,$$

which, in turn, releases oxygen to allow a cyclic reaction and that, when iron is present the $FeSi_2$ liquid phase actively oxidises, releasing SiO and eating through the solid Si to maintain its constant composition. This allows the fast nitriding kinetics observed with impure (Fe containing) Si powder.

Riley: There is no doubt that, under conditions where one would expect $FeSi_{2(1)}$ to be formed then acceleration of reaction rate can be observed. I would not say that nitridation necessarily involves SiO, however, and it is unreasonable to ignore the existence of the high vapour pressures of Si_2N or Si. Of course, one would not expect to get fast nitridation in the presence of oxide films and these have probably been responsible for slow reactions observed with high purity silicon.

Singhal: Has nitridation been attempted with atomic nitrogen or nitrogen plasma? Would you expect the rates of nitridation to be much different in atomic nitrogen than in molecular nitrogen?

Riley: Atomic nitrogen appears to be a more effective nitriding species, as illustrated by experiments with low pressure nitrogen in an HF glow discharge (74). It may also account in part for the faster nitridation rates observed with ammonia.

Jelacic: Do you find different sorts of whiskers as a function of the nitriding atmosphere? Because we obtained a layer of a kind of whiskers that cover the surface of the nitrided silicon, when we use NH_3 as nitriding agent.

Riley: The use of hydrogen/nitrogen mixtures has been associated with enhanced whisker production under some conditions (88), but this observation is open to more than one interpretation.

THE ROLE OF IMPURITIES IN THE FORMATION OF REACTION-BONDED SILICON
NITRIDE

A. de S. Jayatilaka* and J.A. Leake

Department of Metallurgy and Materials Science,
University of Cambridge, Pembroke Street,
Cambridge, CB2 3OZ, U.K.

ABSTRACT. The influence of impurities, notably iron, on the
nitridation of silicon powder is briefly reviewed and new data
presented. The most effective are metals which form silicon-rich
eutectic liquids melting not far below the nitridation temperature.
Aluminium and the alkali metals are ineffective. Silicate liquids
do not appear to be necessary. The relative amounts of the α and
β phases formed in the presence of certain impurities appear to be
related to α' or β' SiAlON formation.

1. INTRODUCTION

The marked effect of a small amount of iron on the nitridation of
silicon is well known [e.g. 1]. The mechanisms by which it
enhances nitridation and its influence on the proportions of α
and β produced are beginning to be understood. Systematic study
of the effects of other impurities has been rare [2]. The
emphasis here is on impurities in the silicon powder, not in the
furnace atmosphere.

2. PREVIOUS WORK

The extent of nitridation and the proportions of α and β formed
are known to be affected by many factors [3,4,5]. Vapour-solid,
vapour-liquid-solid and solid state mechanisms of growth have
been suggested; vapour transport of silicon occurring by SiO or

---

*Now at: Applied Sciences Laboratory, University of Sussex,
          Falmer, Brighton, BN1 9QT, U.K.

Si species. Definite crystallographic relationships in the microstructure may arise [4,6,7]. Transformation of α to β occurs at
higher temperatures and/or pressures but evidence for the reverse
transformation is scanty. A detailed study of the rôle of
impurities [2,3] revealed two stages in the presence of iron;
firstly the disruption and/or removal of the silica layer on the
silicon and secondly a presumed changed morphology significantly
reducing the sealing-off of the nitrogen channels, consistent with
the observation that iron promotes the growth of α needles rather
than fine-grained material [4]. Iron reacts very rapidly with
silicon at nitriding temperatures despite the oxide layer [2] and
there have been suggestions that eutectic liquid formation is
important [8]. The $FeSi_2$-Si eutectic temperature is 1208°C [9]
and the effect of iron becomes negligible below this [2]. The
formation of $FeSi_2$ in beads of unreacted silicon is known [1].
Additions of sodium fluoride and aluminium fluoride [10] and
aluminium [11] to silicon and more recently [2] additions of about
1wt% of (a) cobalt, chromium, nickel or manganese and (b) magnesium,
aluminium, zinc or tantalum showed that the first series transition
metals enhanced nitridation and modified the form of the kinetics
whereas other additives did not. Aluminium favours the formation
of β [11] and the common preponderance of β-formation in the later
stages of nitridation may partly be due to the absorption of
aluminium-bearing species [12]. There has been but little success
in locating impurities in the micostructure or in detecting liquid-
formation below the melting-point of silicon [13].

3. EXPERIMENTAL

Powder from semiconductor-grade silicon (Monsanto, Belgium) ground
in a tungsten carbide mill and showing no appreciable traces of
metallic impurities spectroscopically was used. Selected metallic
impurities (½wt%) in a variety of forms were thoroughly manually
mixed into samples; Fe as $FeSi_2$, Cr as $CrSi_2$ or Cr_2O_3, Al as the
metal or Al_2O_3, Na as Na_2CO_3 or by soaking the silicon powder in
caustic soda and Li as Li_2CO_3. All starting powders were finer
than 200 mesh.

Cylindrical compacts about 2cm diam x 6cm formed by isostatic
pressing, 100 MPa, initial green density $1.50Mgm^{-3}$, were nitrided
on alumina boats in an externally heated mullite tube. After
evacuation to less than 10^{-2}Pa, 'white spot' nitrogen was supplied
to maintain a nearly constant pressure slightly above atmospheric.
After pre-nitriding at 1300°C disc specimens 5mm thick were cut
from the compact. Firstly pure and iron-doped discs were nitrided
at 1350°C for a sequence of times from which it was found in each
case that nitridation had almost ceased after 24h. Subsequent
nitridations were all for 24h.

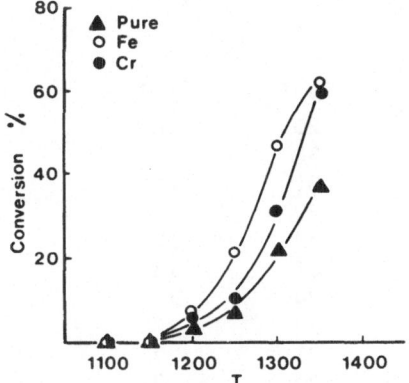 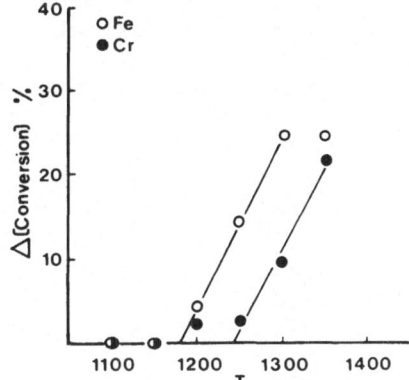

Fig. la. Conversion in 24h
as a function of temperature
($^{\circ}$C) for pure, Fe- and Cr-
doped silicon.

Fig. lb. Difference in
conversion between pure and
doped samples.

    The conversion was determined via the weight gain assuming
no loss of material. The proportions of α and β were determined
for powdered specimens by X-ray diffractometry and, for selected
specimens, the final impurity content was measured by X-ray
fluorescence analysis; Tables 1 and 2. More alumina was detected
in the lower half than in the upper half of one specimen
indicating uptake from the alumina boat [14]. Optical and
electron optical techniques failed to reveal clearly either the
location of the additives after nitridation or definite microstructural features depending on particular additives.

    In view of the interest in eutectics further specimens were
nitrided for 24h at selected temperatures from 1100°C to 1300°C,
Figure la. In Figure lb the difference between doped and pure
specimens is plotted to emphasise the change between iron- and
chromium-containing material.

    Further experiments were performed to confirm the interaction
between impurities and the oxide layer on the silicon by increasing
the oxidation. Firstly pure powder was oxidised for 0.25h at
1000°C in air. Compacts of this powder with selected additives
were nitrided for 24h at 1350°C. High conversions, not quite so
high as with unoxidised powder, were obtained with iron and
chromium but a low conversion, approximately the same as for the
undoped oxidised powder, was observed with aluminium. Secondly
discs oxidised at 1000°C in air after pre-nitriding to a fixed
amount were nitrided for 24h at 1350°C. The conversions showed
that once some silicon nitride has been formed subsequent oxidation
has relatively little effect [14].

TABLE 1.   Nitridation of pure silicon powder with $\frac{1}{2}$wt%
           additives for 24h at 1350°C

Additive	Form	Conversion %	α:β	c_α %	c_β %
nil	–	36	2.7	26	10
Fe	$FeSi_2$	60	1.7	38	22
Cr	$CrSi_2$	57	1.9	37	20
Cr	Cr_2O_3	51	1.1	27	24
Ca	CaO	46	2.6	33	13
Ca	$CaSi_2$	49	2.8	36	13
Al	Al_2O_3	36	2.1	24	12
Al	Al	36	1.9	23	13
Na	Na_2CO_3	28	2.8	21	7
Na	$(NaOH)$	27	3.0	20	7
Li	Li_2CO_3	26	6.1	22	4

TABLE 2.   Additive concentration after nitridation

Additive	Conversion %	Final concentration, wt%	
		If no loss*	Observed
Fe	60	0.36	0.36
Cr	57	0.36	0.32
Ca	49	0.38	0.20
Al	36	0.40	0.50

* After allowance for the weight gain.

4.   DISCUSSION

Although the nitridation system used was less precisely controlled
than ideal all the important parameters were approximately constant
throughout.   Confidence can therefore be placed in trends displayed
by the data although in a different nitriding system the absolute
values obtained could be different.   The results show that
nitridation is greatly enhanced by iron, chromium and to a lesser
extent calcium.   The enhancement is a little greater when the
metal is added as its silicide rather than its oxide.   Aluminium
has no effect, either as oxide or metal, and sodium and lithium
cause a modest retardation.   Iron, chromium and calcium all form
silicon-rich eutectics melting at 1208°C, 1320°C and 980°C
respectively [9].   The Al-Si eutectic occurs at a low temperature
and is very aluminium-rich.   Thus it is less likely to be of
assistance to this reaction.   Silicate liquids do not appear
relevant because there are no silicates in the Cr_2O_3-SiO_2 system
and because sodium and lithium would then be expected to accelerate
nitridation rather than to retard it.   Possibly the observed
retardation is due to the presence of a low viscosity silicate melt
which impedes the flow of nitrogen to reaction sites.   The nul

effect of aluminium would be explicable since it can replace silicon
in a network without disruption. The temperature dependences of
iron- and chromium-doped specimens show by extrapolation of the
linear parts (Figure 1b) that the effect of the additives goes to
zero at about 1180°C and 1240°C respectively. A rapid variation of
conversion with temperature is expected but the correlation with
eutectic temperature is noteworthy. The agreement between
eutectic temperature and the onset of nitridation is satisfactory
for iron, bearing in mind possible temperature variations within
the specimen, but not for chromium. Nevertheless the displacement
of the temperature intercept for chromium towards higher
temperatures is in the correct direction although less than
expected if the binary eutectic is the only significant factor.

Iron and chromium showed no appreciable change in concentration.
The increase for aluminium probably arises by transport from the
alumina boat but some contamination from the furnace tube is
possible [12]. The decrease for calcium is due to its volatility
at the nitridation temperature [15]. Sodium and lithium will be
lost rapidly for the same reason. Compensation for this in the
latter case by placing additional lithium carbonate in the furnace
led to very low conversion. The surface of the specimen was
rapidly covered by a glassy-looking layer including crystalline
lithium disilicate and lithium silicon nitride (Li_2SiN_3).

For pure silicon nitridation is impeded by oxidation. Iron
and chromium rapidly disrupt the oxide layer whilst aluminium
does not.

Iron and chromium promote β-formation, possibly influencing
growth mechanisms although no change in morphology was detected.
Neither calcium nor sodium affects α:β although they have opposite
effects on the kinetics. This may indicate that the mechanism of
acceleration by calcium is not identical to that for iron or
chromium. Aluminium, which has no effect on kinetics, produces a
reduction in α-formation and an enhancement of β whilst lithium,
which produces a modest deceleration of the reaction, suppresses
β-formation almost entirely. Prompted by Professor Jack it is
interesting to speculate on the relationship between the behaviour
of aluminium and lithium here and in SiAlON systems. Aluminium
occurs with wide composition variations in β'-SiAlON structures
whilst the lithium SiAlON system is one of the few to show
extensive α'-formation.

5. CONCLUSIONS AND ACKNOWLEDGEMENTS

The results reported in this paper together with published data
lead to the following main conclusions:
1. First transition series metals are particularly effective in
    initiating nitridation in the presence of a thin oxide layer

and in subsequently promoting more complete conversion possibly by favouring a particular growth mechanism which may also account for the enhanced production of β.

2. The temperature dependence of accelerated nitridation indicates the possibility that effective additives are those which form silicon-rich eutectics.

3. The formation of silicate melts does not appear to assist nitridation and may indeed hinder it as is shown by the results for sodium and lithium.

4. Aluminium does not influence conversion but favours β-formation. Pick-up of aluminium from the furnace tube or boat is significant.

5. The tendencies of aluminium to favour β-formation and lithium α-formation may be correlated with their behaviour in SiAlON systems.

The authors wish to thank Professor R.W.K. Honeycombe for the provision of laboratory facilities. A. de S.J. is indebted to the Commonwealth Scholarship Commission for financial support.

REFERENCES

1. D.R. Messier and P. Wong, J. Amer. Ceram. Soc. 56, 480 (1973).
2. A. Atkinson and A.J. Moulson, private communication.
3. A. Atkinson, A.J. Mouslon and E.W. Roberts, J. Amer. Ceram. Soc. (in press).
4. H.M. Jennings and M.H. Richman, J. Mat. Sci. (in press).
5. D.P. Elias and M.W. Lindley, J. Mat. Sci. 11, 1278 (1976).
6. R.B. Guthrie and F.L. Riley, Proc. Brit. Ceram. Soc. 22, 275 (1973).
7. W.P. Clancy, The Microscope 22, 279 (1974).
8. K.H. Jack, Special Ceramics 5, ed. by P. Popper, Brit. Ceram. Res. Ass., Stoke-on-Trent, 1972, p.286.
9. M. Hansen, Constitution of Binary Alloys, McGraw Hill, New York, 1955.
10. K. Suzuki and T. Yamauchi, Chemical and Thermodynamic Properties at High Temperatures, A Symposium, Montreal, August 1961, p.213.
11. A. Atkinson, P.J. Leatt and A.J. Moulson, J. Mat. Sci. 7, 482 (1972).
12. R.B. Guthrie and F.L. Riley, J. Mat. Sci. 9, 1363 (1974).
13. D.J. Godfrey, Proc. Brit. Ceram. Soc. 25, 325 (1975).
14. A. de S. Jayatilaka, Dissertation for the Degree of Ph.D., University of Cambridge, 1975.
15. B.C.H. Steele and M.A. Williams, J. Mat. Sci. 8, 427 (1973).

DISCUSSION (Goursat, Krauth, Smeltzer)

Morgan: As we see for each rather simple explanation for a particular effect, e.g. the ability of Al to stimulate β, or Li, α;
we see that other ions behave in an unexpected way. We see that
Cr is just as effective as Al in stimulating β even though the
likelihood of similar action is very small. One suspects then
that, in general, unless the additive specifically affects the
initial formation of α from silicon, the general tendency will be
for the additive to increase the α-β effect by providing a transport mechanism. In addition, when one constructs a model of α-
silicon nitride, it is easy (it happened to us accidentally) to
produce very complex defects in the structure including strange
twist possibilities and stacking faults. This suggested to us
that a close study of α is required to determine whether α is
rather variable.

Godfrey: The microstructural effects of iron contamination can
be seen in the production of large holes, characteristic of the
microstructure of commercial purity RBSN, which we found were
absent in 99.95% (cation) purity RBSN. Regarding the fate of
iron, it may be seen easily by atomic number contrast in SEM
examination to be present in the 'unnitrided silicon' phase, although it is often difficult to distinguish the slightly more
yellow colour of $FeSi_2$ by optical micrography. This iron disil-
icide is often very remote from the large holes formed by the
initial eutectic melting.

Jack: Taking Dr. Morgan's point a little further, and without
wanting to prolong the unresolved controversy about the α-β relationship, there certainly are many different "α-silicon nitrides".
Dr. Thompson has measured the unit-cell dimensions of 20 or more
so-called pure "α-silicon nitrides" from different sources and
finds very wide variations in unit-cell dimensions depending on
their method of preparation, e.g. by vapour deposition on graphite
from $SiCl_4$ and NH_3, or by reaction of SiO with N_2, or by nitriding
Si with N_2. The cell dimensions of α-whiskers prepared from SiO
and N_2 depend upon the size of the whiskers and so the existence
of facetting as suggested by Dr. Morgan is not improbable.

Jelacic: We have worked with both impure and pure silicon. The
impure had about 1.5% Fe, and we found too that Fe enhances the
conversion to nitride and the amount of β-silicon nitride at
1250-1300°C, but the difference between pure and impure silicon
became less at higher temperature (1400°C).

THE EFFECTS OF IRON ON THE NITRIDATION OF SILICON

S.M. Boyer[*], D. Sang and A.J. Moulson

Department of Ceramics, University of Leeds,
Leeds, U.K.

ABSTRACT. It is known that an oxide film covering silicon inhibits nitridation whilst iron impurity accelerates the reaction. As part of a study to explain these effects experiments have been carried out to determine the effect of iron on the oxidation kinetics of high purity silicon. Below a critical oxygen pressure ($\sim 3 \times 10^{-3}$ bar @ 1370°C) iron induced a transition from passive to active oxidation. Above this critical pressure iron was found to introduce a short-lived enhanced oxidation rate. The passive-active oxidation transition is thought to be due to iron induced devitrification of the native oxide layer thus exposing free silicon surface and allowing a Wagner-type active oxidation mechanism to operate at low oxygen pressures.

This oxide removal mechanism is invoked in an analysis of the effect of iron on the nitridation of silicon powder compacts. A model for the nitridation of silicon is proposed involving two parallel and independent reactions: nucleation and growth through the native oxide layer, and oxide removal throughout the whole compact followed by secondary, iron-induced, nitride nucleation and growth. It is shown that the extent of this secondary growth is determined by the iron concentration at the p.p.m. level.

1. INTRODUCTION

Silicon which has been exposed to air at room temperature is always associated with a thin amorphous oxide film (1). Further oxidation, at higher temperatures, is passive occurring by diffusion of the oxidising species through the coherent film, the kinetics of which are parabolic (2).

[*]One of us, SMB, wishes to acknowledge the financial support of A.M.L. (Ministry of Defence P.E.).

It is known that the 'native' oxide film (~ 3 nm) retards, and that a thick oxide film (~ 1 μm) can effectively prevent a reaction between pure silicon and nitrogen (1). Iron has long been known to accelerate this reaction, presumably by some interaction with the inhibiting oxide film and possibly involving its removal as $SiO(g)$ (1). Such an interaction could be expected to affect the well established oxidation kinetics, and so experiments have been carried out to determine the oxidation rate of silicon when contaminated by iron.

2. EXPERIMENTAL

2.1 Silicon Oxidation

Spectrographically pure silicon powder, prepared in the manner described by Atkinson, Moulson and Roberts (3) was oxidised at 1370°C at various pressures of air and the oxidation rates followed by thermogravimetry. In each case the sample was removed from the balance when the 'zero-rate' region was approached and ~ 3000 ppm iron as an alcoholic nitrate solution added to the oxidised silicon. The samples were then returned to the balance and further heated in air. Below a critical air pressure of $\sim 1.5 \ 10^{-2}$ bar ($P_{O_2} = 3 \ . \ 10^{-3}$ bar) enhanced oxidation rates of two or three orders of magnitude greater than the passive oxidation rates were observed. Above this critical pressure short lived enhanced rates were observed, the rate of enhancement decreasing with increasing pressure to zero at ~ 0.1 bar. The results are shown schematically in Fig. 1. X-ray analysis of the reaction products below the critical oxygen pressure ($P_{O_2}{}^c$) showed only silicon remaining at the bottom of the crucible and α-cristobalite to have formed predominantly at the top of the crucible and also in other parts of the reaction system. Above $P_{O_2}{}^c$ the silicon particles were covered with a layer of α-cristobalite. The difference in location of reaction product indicates a change from vapour phase to a solid-gas reaction on exceeding $P_{O_2}{}^c$.

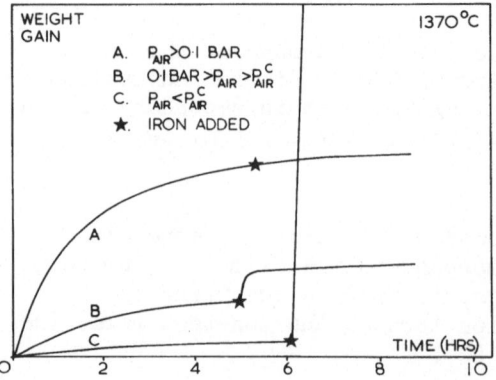

Fig. 1. Pressure dependence in the oxidation kinetics of iron-contaminated silicon.

2.2 The effect of iron on vitreous silica

The means by which iron affects the surface coating of vitreous silica on silicon was investigated further. A film of metallic iron was vacuum evaporated onto the surface of transparent vitreous silica slabs and the samples heated under nitriding conditions (P_{N_2} = 1 bar; P_{O_2} ~ 10^{-5} bar; 1150 to 1400°C). In all cases the contaminated surfaces were extensively microcracked and optical analysis under crossed polars showed small anisotropic crystals. X-ray analysis showed α-quartz and α-cristobalite as the only crystalline phases both above and below the temperature at which melt formation could be expected (~ 1230°C) (4). Vitreous silica which was uncontaminated remained unaffected under the above conditions.

3. DISCUSSION

3.1 Silicon Oxidation

The oxidation kinetics of iron-contaminated silicon may be interpreted in terms of the Wagner oxidation model (Fig. 2) in which oxidation of a free silicon surface, i.e. active oxidation, occurs by the formation of volatile silicon monoxide (5). This free silicon surface will be maintained in a partial pressure of oxygen provided that P_{SiO} at the surface is greater than P_{O_2}. The flux of silicon, as SiO, away from the surface increases linearly with P_{O_2} until a critical SiO pressure is reached, determined by the equilibrium:

$$Si + SiO_2 = 2SiO \qquad \ldots\ldots (1)$$

when $P_{O_2} \geqslant P_{SiO}{}^c$ surface coverage with a passivating layer of vitreous silica occurs. ($P_{O_2}{}^c$ = 5 . 10^{-3} bar @ 1370°C.)
In the case of an already oxidised silicon surface an oxygen

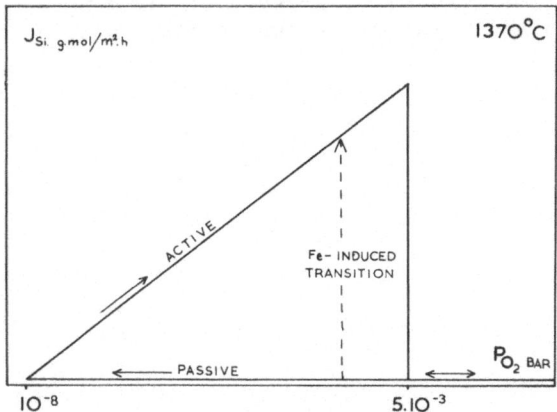

Fig. 2. The Wagner oxidation model (Schematic)
N.B. Linear scales.

pressure of less than half the corresponding P_{SiO} for the reaction:

$$SiO_2 = SiO + \tfrac{1}{2} O_2 \qquad \qquad \cdots (2)$$

(i.e. $P_{O_2} \approx 10^{-8}$ bar @ 1370°C) must be achieved before loss of the oxide layer can expose free silicon surface. In the present work it is believed that iron, by devitrification and subsequent disruption of the surface silica layer, exposes free silicon and that when $P_{O_2} < P_{O_2}{}^c$ the surface silica is lost as SiO (Eq.1). A transition from conditions of passive to those of active oxidation is thus induced, oxygen impinging on the silicon surface forming volatile SiO which is further oxidised to SiO_2 elsewhere in the reaction system (Fig. 1 curve 'c'). When $P_{O_2} > P_{O_2}{}^c$ the free silicon surface exposed is rapidly re-covered with a passivating silica layer by a recurrent disruption/re-oxidation process, passivation at $P_{O_2} > 2 \cdot 10^{-2}$ bar ($p_{air} = 0.1$ bar) being too rapid to produce any measurable rate enhancement (Fig. 1 curves 'b' and 'a').

3.2 Silicon Nitridation

The kinetics of nitridation ($P_{N_2} = 1$ bar; $P_{O_2} \sim 10^{-6}$ bar; 1370°C) of pure silicon with an inhibiting 'native' oxide film have been studied by Atkinson and Moulson (1) who found only a small fraction ($\sim 2\%$) of silicon to convert to silicon nitride. On doping with controlled low amounts of iron they observed enhanced nitridation rates after an initial induction time. These results have now been interpreted as follows:

The initial kinetics for iron-doped samples closely follow those of the undoped sample. It therefore seems likely that, in the initial stages of reaction, nucleation and growth through the oxide layer is taking place and is largely unaffected by the presence of iron. For this reason the kinetics curve, taken from the original data, of the undoped sample was subtracted from those of the doped samples. The results, expressed in terms of weight gain per unit area against time, are shown in Fig. 3.

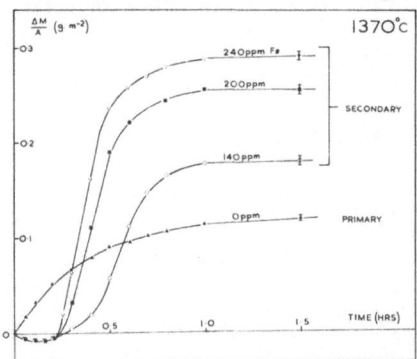

Fig. 3. 'Primary' and 'Secondary' nitridation kinetics for iron-doped silicon.

It can be seen that the resulting curves show three distinct
regions:

(a) During the first 20 minutes there is a slow weight loss, in-
     dependent of the iron concentration and of the same order as
the total weight of oxygen, as surface oxide, in the sample.

(b) After this induction period there is a rapid increase in
     weight proportional to the iron concentration, all samples
reaching maximum weight simultaneously.

(c) Finally, there is a region of zero weight gain when the un-
     doped sample is still increasing in weight.

The above suggests that the 'sigmoidal' kinetics generally
observed during nitridation in the presence of iron may be understood in terms of two parallel and independent reactions: 'primary-
nitridation'; a surface reaction of nucleation and growth through,
and possibly incorporating as oxynitride, the native oxide layer
as has been suggested by Hüttinger (6), and Messier and Wong (7);
and 'secondary-nitridation'; oxide removal followed by further
nitride nucleation induced by the presence of iron. The loss of
surface oxide can occur either by exposing the underlying silicon
or by rendering the oxide film 'transparent' to diffusing species,
either of which allows the reaction described in equation (1) to
proceed. Present evidence strongly suggests devitrification and
disruption as the relevant mechanism. (Measurable weight loss at
this stage indicates that the SiO formed does not react within the
compact and so vapour phase reactions involving SiO play no significant part in this particular reaction system. It is supposed
that SiO condenses, possibly as $\alpha-Si_3N_4$, elsewhere in the reaction
system.) 'Secondary' nucleation then occurs, the nucleation
density being proportional to the iron concentration. This is
evidenced by the fact that both the maximum nitridation rates and
final weight gains are directly proportional to the iron concentration (Figs. 4 and 5). It is likely that the nuclei form at

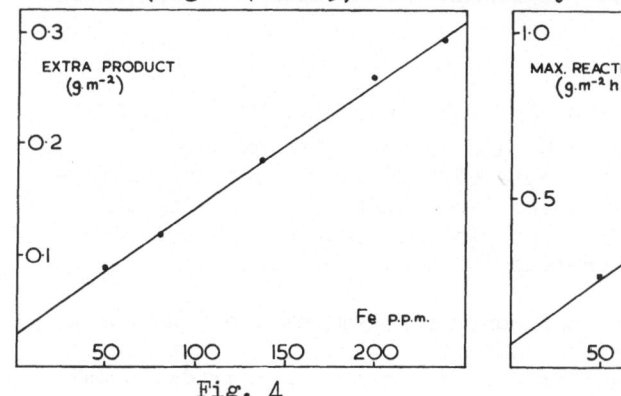

Fig. 4

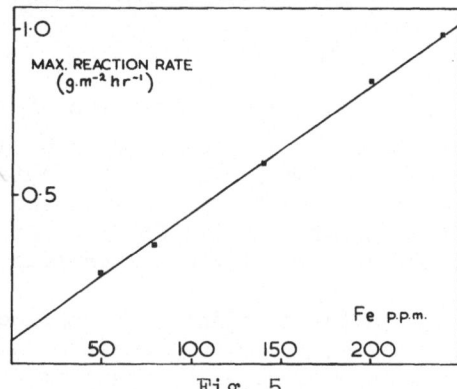

Fig. 5

The dependence of 'Secondary' nitridation on iron concentration.

Fe-Si melt sites, since it has been shown that Fe-induced enhanced nitridation ceases below 1200°C (8,9), the Fe-Si solidus
temperature. The morphology of this secondary growth has yet to
be fully investigated although there is some evidence that growth
may be controlled by surface diffusion of silicon, possibly an
enhanced diffusion through a film of an Si-Fe liquid which has
been shown to wet silicon nitride (10). An analysis of the
reaction kinetics shown in Fig. 3 suggests that the 'primary-
nitridation' occurs throughout and after the iron-induced reaction
and that the reacting surface is finally closed off by coalescence
of primary nitride growths.

A preliminary study of the effects of certain other metals
(Al, Ta, Ni) on vitreous silica under nitriding conditions showed
only Ni to devitrify the silica, X-ray analysis showing both Al
and Ta to have nitrided. It was also found that the metals which
have been demonstrated to enhance nitridation (Fe, Cr, Mn, Ni)
and those which were seen to be ineffective (Mg, Ca, Al, Zn, Ta)
(1) could be classified in terms of the free energies of formation
of their nitrides (11), those which were ineffective nitriding
preferentially to silicon. The above suggests that it is the
metallic contaminant or its oxide, rather than the nitride, which,
when in contact with the silica, promotes devitrification.

Although in the present study of nitridation to low levels
of conversion and where gaseous impurity levels were carefully
controlled SiO was found to play no significant part, its role in
the complete nitridation of silicon must be explained. Three considerations are important here: Firstly, whenever free silicon
surface is available, and the oxygen pressure is below $P_{O_2}{}^c$,
active oxidation will occur. Secondly, nitridation of the SiO
generated by this reaction will further release oxygen, and so
may form the basis of a cyclic reaction route. Finally, SiO may
diffuse out of the compact and become unavailable for reaction.

4. CONCLUSIONS

(1) At low oxygen pressures ($P_{O_2} < 5 \cdot 10^{-3}$ bar, @ 1370°C) iron
     induces a transition from passive to active oxidation of
silicon. The transition is achieved by devitrification and subsequent disruption of the protective silica film allowing the
SiO generated at the Si/SiO_2 interface to escape, the reaction
continuing until the silica film is removed completely.

(2) In the nitridation of silicon, iron removes the inhibiting
     oxide layer in the manner described above. The exposed
silicon surface will remain free of oxide during nitridation at

oxygen partial pressures of up to $\sim 5 \cdot 10^{-3}$ bar (~ 3 torr), any oxygen up to this pressure forming SiO within the compact. After removal of the oxide film, iron nucleates extra nitride growth at Fe-Si melt sites.

(3) From the evidence on the effects of iron and of other con-
     taminants a set of criteria has been formulated. It is suggested that in order to enhance nitridation, a metallic con-taminant should:

   (a) devitrify silica under nitriding conditions,

   (b) form a liquid with silicon at nitriding temperatures,

 and (c) nitride less readily than silicon.

REFERENCES

1. A. Atkinson and A.J. Moulson, Science of Ceramics 8,
    Publ. Brit. Ceram. Soc., 1976, p.111.

2. A.G. Ravesz, J. Non-Cryst. Solids (11), 309, 1973.

3. A. Atkinson, A.J. Moulson and E.W. Roberts, J. Am. Ceram.Soc.
    (59), 285, 1976.

4. L.S. Darken, J. Am. Ceram. Soc. (70), 2051, 1948.

5. C. Wagner, J. Appl. Phys. (29) 9, 1295, 1958.

6. K.J. Hüttinger, High Temperatures-High Pressures (2), 89,
    1970.

7. D.R. Messier and P. Wong, J. Am. Ceram. Soc. (56) 9, 480,
    1973.

8. A. Atkinson and A.J. Moulson, S.R.C. Project B/SR/7943,
    Progress Report 5, Dept. of Ceramics, University of Leeds.

9. J.A. Leake and A. de S. Jayatilaka; This volume.

10. T.J. Whalen and A.T. Anderson, J. Am. Ceram. Soc. (58), 396,
     1975.

11. M. Olette and M.F. Ancey-Moret, IRSID, 1962.

DISCUSSION (Goursat, Krauth, Smeltzer)

Jack: The role of impurities, including iron, is to promote the reactions:

$$6SiO + 2N_2 \longrightarrow Si_3N_4 + 3SiO_2 \qquad (1)$$

$$3SiO_2 + 3Si \longrightarrow 6SiO \qquad (2)$$

by forming one or more liquid phases, by reaction with Si to form a liquid eutectic or with SiO_2 to form a silicate, or both. The reaction to form SiO is then a liquid/solid or a liquid/liquid one and so much faster than solid/solid. The initial step when the Si surface is covered with a 30 Å layer of SiO_2 might be

$$2Fe + 2SiO_2 \longrightarrow 2FeO + \uparrow 2SiO \quad \text{followed by}$$

$$2FeO + SiO_2 \longrightarrow Fe_2SiO_4, \text{ or } 2FeO + 2Si \longrightarrow 2Fe + \uparrow 2SiO.$$

Leake: Whilst iron silicates might well be formed at these temperatures, the Cr_2O_3-SiO_2 diagram is very different and appears to show no chromium silicates.

Messier: It seems to me that the amounts of iron added (300 ppm maximum) are too small to justify proposing a mechanism involving iron silicate formation as a means of promoting active oxidation and consequently an enhanced rate of nitridation.

Boyer: We agree with the likelihood of the cyclic reactions suggested by Professor Jack, but we feel it is unlikely that iron-silicates play a part in this reaction. Our evidence suggests that the silica layer covering the silicon is completely removed in the early stages of reaction, leaving none available for silicate formation and hence SiO must be formed by direct reaction with silicon. X-ray analysis has never revealed the presence of silicates in any of our iron contaminated samples.

Schaeffer: I would like to know why an alkali silicate layer apparently does not increase the conversion to nitride.

Leake: Possibly the molten silicates formed when alkalis are added restrict nitrogen permeation into the compact.

Boyer: The amount of molten alkali silicate formed would be insufficient to restrict nitrogen flow into a compact and would be more likely to form a viscous film around each particle.

Jelacic: Is the role of iron that of influencing surface tension?

Moulson: The formation of local regions of an iron silicate liquid which did not wet the silicon surface would, by exposing the silicon, have the same effect on initial oxide removal as would its disruption by devitrification. We believe that present evidence favours devitrification as the mechanism of oxide-removal although further work will be required to establish one particular mechanism as finally correct.

Singhal: In calculating the critical oxygen partial pressure for active-passive oxidation of silicon, you have assumed the diffusion coefficients of SiO and O_2 molecules to be equal. Is there any evidence or justification for this assumption?

Moulson: We have not assumed the diffusion coefficients of SiO and O_2 molecules to be equal, the ratio D_{SiO}/D_{O_2} having been estimated by Wagner (J. App. Phys. 29 (9) 1295-7 (1958)) to be 0.64. Our values for the critical oxygen pressure have been calculated on this basis.

NITRIDATION OF Si AND Si + Al COMPACTS UNDER DIRECTED GAS FLOW CONDITIONS

J.P. Torre and A. Mocellin

Ecole des Mines de Paris, Centre des Matériaux
B.P.87, 91003 Evry Cedex, France

ABSTRACT. Both silicon powder (99.96% pure) and silicon-aluminium blends have been isostatically compacted and nitrided under directed gas flow conditions, in an attempt to evidence the steps of the overall reaction mechanism when residual oxygen is present in the nitriding agent. The early stages only have been investigated, as the experiments typically consisted of a hold at 1250°C for no longer than 20 hours, followed by one hour at 1450°C.
Micrographic observations and X-ray diffraction analyses clearly suggest that oxydation reactions of the initially formed Si_3N_4 (and eventually AlN) to give silicon monoxide (and Al_2O_3) are taking place : depending upon the local availability of "fresh" gaseous molecules within the porous compact, nitrogen consuming and nitrogen producing (oxidation) reactions take place at alternate places and times. These can either promote the development of large size pores, or conversely facilitate their elimination. Monitoring of the oxygen partial pressure during a nitriding schedule may therefore prove helpful for a better control of microstructure development in reaction bonded silicon nitride.

1. INTRODUCTION

The inhibiting effect of oxygen on the nitridation of silicon compacts has been pointed out |1,2| and shown to be related to the presence of protective silica layers on the silicon particles |3|. Similarly, the apparent catalytic influence of some impurities of the starting powder, e.g. transition metals, has been noted |4|, and recent work suggests that they facilitate the removal of surface silica as volatile SiO |3,5|.

It seems therefore that the preparation of highly nitrided silicon parts requires either as low as possible an oxygen partial pressure, or the presence of a suitable oxygen getter within the compacts, so that the corresponding oxygen pressure in the nitriding gas can be raised in practice, and kept at easily controllable levels. Aluminium has recently been shown to play such a role and to lead to the formation of an alpha-free β'-Si-Al-O-N solid solution |6|. It was also suggested that under such conditions, one of the reaction steps of the overall process, was the oxydation of some of the initially formed Si_3N_4 |6| according to :

$$Si_3N_4 + \frac{3}{2} O_2 \rightarrow 3 \ SiO + 2N_2$$

The work to be reported here was undertaken to further clarify the nitridation mechanism in the presence of residual oxygen. In an attempt at experimentally evidencing some of the reaction steps, it was chosen to submit the powder compacts to a macroscopically directed gas flow, in such a way that permeation would take place through one face only, and the reaction front would progress more or less parallel to this interface.

2. SAMPLE PREPARATION

Chemical analyses and particle sizes of the starting powders were previously indicated |6|. Isostatically pressed compacts ($196 \ MN/m^2$) were prepared, either from pure silicon or from Si + 6 weight percent Al, dry blended for 24 hours in a "Turbula" mixer. Cylinders about 15 mm high, were machined from the green compacts, such that their diameters closely fitted the inside diameter (20 mm) of the mullite tube of a SiC resistor furnace. The operating conditions were similar to those previously described |6|. The temperature schedule consisted in a 20 hour hold at 1250°C, followed by one hour at 1450°C. The oxygen content of the nitriding agent (70% N_2, 30% Argon mixture) was set to 500 ppm by volume and the flow rate to one renewal per hour corresponding to a 0.02 $cm.s^{-1}$ initial velocity at the inlet face.

The nitrided samples were cut parallel to the gas flow direction and polished with an alumina slurry for microstructure observation. X-ray diffraction patterns were recorded (Co $K\alpha_1$ radiation) on polished surfaces perpendicular to the gas flow directions at various depths within the samples (Figure 1).

3. RESULTS AND DISCUSSION

Microstructure observations and X-ray analyses show that Si and Si + Al behaved differently. In the former samples (Figure 2) a uniformly porous layer developed between the relatively pore-free

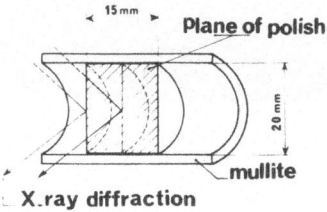

Fig. 1. Experimental set-up.

inlet zone and the silicon rich core. In the Si + Al compacts
(Figure 3), several more or less porous regions can be identified
from the external surface to the molten silicon core. Both α- and
β- Si_3N_4 are present in the Al free samples. Weak traces of α-
cristoballite are also noticeable locally. Neither α- Si_3N_4 nor
silica can be detected in the nitrided Si + Al compacts, but minor
amounts of Al_2O_3 and AlN are found, and the intensities of the
corresponding lines vary in opposite directions with position
within the sample. Plots of % Si nitrided versus position show
alternating decreases and increases in both cases. No definite
trend is visible on the α/β plot (Figure 2) : the amplitude of
variations of this ratio can hardly be claimed to exceed the
experimental uncertainties.

As shown on the thermochemical diagram of Figure 4 |7-10|,
the following four compounds : Si_3N_4, SiO_2, Al_2O_3, AlN may be
formed when both the N_2 and O_2 potentials are high in the furnace
atmosphere or within a pore. It is also clear from this diagram
that a modification in either one of the oxygen or nitrogen potentials may be sufficient to render some of the above phases thermodynamically unstable. In other words, domains may be theoretically
indentified in the (P_{N_2} ; P_{O_2}) plane, such that individual condensed phases should alternatively appear or disappear as the gas
composition in the system crosses the corresponding boundaries.
From a practical point of view then, changes in the phase composition of a given sample, may be correlated by means of Figure 4,
with changes in the gas composition in contact with it ; whether
these are controlled by an outside operator or any physico-chemical
constraint internal to the system. It is also of interest to note
that some of the reactions mentionned on Figure 4 are gas consuming, whereas others are gas producing. Increases in the amount
and volume of pores are of course to be related to the latter kind
of reactions.

In our pure Si compacts, nitride maxima approximately correspond to SiO_2 minima, whereas in the Si + Al samples, the frequency

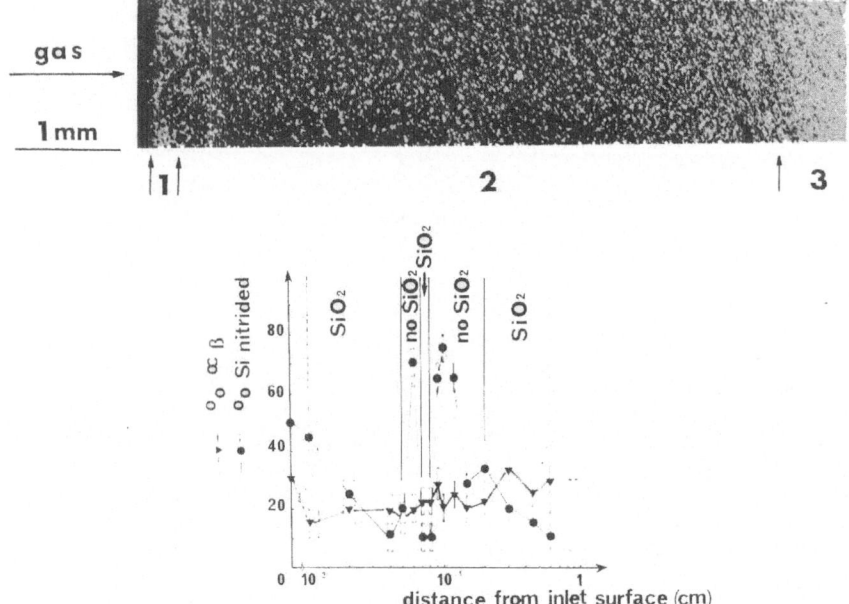

Fig. 2. Si compacts : results.

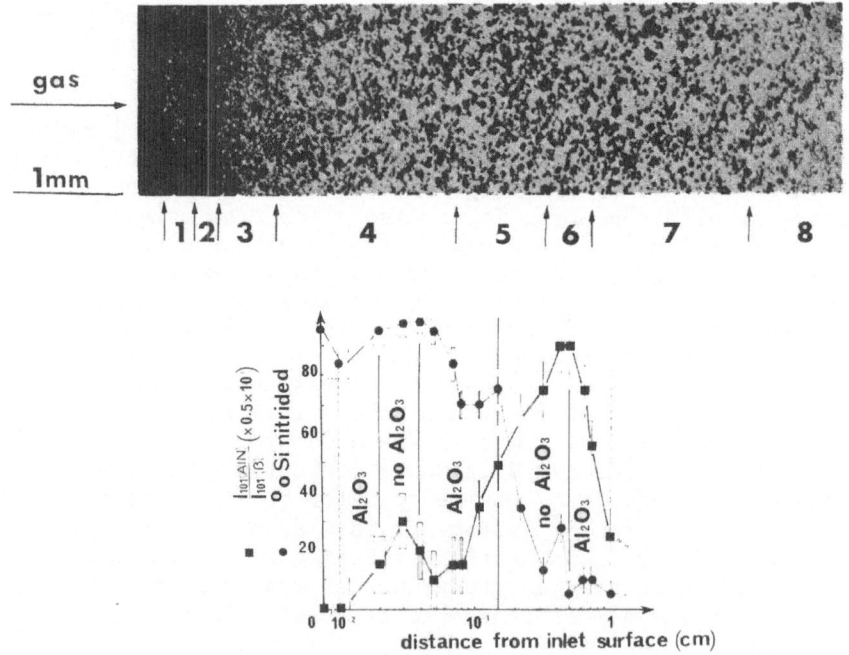

Fig. 3. Si + Al compacts : results.

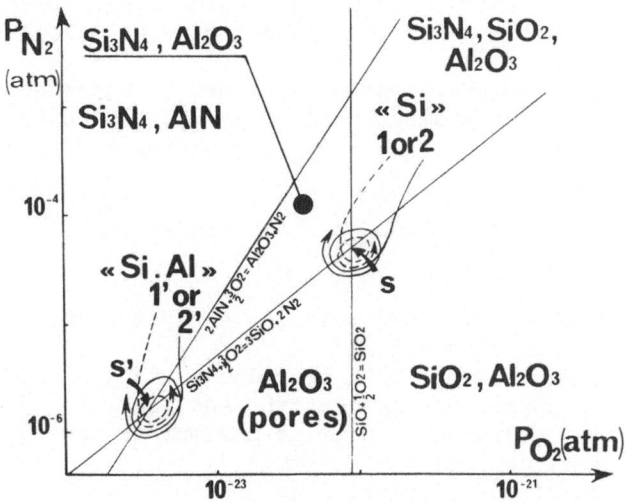

Fig. 4. Thermochemical diagram (1250°C, P_{SiO} = 8.9 x 10^{-3}atm
corresponding to the following equilibrium :
    Si + SiO$_2$ = 2 SiO |3|)

of fluctuations of the Si$_3$N$_4$ content is about twice that for AlN
(Figures 2 and 3). In terms of the above argument, such results
appear consistent with changes of the gas compositions with time
and/or position in the compacts, which brings further support to
a previous finding |6| according to which the overall nitridation
process is gas flow controlled in the present kind of samples.
Qualitative gas composition paths have also been sketched on
Figure 4, which are in agreement with the recorded fluctuations
of Figure 2 and 3. The system has a tendency to evolve, either
clockwise or counter clockwise, around point S in the Al-free
compacts and around point S' in the Si + Al samples. These respectively correspond to the following simultaneous equilibria :

$$S \begin{cases} Si_3N_4 + \frac{3}{2}O_2 = 3\ SiO + 2N_2 \\ SiO + \frac{1}{2}O_2 \quad = SiO_2 \end{cases}$$

$$S' \begin{cases} Si_3N_4 + \frac{3}{2}O_2 = 3\ SiO + 2N_2 \\ 2\ AlN + \frac{3}{2}O_2 = Al_2O_3 + N_2 \end{cases}$$

The nitrogen pressures within the compacts therefore appear to remain close to 10^{-4} atm and 10^{-6} atm respectively, so that new gas molecules are continuously introduced from the outside reservoir and a "reaction front" may proceed inward. Such pressure levels are not unrealistic as the following rough calculation suggests.

Consider, for example, zone 5 on Figure 3. The excess N_2 pressure necessary for raising the upstream $13 \pm 5\%$ nitride content to its next maximum value of $28 \pm 5\%$ (neglecting AlN formation) is approximately :

$$P_{N_2} = \frac{n}{\frac{e}{v} \times S \times P} \quad x \quad \frac{1}{\frac{D^2}{8}\sqrt{\frac{2\,RT}{M}}} \qquad \text{with :}$$

n : number of involved nitrogen moles, i.e. approximately 6×10^{-4}
e : thickness of zone 5 (= 0,6 mm)
V : displacement rate of the % Si nitrided maximum in zone 5, assumed to be approximately constant and equal to $\frac{\Delta x}{\Delta t}$, with $\Delta x = 4$ mm i.e. position of nitride maximum in zone 5, after $\Delta t = 20$ hours ; hence : $v = 0.2$ mm.h^{-1}.
S : cross section of the compact, i.e. 314 mm^2
P : relative porosity of zone 5, i.e. 45% estimated by quantitative metallography
D : mean pore diameter in zone 5 (= 100μ)
M : molar weight of nitrogen (28g)

Hence :

$$P_{N_2} = 0.1 \text{ Pa} \approx 10^{-6} \text{ atm}$$

4. CONCLUSION

Nitridation of Si or Si + Al compacts under directed gas flow conditions suggests that oxidation reactions of some initially formed Si_3N_4 (and also AlN) to give silicon monoxide (and Al_2O_3) are taking place locally within the samples. It appears that, depending upon the local availability of "fresh" gaseous molecules within the pores, nitrogen consuming and nitrogen producing (oxidation) reactions take place at alternate places and times. The latter promote the development of large size pores, whereas the former facilitate their elimination. Monitoring of the operating conditions during a nitriding schedule (oxygen partial pressure, gas flow rate...) may therefore prove helpful for a better control of microstructure development in reaction bonded silicon nitride.

ACKNOWLEDGEMENTS

Financial support for this work was provided by D.R.M.E. ; Contract n° 74/531.

REFERENCES

1.   K.J. Hüttinger, High Tem. High Pressures 2 (1970) 89.
2.   D.R. Messier, P. Wong, A.E. Ingram, Journ. Am. Ceram. Soc. 56 (1973) 17.
3.   Shi-Shong Lin, Journ. Am. Ceram. Soc. 58 (1975) 21.
4.   D.R. Messier, P. Wong, Journ. Am. Ceram. Soc. 56 (1973) 480.
5.   A. Atkinson, A.J. Moulson, Sci. Ceram. 8(1976)111.
6.   J.P. Torre, A. Mocellin, J. Mat. Sci. (to be published).
7.   JANAF Thermochemical Tables (1965).
8.   K. Blegen, Special Ceramics 6, B.C.R.A. (1974) 223,P. Popper ed.
9.   O. Kubaschewski, Ll.E. Evans, C.B. Alcock, Metallurgical Thermochemistry, Pergamon Press (1967).
10. J. Drowart, G. de Maria, R.P. Burns, M.B. Inghram, J. Chem. Phys. 32 (1960) 1366.

DISCUSSION (Goursat, Krauth, Smeltzer)

Moulson: This contribution stresses the need to give consideration to the possibility of 'redox' reactions occurring at the microscopical level within a compact.

It should be remembered that any volatile species liberated from hot (c. 1500°C) ceramic tubes will tend to condense on the specimen because of the inevitable temperature gradient.

Cannon: When an Al-Si powder mixture is heated the Al melts well below the nitriding temperature. One might expect segregation of the Al to give regions of higher Al concentration and lower permeability. In liquid phase sintering of $NaF-CaF_2$ compacts we find significant variability in liquid concentration throughout the body after melting; that is the liquid tends to segregate into regions of high liquid concentration. This tendency to segregate would be even greater in a system with a high contact angle than in the $NaF-CaF_2$ system in which the liquid wets the solid with a very low contact angle.

Mocellin: Such an effect as described is probably more likely to occur when the overall compact dimensions are increased and/or the gas supply to its interior is reduced so that neither AlN nor Al_2O_3 forms on Al melting.

Klomp: Since redox-reactions or dissociation reactions appear to occur, it seems to be more valuable to control the partial pressures of the gas (p_{O_2} and p_{N_2}) than the gas flow. There is a gas take-up that differs over the length of the compact, so an excess of the gas regarding the partial pressures is required.

Mocellin: It is the author's opinion that both the gas flow rate and composition (N_2 and O_2 partial pressures) are very important parameters to keep in close control in a practical situation, at least because one may or may not want to keep flushing the nitriding compact with an outside gas not necessarily having at all times the most desirable compositions.

Desmaison: Possibly you could gain more information on the reaction mechanism by taking into account the rate of each elementary process appearing on your thermodynamic diagram.

Mocellin: Obviously, you are right. We believe that sophisticated kinetic studies are of little value as long as the overall mechanism under consideration is not sufficiently understood as to its elementary steps. Our objective here was to call attention to some still unanswered questions that we think should be tackled prior to very involved kinetic work.

THE EFFECTS OF ALUMINIUM AND ALUMINA ADDITIONS ON THE FORMATION OF β-SILICON NITRIDE

C. Jelacic and H. Dervisbegovic

Department of Inorganic Chemistry,
Technological Faculty Tuzla, Yugoslavia.

ABSTRACT

Technical silicon containing about 1.5% Fe, nitrided in pure nitrogen at a temperature of 1400°C for 8 h, gives predominantly α-silicon nitride. While 3% of aluminium has no effect on the formation of β-silicon nitride, the addition of 4% of either aluminium powder, or alumina, to the silicon, followed by pressing, rapidly increases the yield of β-silicon nitride. The larger the quantity of additive, and the higher the pressure, the greater is the yield of β-silicon nitride. With the addition of 6% of aluminium, and with a pressure of 150 MN m^{-2} (1.5 ton cm^{-2}), the predominant product is β-silicon nitride, with insignificant amounts of α-silicon nitride.

Nitridation of mixtures of silicon and alumina in the ratios 5:3 and 1:1 yields β-sialon at 1400°C in pure nitrogen (with insignificant amounts of α-silicon nitride and alumina). In nitrogen containing 1% of oxygen a large quantity of α-alumina, which has not been dissolved in the β-silicon nitride lattice, remains.

d-spacings for a 37% solution of alumina in β-silicon nitride are increased by 0.04 to 0.05 A°, in relation to pure β-silicon nitride. Spectroscopically pure silicon behaves in the same way as technical silicon.

At temperatures between 1400 and 1550°C solid solutions of alumina in silicon nitride are not obtained if pre-formed silicon nitride, and alumina, is used. It is supposed that inclusion of α-Al$_2$O$_3$ into the β-Si$_3$N$_4$ lattice is specially facilitated while the lattice is in the course of formation ('in statu nascendi'). That aluminium powder behaves in an identical manner to alumina is explained partly in terms of its conversion to alumina by extraction of oxygen from surface silica on the silicon nitride.

DISCUSSION (Goursat, Krauth, Smeltzer)

Godfrey: Two years ago we reported similar work on the reaction-bonding of Si-Al_2O_3 compacts. Reaction bonded sialons were produced, but showed decreased strength as the amounts of Al_2O_3 were increased, probably due to the volatilization of SiO increasing defect size. Did you measure the strength of your materials?

Jelacic: I did not make systematic measurements of the mechanical strength of compacts because I had a small number of them. The strength was about 1.5-2 ton/cm^2 for resistance to pressure.

Riley: An effect of compaction pressure, apparently increasing the proportion of β-phase produced, has been reported by Parr and colleagues (Powder Met., 1961, 8, 152). This could be the result of a higher compact density retarding an initial α-forming process, leaving more silicon available for conversion to the β-phase subsequently as full nitridation temperature is reached.

PREPARATION AND CHARACTERIZATION OF SILICON NITRIDE POWDERS
FOR HOT PRESSING

H. Haag, W.D. Glaeser and B. Krismer

Hermann C. Starck Berlin

ABSTRACT

It is well established that silicon nitride powder suited for hot pressing needs to be predominantly α-phase Si_3N_4. Not more than about 5% β-phase can be tolerated. β-phase formation is favoured by temperatures running high, but lowering reaction temperatures necessitates prolonged nitridation times. In a research project sponsored by the German Ministry of Research and Technology (BMFT, Project No. 01ZA 075-Z13 NTS 1003) the manufacture of α-phase silicon nitride powder has been investigated with a view firstly to optimising reaction times, and secondly to establishing conditions for the production of high yields of α-phase.

Experiments have shown that:

1. nitriding in pure ammonia in the temperature range 1280-1330°C increases the reaction rate by at least a factor of two as compared with a nitrogen-hydrogen atmosphere;

2. the particle size range 0-53 μm is suitable for complete nitridation;

3. a two-stage process (with second stage temperatures of 1355 and 1400°C)leads to a considerable reduction in the reaction time, without risk of pronounced β-phase formation.

The nitriding is best done in stages, with a first run at 1300°C for 10 h where most of the silicon is reacted to silicon nitride. Very little β-phase is formed. After the first nitriding the reaction mass is milled, and then the nitridation completed. This is done in two steps, the first lasting 1 h at 1330°C and the second step lasting 7 h at 1400°C. Only α-phase forms from the residual unreacted silicon, and the α/β-phase yield consequently increases. The complete nitriding can therefore be done in far less than 72 h if a two-stage reaction in 100% ammonia is used. The final amount of β-phase is typically less than 6%, whereas the amount of α-phase reaches 90%.

DISCUSSION (Goursat, Krauth, Smeltzer)

Jack: To what extent is the NH_3 cracked under your nitriding conditions - is it not just nitrogen plus hydrogen?

Glaeser: What I intended to say was that we use NH_3 gas at the furnace input - flushing the furnace and operating in a flow of NH_3. The gas heats up in the furnace and gets cracked, the amount of which has not been determined.

Engel: In which form is the silicon contained in the furnace during nitriding? What kind of boats do you use?

Glaeser: We heap up loose powder in boats. Quite a range of materials is suitable, silicon nitride being one of them. The furnace is a commercial one, consisting of molybdenum elements behind muffles, but in the reaction atmosphere.

NITRIDATION OF CALCIUM HEXABORIDE

K. A. Schwetz

Elektroschmelzwerk Kempten GmbH., 896 Kempten,
W. Germany

ABSTRACT. The experimental findings regarding the formation of
boron nitride from calcium boride and gaseous nitrogen at temperatures varying between 900 and 2100 °C are in conformity with
the results established through thermodynamic calculation. The
reaction in nitridation is exothermic. The heat of reaction calculated by approximation is bigger than that in the formation of
boron nitride from the elements. Thermogravimetric measurements
reveal that at temperatures below the boiling point of calcium
metal the reaction quickly comes to rest. At reaction temperatures of 1550 °C and higher the kinetic inhibition is, however,
very short-lived. The nitridation of calcium hexaboride with a
30 % addition of boric acid proceeds much faster than that of
elemental boron and calcium boride without addition of boric acid.

1. THERMODYNAMIC CALCULATIONS

As can be ascertained from the literature (1 - 2) calcium hexaboride is stable towards nitrogen up to 2000 °C. The author is
therefore going to check the result of his recent study (3) according to which the nitridation of ultrafine CaB_6-powders to boron
nitride was observed in a temperature range of 2100 °C to less
than 1000 °C, through thermodynamic computation as follows:
Unlike the formation of boron nitride from the elements

$$B + 1/2\ N_2 = BN \quad \ldots \ldots \ldots \ldots \ldots \ldots \text{eq. 1}$$

the nitridation of CaB_6 is proceeding at temperatures above the
decomposition temperature of calcium nitride Ca_3N_2 (1250 °C) with
elemental calcium being liberated simultaneously:

$$CaB_6 + 3\ N_2 = 6\ BN + Ca\ \ldots\ldots\ldots\text{eq. 2}$$

with B_2O_3 being present calcium is able to form additional boron nitride according to

$$B_2O_3 + 3\ Ca + N_2 = 3\ CaO + 2\ BN\ \ldots\ \text{eq. 3}$$

to result in the overall equation

$$3\ CaB_6 + B_2O_3 + 10\ N_2 = 20\ BN + 3\ CaO\ .\ \text{eq. 4}$$

As there was no reference in literature as to the free formation enthalpy of CaB_6, the free enthalpy change searched for the new reaction 4 was computed on the basis of the free enthalpy change of the known reaction 5, i. e. nitridation of B_2O_3 in the presence of carbon, and 6, i. e. carbothermic production of CaB_6:

$$B_2O_3 + 3\ C + N_2 = 2\ BN + 3\ CO\ \ldots\ldots\ \text{eq. 5}$$
$$CaO + 3\ B_2O_3 + 10\ C = CaB_6 + 10\ CO.\ \ldots\ \text{eq. 6}$$

When adding equation 6 multiplied by the factor 3 to equation 4, equation 5 multiplied by the factor 10 is obtained. The data on the free formation enthalpies of boron nitride, boric oxide, calcium oxide, and carbon monoxide were extracted from the JANAF - Thermochemical Tables (4). For ΔG_T° of reaction 6 the following approximation according to Peshev (5) can be established:

$$\Delta G_T^\circ = 775140 - 429{,}29\ T\qquad\text{(cal/mol).}$$

The free enthalpy change and logarithms of the equilibrium constants calculated per mole boron nitride for reactions 1 to 5 are summarized in table 1. The obtained and in figure 1 plotted ΔG values prove that the formation of boron nitride from calcium boride may take place even at lower temperatures. Furthermore, it is evident that the nitridation of CaB_6 in the presence of B_2O_3 is favoured thermodynamically as compared to the nitridation of pure CaB_6. The zero line crosses the ΔG curve of reaction 4 at 2400 °K, i. e. that the formation of BN is theoretically possible up to this temperature which was also confirmed by the author's own experiments (3). Table 2 shows the enthalpy values ΔH_T° per mole BN calculated by approximation from the equilibrium constants at various temperatures according to Van't Hoff.

Table 1. Equilibrium data for the reactions 1 to 5.
ΔG_T° values in kcal/mole BN

Temp.	Reaction 1		Reaction 2		Reaction 3		Reaction 4		Reaction 5	
°K	ΔG°	$\lg K_p$	ΔG°	$\lg K_p$	ΔG°	$\lg K_p$	ΔG°	$\lg K_p$	ΔG°	$\lg K_p$
1000	−38,89	8,499	−33,68	7,40	−108,1	23,65	−41,11	8,99	+10,76	−
1500	−28,39	4,137	−19,72	2,88	−103,6	15,12	−26,87	3,92	− 7,18	1,05
2000	−18,03	1,970	− 5,9	0,64	− 66,7	7,30	−12,00	1,31	−24,5	2,68
2500	− 7,59	−	+10,98	−	− 30,8	2,69	+ 6,80	−	−37,91	3,32

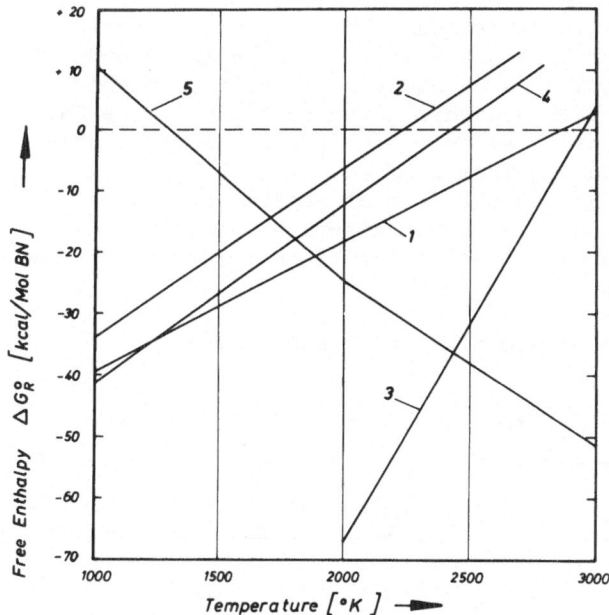

Figure 1. The free reaction enthalpy as a function of the absolute
temperature for the reactions 1 to 5.

Table 2. Heat of reaction per mole boron nitride.

Reaction	ΔH_T^o kcal/mole BN	Temp.Range °K
Eq.1 Formation of boron nitride from elements	− 59,7	1000 − 2000
Eq.2 Nitridation of pure Ca-Hexaboride	− 61,8	1000 − 2000
Eq.3 Nitridation of boron oxide in the presence of Ca	− 210	2000 − 2500
Eq.4 Nitridation of Ca-Hexaboride with H_3BO_3 addition	− 70,2	1000 − 2000
Eq.5 Nitridation of boron oxide in the presence of C	+ 38,9	1500 − 2500

Consequently the nitridation of CaB_6 is exothermic, the heat of
reaction per mole boron nitride being even higher than that in the
formation of boron nitride from the elements.

2. KINETIC INVESTIGATIONS

The kinetics of the nitridation of calcium hexaboride and boron
submicron powders – the latter serving as a comparative substance
– were studied thermogravimetrically in isothermic conditions at
1350, 1450 and 1550 °C. The boron submicron powder was prepared
from a pure, vacuum grade – 325 mesh powder fraction; the calcium
hexaboride was a commercial grade, such as it is obtained today in

large scale production in an electric furnace according to the carbothermic reduction melting process. This boron rich boride is very pure (metallic impurities 0,1 % max.), taking no account of minor contents in carbon and adherent hydrated B_2O_3. The characteristic data of the materials used are given in table 3. The CaB_6 powders A and B differ mainly as to their specific surface area (BET). Powder A' was produced from A by adding 30 % H_3BO_3. All submicron powders were used in the form of agglomerates with 0,4 to 1,0 mm diameter and 50 - 70 % theoretical density. Linde-nitrogen, purified with a molecular sieve (Oxygen $<$0,1 vpm), was used as nitriding gas.

Table 3. Characteristics of starting materials.

Starting Material	Vendor	Surface Area, m^2/g	total B %	B_2O_3 %	H_2O %	Ca %	C %
Cryst.Boron	H.C. Starck,Berlin	9,5	$>$99	0,86	–	–	–
Ca-Hexaboride A	} ESK/Kempten	7,1	57,38	3,20	1,5	34,11	4,26
Ca-Hexaboride A'		–	49,34	16,20	10	26,78	3,40
Ca-Hexaboride B	}	15,1	58,31	3,03	–	35,47	4,17

Experimental

The TG-measurements were carried out with a thermobalance from Netzsch, Selb/Bavaria. Electrically heated vertical tube furnace, maximum temperature: 1550 °C, sample weight: 100 mg, heating rate: 50 °C per min., nitrogen flow: 30 l per hour, crucible material: hot pressed, methanol treated boron nitride (weight constant at 1550 °C), height of the powder filling: 2 mm, prior to heating up the alumina tube was flushed 1 hour with nitrogen.

Results

Fig. 2 and 3 show the weight gain versus time curves obtained at 1350, 1450 and 1550 °C.

Table 4. Nitridation of B and CaB_6-Submicron powders at 1 atm. N_2.

Sample	Temp. °C	Hold Time min	Weight Gain*) %	Weight Loss*) %	Nitrogen Content %	Boron Conversion %
Boron	1550	510	105	nil	52,5	79
"	1450	310	60	nil	35,2	45
"	1350	330	33	nil	24,6	25
CaB_6- A	1550	255	30	$\sim$ 8	29,0	50
" - B	1550	340	32	$\sim$ 14	34,6	60
" - B	1450	160	14	nil	12,2	18
" - A'	1550	350	46	$\sim$ 6	36,2	79
" - A'	1350	145	16	nil	13,7	22

Latent weight loss calculated from: WL = f_N. E - WG
Letters of reference: WL ... weight loss   E ... final weight (mg)
     f_N ... nitrogen content (weight fraction, %/100)
*) ... based on dehydrated starting material   WG ... weight gain (mg)

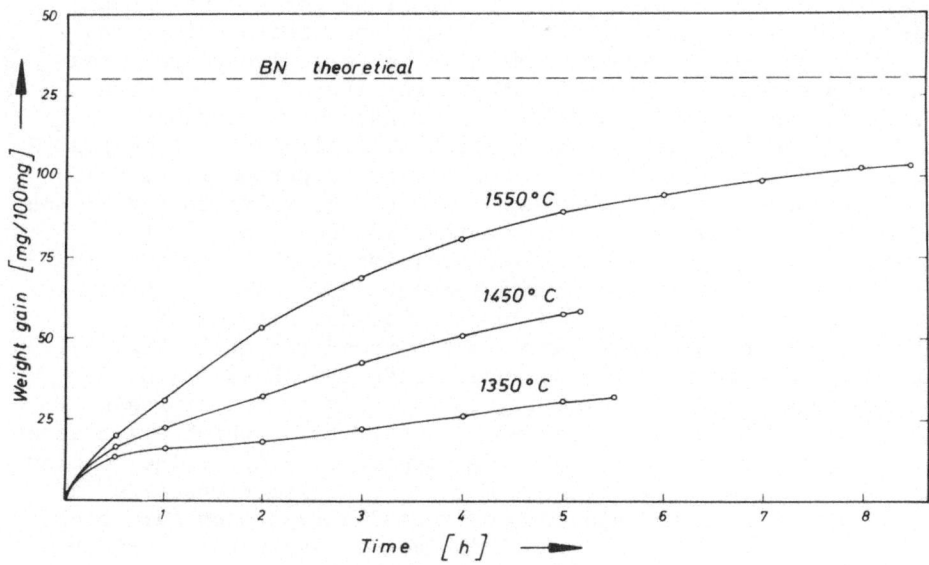

Fig. 2. Nitridation isotherms for crystalline boron (BET: 9,5 m²/g) in pure nitrogen at 1 atm.

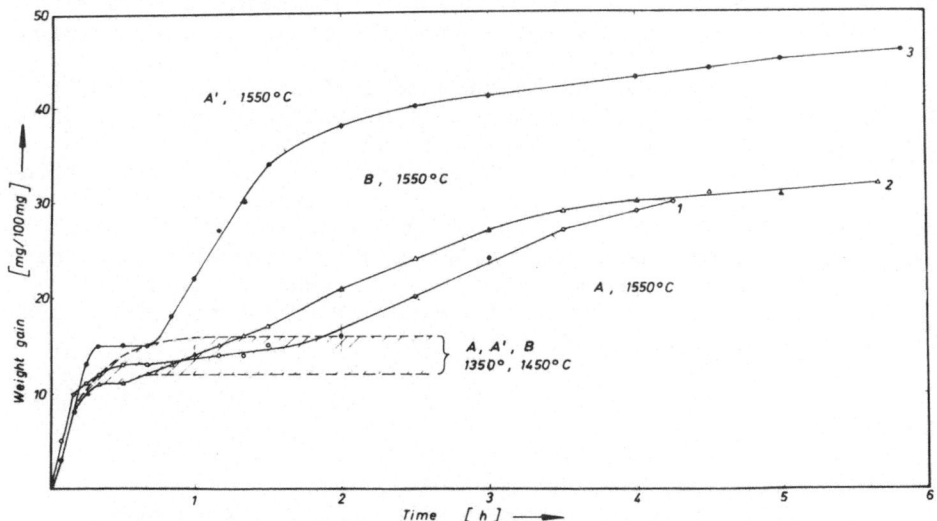

Fig. 3. Nitridation isotherms for technically pure CaB_6
   A : no H_3BO_3 addition, BET: 7,1 m²/g
   A': powder A with an addition of 30 % H_3BO_3
   B : no H_3BO_3 addition, BET: 15,1 m²/g

The final weight gains as well as the final boron conversions, determined on the basis of the nitrogen analyses and the final latent weight losses are presented in table 4. The weight loss occurring simultaneously with the weight increase in calcium hexaboride is due to volatilization of gaseous by-products (Ca, $CaCN_2$, CO) during the nitriding process. As indicated from Fig. 2 the nitridation of elemental boron follows a parabolic rate law at temperatures between 1350 and 1550 °C. According to Fig. 3 the CaB_6 submicron powders show no further increase in weight – irrespective of their chemical composition and BET surface area – after being held at temperatures between 1350 and 1450 °C for about one hour, which seems to prove that the nitridation is inhibited at these temperatures (see hatched region in Fig. 3). The curves are horizontal, because beside the BN also liquid calcium metal is formed. The liquid Ca inhibits the nitrogen gas transport to the non reacted CaB_6-particles, which are covered with the liquid phase. At 1550 °C, consequently above the boiling point of calcium (1487 °C), the kinetic inhibition observed is relatively short-lived and the horizontal curves then turn back to a weight gain course. The calcium metal evaporizes, further nitridation can occur, resulting in a boron conversion of 60 % after almost 6 hours (for CaB_6-B powder). An interesting feature is the strong acceleration of the reaction rate of the sample with 30 % H_3BO_3 addition (see 1550 °C curve of CaB_6-A'). As soon as the kinetic inhibition comes to an end the curve is ascending linear for one hour. In this case the calcium predominantly reacts with B_2O_3 forming additional BN and CaO and the final boron conversion after 6 hours at 1550 °C is 79 %.

As a consequence the latent weight loss in CaB_6-A' with a 16,2 % B_2O_3 content amounts to less than half of that in CaB_6-B powder with 3 % B_2O_3 when held at 1550 °C for almost 6 hours.

A comparison between the rate of the nitridation of calcium boride and that of elemental boron at 1550 °C reveals the following: CaB_6-A and -B react slower, whereas CaB_6-A' with a 16,2 % boric oxide content reacts perceptibly faster with nitrogen than elemental boron. To give an example: in CaB_6-A' powder a boron conversion of approx. 79 % is obtained already after 350 min, in elemental boron, however, only after 510 min.

REFERENCES

1. M. L. Andrieux, Rev. Met <u>32</u> (1935), 487.
2. N. N. Greenwood, in: Comprehensive Inorganic Chemistry, Pergamon, 1973, Vol. 1, 729.
3. K. A. Schwetz, paper presented at the DKG-Annual Meeting, Düsseldorf, Oct. 1976, Ber. Dtsch. Keram. Ges., in press.
4. JANAF-Thermochemical Tables, 2nd Edition 1971.
5. P. Peshev, Rev. Int. Hautes Temper. et Refract. <u>4</u> (1967), 289.

DIRECT PREPARATION OF SIALON-TYPE MATERIALS FROM NATURALLY OCCURRING MINERALS

S. Umebayashi

National Industrial Research Institute of Kyushu, Shuku-machi, Tosu City, Saga Pref., Japan.

Dense sintered solids with compositions in the Si_3N_4-Al_2O_3-AlN system were obtained by hot-pressing a mixture of naturally occurring silica sand (SiO_2 99.9; Al_2O_3 0.018, TiO_2 0.0023; Fe_2O_3 0.0031; Na_2O 0.0022; K_2O 0.001; loss on ignition 0.03($^w/o$)) and Al powder (99.7%) in a N_2 atmosphere. Mixed powder compacts of the silica and 40-60 wt% Al powder were formed into discs 25 mm in diameter and 25 mm long under a pressure of 400 kg/cm². Samples were heated at 1400°C for 10 h in a N_2 atmosphere. After reaction sintering at this temperature, the material was powdered again by vibratory ball milling to 14 μm average particle size for hot pressing. Hot-pressing was carried out at a pressure of 200 kg/cm² and temperatures from 1600 to 2000°C, for 1 h in a N_2 atmosphere, using graphite dies 30 mm in diameter. The composition of the crystalline phases was identified by X-ray diffraction using Cu Kα radiation.

Density and phase composition of the sintered solids varied with Al content and hot-pressing temperature. Table 1 gives the bulk specific gravity, apparent porosity and weight-loss of the products hot-pressed at various temperatures. High density solids were obtained between 1700° and 1800°C. Above these temperatures, weight loss of the samples increased due to vaporization during hot-pressing. The temperature for densification and that for vaporization has a tendency to increase with increasing Al content in the starting material.

Table 2 gives the phase composition of the sintered solids prepared with different Al contents and different hot-pressing temperatures. β'-Si_3N_4 is considered to be a β-Si_3N_4-Al_2O_3-AlN solid solution (1), and its unit cell dimensions, a_0 and c_0, increased with increasing hot-pressing temperature and increasing

Hot-pressing Temp.	Apparent porosity (%)	Bulk sp. gravity	Weight loss (%)
	HQ40		
1600°C	19.20	2.73	0.00
1700°C	2.10	3.28	0.00
1750°C	0.04	3.28	0.48
1800°C	0.08	3.19	30.00
	HQ50		
1600°C	30.80	2.18	0.00
1700°C	7.80	2.70	0.00
1800°C	0.05	3.25	12.50
1850°C	0.10	3.21	18.50
1900°C	0.33	3.17	37.34
	HQ60		
1600°C	31.2	2.39	0.00
1700°C	29.20	2.47	0.00
1800°C	3.00	3.40	4.20
1850°C	0.17	3.46	8.50
1900°C	0.29	3.43	13.50
2000°C	0.66	3.35	22.30

Table 1: Apparent porosity, bulk specific gravity, and weight loss of the hot-pressed solids.

Al content as shown in Fig. 1. X-phase (2) is a minor constituent and its existence is observed only in solids prepared with 40% Al. The X-phase is considered to be formed by reaction of Si_2N_2O formed from SiO_2 and Si_3N_4. The SiO_2 was due to the low Al content. The 15R-AlN polytype phase (2) appeared and the X-phase disappeared with increasing Al content. Only 15R polytype phase was observed as a component of the products when 60% Al was used, and the hot-pressing temperature was above 1950°C. Table 3 gives X-ray data on the 15R polytype phase. The d values of the 15R polytype increased slightly with increasing hot-pressing temperature.

It is not yet clear if a glass phase is present in the products. Jack (2) reported that X-phase crystallizes from a silica-rich nitrogen glass on cooling from 1650°C, and some glass is always retained. Hot-pressing of the mixture containing 40% Al above this temperature may therefore give products containing a glassy component in addition to β'-Si$_3$N$_4$, α-Al$_2$O$_3$ and X-phase. However, it may be that in the products prepared from 50 and 60% Al, excess Al prevents or reduces the formation of a glass phase, by tying up oxygen with Al in solid solution, β'-Si$_3$N$_4$ or 15R polytype lattices.

	HQ40	HQ50	HQ60
1600°C	α-Al$_2$O$_3$ > β'-Si$_3$N$_4$ >> X	α-Al$_2$O$_3$ > β'-Si$_3$N$_4$ > 15R	15R > α-Al$_2$O$_3$ > AlN > β'-Si$_3$N$_4$
1700°C	α-Al$_2$O$_3$ > β'-Si$_3$N$_4$ >> X	β'-Si$_3$N$_4$ > α-Al$_2$O$_3$ > 15R	15R > α-Al$_2$O$_3$ > AlN > β'-Si$_3$N$_4$
1800°C	β'-Si$_3$N$_4$ >> α-Al$_2$O$_3$ >> X	β'-Si$_3$N$_4$ > 15R > α-Al$_2$O$_3$	15R >> β'-Si$_3$N$_4$ > α-Al$_2$O$_3$ (trace)
1900°C		β'-Si$_3$N$_4$ > 15R >>> α-Al$_2$O$_3$ (trace)	15R >>> α-Al$_2$O$_3$ = β'-Si$_3$N$_4$ (trace)
1950°C			15R
2000°C			15R

Table 2: Phases present in hot-pressed solids.

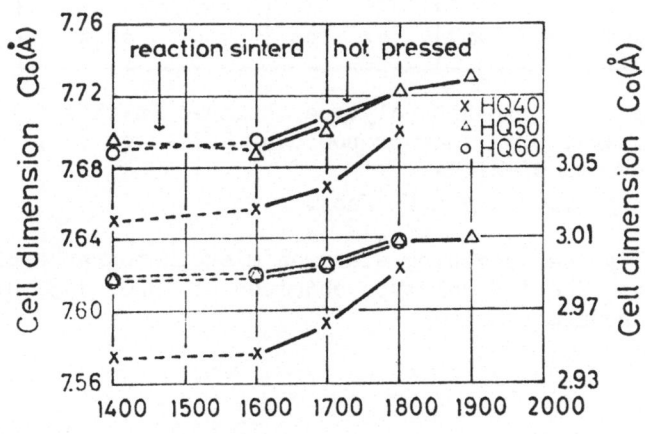

Fig. 1: Variation of unit cell dimensions of β'-Si$_3$N$_4$ with hot-pressing temperature.

Index	15R-AlN**		HQ60-18		HQ60-19		HQ60-20	
	d	I	d	I	d	I	d	I
0015	2.787	s	2.790	250	2.795	300	2.796	375
101	2.602	s	2.604	215	2.606	230	2.607	310
012	2.587	ms	2.591	155	2.591	170	2.594	210
104	2.529	w	2.529	26*	2.531	22	2.535	28
015	2.489	ms	2.492	94*	2.492	106	2.495	112.5
107	2.389	s	2.392	182	2.392	200	2.396	210
018	2.333	mw	2.332	30	2.333	38*	2.335	55*
0018	2.323	mw						
1010	2.212	mw	2.215	32*	2.216	34	2.219	50
0111	2.150	m	2.152	35	2.152	40	2.156	37.5
0021	1.991	w	1.989	8	1.995	12	2.002	17.5
0114	1.964	mw	1.967	26	1.968	28	1.970	35
1016	1.845	ms	1.848	70*	1.849	80	1.852	95
0117	1.789	ms	1.792	20*	1.793	20	1.795	22.5

Table 3: X-ray data for 15R-AlN polytype obtained in this work compared with that proposed by K.H. Jack (2). *Overlap with β'-Si_3N_4 **15R-AlN polytype by K.H. Jack.

Fig. 2 shows typical scanning electron microphotographs of fractured surfaces. The 15R polytype obtained at 1900, 1950 or 2000°C shows a fibrous morphology.

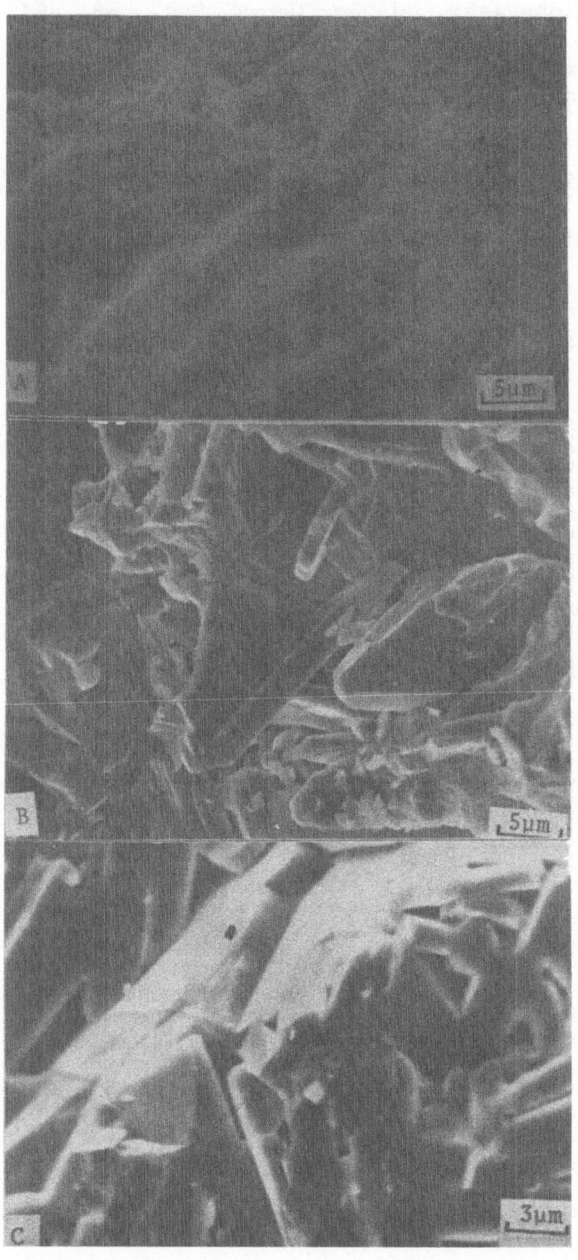

Fig. 2: Scanning electron micrographs of fractured surfaces.
a) 40% Al, 1800°C; b) 50% Al, 1900°C; c) 60% Al, 1200°C.

REFERENCES

1. S. Umebayashi and K. Kobayashi. J. Am. Ceram. Soc., 58, 464 (1975).
2. K.H. Jack. "Sialons and Related Nitrogen Ceramics", preprints; A Symposium, Tokyo, November 1975.

DISCUSSION (Goursat, Krauth, Smeltzer)

Billy: What is the oxidation resistance of your sintered materials? The 15R phase (sialon phase of Goursat and Billy) has a relatively poor oxidation resistance.

Umebayashi: Results are as follows (1200°C, 40 h):

	Weight gain (mg/cm^2)	Phases observed in the specimen
HQ40-18(1)	0.07	$\beta'-Si_3N_4 \gg \alpha-Al_2O_3 \gg X$
HQ40-18(2)	≈ 0	$\beta'-Si_3N_4$
HQ40-40Si-18(1)	0.04	$\beta'-Si_3N_4$
HS50-18(1)	0.23	$15R > \alpha-Al_2O_3 > \beta'-Si_3N_4$
HS60-18(1)	0.30	$15R \ggg \alpha-Al_2O_3$

Key:

Q	40	18	(1)	S
silica sand	% Al content	Temperature (°C)	Time (h)	Volcanic ash

HQ40 - 40 Si

silica sand:Al:Si

(60   :40:40)

The oxidation resistance of fully dense 15R-polytype is better; the oxidation resistance of 27R-polytype was 0.50 mg/cm^2 at 1200°C for 200 h in air.

Mangels: What is the strength of your 15R polytype?

Umebayashi: Room temperature strength is 50-60 Kg/mm^2.

Section F

DENSIFICATION PROCESSES

BASIC AND PRACTICAL ASPECTS IN SINTERING NITROGEN CERAMICS

A.L. Stuijts

Philips Research Laboratories
Eindhoven, The Netherlands

ABSTRACT. The present state of the science of sintering is
reviewed. The consequences for covalently bonded materials such
as nitrogen ceramics, which are known generally for their poor
sintering behaviour, are considered. Problems are posed by the need
for very fine starting powders because of their very high surface
reactivity. The recent development in sintering silicon carbide
to a high density is discussed in more detail. Liquid-phase
sintering and pressure sintering are an effective means of
obtaining densification, but both have their disadvantages for
practice. Some thoughts on "activated sintering" are presented.

1. INTRODUCTION

The interface between two phases represents an extra amount of free
energy. The aspect of interfacial free energy comes very strongly
to the fore when powders are used as starting materials for forming
products by sintering. Powder compacts used in normal sintering
practice of ceramics have a very high specific surface area,
usually 1-10 m^2/g, which corresponds to an extra free energy of
about 1 cal/g. This relatively small amount of energy provides the
driving force for the sintering process.

A powder compact can lower its overall free energy by lowering
the interface area and/or by changing the kind of interfacial
energy. In both cases transport of matter has to occur, which is
facilitated by heating the compact to a sufficiently high
temperature. There are a great number of possible mechanisms by
which transport of matter may take place. They may cause one or
more of the following effects on the scale of the particles:

smoothing	rounding off of the pores
strengthening	growth of a neck between the particles
shrinkage	increase in density of the compact

The extent to which a certain mechanism contributes to one or other of these phenomena depends on the type of transport mechanism and on the material under consideration.

For a material flux to occur a gradient in the chemical potential has to be present. The problem is to show the origin and magnitude of this gradient and to find the possible paths for transport of matter. The principal material transport mechanisms are understood in a very general way. If these principal mechanisms are to be applied in order to understand or predict the sintering behaviour of a certain material, then problems arise. These problems are caused by the fact that every material is characterised by its chemical composition and crystalline structure, which often results in a unique and specific high-temperature behaviour.

In this paper a review is given of the current views on the mechanisms of transport of matter and the main diffusion paths in crystalline solids. This paper cannot treat the sintering process in all details. A choice of review papers is therefore included in the list of references for further reading (1-8).

After treating the basic aspects of sintering models and the thermodynamics of the process, an attempt is made to make some generalisations on the sintering behaviour of classes of materials according to the type of atomic bonding: metallic--ionic-covalent.

Processes on the particle scale other than those previously mentioned do occur at the sintering temperature. In crystalline materials, grain growth and the interaction between pores and grain boundaries are of paramount importance. This occurs at the later stage of the sintering process, when a great part of the porosity has disappeared. Pore growth may occur. The effect of gases trapped within the pores has also to be considered. It is essential to understand these phenomena where one needs a high density in combination with a certain microstructure. This is particularly true for a material which shows an intrinsically poor sintering behaviour, such as the covalently bonded materials to which the nitrogen ceramics belong.

Apart from the basic sintering aspects a number of practical aspects are considered. Liquid-phase sintering, pressure sintering, or a combination of both, can offer a solution when theory predicts a low shrinkage rate but a high rate is required. Pressure sintering offers an especially effective means by which high densities can be obtained in materials with poor sintering behaviour.

The aspects of grain growth, grain boundary-pore interaction, and pressure sintering are treated in other papers at
this conference. Although very little is known fundamentally about
the sintering behaviour of covalently bonded materials, some
thoughts will be presented in this paper on specific problems in
sintering nitrogen ceramics. Some preliminary and critical
thoughts on liquid-phase and "activated" sintering will also be
put forward. The recent success in sintering silicon carbide to
a high density (9,10) forms a reference for the interpretation
of sintering behaviour.

2. SINTERING MODELS

The curvature of the particle surfaces on micron scale in a
consolidated powder mass is the important factor in the chemical
potential. The driving force for the flow of matter to fill the
pores results from differences in the chemical potential for
places with different surface curvatures. The flow of matter occurs
between places which act as sources and sinks (11). The complex
geometry of a powder mass cannot be described with a few parameters
and thus simplifications have to be made. An idealized situation

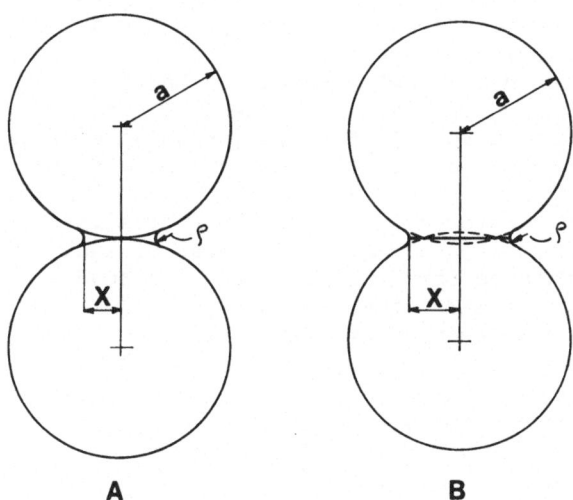

Fig. 1. Two spheres representing the initial stage of sintering.
Neck growth occurs A) with the surface or B) with the grain
boundary between the spheres as the source of matter to fill the
neck. Radius of neck is x, radius of curvature of the neck surface
is ρ.

is represented in the model of two coalescing spheres as shown in
Fig. 1. This model has often been dealt with in the literature and
has become the basis of the science of sintering. This basis was
laid by Frenkel (12) in considering Newtonian viscous flow as the
transport mechanism for filling of the neck. Later this model was
extended into a generalized theory by Kuczynski (13). He
introduced the model of a sphere on a plate or a bundle of wires,
which are experimentally easier to handle. Their work and that of
many others has brought a substantial increase in our understanding
of the transport mechanisms involved in sintering.

In most cases in practice the intention is that at the end of
the sintering process most of the porosity should have disappeared.
Three stages are distinguished in the sintering process, the
initial, the intermediate, and the final stage, roughly corresponding
to a shrinkage of a few %, up to about 90%, and 90 to 100%
respectively. The two-sphere model only refers to the initial stage
to which most work on models has been devoted.

From his studies on the rate of neck growth Kuczynski (13)
derived for the rate of neck growth a general equation

$$\left(\frac{x}{a}\right)^n = \frac{F(T)}{a^m} t$$

In this equation t is the time and $F(T)$ is a temperature-dependent
constant. x and a are defined in Fig. 1. The transport mechanisms
are characterized by the exponents n and m. In his recent review
paper (6) Kuczynski lists the neck-growth mechanisms and the
corresponding values for n and m as given in Table 1. The constant
$F(T)$ will contain the relevant materials constant, such as diffusion
coefficient in the case of volume diffusion, the vapour pressure
in the case of evaporation-condensation, etc.

In this paper we are not interested in the exact details of
the different models, but more in their general character and,
especially, in their significance for densification of a powder
compact. In order to do, use is made of a recent report by
Ashby (11) on sintering pure one-component systems in the
absence of an applied stress.

Table 1   Neck-growth mechanisms (6)

m	n	Mechanism
1	2	Newtonian viscous flow
p	2p	Viscous flow
2	3	Evaporation and condensation
3	5	Volume diffusion
4	6	Grain-boundary diffusion
4	7	Surface diffusion

Table 2   (After (11))

Mechanism No	Transport path	Source of matter	Sink of matter
1	Surface diffusion	Surface	Neck
2	Lattice diffusion	Surface	Neck
3	Vapour transport	Surface	Neck
4	Boundary diffusion	Grain boundary	Neck
5	Lattice diffusion	Grain boundary	Neck
6	Lattice diffusion	Dislocations	Neck

In treating the mechanisms of transport of matter, mostly several mechanism are excluded a priori. In Ashby's paper amorphous materials are not considered, which means that Newtonian viscous flow is excluded. It is also argued in this paper, as done before by Kuczynski (6) and others, that plastic flow, or better dislocation creep, cannot contribute significantly to densification. For the present paper we will follow Ashby and exclude these two mechanisms.

Ashby considered six distinguishable mechanisms for the flow of matter, as shown in Table 2. In all cases mentioned here matter flows into the neck between the powder particles and in Fig. 2 the possible diffusion paths are shown. All mechanisms mentioned in Table 2 contribute to neck growth, but not all lead to densification of a powder compact. In densification the distance between the

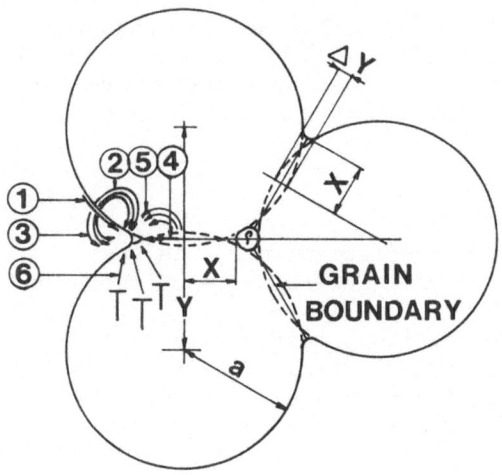

Fig. 2. Three-sphere model representing a multi-particle array. The six transport paths from Table 2 are shown. In all cases neck growth occurs, but only paths 4 to 6 lead to densification. After (11).

centres of the particles has to decrease. Those mechanisms where the surface is the source of matter only lead to rounding off of the pores (smoothing). This can easily be derived from Fig. 2. Only the mechanisms 4 to 6 can cause densification of a powder compact.

In a powder compact there will be formation of a neck to some extent already (14). This contribution to neck formation will be much less in brittle ionic or covalently bonded materials than in ductile metals like pure silver or copper.

The basic equations for neck growth during the different stages have to be transformed to equations for powder compacts, where the linear shrinkage $\Delta L/L_0$ can be measured as a function of time. This is necessary as the two-sphere model can only be applied to sizes which are relatively large compared to the powder particle sizes used in real sintering practice.

Derivations have been made by a great number of investigators, starting with the work of Coble (1, 15). It is necessary to make number of geometrical assumptions and therefore the relations derived differ in detail (16, 17, 18). As several transport mechanisms can contribute simultaneously, shrinkage equations have also been derived where different mechanisms are combined (19 - 22).

A typical equation for shrinkage by lattice diffusion from the grain boundary to the neck surface is (23):

$$\frac{\Delta L}{L_0} = (\frac{2D_t \gamma_{sv} \Omega_0 t}{kTa^3})^{1/2} \qquad (2.1)$$

D_T is the tracer diffusion coefficient, $\Delta L/L_0$ the shrinkage ratio, γ_{sv} the surface energy.

If the transport path is from the grain boundary by grain-boundary diffusion to the neck, the equation can be (17):

$$\frac{\Delta L}{L_0} = (\frac{50D_b \delta\gamma_{sv} \Omega_0 t}{7\pi kTa^4})^{0.31} \qquad (2.2)$$

D_b is the grain-boundary diffusion coefficient and δ is the thickness of the grain boundary. δ will generally be of the order of a few atom layers.

A shrinkage equation with which it is possible to calculate the volume, grain-boundary and surface diffusion coefficients, provided vapour transport and plastic flow are negligible, has been derived by Johnson (19):

$$\frac{X^3 R}{X+R\cos\alpha}\,\dot{Y} = \frac{2\gamma_{sv}\Omega_o D_t}{\pi kTa^3}\frac{A_v}{ax} + \frac{4\gamma_{sv}\Omega_o D_b\delta}{kTa^4} \qquad (2.3)$$

where $X = x/a$
$R = \rho/a$
$Y = L/Lo$
$\dot{Y}$ = shrinkage rate
A_v = neck surface area
With this equation it is a straight-forward matter to derive the volume and grain-boundary diffusion coefficients. In measuring Y, $\dot{Y}$ and x it is also possible to estimate the surface-diffusion coefficient.

The application of these quantitative relations to conclude about the matter-transport mechanisms and the derivation of diffusion coefficients is only possible if a good model system is used. A carefully prepared compact of spheres of narrow size distribution has to be used. If care is taken in this respect good agreement with known diffusion data can be found.

From the relations conclusions can be drawn about the relative importance of each of the mechanisms as a function of the important parameters: particle size, temperature, time, and, in the case of relation (2.3), the instantaneous geometry. They show that compared to grain-boundary diffusion, volume diffusion will be dominant at high temperatures and for larger particles. Another useful generalization is that, at a given temperature, the flux ratios of volume to grain-boundary diffusion and grain--boundary to surface diffusion will increase with sintering time and particle size (24).

It is not possible to apply these relations to real powder systems (7, 24, 25) because many factors play a role violating the basic assumptions in the geometric models. To mention a few: non-ideal packing of the particles, non-spherical particles, anisotropy in the surface energy, grain growth already at the beginning of the sintering process, etc. However, it is indeed possible to derive from these equations some consequences affecting the practice of sintering different classes of materials.

Ashby (11) has combined equations selected from literature for the rate of neck growth, $\dot{x}$, or has derived equations himself for the different stages of the sintering process. Using known physical constants and diffusion data it was then possible to construct sintering diagrams (the word sintering maps would probably be a better choice) for different materials. In Fig. 3 a schematic representation is given of such a sintering diagram. The different fields indicate the dominant mechanisms for neck growth. At the boundaries of the fields, represented by the heavy lines, two mechanisms contribute equally to the sintering

338

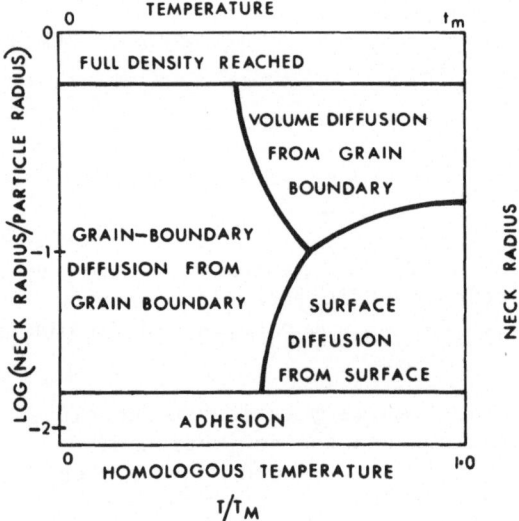

Fig. 3. Schematic representation of a sintering diagram as proposed by Ashby (11). T_m and t_m are the melting temperatures in °K and °C, respectively.

rate. Other transport mechanisms, e.g. evaporation and condensation may become dominant, depending on the material.

In these sintering maps it can nicely be displayed what is known from model sintering studies. Mechanism No. 4 from Table 2, grain-boundary diffusion, plays an important role. At higher temperatures, volume diffusion with the grain boundary as the source of matter (mechanism No. 5) becomes dominant.

3. THERMODYNAMIC ASPECTS

3.1. Influence of the surface curvature

As already mentioned the driving force for the flow of matter arises from the differences in surface curvatures between sources and sinks (11,26). The neck surface which forms between two spheres is characterized by the two principal radii of curvature, ρ and x (Fig. 1), having opposite signs. The difference in free energy between the neck surface and the particle surface then becomes:

$$\Delta G = \gamma_{sv}\Omega(\frac{1}{\rho} - \frac{1}{x} + \frac{2}{a}) = \frac{\gamma_{sv}\Omega}{\rho} \quad (a > x > \rho)$$

where Ω is the molar volume. This energy difference causes a material flux. Often it can be translated into a concentration gradient of a mobile component. This will be illustrated for the simple case of vapour phase transport.

The relation between the differences in vapour pressure and free energy is given by the Kelvin or Thomson-Gibbs equation:

$$\ln \frac{P_\rho}{P_a} = \frac{\Delta G}{kT} \cong \frac{\gamma_{sv}\Omega_0}{\rho kT}$$

where P_ρ and P_a are the vapour pressures above the neck surface and the particle surface respectively, and Ω_0 is the atomic or molecular volume. When $P_a \cong P_0$. the vapour pressure above a flat surface, and $\Delta P \ll P_0$ we obtain:

$$\Delta P \cong P_0 \frac{\gamma\Omega}{\rho kT}$$

This is the basis for the evaporation and condensation mechanism when the material has a sufficiently high vapour pressure. The convex particle surface will evaporate material that condenses into the concave neck surface.

In most cases transport of matter has to proceed by surface, grain-boundary, or volume diffusion. The surface of a solid forms a high-diffusivity path. Surface diffusion can act as a mechanism to lower the free energy by smoothing the pores without densification of the compact. With small particles its contribution will become more prominent.

The grain boundary, a two-dimensional defect in a solid, can provide a high-diffusivity path for atoms. The grain-boundary diffusion coefficients are mostly orders of magnitude larger than the lattice diffusion coefficients. The activation energies generally lie between those for surface and volume diffusion.

Diffusion transport in the bulk of a solid is only possible in the presence of defects. Vacancies are the most dominant point defects making diffusional flow possible in a solid. In some materials, however, interstitials can play a dominant role. It is of great importance to have knowledge of the point defects in the material investigated, as volume diffusion contributes to densification.

In analogy with the calculation of the vapour pressure it can be derived that the difference in vacancy concentration between the neck surface and the particle surface is given by

$$\Delta C = C - C_0 = \frac{2C_0\gamma_{sv}\Omega_0}{\rho kT}$$

where C_0 is the concentration of vacancies below a flat surface. For a simple metal this is

$$C_0 = \exp(-\Delta H/kT)$$

where ΔH is the enthalpy for the formation of a lattice vacancy.

As a result of the gradients in the vacancy concentration (which corresponds to the gradients in the chemical potential) diffusional fluxes of vacancies are set up, with a counterflow of atoms. Assuming rapid equilibration between solid and atmosphere, a quasi steady state condition results where vacancies are generated at the neck surface and disappear at a vacancy sink. The counterflow of matter fills up the neck.

In Table 2 the sources of matter, corresponding to the vacancy sinks, listed are the convex surface, the grain boundary, or dislocations. It has become clear from the theoretical work of Nabarro (27) and Herring (28) that grain boundaries are very effective sources and sinks for vacancies. This means that atoms move radially through the grain boundary to the neck (mechanism No. 4 in Table 2), or from the grain boundary through the bulk to the surface of the neck (mechanism No. 5 in Table 2). At the same time the removal of material from the grain boundary causes the distance between the particle centres to decrease, which is essential for densification.

Comparing the rates of neck growth in a two-sphere model with and without a grain boundary as a source of matter results, in the first case, in a rate faster by a factor of 3 (29).

3.2. Effect of dihedral angle

As pointed out by Herring (26) the equilibration of surface tensions requires a neck between two crystalline particles to have a sharp angle where the grain boundary meets the surface, as shown in Fig. 4 (after (30)). Kingery and Francois (31) have indicated the importance of an equilibrium dihedral angle less than $180°$ for the interaction between grain boundaries and pores. Under certain circumstances it can even lead to pore growth (31,32).

When an equilibrium dihedral angle ϕ is formed at the neck the following relation will hold:

$$\frac{\gamma_{ss}}{\gamma_{sv}} = 2\cos\frac{\phi}{2}$$

where γ_{ss} is the grain boundary energy. For $\gamma_{ss} \ll \gamma_{sv}$ a dihedral angle $\phi = 180°$ will be approximated and a spherical neck surface formed. The minimum value of $\phi = 0°$ gives $\gamma_{ss} \geq 2\gamma_{sv}$.

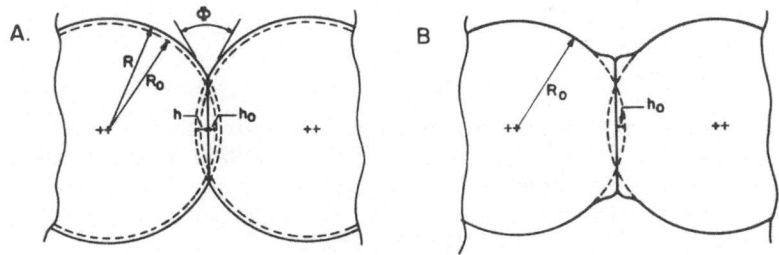

Fig. 4. Two-sphere densification model with a sharp dihedral angle φ (After Pask and Hoge (30)). A. No neck surface is formed; all matter is redistributed over the surface and fractional linear shrinkage is given by h_o/R_o. B. The classical neck region in combination with a sharp dihedral angle.

Normally in the sintering models the grain-boundary energy is neglected compared to the surface energy. It will be clear that for a ratio γ_{ss}/γ_{sv} above 2 neck growth is impeded. It has recently been postulated that a relatively high value of the ratio γ_{ss}/γ_{sv} could be the reason for the poor sinterability of silicon carbide (10). This is a covalently bonded material, having interatomic forces which are strongly directional and working only at a short distance. Therefore, one might suspect that the interaction between two solid surfaces is small, possibly resulting in a relatively large value of γ_{ss}/γ_{sv}. As nitrogen ceramics are also covalently bonded materials, it is important to treat the case of a sharp dihedral angle in more detail here.

The effect of a sharp dihedral angle has been discussed by Herring (26), White (33,34), and recently by Pask and Hoge (30). The model treated by Pask and Hoge uses the geometric configuration shown in Fig. 4A. A situation as shown here can only be maintained when a rapid redistribution of material over the free surfaces takes place. In the case of Fig. 4B the matter transport to the neck surface is faster than the rate at which matter is redistributed, but this is assumed to be a non-equilibrium perturbation of the first case.

The change in free energy of the system is given by

$$\delta G = \delta \int \gamma_{sv} dA_{sv} + \delta \int \gamma_{ss} dA_{ss}$$

where A_{sv} and A_{ss} are the surface area of the solid-vapour interfaces and the grain boundaries respectively.

In a stacking of particles the dihedral angle is governed by geometrical factors. This can easily be seen in a two-dimensional model where the pores assume the shape of a regular polygon. In that case is

$$\phi' = 180°(\frac{n-2}{n})$$

where n is the number of grains surrounding the pore and ϕ' the actual dihedral angle. Only if $\phi' > \phi$, where ϕ is the equilibrium dihedral angle, sintering occurs.

From this a critical value for γ_{ss}/γ_{sv} can be derived, for each type of stacking, defined by the relation:

$$\cos\frac{\phi'}{2} = \frac{\gamma_{ss}}{2\gamma_{sv}}$$

The value of this model is that it points to the importance of interfacial energies for sintering practice. From Pask and Hoge's model it can be learned that a high green-density is a useful means of increasing the critical value of γ_{ss}/γ_{sv}.

Next, one has to realize that adsorbed or segregated species very drastically affect the interfacial energies and thus influence the sintering process. A clean surface of a covalently bonded material is extremely reactive as a result of the strongly directional "dangling" bonds. In practical powders their surfaces will not be clean and thus drastically affect neck growth and sintering.

Prochazka (18) has been able to sinter silicon carbide to a high density. In his procedure the starting powder has to be the cubic β-silicon carbide and to have a high specific surface area ($\sim$ 10 m^2/g). By adding several tenths of one percent boron and carbon, densities of well over 95% of theoretical can be reached. His explanation is that carbon increases the effective surface energy by removal of (adsorbed) SiO_2 and Si. This seems a reasonable explanation as such fine powders will be extremely reactive to oxygen. Reaction between silica and carbon produces the volatile SiO.

SiC can lose Si at high temperatures; to what extent the carbon has a role to suppress this effect is not clear.

The role attributed to boron by Prochazka is the decrease of grain-boundary energy by selective segregation in grain boundaries. This hypothesis has to be examined critically as it is not based on other experimental evidence than that boron provokes shrinkage. There is relatively little information available in the literature on the ratio of the grain boundary to the surface free energies. Most measurements known have been made on metals of cubic or near-cubic symmetry. All known data show the ratio γ_{ss}/γ_{sv} to be smaller than 1.

As to segregation there is likewise very little information about the ratio by which the surface and grain boundary energies will be lowered. As stated by Seah and Lea (35), this difference will depend on the particular system considered.

4. PRACTICAL ASPECTS IN SINTERING

It is useful to start this section on practical aspects by reproducing a plot by White (25) on a comparison of sinterability of oxides compared to metals. Fig. 5 shows the shrinkage after 1 hr as a function of the homologous temperature. It compares isochronal sintering curves of a number of oxides with those of copper particles of comparable particle size, but differing in shape. The conclusion is, as generally found in practice, that oxides show a greater sinterability than metals although the volume diffusion coefficients of the slowest-diffusing ion are generally much lower than those for most metals.

The increased sinterability of ionic compounds also follows from the diffusion coefficients derived from sintering data. In alumina, Coble found (15) that the apparent diffusion coefficient lies nearer to that of the tracer diffusion coefficient of the Al^{3+} ion, which is about 10^3 times larger than that of the O^{2-} ions.

Several explanations have been put forward. Coble suggested that rapid transport of O^{2-} ions takes place at the grain boundaries, while Al^{3+} ions are transported via the bulk. Kuczynski (36) suggested the existence of a layer with an increased vacancy concentration as a result of interaction with

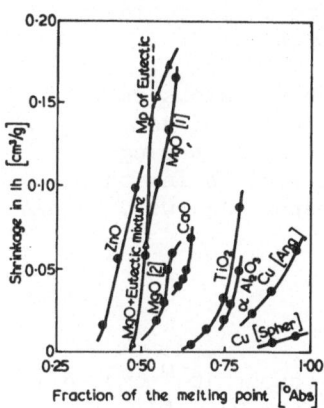

Fig. 5. Shrinkage as a function of homologous temperature of a number of oxide and metal powders. MgO [1] and MgO [2] were calcined from a carbonate at 900°C and 1450°C respectively. Cu [Ang] and Cu [Spher] refer to angular and spherical copper powders. After White (25).

the sintering atmosphere. Reijnen (37) explains the large, apparent diffusion coefficient by the effect of a deviation from the stoichiometric composition. He also showed that there are ionic compounds, such as nickel aluminate with an excess of Al_2O_3 over the stoichiometric composition, which have intrinsically a very low shrinkage rate.

4.1. Sintering of covalently bonded materials

Covalently bonded materials do not appear in Fig. 5. It is generally stated that (pure) covalently bonded materials, such as silicon nitride, do not sinter (38). To check this it is useful first of all to check whether the relations given in the previous section predict a low sintering rate. A typical example of a covalently bonded material where the self-diffusion coefficient is known is silicon. A reasonable sintering rate, e.g. 5% shrinkage in 1 hour, follows from equation (2.1.) when the ratio of D (cm^2 sec^{-1})/a^3 (cm^3) ~1 (39).

At a temperature of $1225^\circ C$, which is about $200^\circ C$ below the melting point, the self diffusion coefficient is about 10^{-14} cm^2/sec (40). From the above ratio D/a^3 it follows that a powder is needed with a particle size of about 0.5 µm. This raises the problem of producing powders of such a fine particle size without contamination, such as formation of an adsorbed oxide layer. Sintering will stop as the transport of matter, silicon in this case, will be seriously hindered. In silicon and silicon compounds it is not possible to remove the oxide layer in a suitable atmosphere. In oxide ceramics, where it is normal to use such fine powders, this is only a minor problem.

Silicon is also a good example with which to demonstrate the problem arising from sintering mechanisms which hinder densification by giving only neck growth when the sintering rate by volume or grain boundary diffusion is low. Taking the neck-growth rate, x, equations from Ashby's paper (11) we obtain for vapour transport (mechanism No. 1)

$$\dot{x} = P_v F \left(\frac{\Omega_0}{2\pi dkT}\right)^{1/2} K_1$$

where P_v is the vapour pressure, $F = \gamma_s \Omega_0 /kT$ (about 10^{-6} cm), d is the density of the material and $K_1 = (1/\rho - 1/x + 2/a).(1 - x/a)$. For volume diffusion according to mechanism No. 5 we have

$$\dot{x} = 4 D_T F K_2^2$$

where $K_2 = (1/\rho - 1/x)$. K_1 and K_2 can be taken to be about equal for $x < a$. Comparing these two rates for the initial stage of sintering at $1225^\circ C$, where P_v for silicon $\approx$ 1 dyn/cm^2, we find a

ratio of about 10^7. This very large factor shows the enormous
problem encountered when the intrinsic diffusional behaviour
of the material is poor.

It can be predicted that the contribution of volume and
also of grain-boundary diffusion to neck growth (and thus
shrinkage) relative to that of evaporation-condensation
increases with decreasing particle size. This suggests the
advantage of using silicon starting powders with a particle size
lower that the value of 0.5 μm calculated above.

The diffusion coefficients in the nitrides and in the sialons
are not known. However, the self-diffusion coefficients of
carbon and silicon have been measured in both p-type and n-type
α-silicon carbide (41). The values reported are even higher than
the values of the self-diffusion of silicon in pure Si, as used
above. Grain-boundary diffusion coefficients are not known, but
can be expected to contribute also in covalently-bonded materials.
Therefore, we can make the conclusion that intrinsically densification
of covalently-bonded materials is possible, provided a
sufficiently small particle size is used. The practical problems
encountered, such as an oxide layer at the particle surface and
dissociation of the compound, have to be solved in a way which
will be specific for the compound investigated. It is remarkable
to find in the literature many examples where it is stated
that covalently-bonded materials do not sinter where particle
sizes of a few up to even 20 μm have been used.

4.2. Practical means to improve sintering

Ceramic practice has long been able to improve sintering when
the intrinsic diffusional behaviour was poor or decrease un-
practically high sintering temperatures. The oldest practice
is to apply a liquid phase at the sintering temperature. The
amount used varies between about 1 to several tens vol %.

There is relatively little fundamental research in the field
of liquid-phase sintering. The reason is obvious: the systems are
chemically very complex. Some reviews have been included in the
list of references (42 - 45).

Again, in case of equilibrium the relation applies

$$\gamma_{ss} = 2\gamma_{sL} \cos \frac{\phi}{2}$$

where γ_{sL} is now the interfacial tension solid-liquid. In case
$\gamma_{ss} \geq 2\gamma_{sL}$ the angle ϕ will be zero and the particles will be
completely surrounded by the liquid.

Part of the role of a liquid phase, especially at higher
volume percentages, is a rearrangement of the particles by
removing bridges and other imperfections of the packing by

capillary forces. The most important densification mechanism, however, is by dissolution of the material, diffusion through the liquid and precipitation at places of lower chemical potential. This mechanism is quite analogous to the sintering of a pure solid phase by grain boundary diffusion. However, because of the much higher diffusion rates in liquids the matter transport will be much faster and can occur at relatively low temperatures compared to the normal sintering temperatures.

The application of a liquid phase has also been successful in attempts to densify nitrogen ceramics. Examples are the addition of Y_2O_3 and SiO_2 to AlN (46), of MgO to Si_3N_4 (47), of Al_2O_3 and MgO to Si_2N_2O (48), and of B_4C to SiC (49). The quantities used generally lie between 5 and 10 wt %.

It is questionable, however, whether liquid phase sintering will result in materials suitable for use in high-performance applications. At high temperatures the second phase which is present increases the creep rate. It is, therefore, necessary to increase the refractoriness of the second phase (38), or apply "transient liquid-phase sintering".

Much more effective for obtaining good high-temperature properties is the application of hot-pressing or pressure sintering. By an external pressure the liquid phase can drastically be decreased or even eliminated. High densities can be obtained in combination with small grain sizes, which is favourable for high-temperature, high-strength materials. The problem for practice is the high cost of the operation and the problem of shaping possibilities. In the latter aspect progress has been made by the so-called "powder vehicle hot-pressing technique" (50). It is outside the scope of this paper to treat this subject further.

Very interesting possibilities lie in the field of "activated sintering". Although the present author is not in favour of the term as normal sintering also requires "activation", the definition of the field as given by W.D. Jones (51) can be accepted: "It is connected with the sintering of powder compacts, where the rate of sintering is modified by some physical or chemical treatment of powder or compact, or by incorporating reactive gases in the sintering atmosphere with a view to enhance densification". Review papers have been included in the list of references (52 - 54).

The appropriate method of activation depends very much on the physical and chemical properties of the material. In this respect it is not in fact possible to give general principles other than that, in view of the very low matter transport velocities in crystalline solids, an increase of structural-disorder and modifying surface or grain boundary properties can give rise to enhanced densification rates.

A typical example is the enhanced reactivity of solids at the temperature of an allotropic transformation. Morgan (55) found enhanced densification in zinc and cadmium sulfide around the sphalerite-wurzite phase transition. Chaklader and Roberts (56) could obtain densification of quartz, normally not sintering, when an intermediate phase developed before the transition to cristobalite. A similar effect could be expected to occur in SiC around the β-α transition.

The field which will be most important to practice is the enhancement of densification by a small amount of an additive which does not dissolve into the lattice, nor form a liquid phase. A typical example is the activated sintering process of tungsten, where the addition of 1 wt % Ni leads to densities of well over 90% after sintering for 30 minutes at 1100°C. Without Ni no densification occurs.

Other examples can be found in the hot-pressing of carbides such as TaC (57), where the addition of 1 wt % of a transition metal leads to full densification which cannot be reached without the addition. It has also been reported (58) that addition of 1-2 wt % of Ni to a fine powder of AlN leads to high densities without the application of an external pressure. Their results could not be explained by liquid-phase sintering.

Although a good explanation is still lacking, the experimental observations up to now tend to support of theories which relate the kinetics to a rapid diffusion at the interface between the powder and a thin film of the additive (59), or through a high-diffusivity grain boundary layer which has resulted from segregation of the additive (60, 61). This could possibly be an alternative explanation for the results of Prochazka (10) on the sintering of boron-doped SiC.

5. CONCLUDING REMARKS

Although very little is known about the relevant properties, such as diffusion coefficients and interfacial energies, this paper has tried to give a general view of the possibilities for sintering nitrogen ceramics. It is found that pure nitrogen ceramics, because of their covalent bonding character, will be difficult to densify. The success in densifying silicon carbide (10) however, gives grounds for optimism. There has been progress in the field of sintering nitrogen ceramics, among other things by the growing awareness of the importance of the use of very fine starting powders(58, 62, 63). Sintering with a liquid phase is very effective for densification, but will cause deterioration of the intrinsically good high-temperature properties of the nitrogen ceramics.

Although reaction-sintered and hot-pressed materials are of great importance for practice at this moment, there is a need for dense-sintered nitrides, such as Si_3N_4 and AlN. They are, however, subject to the specific problem of loss of nitrogen at high temperatures (64). It has been proposed (47) to apply an increased nitrogen pressure during sintering. However, one then has to take into account the effect of the gases trapped within the pores. They will affect the high-temperature properties.

Sialons can be sintered to a high temperature without addition of a liquid phase. As their properties lie closer to Al_2O_3 (38) one can expect their type of bonding to have changed from pure covalent to a mixed covalent-ionic, resulting in better sintering properties.

6. ACKNOWLEDGEMENTS

The author wants to thank his colleagues who have contributed much to several of the matters treated in this paper. He especially wishes to thank Dr. D. Veldkamp, Mr. F. Kools, Dr. M. Sparnaay and Dr. A. Broese van Groenou for their comments and the fruitful discussions.

REFERENCES

1. R.L. Coble and J.E. Burke, Sintering in Ceramics, Progr.Ceram. Sci. 3 (1963) 197.
2. H.F. Fischmeister und H.E. Exner, Theorien des Sinterns, Metall 18 (1964) 932, 19 (1965) 113, 19 (1965) 944.
3. F. Thümmler and W. Thomma, The Sintering Process, Met.Rev. 115, J.Inst.Metals 12 (1967) 69.
4. D.L. Johnson and I.B. Cutler, Use of Phase Diagrams in the Sintering of Ceramics and Metals in: Phase Diagrams, ed. A.M. Alper, 2 (1970) 265. New York: Academic.
5. J.R. Moon, Sintering of Metal Powders, Powder Met. Int. 3 (1971) 147, 3 (1971) 194.
6. G.C. Kuczynski, Physics and Chemistry of Sintering, Adv.Coll.Interf.Sci. 3 (1972) 275.
7. A.L. Stuijts, Synthesis of Materials from Powders by Sintering, Ann.Rev.Mat.Sci., Vol. 3 (1973) 363.
8. Ja. E. Geguzin, Physik des Sinterns (1973) Leipzig: VEB Deutscher Verlag für Grundstoffindustrie.
9. S. Prochazka, Ceramics for High Performance Applications, ed. J.J. Burke, A.E. Gorum and R.N. Katz (1974) pag. 239, Chesnut Hill, Mass.:Brook Hill Publ. Co.
10. S. Prochazka, Mat.Sci.Res., Vol. 9 (1975) 421.
11. M.F. Ashby, Acta.Met. 22 (1974) 275.
12. J. Frenkel, J.Phys. (USSR) 9 (1945) 385.
13. G.C. Kuczynski, Trans.AIME 185 (1949) 169.
14. K.E. Easterling and A.R. Thölén, Mod.Developm.Powder Met. 4 (1971) 221.
15. R.L. Coble, J.Appl.Phys. 32 (1961) 787, 793.

16. W. Beere, Acta Met. 23 (1975) 139.
17. D.L. Johnson and I.B. Cutler, J.Amer.Ceram.Soc. 46 (1963) 541.
18. G.C. Kuczynski, Mat.Sci.Res. Vol. 10 (1975) 325.
19. D.L. Johnson, J.Appl.Phys. 40 (1969) 192.
20. D.L. Johnson, J.Amer.Ceram.Soc. 53 (1970) 574.
21. G.H. Gessinger, Scripta Met. 4 (1970) 673.
22. R.L. Eadie and G.C. Weatherly, Mat.Sci.Res.Vol. 10 (1975) 239.
23. R.L. Coble, J.Amer.Ceram.Soc. 41 (1958) 55.
24. D.L. Johnson, ref. 9, p. 173.
25. J. White, Proc.Brit.Ceram.Soc. No. 3 (1965) 155.
26. C. Herring, Physics of Powder Metallurgy, ed. W.E. Kingston,
    (1951) 143. New York: McGraw-Hill.
27. F.R.N. Nabarro, Rep.Conf. Strength on Solids, Phys.Soc.London
    (1948) 75.
28. C. Herring, J.Appl.Phys. 21 (1950) 437.
29. W.D. Kingery and M. Berg, J.Appl.Phys. 26 (1955) 1205.
30. J.A. Pask and C.E. Hoge, Mat.Sci.Res. Vol. 10 (1975) 220.
31. W.D. Kingery and B. Francois, Sintering and Related Phenomena,
    Proc. 2nd Conf. Notre Dame, eds. G.C. Kuczynski, N.A. Hooton
    and C.F. Gibbon (1967) 471. New York: Gordon and Breach.
32. P.J.L. Reijnen, Sci.Ceram. 4 (1968) 169.
33. J. White, ref. 20. p. 245.
34. J. White, Sci.Ceram. 2 (1965) 305.
35. M.P. Seah and C. Lea, Phil.Mag. 31 (1975) 627.
36. G.C. Kuczynski, Pulvermetallurgie in der Atomkerntechnik,
    4. Plansee Seminar, ed. F. Benesovsky (1962) 166. Reutte:
    Metallwerk Plansee A.G.
37. P.J.L. Reijnen, Proc. 6 Int.Symp.Reactiv.Solids, eds.
    J.W. Michell, R.C. De Vries, R.W. Roberts, P. Cannon,
    (1969) 99. New York: Wiley.
38. K.H. Jack, J.Mat.Sci. 11 (1976) 1135.
39. J.E. Burke, Chemical and Mechanical Behavior of Inorganic
    Materials, eds. A.W. Searcy, D.V. Ragone, U. Colombo (1970)
    413. New York: Wiley.
40. D. Shaw, Atomic Diffusion in Semiconductors (1973)
    London/New York: Plenum.
41. R.N. Ghoshtagore and R.L. Coble, Phys.Rev. 143 (1966) 623.
42. W.D. Kingery, Ceramic Fabrication Processes, ed. W.D. Kingery
    (1958) 131. New York: Wiley.
43. V.N. Eremenko, Yu.V. Naidich, and I.A. Lavrinenko,
    Liquid-Phase Sintering (1970). New York: Plenum.
44. T.J. Whalen and M. Humenik, ref. 20. p. 715.
45. W.J. Huppmann, Mat.Sci.Res. No. 10. (1975) 359.
46. K. Komeya, H. Inoue and A. Tsuge, J.Amer.Ceram.Soc. 57 (1974) 411.
47. M. Mitomo, J.Mat.Sci. 11 (1976) 1103.
48. J.P. Mary, P. Lortholary, P. Goursat, M. Billy and J. Mexmain,
    Bull.Soc.Fr.Céram. 105 (1974) 3.
49. S.R. Billington, J. Chown and A.E.S. White, Special Ceramics 1964
    (1965) 19.
50. F.F. Lange and G.R. Terwilliger, Amer.Ceram.Soc.Bull. 52 (1973) 563

350

51. W.D. Jones, Fundamental Principles of Powder Metallurgy (1960), London: Arnold (Publishers) Ltd.
52. M. Eudier, Symp.Powder Metallurgy, Spec.Rep. No. 58, Iron and Steel Institute London (1956) 59, 345.
53. A.J. Shaler, ref. 20, p.807.
54. A.S. Reshamwala and G.S. Tendolkar, Powder Met.Int. 1 (1969) 58, 2 (1970) 15.
55. W. Schintlmeister and K. Richter, Planseeber. Pulvermet. 18 (1970)3.
56. G.V. Samsonov and V.J. Yakovlev, Z.Metallk. 62 (1971) 621.
57. B. Lersmacher and S. Scholz, Archiv.Eisenhuttenwesen 32 (1961) 421.
58. M. Trontelj and D. Kolar, J.Mat.Sci. 8 (1973) 136.
59. G.E. Rhead, Scr.Met. 6 (1972) 47.
60. F.J.A. Den Broeder, private communication. Ref. 7., p.379.
61. G.H. Gessinger and Ch. Buxbaum,Mat.Sci.Res. Vol. 10 (1975) 295.
62. K. Komeya and H. Inoue, J.Mat.Sci. 4 (1969) 1045.
63. K.S. Mazdiyasni and C.M. Cooke, J.Amer.Ceram.Soc. 56 (1973) 628.
64. R.D. Pehlke and J.F. Elliott, Trans.AIME, 215 (1959) 781.

DISCUSSION (Cannon, Moulson, Lange)

Cannon: Although the Ashby sintering maps are conceptually useful for visualizing the operative processes, there are two particular dangers which can result from using those for metals as guides for ceramics. One is that the relative temperature dependencies of lattice, surface and grain boundary diffusion are generally quite different and so the relative positions of various regions on the maps are different. For instance the widely accepted notion of $\Delta H_{gb} \approx \frac{1}{2} \Delta H_l$ appears not to be applicable to oxides. More importantly the ambipolar effects which must result for sintering compounds have not been incorporated into any of the sintering equations. Thus Johnson's formalism for separating boundary and lattice diffusion contributions is entirely incorrect except for the simple case where transport of one species by either boundary or lattice is rate controlling over the entire range of densification.

The Sintering Behavior of Covalently-Bonded Materials*

C. Greskovich, S. Prochazka and J. H. Rosolowski

Physical Chemistry Laboratory, General Electric
Corporate Research and Development, Schenectady,
New York, 12301

ABSTRACT. Covalently-bonded solids do sinter under appropriate experimental conditions. Si and β-SiC (doped with B+C) appear to sinter in the absence of liquid phase whereas the sintering mechanism of doped Si_3N_4 used in this study has not been yet identified. High temperatures >1800°C are required to promote sufficient sintering of submicron powders of β-SiC doped with 0.6%B+0.8%C so that relative densities greater than 95% can be achieved. High nitrogen pressure promotes intergranular bonding and densification in Si_3N_4.

1. INTRODUCTION

Sintering is a fundamental consolidation process for many metal and ceramic oxide powders, but this method has not been fully explored for the nominally unsinterable covalently-bonded crystalline solids. This paper presents some of the major factors responsible for the unsinterability of Si, β-SiC and Si_3N_4. Under proper conditions these substances do sinter to near theoretical density. For Si_3N_4, this is achieved through the use of non-oxide additives.

---

* The research was supported by the Advanced Research Projects Agency of the Department of Defense and monitored by AFML under Contract No. F33615-76-C-5033.

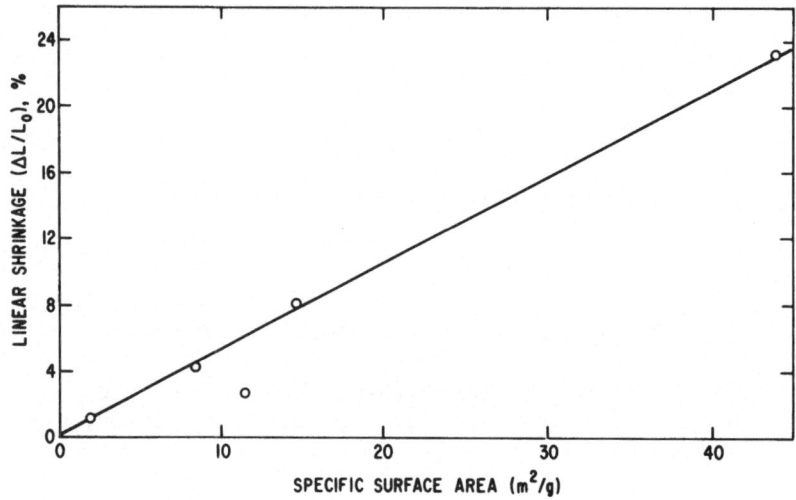

Fig. 1. Linear shrinkage of silicon powder compacts of about the same green density (∿50%) fired at 1350°C for 60 min. in Argon as a function of specific surface area for the starting powders[1].

2. SOME ASPECTS OF THE SINTERING BEHAVIOR OF Si and β-SiC

Contrary to popular belief, covalently-bonded solids do sinter by diffusional transport mechanisms, but because of low self-diffusion coefficients even at high temperatures, ultrafine particle sizes are required. The pronounced effect of particle size on the sintering behavior of silicon is shown in Fig. 1. By using a Si powder with a specific surface area of 44 m^2/g (∿0.06µ), relative densities of 92% and 99% were achieved at 1350 and 1380°C, respectively. It can be shown [1] that the observed increase in shrinkage of Si compacts with decreasing particle size implies that the amount of matter transported into the interparticle "neck-region" by volume and/or grain boundary (g.b.) diffusion increases relative to that transported by the vapor.

The effect of particle size and the addition of boron and carbon dopants on microstructural development in Si and β-SiC, respectively, are shown in Figs. 2 and 3. For nondensifying Si powders the characteristic microstructure of fired compacts (Fig. 2A) is composed of large, dense regions spaced by large pores and is caused by the domination of matter transport through vapor. By reducing particle size the microstructure of the resulting dense, fired compacts (Fig. 2B) is composed of finer grains and pores due to the greater transport of matter by volume

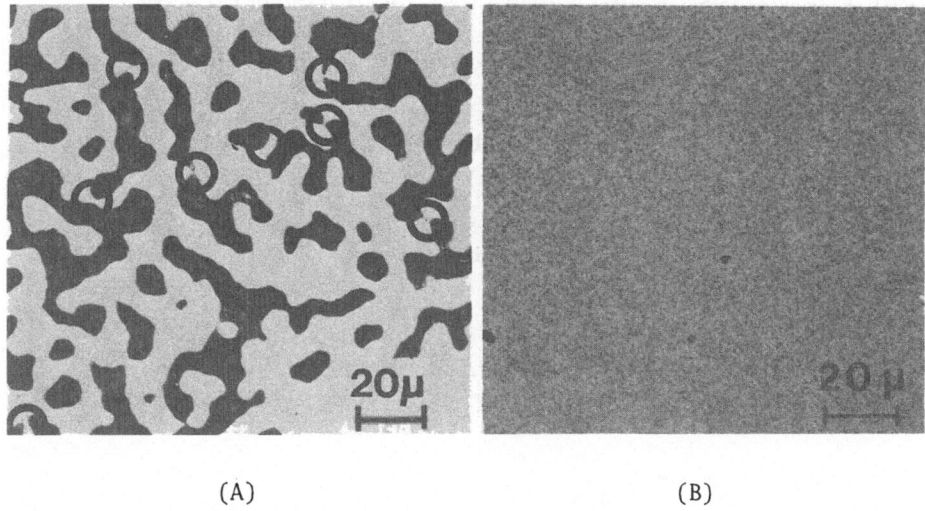

(A)   (B)

Fig. 2.   Reflected-light photomicrographs of polished sections of
sintered silicon compacts prepared from (A) 0.23μ powder and
(B) 0.06μ powder.   Compacts fired at 1350°C for 60 min. in Argon.
In (A) the solid phase is white and pores are black (relative
density ∿60%).   Small "necks" connecting the solid phase shown inside the black circles are undergoing disconnection.   In (B) the
compact sintered to a relative density of 92%.

(A)   (B)

Fig. 3.   SEM photomicrographs of fractured surfaces of fired compacts of β-SiC containing (A) 0.8 wt%C and (B) 0.8 wt%C and 0.6wt%B.
Compacts fired at 1900°C for 60 min. in Argon.

and/or g.b. diffusion which prevents the disconnection of grains[1].

In Fig. 3, the effect of the simultaneous addition of B+C on the development of microstructure in β-SiC during the early stages of solid-state sintering of compacts of 0.2μ powder particles is illustrated. β-SiC containing 0.8 wt% C does not densify even at temperatures up to 2000°C, and exhibits a microstructure characterized by randomly distributed dense regions separated by large pores. The addition of 0.6 wt% B to this same powder permits appreciable densification at 1800 to 1900°C and, again, a fine pore-grain microstructure develops. Boron is found to reduce the relative rate of matter transport at low temperatures by surface diffusion which is responsible for grain and pore growth, thereby permitting densification to occur at high temperatures by volume and/or g.b. diffusion[1].

3.   THE SINTERING OF Si_3N_4

3.1 Obstacles to Sintering

The application of some of major findings described above for sintering Si and β-SiC combined with the high susceptibility of Si_3N_4 to thermal decomposition leads to the speculation that primarily four obstacles hinder the solid state sintering of Si_3N_4 without the use of additions which introduce property degrading phases. First the thermal decomposition temperatures of Si_3N_4 is about 1875°C in 0.1 MPa of N_2 and thus large weight losses by vaporization and thermal decompositon are anticipated and observed. This characteristic suggests that submicron Si_3N_4 powders must be prepared to enhance the relative volume and/or g.b. diffusion at lower temperatures. Secondly, further control of thermal decomposition must be accomplished by the use of high nitrogen pressures. Next, the atomic mobility in pure Si_3N_4 is essentially negligible below 1550°C (this being demonstrated by attempts to densify Si_3N_4 by using diamond-forming conditions[2]), indicating that the use of perhaps solute additives is probably necessary to enhance sinterability by promoting grain boundary (neck) formation but yet insufficient to significantly degrade high temperature properties. Finally, the α→β-Si_3N_4 transformation, which may lead to high aspect ratio grain morphologies and lower final densities, should be controlled, if necessary.

3.2 Results and Discussion

The thermal decomposition and sintering behavior of "pure"

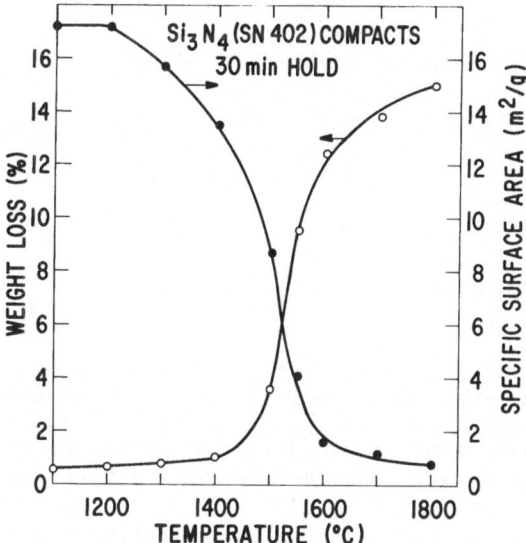

Fig. 4. Weight loss and specific surface area as a function of
temperature for Si$_3$N$_4$ compacts fired in 0.1 MPa of N$_2$.

Si$_3$N$_4$ are illustrated in Fig. 4 and Table 1. Most experiments
were done on two amorphous Si$_3$N$_4$ powders with metallic purity
>99.99 wt%, one being purchased from GTE Sylvania (SN-402) with
an oxygen content ∿ 3 wt% and particle size ∿0.15μ and the other
produced "In-House" with comparable oxygen content and particle
size. Most of the weight loss and specific surface area reduction
(Fig. 4) occurs between 1400 and 1600°C in 0.1 MPa of nitrogen.
Since essentially no macroscopic shrinkage is observed even up to
1800°C, the major portion of material loss and surface area reduction is caused by the combined effects of thermal decomposition
into Si and N$_2$ (and recombination) and vapor phase transport. The
fired compacts were always crumbly because matter is removed faster
from the neck regions between grains by vapor phase and/or thermal
decomposition processes than can be supplied by any of the available intergranular bonding mechanisms.

Under high nitrogen pressure (Table I) a compact of SN-402
Si$_3$N$_4$ fired at 1850°C in 4.7 MPa of N$_2$ (Exp. 1(exhibited little
macroscopic shrinkage (1-2%) and a weight loss ∿10%, which is much
lower than ∿25% observed for the same compact fired at the same
temperature in 0.1 MPa of N$_2$. For Exp. 2 and 3 (In-House Si$_3$N$_4$
powder was used) 10 to 20% linear shrinkage was observed even
though the weight loss was relatively high, ∿20%. Whether or
not the sintering observed is characteristic of "pure" Si$_3$N$_4$ or is
caused by impurity contamination from a pressurized SiC tube during

356

TABLE I. RESULTS ON SINTERING Si_3N_4 UNDER HIGH NITROGEN PRESSURE

Exp. No.	Temp.(°C)	Time(sec.)	P_{N_2}(MPa)	$\frac{\Delta L}{L_o}$(%)	$\frac{\Delta W}{W_o}$(%)
1	1850	2700	4.7	1-2	10
2	2000	1200	8	10	20
3	2100	900	8	20	21

firing is as yet undetermined. Nevertheless, the highest final density obtained was 1.9 g/cc for Exp. 3. A SEM of a fractured surface (Fig. 5) of the specimen prepared in Exp. 3 showed that high nitrogen pressure appeared to enhance bonding between Si_3N_4 grains primarily in the beta form.

The above results indicate that chemical additives are necessary to promote full densification of Si_3N_4. Consequently a number of non-oxide additives were selected for investigation to avoid grain boundary, oxide phases deleterious to high temperature strength and creep.

The sintering results for doped-Si_3N_4 (In-House powder) fired in nitrogen showed that nearly theoretically-dense Si_3N_4 could be produced with several non-oxide additives. Grain boundaries were difficult to chemically-etch, possibly indicating the absence of much grain boundary phase. Fractured surface exhibited primarily intergranular fracture. Mechanical property measurements of these dense Si_3N_4 ceramics are currently in progress.

Fig. 5. SEM photomicrograph of a fractured surface of Si_3N_4 prepared in Table I, Exp. No. 3.

REFERENCES

1. C. Greskovich and J. H. Rosolowski, "Sintering of Covalent
    Solids," J. Amer. Ceram. Soc., 59 (7-8) (1976).
2. C. Greskovich, S. Prochazka and J. H. Rosolowski, "Basic
    Research on Sintered Ceramics," General Electric Report
    SRD-76-036, March 1976.

DISCUSSION (Cannon, Moulson, Lange)

Riley: Why is it that you don't seem to have the same problems
with loss by evaporation in silicon sintering that you get with
silicon nitride sintering?

Greskovich: You do have similar problems in both materials but to
different degrees. In the sintering of Si, the relative ratio of
matter transport (into the interparticle necks) by volume and/or
grain boundary diffusion as compared to vapour phase transport
apparently increases markedly with decreasing particle size so
that densification proceeds. In the case of sintering silicon
nitride in nitrogen, thermal decomposition or vaporisation largely
dominates any enhancement in volume and/or grain boundary diffusion
provided by reduction in particle size. Hence, densification
does not occur.

Morgan: It does appear that both the weight loss and the
reduction in surface area of the silicon nitride powders is
coincident with the amorphous $\rightarrow \alpha$ transformation. Is this so?

Greskovich: Most of the weight loss and surface area reduction in
silicon nitride compacts takes place when the material is
crystalline (α-Si_3N_4). However surface area reduction can and
does occur at temperatures below $1450°C$ and is probably caused
by surface diffusion on amorphous silicon nitride particles.

Lange: We find β-rich second phases in sintered SiC and suggest
liquid phase sintering, i.e. SiC decomposed at high temperature
into Si + C. Liquids can form by reaction of Si + β.

Greskovich: Our work at GEC shows no detectable evidence of any
liquid phase in the microstructure of sintered dense SiC containing
excess carbon and boron. If your claim were true, then SiC could
be liquid phase sintered without the presence of the excess carbon.
Since we find it essential to have excess carbon to promote
densification during sintering, we believe that the densification
mechanism is via solid state rather than liquid phase sintering.

Schwetz: There is a lot of oxygen in the amorphous Sylvania submicron silicon nitride powder (approx.3%). Did you remove the
oxygen before sintering?

Greskovich: We did in some cases. However, we found no effect
during the sintering of undoped silicon nitride powder.

PRESSURELESS SINTERED SILICON NITRIDE

I. Oda, M. Kaneno and N. Yamamoto

Research and Development Laboratory,
NGK INSULATORS, LTD., Mizuho, Nagoya, Japan

1. INTRODUCTION

The influence of various kinds of additives on the densification
during pressureless sintering (PS) of silicon nitride was investigated. The combination of three kinds of additives, $MgO-BeO-CeO_2$,
has been found to densify silicon nitride to 97% of theoretical
and its mechanical properties is superior to the sintered body
with MgO only.
This paper summarizes the work concerned with the relationship
between the additives and properties of silicon nitride, when
MgO, BeO and CeO_2 are used.

2. EXPERIMENTAL

The Si_3N_4 powder used for the pressureless sintering studies was
AME's controlled phase 85. X-ray diffraction analysis revealed
that the powder was composed of about 90% α-phase and remains were
β-phase, unreacted Si and other impurities. According to the direct
examination in the electron microscope, the powders are characterized by the needle like or elongated particles. Chemical analysis
of the powders are reported in Table 1; this Table shows the
principal cation impurities in the powders of starting, after
grinding and after acid treatment, respectively. This chemical
analysis indicates that the powder after acid treatment contains
about 0.1 - 0.2% of Fe, Al and Ca, respectively, thus totally
about 0.5% of impurity level
The standard procedure was used to prepare the powders for pressureless sintering. The starting powders were ground for 100 hrs in
iron ball mill to be less than 10 µ and then treated by HCl to

Table 1. Chemical analysis of AME powder

(wt%)

	Fe	Al	Ca	Mg	Na	K
Starting powder	0.442	0.301	0.157	0.018	0.004	0.006
After grinding in iron ball mill	5.291	0.457	0.255	0.007	0.007	0.002
After iron removal by acid treating	0.112	0.234	0.072	tr.	tr.	tr.

remove iron impurity. The average particle size of the powders milled for 100 hrs was about 0.5 μ. This treated Si_3N_4 powders were added with the desired amounts of additives in a plastic bottle. The mixed powders were hydrostatically pressed into disks at 40,000 psi. Sintering was done at 1,600 - 1,800°C in nitrogen atmosphere without pressure.

3. RESULTS AND DISCUSSION

3.1 The effect of additives

Table 2 shows the open porosity and relative amount of β-Si_3N_4 of the sintered body, when Si_3N_4 compact with 5 wt% of each additive was sintered at 1,800°C for 0.5 hrs. The relative amount of β-Si_3N_4 was determined by comparing the peak height of X-ray diffraction of (210),(101) of β-Si_3N_4 with (210),(102) of α-Si_3N_4. X-ray diffraction patterns of sintered body are shown in Fig. 1, when 5 wt% BeO and 5 wt% MgO were added, respectively. According to X-ray analysis, α-phase silicon nitride almost transformed to β-phase during sintering in case of BeO, while about half of phase remained in case of MgO. This fact means that BeO has an important role of accelerating α to β phase transformation of silicon nitride. It is also suggested, however, that MgO has the

Table 2. Influence of Pressureless Sintering Silicon nitride

Additive	Open porosity (%)	Relative amount of β-phase
MgO	11.7	50
BeO	30.7	100
CeO_2	33.2	30

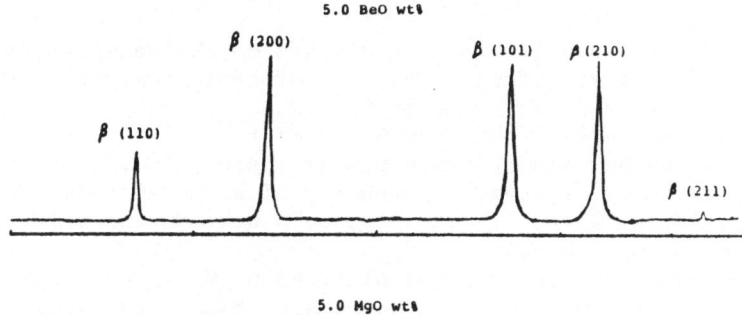

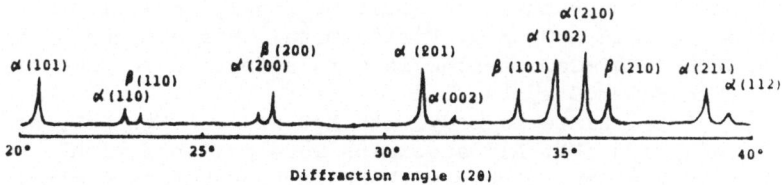

Fig. 1. X-ray diffraction patterns of PS-Si$_3$N$_4$

most effective to densify silicon nitride by forming a liquid
phase with SiO$_2$ present in Si$_3$N$_4$ powder as shown in Table 2.
The effect of additives of MgO and BeO was studied in more detail
in order to obtain a highly densified pressureless sintered silicon nitride.

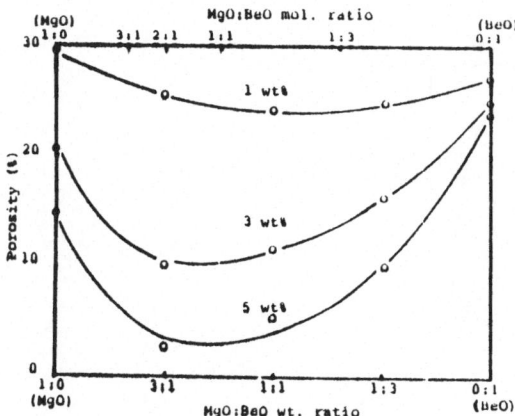

Fig. 2. Influence of amounts of BeO and MgO on the densification
of PS-Si$_3$N$_4$

362

3.2 The effect of additives in the system of BeO-MgO-CeO$_2$

To determine the combined effect of additives on the densification
behavior, experiments at different amounts of additives were done
in the system of MgO-BeO, as shown in Fig. 2.
The most densified sintered body was 96.9% of theoretical, when
added with 1.25 wt% BeO and 3.75 wt% MgO at a same time.
In case of BeO-MgO-CeO$_2$ system, the density of sintered body was
achieved to be 97.1%, when 1.25 wt% BeO, 3.75 wt% MgO and 5.0 wt%
CeO$_2$ were added simultaneously.
The firing shrinkage of pressureless sintered Si$_3$N$_4$ in non-iso-
thermal condition is shown in Fig. 3. The shrinkage curve indicates that Si$_3$N$_4$ compact with BeO-MgO or BeO-MgO-CeO$_2$ system was
densified at lower temperature and could be highly densified
finally, comparing with MgO only. Furthermore, this shrinkage curve
also shows that BeO-MgO-CeO$_2$ system is more effective in producing
dense bodies than BeO-MgO system.
Microstructure of the sintered body with MgO and BeO-MgO-CeO$_2$ are
shown in Fig. 4(A) and (B). All specimens were polished with
alumina powder (0.2 μ) and etched with HF etchant before electron
microscopic examination.
When only MgO is used as an additive, elongated grains are observed with wide distributions of grain sizes and shapes, and glassy
phase is also observed at the grain boundaries as shown in Fig.
4(A). On the contrary, when BeO-MgO-CeO$_2$ are used as additive,
the grains had a more uniform size distribution and the amount of
the glassy phase was small. This fact suggests that BeO accelerates
the sintering of silicon nitride by inhibiting the grain growth
and solving partially into the silicon nitride grains, when being
coexisted as a sintering aid with MgO.
The effect of the phase transformation by BeO on the sintering
mechanism of silicon nitride was not examined in this experiment.

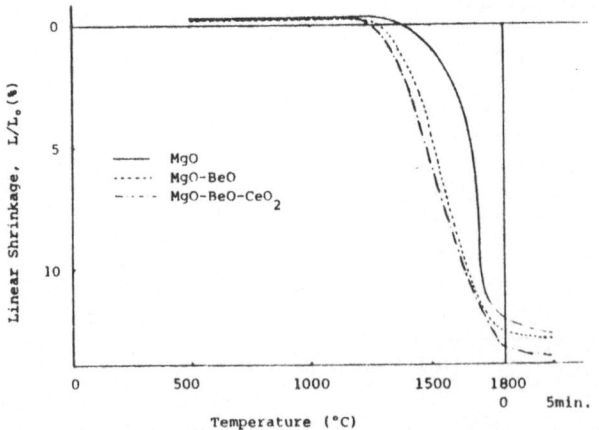

Fig. 3. Firing shrinkage of PS-Si$_3$N$_4$

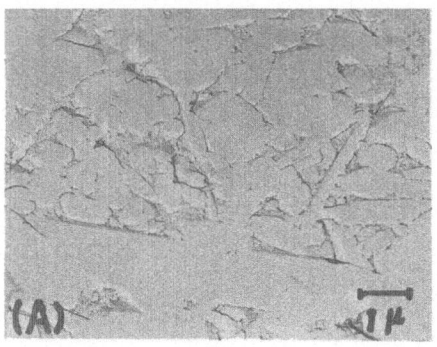 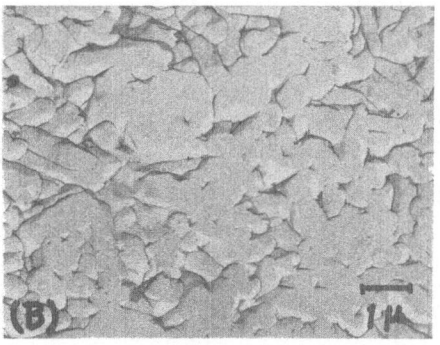

Fig. 4. Electron micrograph of PS-Si$_3$N$_4$ added with MgO(A) and BeO-MgO-CeO$_2$(B)

In case of BeO-MgO-CeO$_2$ system, the densification rate was accele-rated comparing with BeO-MgO system, however it seems to be caused by increasing the total amounts of additives. Nevertheless, CeO$_2$ addition has an effect on improving the mechanical strength report-ed below. The unit cell dimensions of sintered body with additions of BeO-MgO and BeO-MgO-CeO$_2$ were not different from that of β-phase in the starting powder, as shown in Table 3.

The bending strength of pressureless sintered silicon nitride are shown in Fig. 5. Test specimens in strength experiment were dia-mond-cut 3 x 3 mm in cross section and they were broken in three point loading on a 25.4 mm span with a cross head speed of 0.05 cm/min.

The strength-temperature curves for both specimens have the same shape, with strength decreasing above 1,000°C. In case of BeO-MgO-CeO$_2$ addition, strength of 118,000 psi have been obtained at room temperature, however it decreases about half above 1,000°C.

This strength decrease at elevated temperature may be derived from the lack of refractoriness of grain boundary phase.

Accordingly, further strength improvement at elevated temperature must be studied hereafter.

Table 3. Unit cell dimensions of β-Si$_3$N$_4$

	a_o (Å)	c_o (Å)
AME starting powder	7.609	2.910
BeO addition	7.607	2.910
BeO+MgO addition	7.607	2.911
BeO+MgO+CeO$_2$ addition	7.609	2.910

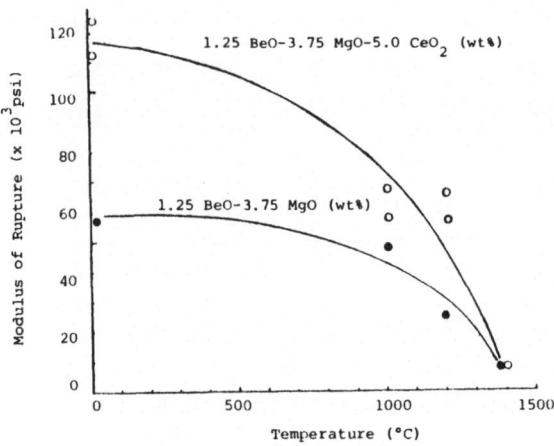

Fig. 5. Strength of PS-Si$_3$N$_4$

The properties of pressureless sintered silicon nitride are summarized in Table 4.

4. CONCLUSION

The α-Si$_3$N$_4$ powder with the addition of BeO-MgO-CeO$_2$ can be densified to 97% of theoretical by pressureless sintering. Densification may be occured through liquid phase sintering. Addition of CeO$_2$ has an marked effect on the improvement of mechanical strength. The problem is that the decrease of high temperature strength is rather large and so strength improvement at elevated temperature is now being studied.

Table 4. Properties of PS-Si$_3$N$_4$

Bulk Density (g/cm^3)	3.10
Porosity(total) (%)	2.9
Modulus of Rupture (psi)	
R. T.	118,000
1,000°C	64,000
1,200°C	63,000
1,400°C	7,000
Modulus of Elasticity (psi)	40 x 10^6
Linear Thermal Expansion Coefficient (/°C)	3.72 x 10^{-6}
Thermal Conductivity (cal/cm.sec.°C)	0.06
Thermal Shock Factor (σ/α.E)	790

DISCUSSION

Billy: It is surprising that you could get such high densification by using temperatures as high as 1800°C, where Si_3N_4 decomposes. How did you prevent decomposition of silicon nitride?

Oda: No special techniques were used. The experimental procedure was to place the silicon nitride sample with MgO, BeO and CeO_2 additions between two billets of reaction bonded silicon nitride in a graphite crucible, and RF heat in flowing nitrogen (1 l/min).

Lange: The pressureless sintering of Si_3N_4 can be viewed as a competition between pore growth by decomposition and pore shrinkage by liquid phase sintering. This indicates a need to optimize the sintering temperature and time in order to minimise decomposition. Did you observe any weight loss?

Oda: We observed between 4 and 7% weight losses during sintering at 1700-1800°C for one hour under nitrogen.

Godfrey: As some of the environments for which these ceramic materials are intended (e.g. gas turbines) contain large quantities of H_2O, what thought is being given to the deleterious effects and toxic hazards of $Be(OH)_2$ volatilization at about 1000°C and above?

Oda: We know that BeO and its compounds are generally toxic as you indicate, but we have not verified whether our Si_3N_4 body containing BeO shows toxic hazards under some conditions. We think that our BeO does not bring a serious problem, because our Si_3N_4 body contains only a small amount of BeO, and moreover the BeO seems to form a stabilized phase in the sintered body.

Jack: Promotion of the $\alpha-\beta$ transformation by BeO is understandable because BeO gives Be_2SiO_4 with the silica on the Si_3N_4 and Be_2SiO_4 (phenacite) has the silicon nitride β structure. However, Mr. Oda kindly provided us with a sample of his material in which we have found unchanged BeO. Has Mr. Oda found the same?

Oda: We could not find unchanged BeO by using a Rotating Anode X-ray Diffractometer. However, we could confirm the existence of phenacite when a mixture of BeO and SiO_2 was fired under the same sintering conditions.

DENSIFICATION OF NITRIDES BY HOT PRESSING

T. Vasilos

Avco Systems Division, Materials Sciences Dept.,
Lowell, Massachusetts

ABSTRACT. The hot pressing of refractory compositions has been
given particular attention in recent years with respect to
achieving dense, fine-grained microstructures. The current
discussion on nitrides is treated in terms of mechanisms and
technology, and comparisons are made with other compounds such as
oxides and carbides. Attention is given to the role of particulate
characteristics, impurities and additives on microstructure development and process mechanism. The significance of such process
parameters as applied temperature, pressure, atmosphere and mold
geometry is treated in relation to composition. Further discussion
is given to the technology of forming special shapes and size
scale-up consistent with state-of-the-art practice.

1. INTRODUCTION

Hot pressing as a fabrication process has found increased use in
the ceramics processing industry in recent years with particular
emphasis on the preparation of materials with improved properties
through composition, microstructure and density control.

Despite the recent success achieved in sintering such compounds
as silicon carbide, silicon nitride, and certain Sialon compositions, hot pressing of these materials is currently conducted to
achieve maximized strength and wear or corrosion resistance
properties.

Table I provides a general listing of ceramic materials fabricated
by hot pressing and their associated applications.

TABLE I. HOT PRESSED CERAMICS AND APPLICATIONS

Material	Characteristics	Application
Aluminum Oxide	Dense, 1-3 μm grain size, high strength	Cutting tools, Gas Bearings
Magnesium Oxide	Dense, transparent	IR Windows
Boron Carbide	Dense, small grain	Ballistic Armor, Sand Blast Nozzles Wear Resistant Parts
Silicon Carbide	Dense, small grain	Special molds and wear parts
Boron Nitride (non-cubic)	Some residual porosity, low strength	Container Vessels Mold Linings Sputtering Targets
Silicon Nitride	Dense, fine grained < 1-5 μm	Advanced Turbine Engine Components, Bearings, Welding Insulators; Potential for radome, IR dome and other thermal protection systems
Various Ferrites	Dense microstructure	High permeability, low loss ferromagnetic materials
Various Ferroelectrics	Dense, translucent structure	Electro-Optical Applications

Also, there are various refractory metal borides and carbides that have also been prepared for special wear resistance applications. Furthermore, the hot pressing process has been employed in the fabrication of special metal-ceramic composites.

The coincident application of temperature and pressure to "dissimilar" components in a composite system provides the basis for an interlocking structure with the prospect for interdiffusional bonding in some instances. This is treated in more detail later in the discussion.

2. PROCESS DISCUSSION

2.1 General

In hot pressing, the pressure can be applied uniaxially, biaxially or isostatically. However, with the exception of hot isostatic pressing, often applied to densify special metal forms, and very high pressure forming as in polycrystalline diamond, uniaxial hot pressing represents the normal technique most often used.

The process consists of applying pressure to the ceramic in powder or cold pressed compact form, in a refractory mold, usually of a form of graphite, which is heated to a specific temperature for densification. Details appear in the literature.[1] Pressures normally employed range between 1000 and 10,000 psi, although in some work pressures well beyond this level have been used with non-graphite dies. Temperatures can range up to 2500°C for densification in graphite molds; however, by proper selection of starting material characteristics, the process, with its relatively short time cycle and lower equivalent densification temperature, can be used to fabricate dense fine grain microstructures with or without the use of additives to control grain growth. This is particularly true of the essentially ionic bonded compounds, e.g., MgO, Al_2O_3, ZrO_2, and BeO, where ion diffusion rates are adequate for relatively low temperature densification. However, for compounds such as silicon nitride, boron nitride, and silicon carbide, additives are required to promote densification independent of grain growth control. Ion mobilities are sufficiently low that temperatures close to the decomposition temperature are required for these compounds if very high pressures are not employed, e.g., well beyond 10,000 psi and the limit of most graphite molds. Details on process mechanisms are provided below.

2.2 Process mechanisms

As in the case of sintering, particulate characteristics play an important role in determining microstructure development, and the parameters of pressure, temperature, and time are selected and modified in accordance with the densification behavior of the particulates under study. There is varying opinion concerning the mechanisms by which densification proceeds during the hot pressing process. As one might expect, the pertinent mechanism is dependent on the characteristics of the compound as well as the level of applied stress. Some investigators have supported plastic or viscous flow models,[1] e.g., $\ln(P/P_0) = K_c t$ (1), where P = porosity at time t, P_0 = initial relative porosity, and K_c = hot pressing rate constant. This was particularly applicable in the densification of viscous glass, which follow first order kinetics. At very low porosities, $\ln(P/P_o)$ tends to infinity and the relation therefore does not adequately describe final densification behavior. Others have cited particle rearrangement by grain boundary sliding and fragmentation as important densification mechanisms, particularly during early compaction stages and at very high pressures with hard materials such as carbides and borides. Recent investigations of hot pressing have recognized the probable occurrence of more than one mechanism, with considerable emphasis given to the significance of diffusion in characterizing the densification behavior of pure ionic solids such as alumina. Further, attempts have been made explicitly to derive diffusion models. Vasilos and Spriggs[2], for example, concluded that the process of densification during conventional graphite-die hot pressing of most ionic ceramics (i.e., for nominal pressures of 1,000 to about 10,000 psi and at temperatures up to 2500°C) was essentially diffusion-controlled. Rossi and Fulrath[3] have also demonstrated that the densification of alumina by hot pressing proceeds by a diffusional process. A densification model employed by these investigators to account for the densification mechanism is a modification of the Nabarro-Herring creep expression. The strain rate, $\dot{\varepsilon}$, in the relation is approximated by use of isothermal rate of change of porosity or linear shrinkage rate, e.g.,

$$\dot{\varepsilon} \approx \frac{13.3 \, D \, \sigma_e \, \Omega_o}{kT \, (GS)^2} \approx - \frac{dL}{Ldt} \qquad (2)$$

where: σ_e = the effective stress

$\dot{\varepsilon}$ = strain rate (sec^{-1})

$\dfrac{dL}{Ldt}$ = linear shrinkage rate

The observed consistency in the calculated diffusion coefficient values for hot pressing, as well as the reasonable comparison between activation energy values, has suggested that for the low to moderate pressures employed, densification beyond the initial stages is a diffusion-controlled process.

Coble[4] recognized that hot pressing by a diffusional process should also incorporate a term due to the vacancy gradient present from surface energy and pore curvature effects, as in normal sintering. For the final stage of pore shrinkage the deviation in vacancy concentration is

$$\Delta C = \frac{2 \, Co \, \gamma \, \Omega}{kTr_p} \qquad (3)$$

where r_p is the pore radius. Combined into a flux equation for the effect of pressure, one obtains

$$J = \frac{4\pi \, DCo\Omega}{kT} \left(\frac{2\gamma}{r_p} + \sigma_e \right) \qquad (4)$$

and an effective creep rate of

$$\dot{\epsilon} = \frac{13.3 \, D\Omega}{kT(GS)^2} \left(\frac{2\gamma}{r_p} + \sigma_e \right) \qquad (5)$$

This equation has not been tested explicitly; however, if the data of Vasilos and Spriggs[2] is corrected for the driving force of porosity, calculated diffusion coefficients for Al_2O_3 hot pressing agree with direct aluminum ion diffusion measurements within a factor of two.

The work of Coble and Ellis[5] on neck growth between single crystal spheres of alumina under stress is consistent with the conclusion that diffusion is a major transport mechanism in hot pressing. They found that the contribution of plastic flow to alumina densification at the pressures normally used in hot pressing (e.g., less than 10,000 psi) was small. They concluded that the final stage of densification of alumina during normal hot pressing occurs by enhanced diffusion under the influence of stress. The contribution of plastic flow at higher pressures (over 10,000 psi) and with "softer" materials (such as MgO and Ni) was not specifically evaluated by Coble and Ellis, but it was recognized that for materials with higher symmetry and multiple primary slip planes, the contribution of plastic flow should be greater. It was suggested that the creep behavior of given materials should serve to elucidate probable mechanisms; those which undergo creep by plastic flow should also undergo final densification by a diffusion mechanism. Although creep data of this type at the stresses and grain sizes of the studies referenced are not yet available, it is of interest to note that creep measurements of fine grain $(1-3 \, \mu)$ dense (98-99.5 percent), pure magnesia and alumina suggest a diffusional mechanism at lower stress levels (up to 4,000 psi) for both materials.

Based on the available information, it is believed that the principal mechanism by which densification of ceramics proceeds during hot pressing at pressures up to about 10,000 psi is essentially pressure-directed diffusion with a number of other operative and perhaps overlapping mechanisms, as outlined below:[6]

1) The first step is similar to one of cold compaction where the degree of densification is proportional to log applied

pressure and covers a densification range up to about 70 percent of theoretical, depending on particle characteristics. Fragmentation as a coincident step has not been found to be significant in the case of sub-micron size particles.

2) The second step is probably one of sliding and rearrangement wherein the degree of densification is further increased and is proportional to the ratio of applied pressure to the powder particle size.

3) A third step involving plastic flow at grain contacts occurs rapidly (within minutes) and may occur in combination or at the same time as step 2, above. However, the possible contribution of plastic flow to densification is apparently limited to less than 84 percent of theoretical, and is dependent on the ratio of applied stress in the die to the stress at particle contact points, i.e.,

$$P_s = \frac{Pso}{(1 - 2\sigma_A/N\sigma_p)^3} \qquad (6)$$

where:

σ_A = stress applied to die

σ_p = contact stress between particles

N = coordination number of particles (approximately 12 for ideal packing)

Pso = initial packed density

Ps = final density

4) The final stage of densification (up to theoretical) occurs by diffusion under the influence of stress, the densification rate being greater than the pressureless sintering case by a factor essentially proportional to $L/R\gamma$, where L is applied load, R is grain radius, and γ is surface energy; see equation (5).

Ultimate densification is limited by pore entrapment. Overall densification mechanisms at the higher pressures are not clearly established, although it is generally conceded that the plastic deformation characteristics of the higher symmetry materials are important.

The densification of covalent bonded compounds such as silicon nitride with additives appears to follow other kinetics. Wild et al[7] proposed a liquid phase densification mechanism for silicon nitride and Terwilliger and Lange[8] tested Kingery's liquid phase sintering model[9],

$$\Delta L/L_o = - K\ell \, t^{1/n} \qquad (7)$$

where $\Delta L/L_o$ equals fractional length change against densification data. The values of both K_ℓ and n depend on the rate-limiting step (i.e., dissolution into the liquid phase or diffusion through it) and the powder morphology. Kingery's equation has been applied to the analysis of densification data[10,11] where K_ℓ is stress-dependent. Its use and the method of analysis have been discussed previously[12].

Terwilliger and Lange[8] discovered a reasonable fit of their densification data, below 95% of theoretical, (plotted as log-log fractional length change vs. time) to n values of 3 and 5 in eq. (7), which indicates densification of prismatic particles where dissolution or diffusion is the rate-controlling step, respectively.

For higher densities, densification was slower than predicted by eq. (7). Also, the liquid phase sintering equation fit data for densification of silicon nitride with only 1% MgO present. Figure 1 provides a plot of their data for the effect of MgO content on densification of AME silicon nitride powder (90% phase) at 1650°C. Undoubtedly, the densification of silicon nitride without liquid phase forming additive proceeds by another mechanism, e.g., stress-directed diffusion, which hardly exists under these conditions, i.e. final density was 68% of theoretical. Densification experiments conducted by investigators[8] at different stress levels revealed that $K_\ell \propto \sigma^{1/5}$ and according to the model for liquid phase hot pressing $K_\ell \propto \sigma^{1/n}$.

Other requirements for liquid phase sintering fulfilled by MgO and evidently by other additives include wetting of the matrix phase and solubility of the matrix phase in the liquid phase. Also, in this connection, it is likely that powdered boron nitride containing B_2O_3 additive densifies by a liquid phase mechanism.

2.3 Process technology

A general description of ranges in hot pressing fabrication is provided below. While there are no sharp distinctions in the pressure ranges employed, it has been possible to identify arbitrary separations for various mold materials employed as seen in Table II.

The available configurations within these listed ranges provides decreasing specimen size capacity as the operating pressure increases. The size limits shown in each range are generally intended as a guide, and actual diametral limits are dependent on specimen thickness, densification temperature, and desired microstructure.

The relatively low strength levels for conventional graphite die

374

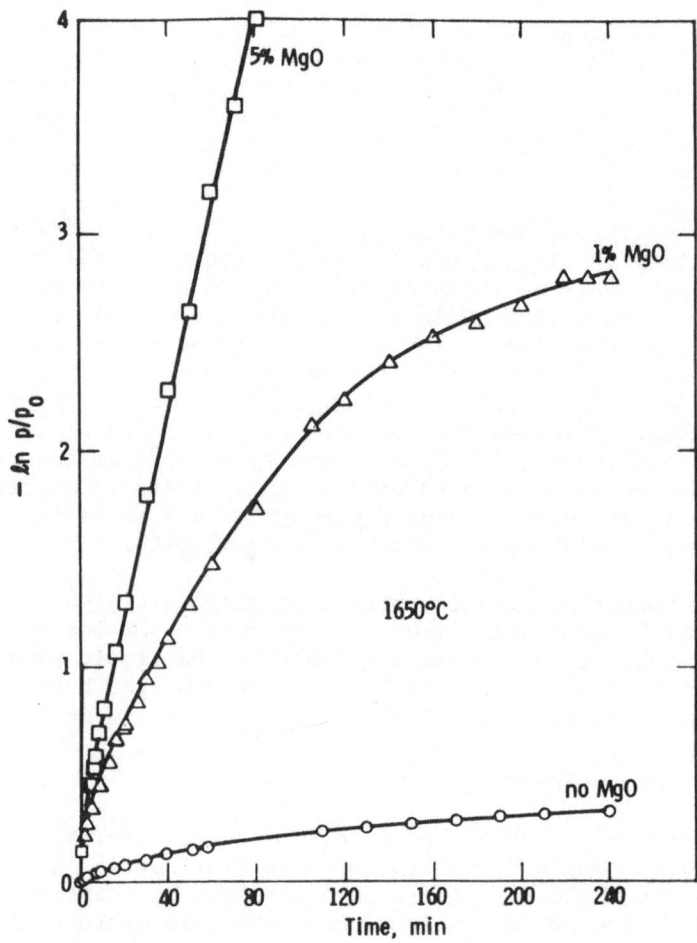

Figure 1. Effect of MgO Content on Densification of AME Powder at 1650°C [8].

TABLE II.   RANGES IN HOT PRESSING FABRICATION

Pressure Range and Application	Mold Materials	Size Limits for Fabricated Parts
1,000 to 5,000 psi uniaxial	Graphites	Up to 2 ft. in diameter (smaller sizes as pressures increase above 2,000 psi)
5,000 to 20,000 psi uniaxial	Special Graphites	
10,000 to 50,000 psi uniaxial	Ceramics (e.g., Al_2O_3, SiC, TiB_2, ZrB_2	Less than 3" in diameter
10,000 to 50,000 psi isostatic	Gas pressure contained in steel	Increased lengths (up to 20") with diameters < 3"
50,000 to 300,000 psi uniaxial	High-strength steel and cemented tungsten carbide at the higher temperatures	Generally 1" in diameter tending to 3" as limit
300,000 to 750,000 psi	Cemented tungsten carbide	$\frac{1}{4}$" to $\frac{1}{2}$" in diameter

parts have been known to limit the minimum attainable grain size at full density for particular ceramic raw materials; however, the recent availability of high-strength graphite and filament-wound graphite has made it possible to attain a level of 10,000-15,000 psi. However, high pressure levels of this sort have not been demonstrated to be useful or necessary to the densification of nitride ceramics in general.

In recent years, modifications to the standard hot pressing process have included the press forging of ceramics, e.g., alumina[13,14] as well as, more recently, silicon nitride[15]. The press forging of alumina provided bodies with high in-line optical transmissivity as a result of nearly complete pore removal and high crystallographic texturing. The textured structure was achieved by plastic deformation primarily on the basal slip system. Further mechanical strength was also enhanced as a result of forging. It was found that the bend strength at -196°C and 1200°C was nearly independent of grain size in the 1-20 μm range[13]. The alumina work included the press forging of cold pressed and/or pre-sintered alumina compacts into thin shell 2-inch and 3-inch diameter hemispheres which retained the crystallographic orientation in the plane of the shell. The optical transparency was

nearly uniform throughout the full hemisphere on the best
samples; however, regions of low transmission or excessive
haze remained a problem in many forgings. During forging,
edge tears often developed in the skirt of the forging and
some of these extended through the 0.4 inch skirt into the
hemispherical portion of the forging. The tearing was thought
to be caused by cavitation resulting from a more rapid
deformation (i.e., strain rates between 2×10^{-4} sec and
1×10^{-3} sec-1) than the available mechanisms could accommodate
at the imposed strain rate.

Inasmuch as hot pressing experiments indicated that a slight
preferred orientation occurs in silicon nitride during a
conventional fabrication cycle, Rhodes et al[15] considered it
worthwhile to investigate developing a stronger texture by
press forging this material. Using similar techniques as
for alumina a crystallographic texture was found which is
opposite that in alumina. The "c" axis for silicon nitride
tended to align itself perpendicular to the pressing direction
and a pronounced microstructure texture was also apparent.
Figure 2 provides a photomicrograph of a cross-section of a
disc shaped forged sample of silicon nitride, featuring a
significant microstructural texture. Inverse pole figure
measurements for the forged sample also reveal development of
a strong { 100} texture. Figure 3 shows a 2-inch base
diameter Si_3N_4 hemisphere forged from a cold pressed cube
shaped preform. The hemispherical surface illustrated is an
"as-pressed" surface and rim cracks were restricted to a
disposable skirt. Limited property data reveal that the stress
rupture life at $2415°F$ was improved by a factor of 6 to 10
times over conventionally hot pressed material.[16] Impact
resistance measurements were inconsistent although a number of
maximum impact resistance values were observed for forged
samples.[15,16] In these instances the sample surface loaded
in tension during testing was the surface perpendicular to the
pressing direction during fabrication, presenting, therefore,
elongated grain shapes on tensile surfaces. One might expect
lower strengths for samples presenting the alternate surface
on the basis of anisotropy. This has not been evaluated yet.

Other modifications to the hot pressing process have included
pseudo-isostatic hot pressing of silicon nitride pre-form
shapes by utilizing a powder bed of BN or C to transmit the
pressure to the shape at temperature. The powder bed is
thought to behave to a first approximation as an isostatic
fluid. Complex air foil and rotor shapes have been produced by
this process for evaluation early in the joint Ford-Westinghouse
program.[17] The approach was later discontinued because of
non-uniform pressure distribution through the powder, in favor
of a "duo-density" method of fabricating a turbine engine rotor

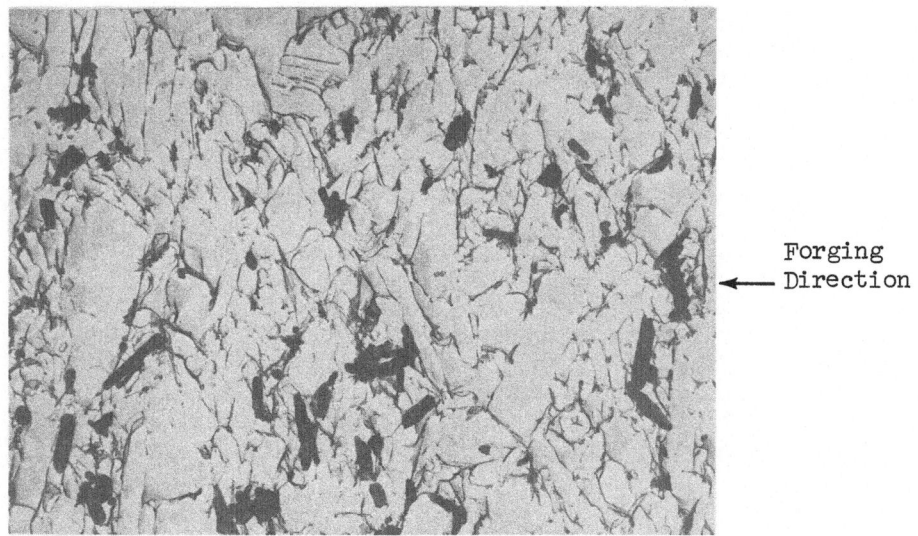

Forging
← Direction

7500X
Figure 2. Microstructural Texture in Press Forged Si₃N₄.

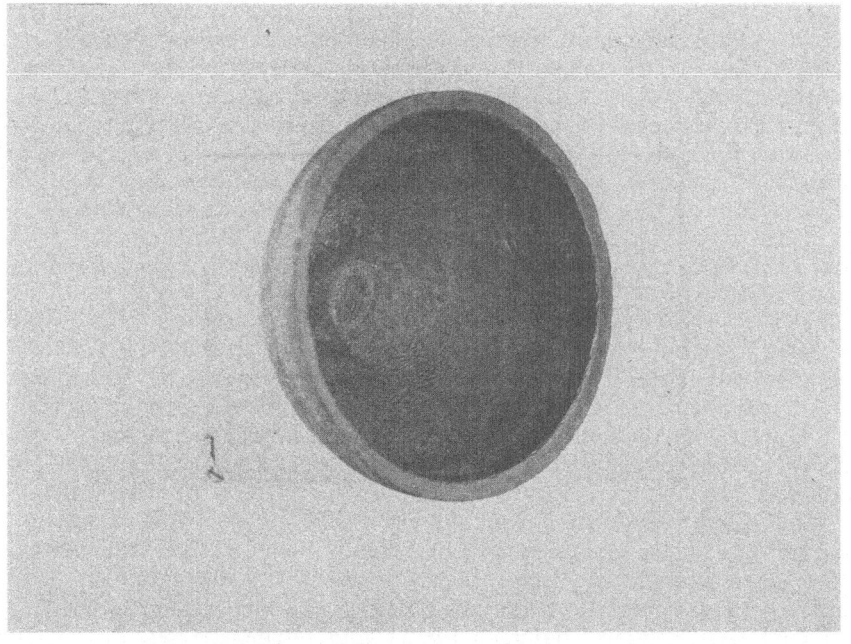

1X
Figure 3. Si₃N₄ Hemisphere Forged from a Cold Pressed Preform.

shape. This concept utilizes the high strength of hot pressed
Si₃N₄ in the hub region where stresses are highest, but
temperatures are moderate and, therefore, creep resulting from
the use of an MgO densification aid is minimized. Reaction-
sintered Si₃N₄, which can be formed into complex airfoil shapes
by injection molding or slip casting, is utilized for the
blade rings. Although the reaction-sintered material is of
lower strength, it appears adequate for the turbine blades
because stress levels in this region are lower than in the hub.
This concept relies on the ability to bond the two Si₃N₄
components (blade ring and hub) by hot pressing techniques
into an integral turbine rotor and is currently the primary
approach for fabricating this shape.[18]   The procedure for
fabricating duo-density turbine rotors involves hot pressing
Si₃N₄ powder to theoretical density for the hub component
while simultaneously bonding it to an encapsulated blade ring.[18]

Currently, emphasis is being given towards improving the
mechanical strength properties of hot pressed silicon nitride
via, 1) the use of higher purity raw materials and, 2) the
substitution of more refractory phase forming additives in
place of magnesium oxide, e.g., yttrium oxide or zirconium
oxide. To date the maxima in strength properties have been
observed through a combination of significantly higher purity
silicon nitride powder with the more refractory additives. In
view of the liquid phase densification mechanism for silicon
nitride it is not likely that a modest adjustment of the
process parameters of temperature and pressure will have
a very significant effect on microstructure and properties.
Unless, of course, the adjustment is disposed toward the
preparation of porous bodies. Rather, it is likely that
significant strength property enhancement will occur via
compositional changes, consistent with grain size and density
requirements.

The fabrication of metal fiber reinforced ceramics by hot
pressing has been studied over a period of years by countless
investigators. In many instances the purposes were related
to desired improvements in thermal shock resistance and
catastrophic fracture behavior. More recently, this same
approach has been applied to silicon nitride with the intent
of improving the impact resistance properties. Hot pressed
samples of Si₃N₄ containing molybdenum or tantalum fibers in
a unidirectional array have been found to vastly improve the
impact resistance,[19,20] and the hot press process provides
for intimate contact between metal and ceramic phases.

Figure 4 shows a 20X cross-section of a unidirectional molybdenum metal fiber reinforced silicon nitride composite. The
rather good match in expansion coefficients between the two

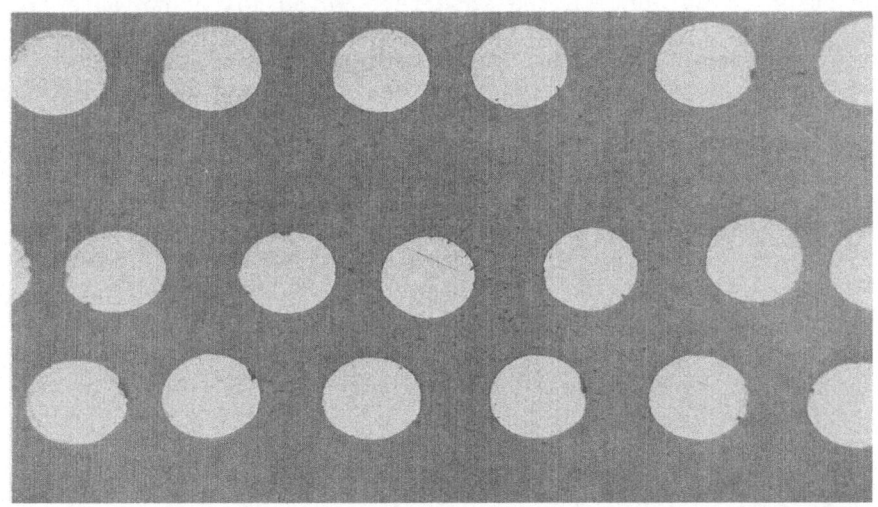

20X

Figure 4. Cross-Section of a Unidirectional Molybdenum Fiber
Reinforced Silicon Nitride Composite.

1.5X

Figure 5. Photograph of a 3-Dimensional Molybdenum Fiber
Reinforced Silicon Nitride Composite.

phases provides for good structural integrity. Some silicide
formation is evident in the interface which may contribute to
the bonding. Further work in the area of fiber reinforced
structures has led to the development of essentially 3-
dimensional metal fiber reinforcement for potential application
in hyperthermal environments.[21] Despite the relatively low
coefficient of expansion of Si_3N_4, thermal stress developing
in aerothermal applications can catastrophically fail this
material. The metal fiber acts as a barrier to continuous
crack propagation under these conditions. Figure 5 is a 1.5X
photograph of such a 3-D composite employing larger diameter
molybdenum filaments in a "Z" direction and smaller diameter
filaments in "X", "Y" directions.[21]

3. SUMMARY

The hot pressing process continues to be employed as a method
for fabricating refractory materials with maximized properties.

Densification mechanisms vary and depend on the nature of the
material being densified, as well as pressure and temperature
levels. In general, silicon nitride, containing additives
to promote densification, does so by a liquid phase mechanism.

It is likely that further improvements in hot pressed
silicon nitride strength properties will be achieved more by
composition control than by adjustment of such process
parameters as pressure and temperature.

Extension of the process to press forging makes it possible
to achieve microstructures with preferred orientation and
some improvement in specific properties. Further extension
of the process to pseudo-isopressing and pressure-bonding
provides a basis for generating complex structures.

The fabrication of metal-ceramic composites is particularly
suited to the hot pressing process and the evaluation of
refractory metal fiber reinforced silicon-nitride is the
subject of current study.

4. REFERENCES

1. P. Murray, E.P. Rodgers and A.E. Williams, "Practical and
    Theoretical Aspects of Hot Pressing of Refractory Oxides,"
    Trans. Brit. Ceram. Soc., 53 (1954), 471-510. See also:
    R.M. Fulrath, "Hot Forming Processes" in Critical Compila-
    tion of Ceramic Forming Methods, Air Force Materials Lab.,
    Tech. Doc. Report No. RTD-TDR-63-4069 (Jan. 1964), 33-43.

2. T. Vasilos and R.M. Spriggs, "Pressure Sintering of Ceramics," Chap. 2 in Progress in Ceramic Science, Vol. 4, J. Burke ed., Pergammon Press (1966).

3. R.C. Rossi and R.M. Fulrath, "Final Stage Densification in Vacuum Hot Pressing of Alumina," J. Am. Ceram. Soc., 48, (1965), 558-64.

4. R.L. Coble, "Mechanisms of Densification During Hot Pressing," in Sintering and Related Phenomena, G.C. Kuczynski et al, eds. Gordon and Breach, N.Y. (1967).

5. R.L. Coble and J.S. Ellis, "Hot Pressing Alumina-Mechanisms of Material Transport," J. Am. Ceram. Soc., 46 (1963), 438-41.

6. T. Vasilos and W. Rhodes, "Fine Particulates to Ultrafine Grain Ceramics," Chap. 8 in Ultrafine-Grain Ceramics, J. Burke ed., Syracuse University Press, Syracuse, N.Y. (1970).

7. S. Wild, P. Grieveson, K.J. Jack and M.J. Latimer, pp. 377-84 in Special Ceramics, Vol. 5, P. Popper ed., The British Ceramic Research Association, Stoke-on-Trent, 1972.

8. G.R. Terwilliger and F.F. Lange, "Hot Pressing Behavior of Si_3N_4," J. Am. Ceram. Soc., 57 (1974), 25-29.

9. W.D. Kingery, "Densification During Sintering in the Presence of a Liquid Phase I," J. Appl. Phys., 30 (1959) 301-306.

10. W.D. Kingery and M.D. Narasimhan, "Densification During Sintering in the Presence of a Liquid Phase: II, ibid., pp. 307-10.

11. W.D. Kingery, J.M. Woulbrown and D.R. Charvat, "Effects of Applied Pressure on Densification During Sintering in the Presence of a Liquid Phase," J. Am. Ceram. Soc., 46 (1963), 391-95.

12. T.J. Whalen and M. Humenik, Jr., pp. 715-46 in Sintering and Related Phenomena. G. Kuczynski et al, eds., Gordon and Breach, N.Y. (1967).

13. A. Heuer, D.J. Sellers and W. Rhodes, "Hot Working of Aluminum Oxide: I", J. Am. Ceram. Soc., 52, 468-74 (1969).

14. W. Rhodes, D.J. Sellers and T. Vasilos, "Hot Working of Aluminum Oxide: II, Optical Properties," ibid., 58, 31-34 (1975).

15. W. Rhodes, P. Berneburg, R. Cannon and W. Steele, "Microstructure Studies in Ceramics," Summary Report Contract N00019-72-C-0298, U.S. Naval Air Systems Command (1973).

16. T. Vasilos and R. Cannon, "Improving the Toughness of Refractory Compounds," Summary Report NASA CR-134813 (1975).

17. A.F. McLean, E.A. Fisher, and R.J. Bratton, "Brittle Materials Design, High Temperature Gas Turbine," Tech. Report AMMRC CTR-72-19 (Sept. 1972).

18. A.F. McLean, R.R. Baker, R.J. Bratton and D.G. Miller, "Brittle Materials Design, High Temperature Gas Turbine," Tech. Report AMMRC CTR-76-12, (April 1976).

19. J.J. Brennan, "Evaluation of Tantalum Fiber Reinforced Si_3N_4" NAVAIR Contract Report, R76-912538-1 (July 1976).

382

20. W. Rhodes and R. Cannon, "High Temperature Compounds for
    Turbine Vanes," Summary Report NASA CR-134531 (1974).
21. T. Vasilos, Current Activity.

DISCUSSION (Cannon, Lange, Moulson)

Katz: Do you see any hope to improve the processing of silicon
nitride by sintering to say 96-98% theoretical density, then
using cladless hot isostatic pressing for final densification to
theoretical density?

Vasilos: Yes; this could be a valuable combination of sintering
and hot pressing.

Stuijts: I can confirm the great value of the technique of hot
isostatic pressing of pre-sintered compacts, as introduced into
the field of electroceramics by Hardtl (Am. Ceram. Soc. Bull.,
1974). On annealing our hot-pressed samples, or on application
at a high temperature, very often bloating occurs. As a possible
reason one may think of gases adsorbed on the powders or included
from the gas atmosphere. Even after taking precautions, such as
pressing in vacuum, we still have bloating.

Greskovich: In the hot pressing of large size ceramic bodies is
there the problem of producing density gradients? If so, what is
a simple method of controlling this?

Vasilos: Indeed there is, which results from both temperature
and pressure gradients. The temperature gradient can be minimised
through appropriate hot press furnace design. The pressure gradient is a function of the part thickness as in the case of cold
pressed ware. This is a strong function too of the compression
ratio of the part being consolidated. A minimum diameter to
thickness ratio of 3:1 is often employed to diminish such
gradients.

Plant: What are the typical tolerances achieved, and what is the
die life, when hot pressing silicon nitride?

Vasilos: In the range of ± .020" for simple square or rectangular
cross section geometries. Through use of special liners the die
life is indefinite while the liner life (non-graphitic) is perhaps 50 runs.

Mangels: What is the additive content of the hot forged silicon
nitride material used in stress rupture tests?

Vasilos: For the stress rupture values I quoted the additive
content was 4% MgO by weight, although samples with $\frac{1}{4}$% MgO have
also been made.

MASS TRANSPORT IN THE HOT PRESSING OF α SILICON NITRIDE

R.J. Brook, T.G. Carruthers, L.J. Bowen, R.J. Weston

Department of Ceramics,
University of Leeds, U.K.

Silicon nitride can be hot pressed to high density by the use of sintering aids, of which magnesium oxide is the best known. It is the objective of the present paper to consider three rate processes in which mass transport can be important, namely, densification, the α-β phase transformation, and creep, to see if a comparison of these can clarify the mechanism by which magnesium oxide accelerates densification in the material.

Hot pressing

Densification data for silicon nitride have generally been analysed using either the liquid phase sintering model proposed by Kingery (1) or the hot pressing equation derived by Murray et al (2). Both models commonly give a good fit for data (3,4) and it is consequently difficult to use them as a basis for identifying the controlling mechanism; they also have the disadvantage that they do not give an explicit value for the controlling diffusion coefficient.

As an alternative approach, the hot pressing model proposed by Coble (5) may be used; the model describes the densification rate for control by grain boundary diffusion and may therefore be also used (6) for liquid phase sintering if the grain boundary thickness is taken as the thickness of the liquid phase boundary film. The densification rate is given by

$$\frac{1}{\rho}\frac{d\rho}{dt} = \frac{47\,\Omega\,WD}{kT\,G^3}(P_T)$$

where G is the grain size, Ω the volume transported by each atom of the slow-diffusing species, W the boundary thickness, D the diffusion coefficient, and P_T the total stress.

384

Using results obtained from work in the Department (6), it is found that the densification rate at a relative density of 0.7 is proportional to P_A, the applied stress. As a consequence, the surface energy contribution to the total stress, which is given in the second term in the relation

$$P_T = \frac{P_A}{\rho} + \frac{\gamma}{r}$$

may be taken as negligible, and a simplified rate equation used:

$$\frac{d\rho}{dt} = \frac{47 \, \Omega \, WD}{kT \, G^3} P_A.$$

The simplified equation can be applied to test the hypothesis that densification occurs by diffusion controlled liquid phase sintering; for under these conditions, W, the thickness of the grain boundary, becomes the thickness of the second phase film present in the boundary and this, in turn, is related to the quantity of second phase present in the sample. If the MgO additive reacts with available silica on the surface of the nitride grains to form a composition close to the $MgSiO_3$–SiO_2 eutectic, then W becomes proportional to the amount of additive. Measurement of densification rate for different amounts of additive shows (Figure 1) that such a linear dependence is indeed found.

Since W may be calculated (from G and the quantity of additive on the basis of uniform boundary film thickness), measurements of densification rate may be directly used (7) to produce values for the process diffusion coefficient. The temperature dependence of

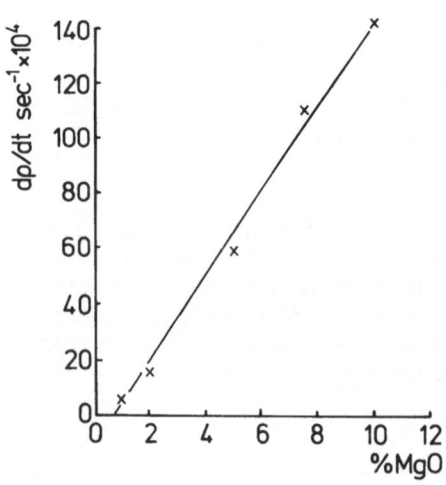

Fig. 1. Variation of strain rate with MgO content at a constant relative density of 0.70. All samples hot pressed at 1650°C and 21 MN m^{-2}. Grain size ~ 0.15 μm.

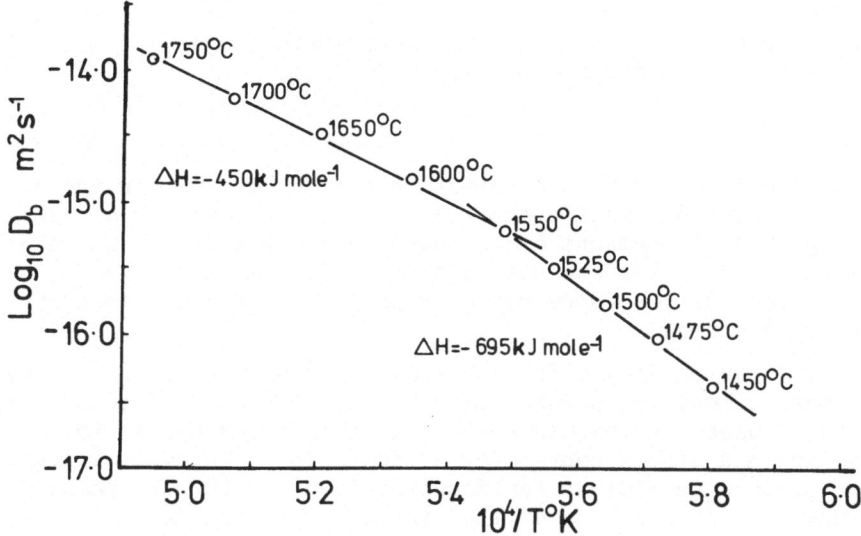

Fig. 2. Arrhenius plot of diffusion coefficients calculated from densification rates of silicon nitride containing 5^w/o MgO, hot pressed at 21 MN m^{-2}.

this coefficient is shown in Figure 2, where it may be seen that the data fall into two temperature regimes divided at a value of approximately 1550°C. This temperature is in good agreement with the solidus in the MgSiO$_3$–SiO$_2$ system (1543°C), and it is tempting to interpret the two enthalpies, above and below 1550°C, as respectively representing diffusion in a second phase boundary region which does or does not contain a liquid phase as one of its elements.

Phase Transformation

The observation (8) that complete densification can be achieved prior to complete transformation from the α to the β form, has been used to suggest that the transformation is not a necessary step in densification. However since removal of 30% porosity can be achieved by the movement of some 15% of the solid material, it is reasonable to suggest that densification and transformation occur simultaneously by the solution-diffusion-reprecipitation mechanism. Such an interpretation is in fact consistent with the degree of transformation reported for the achievement of complete density in the earlier data (8).

To test this hypothesis, measurements of the kinetics of transformation have been made in the Department (7). Data can be fitted to first order kinetics over the entire transformation according to the relation:

$$\frac{1}{\alpha}\frac{d\alpha}{dt} = -K.$$

In the early stages of the transformation, where comparison with densification data is sought, $\alpha \longrightarrow 1$, and

$$\frac{d\alpha}{dt} = -K.$$

For the proposed model, K is a rate constant which includes the appropriate diffusion coefficient and geometrical terms which will in turn depend on the amount of boundary second phase. Although the diffusion coefficient cannot be explicitly evaluated, its temperature dependence can be noted by plotting the dependence of K on the temperature.

This is done in Figure 3; the similarity between this figure, based on measurements of phase compositions by X-ray diffractometry, and Figure 2, based on measurements of densification rates, is most striking and offers convincing evidence that the controlling diffusion process is similar for densification and for the phase transformation. It can be seen that the same two regimes and the same transition temperature are found for the transformation data.

The geometrical features of the transformation can be found by noting the dependence of K on the amount of additive. If the

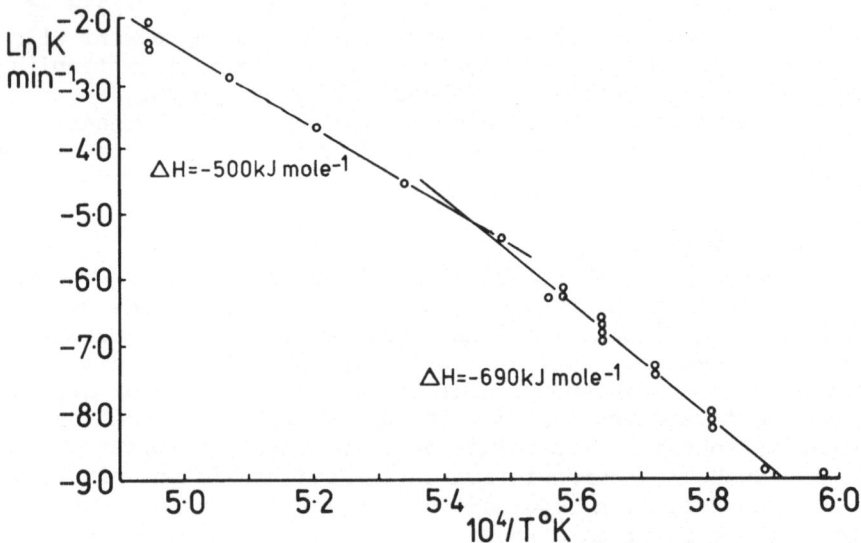

Fig. 3. Arrhenius plot of first order rate constants calculated from phase analysis of silicon nitride containing $5^{W}/o$ MgO and hot pressed at 21 MN m^{-2}.

transformation occurs by diffusion from an α grain across the boundary to a β nucleus, then an inverse dependence on the amount of additive (the diffusion distance, L) is expected. In fact, as indicated in Figure 4, a linear dependence is found, suggesting diffusion parallel to the grain surfaces with the additive amount affecting the cross-section, A, of the diffusion path.

The picture that emerges, therefore, is that densification and the transformation have the same dependence on D and on W. Since $\frac{d\alpha}{dt}$ and $\frac{d\rho}{dt}$ are both dependent on the atom flux, J, one can write a combined flux equation

$$J = - \frac{A\,c\,D}{L\,R\,T}\,(\Omega\,\Delta P + \Delta G_{tr})$$

where A and L are the geometrical factors, c the concentration of the diffusing species, ΔP the stress difference between the point of solution and reprecipitation, and ΔG_{tr} the free energy for the α–β transformation.

Since diffusion is predominantly parallel to the boundaries, the first term dominates; atoms diffusing down the stress gradient imposed by the applied load, take the opportunity to transform, on reprecipitating, to the more stable β phase. The known applied

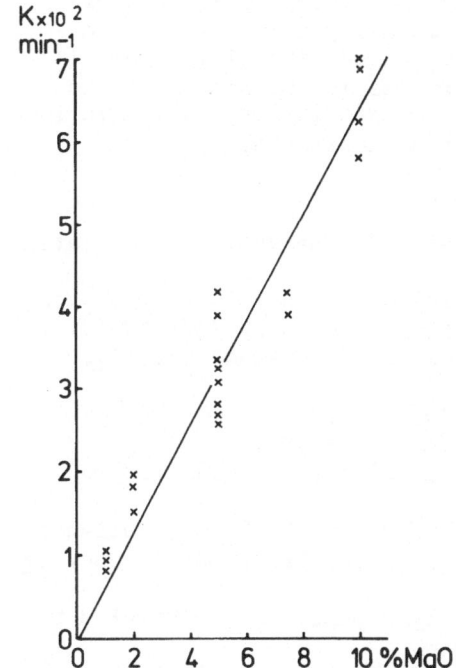

Fig. 4. Dependence of first order rate constant (α–β transformation) on the MgO content of silicon nitride.

loads suggest, therefore, that ΔG_{tr} is substantially less than 800 J.mole^{-1}. Such values are representative of the polytype form of phase change (9).

Creep

Although creep measurements have not been made in the Department, the values for the diffusion coefficient can be used to calculate a minimum value for the creep rate by way of the expressions for diffusional creep. Comparisons with the literature (10) are difficult in part because of the differences between the materials and the conditions employed (notably temperature), and in part because of the scatter of the creep data themselves; however, experimental values are in general higher than those predicted, suggesting that processes other than simple diffusional creep are involved.

Conclusions

The observed consistency between data for hot pressing and for the $\alpha-\beta$ phase transformation in MgO-doped Si_3N_4 suggests that both processes are controlled by diffusion through a grain boundary phase. Further, although transformation occurs together with densification, it is not seen as a necessary factor in bringing about densification.

The model is helpful in showing the linear dependence of densification on the additive, thus allowing selection of the appropriate amount of additive for required sintering rates; it also shows the sensitivity of the process to the nature of the boundary phase. Both of these features should prove valuable in the assessment of alternative sintering additives.

References

1. W.D. Kingery, J.M. Woulbroun and F.R. Charvat. J. Am. Ceram. Soc. 46, 391 (1963).
2. P. Murray, D.T. Livey and J. Williams. Ceramic Fabrication Processes, ed. W.D. Kingery, Wiley, New York, 147 (1958).
3. G.R. Terwilliger and F.F. Lange. J. Am. Ceram. Soc. 57, 25 (1974).
4. J. Mimoto. J. Mat. Sci. 11, 1103 (1976).
5. R.L. Coble. J. Appl. Phys. 41, 4798 (1970).
6. L.J. Bowen, R.J. Weston, T.G. Carruthers and R.J. Brook. Ceramurgia Int. 2 (1976).
7. L.J. Bowen, T.G. Carruthers and R.J. Brook. To be published.
8. R.J. Weston and T.G. Carruthers. Proc. Brit. Ceram. Soc. 22, 197 (1973).
9. A.R. Verma and P. Krishna. 'Polymorphism and Polytypism in Crystals', Wiley, New York, 1966.
10. R. Kossowsky, D.G. Miller and E.S. Diaz. J. Mat. Sci. 10, 983 (1975).

DISCUSSION (Cannon, Lange, Moulson)

Jack: I should like to congratulate the Leeds group on a very fine piece of work but wonder whether the agreement with the model is not a little too good. Would you not expect the activation energy for densification and for transformation to be much more different above and below 1550°C than you have observed? Could you not draw a smooth curve through all your points and not two straight lines intersecting at 1550°C? The liquid phase is not a magnesium silicate but contains nitrogen and so the eutectic will not be at 1550°C - it will vary with concentration of dissolved Si_3N_4 and also impurities content.

Brook: It would be dangerous to infer from an analogy with sialon work for example that experiments with ceramics are necessarily imprecise. We could quote for instance Cutler's sintering work on TiO_2 doped Al_2O_3 or Peelen's grain growth work on MgO doped Al_2O_3; in many respects the Si_3N_4-MgO system is no more difficult than these other systems and we should therefore start to look for agreement with specific models.

We had no clear expectation about the relative magnitudes of diffusion enthalpies in the glassy and crystalline zones since coefficients for N or Si in such phases have not been explored.

It is of course the reader's privilege to decide whether a smooth curve or two lines give a better fit to the data; our own view is that a smooth curve gives unsatisfactory fit particularly to the densification results. We agree that the presence of nitrogen and impurities will cause a deviation from the exact eutectic temperature of the $MgO-SiO_2$ pseudo-binary (1543°C). The results suggest however that the deviation is not large.

Stuijts: You mention the formation of a eutectic at about 1550°C which gives rise to liquid-phase sintering. However, also below this temperature you obtain densification. If one can exclude a lower-melting eutectic as mentioned by Lange, you then have enhanced densification over a high-diffusivity "solid" grain boundary phase. Thus, it is not purely liquid-phase sintering.

Brook: We agree; in the more exact terminology we have sintering in the presence of a second phase.

Clarke: What average thickness does the second phase (amorphous, crystalline, whatever) have on calculation from your MgO content and from the grain size of the Si_3N_4? As I shall be reporting, we do not find any amorphous grain boundary phase, or indeed any second crystalline phase between the grain examined by direct lattice imaging. This would suggest that you would have to modify your very attractive model to include regions where there is no second phase between grains. Would that be possible?

Lumby: How much does your model rely on maintaining a constant grain boundary thickness over a range of temperature?

Brook: The boundary thickness varies between 50-100 Å for the
5-10% MgO additions. The occurrence of boundaries free from
second phase, i.e. which are clean and free of porosity, would
mean an increased thickness of the second phase in other regions
and hence a corresponding reduction in the derived diffusion
coefficients. If the clean boundaries are sub-boundaries between
the crystallites of the original dense powder particles, then no
adjustment is needed. The occurrence of clean boundaries containing pores would be difficult to explain in terms of the theory;
so far as we are aware, these have not been observed.

The model assumes a constant boundary thickness over the
temperature range of the experiments (1450-1750°C); the thickness being related to the amount of boundary second phase. This
is, I believe, reasonable provided the silicon nitride grains do
not have a strongly temperature-dependent solubility in the
boundary phase.

HIGH STRENGTH SILICON NITRIDE PREPARED WITH EUTECTIC FLUX ADDITIONS

J.D. Venables, D.K. McNamara and R.G. Lye

Martin Marietta Corporation, Martin Marietta Laboratories, Baltimore, Maryland 21227, U.S.A.

ABSTRACT

We are investigating the use of mixed oxide eutectic fluxes to promote sintering of hot-pressed silicon nitride. Significant improvements in microstructure and mechanical properties were observed when a eutectic mixture of alumina and yttria (chosen because the eutectic temperature, 1760°C, is convenient for hot-pressing) was employed rather than single oxide additions. For example, the room temperature (biaxial) flexure strength of this material, 1330 MN m^{-2} (190 k spi), appears to be the highest value yet reported for a bulk polycrystalline ceramic.

DISCUSSION (Cannon, Lange, Moulson)

Jelacic: Did you determine the proportions of α- and β-silicon nitride phases?

Venables: The material is composed only of β'-sialon and glass.

Jack: Is the macrostructural inhomogeneity observed a result of non-homogeneous distribution of liquid/glassy phases as suggested by Clarke?

Venables: It has yet to be determined whether the structure results from an inhomogeneous distribution of the liquid/glass phase, inhomogeneous sintering, or both.

Bowen: Polished sections of partially densified compacts of commercial silicon nitride hot-pressed with magnesia show that densification occurs inhomogeneously throughout the compact. The distribution of densification seems to correlate with the mottled colouring often observed in hot-pressed silicon nitride.

Katz: Exactly where does the strength of your high room temperature strength material fall off with temperature?

Venables: We tested at 1400 and 1200°C and found generally poor properties consistent with the presence of a glass boundary phase.

Godfrey: A.D. Sivill, working with Stanley and Tessler at Nottingham University, studied the NBS test extensively 3 years ago; results are given in Sivill's Ph.D. Thesis. He compared 4 ball flexure with ring and ball, and ring and ring flexure, along with 4 point bend tests on square and circular section RBSN beams. A stress analysis and a Weibull analysis, corrected for frictional effects, were produced. The unit volume strength derived from the NBS test was, disquietingly, larger than the unit volume strength derived from the beam tests.

THE FORMATION OF SINGLE PHASE SIALON CERAMICS

R. J. Lumby, B. North and A. J. Taylor

Lucas Group Research Centre, Monkspath, Solihull,
West Midlands.

ABSTRACT

An account of the reactions of silicon nitride with aluminium
nitride and silica (with and without magnesium oxide) is presented, together with an analysis of the resulting structure and
its effect on properties.

Some reactions occurring between silicon nitride, aluminium
nitride and silica are suggested from XRD evidence collected from
specimens of sialons (approximately $Z = 0.8$) hot pressed at temperatures between 1500°C and 1840°C. In a material not containing magnesium oxide, X phase and O' are detected as intermediate
phases, resulting from the reaction of aluminium nitride with
silica and silicon nitride with silica. An addition of 1%
magnesium oxide catalyses these reactions at lower temperatures
by forming a lower melting point liquid phase which assists dissolution of reactants. It is suggested that below 1500°C further
reactions have removed the intermediate phases, and large
quantities of β' have formed. Finally α and β silicon nitride
react with the highly substituted β' and slowly disappear as the
degree of substitution of the β' reduces.

The final structure by inference must contain not only β' but
also a glassy phase containing some silica, magnesium oxide and
also any of the contaminant elements originating from the silicon
nitride. Variation of the proportions of the constituents in the
starting mixture as reported earlier causes changes in the properties, especially creep. Some attempts are made to relate the
compositional changes to possible structural changes. Actual
glass quantities are illustrated with transmission electron

micrographs. Finally, properties of the range of hot pressed sialons, and the composition of the intergranular phase present in these materials, are described.

1. INTRODUCTION

The influence of microstructure on the properties of both silicon nitride and sialon materials has been investigated by many workers using a variety of techniques. By careful correlation of composition, structure and properties it has been possible to develop stronger materials with better high temperature capabilities.

A study of properties and starting constituents led to the preferred charge for hot-pressing of high strength silicon nitride to be defined [1,2] as a fine particle, high α powder. The influence of the glassy intercrystalline phase on the high temperature properties of hot-pressed silicon nitride (H.P.S.N.) has emerged from studies of composition, microstructure, and properties [3,4,5]. The role of magnesia in the densification of Si_3N_4 was identified by X-ray diffraction studies at Newcastle [6] and a solution reprecipitation mechanism was identified by Drew and Lewis at Warwick [7] using transmission electron microscopy.

A combination of X-ray diffraction and optical microstructural studies led to our first questioning the original formula for the composition of expanded β Si_3N_4 [8]. A combination of these techniques with a knowledge of weight loss incurred during sialon formation allowed the so-called 'Z' formula to be confirmed [9] as the correct description of the composition of β'. Latterly, the same combination of techniques has led to spectacular improvements in the properties of sintered sialons. Fully dense sintered materials have been developed which exhibit mean flexural strengths at room temperature of 821 MNm^{-2} (117.5 k.p.s.i.).

The present work, described in more detail elsewhere [10], is restricted to a description of the reactions occurring during the formation of Z = 1 sialon materials from Si_3N_4/AlN/SiO$_2$ mixtures.

Additionally a more complete explanation is presented for the properties and microstructure of the so-called 101-121 series of sialon materials. This series containing 1% MgO were prepared in early 1973, and described in detail at the Special Ceramics Symposium in July 1974 [9]. Additions of AlN to a balanced X formula composition were shown to give rise to massive improvements in the creep properties of the hot-pressed products.

2.   EXPRESSION OF SIALON PHASE BEHAVIOUR

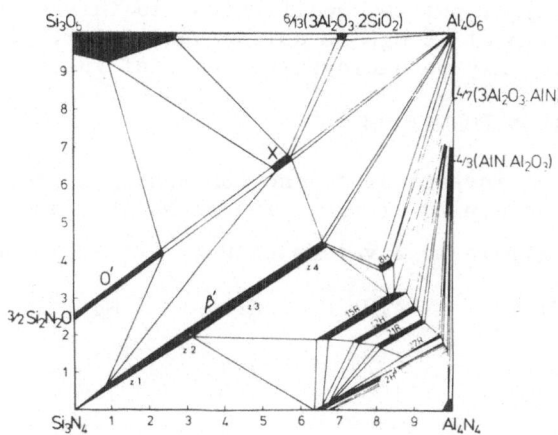

Figure 1.   The Si_3N_4–A1N–A1$_2O_3$–SiO$_2$ behaviour diagram [11].

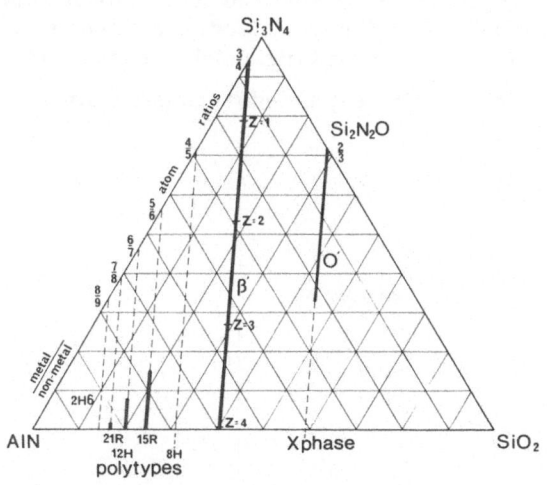

Figure 2.   The Si_3N_4–A1N–SiO$_2$ ternary system.

The currently accepted [11] mode of expression for the phases
present in the quaternary system containing Si_3N_4/A1N/A1$_2O_3$ and
SiO$_2$ is shown in Figure 1. In order that compositions can be
expressed on a weight percentage basis, (as opposed to the equiva-

lent percentage shown in Figure 1), Figure 2 shows the same combination of phases which appear in the ternary system containing
Si_3N_4/A1N/SiO_2. Allowance is made for the typical oxide contents
of both nitride materials, but the modification to this diagram
produced by the addition of 1% MgO is not accounted for. This
diagram will therefore only be employed as a useful guide.

3. REACTIONS AND SIALON FORMATION

Two sialon charges were investigated, one containing no additional
metal oxide and one containing 1% MgO. The Si_3N, A1N, and SiO_2
contents, which approximate to Z values of 0.8, are shown below.

Material	Si_3N_4	A1N	SiO_2	MgO
A	80.2	12.4	7.4	–
B	83.0	10.0	6.0	1.0

Materials A and B were hot-pressed for 1 hour using 15.4 MNm^{-2} at
a variety of temperatures shown in Figures 3 and 4. The phases
present in each material at each temperature were determined by
X-ray diffraction. At temperatures between 1500 and 1600°C, in
Material A, O' and X phase were observed and the A1N content was
seen to decrease. Above 1600°C, O', X phase, and A1N disappear,
and the α and β Si_3N_4 present in the starting powder reduce in
quantity whilst a rapid growth in the expanded β phase content
occurs.

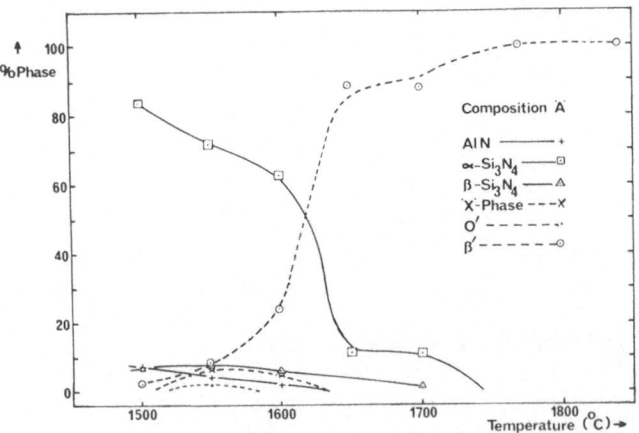

Figure 3. Variation of crystalline phase content with hot pressing temperature – Material A – No MgO [10].

Possible reactions in the absence of MgO.

1500–1600°C

To form: .... O'

$$Si_3N_4 + SiO_2 \longrightarrow 2\ Si_2N_2O \tag{1}$$

.... X phase

$$AlN + SiO_2 \longrightarrow SiAlO_2N \tag{2}$$

Above 1600°C

To remove: ... O'

$$2AlN + 2Si_2N_2O \longrightarrow Si_4Al_2O_2N_6 \tag{3}$$

.... X phase

$$2AlN + 2SiAlO_2N \longrightarrow Si_3Al_4O_4N_4 \tag{4}$$

and Si_3N_4

$$Si_2Al_4O_4N_4 + 6Si_3N_4 \longrightarrow 4\ Si_5AlON_7 \tag{5}$$

$$Si_4Al_2O_2N_6 + 2Si_3N_4 \longrightarrow 2\ Si_5AlON_7 \tag{6}$$

Since none of reactions 3–6 are discrete Z = 4 sialon is rarely observed except where SiO_2 and AlN are reacted together alone as follows

$$2\ SiO_2 + 4\ AlN \longrightarrow Si_2Al_4O_4N_4 \tag{7}$$

Equations (3) to (7) may well oversimplify the succession of reactions. It would appear that a small amount of Si_3N_4 also reacts and results in, for example, the β' sialon formed in equation (4) having a Z value of less than 4. A better possible expression of this reaction is:

$$(2 - Z/2)\ Si_3N_4 + Z/2\ AlN + Z/2\ SiAlO_2N \longrightarrow$$

$$Si_{6-Z}Al_ZO_ZN_{8-Z}$$

The effect of the addition of 1% MgO (Figure 4) is either to catalyse the formation and removal of O' and 'X' at temperatures lower than 1500°C or to bypass or suppress their formation. Observation of the phases present with the time of hot pressing at 1650°C (Figure 5) shows that the 1% MgO addition does not appear

398

to accelerate the reaction between highly substituted β' and the α and β in the charge (i.e. reactions (5) and (6)). The rate of disappearance of α and β appears to be identical for each material.

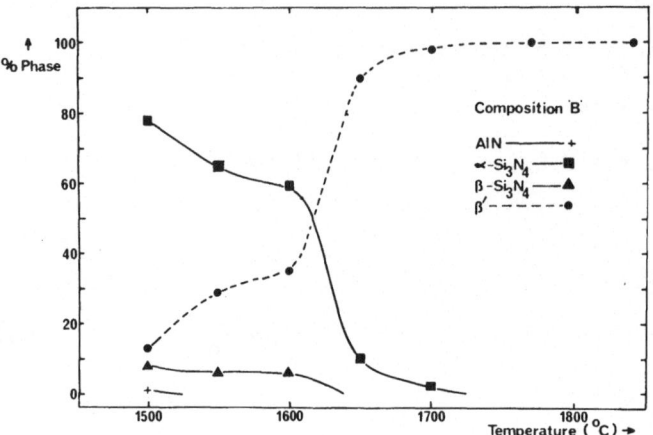

Figure 4.   Variation of crystalline phase content with temperature. Material B - 1% MgO [10].

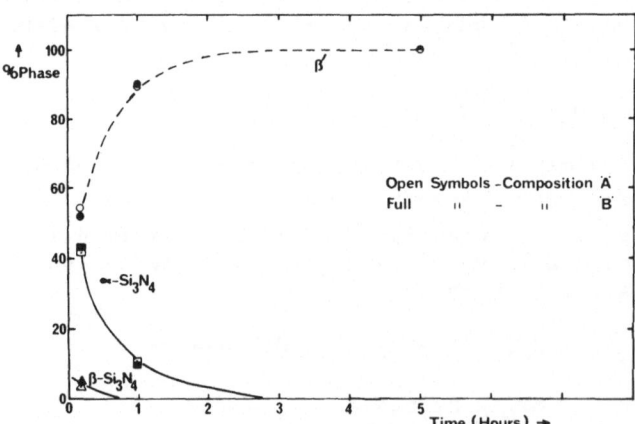

Figure 5.   Variation of crystalline phase content with time of hot pressing at 1650°C [10].

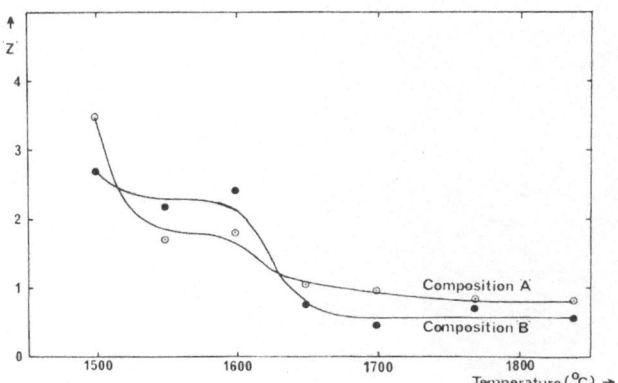

Figure 6. Variation of Z, the degree of substitution of the β' sialon with hot pressing temperature. [10].

The change of the degree of substitution Z with temperature (Figure 6) for these materials shows the expected reduction in Z as the reaction proceeds to a value dictated by the overall nominal composition.

At the end of the reaction materials, which are typically oxygen-rich with respect to the formula $Si_{6-Z}Al_ZO_ZN_{8-Z}$, and materials containing 1% MgO, will be β' sialon plus a glassy phase.

4. 101–121 SIALON SERIES [9, 10]

This series, similar in composition to Material B, ranged in composition from an X = 0.8 material (101) to the formula $Si_{6-3/4X}Al_{2X/3}O_XN_{8-X}$ to a Z = 0.8 (approximately) material made to the formula $Si_{6-Z}Al_ZO_ZN_{8-Z}$. The compositional adjustment was made by discrete additions of AlN, and material contained 1% MgO. All materials in the fully reacted condition would be expected to be β' sialon plus a glassy phase.

400

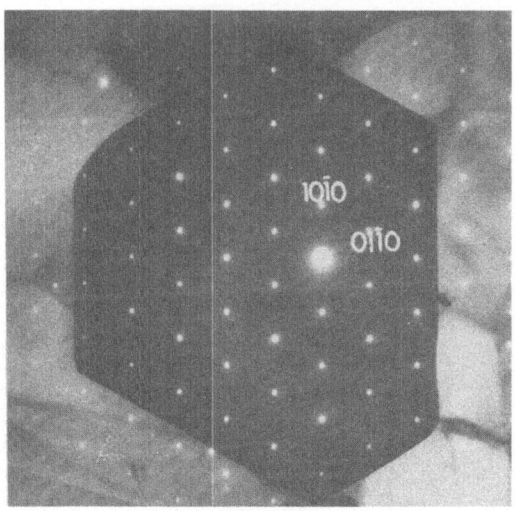

Figure 7. Selected area electron diffraction on a β' sialon grain in Material 101. [10].

Transmission electron microscopy by Dr Lewis of the University of Warwick has clearly identified β' sialon by electron diffraction (Figure 7) and the glassy triple points in an oxygen rich material like 101 (Figure 8). The distinction between a glassy material like 101 and a relatively glass free material like 121 is obvious (Figure 9).

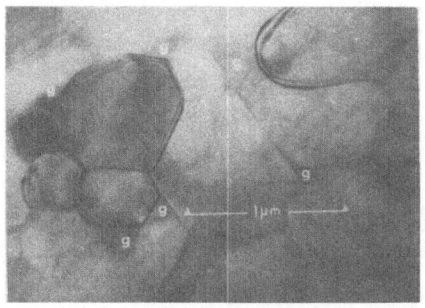
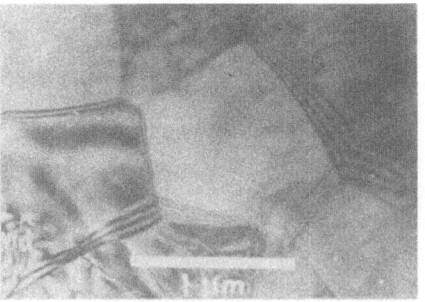

Figure 8. Material 101 showing intercrystalline glass phase (g) [10].   Figure 9. Material 121 showing the absence of an inter-crystalline phase [10].

Proceeding through the series 101 to 121 there was a progressive reduction in the amount of the intercrystalline phase.

5.   101–121 PHASE BALANCE

It is of interest to attempt to analyse this range of compositions
with respect to the phase diagram.   To do this involves making
many assumptions concerned with

a)   The accuracy of the nominal composition.

b)   The compositional changes occurring during the reaction.

c)   The validity of the phase diagram.

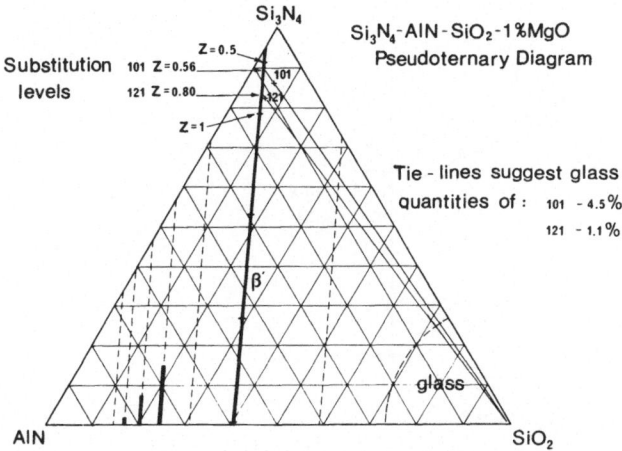

Figure 10.   The Si_3N_4–A1N–SiO_2–1% MgO Pseudoternary diagram.

Putting all of these objections aside allows the following somewhat tenuous observations to be made.   Figure 10 shows the relative
positions of materials 101 and 121, and from these:

1.   101 would be expected to be a more glassy material.

2.   101 would be expected to contain β' sialon of Z value = 0.56
as opposed to 121 which would have a Z = 0.80.   Although it is not
possible to accurately measure glass contents from electron micrographs, the quantities observed and expected are close.   Lattice
parameter measurements for 101 and 121 materials allowed Z values
to be estimated as Z_{101} = 0.53 and Z_{121} = 0.86, again significantly
close.

Glass composition will significantly affect its viscosity at high
temperatures and hence the properties of these materials.   Auger
electron spectroscopy performed by Lewis and Coworkers has shown
that:

1) Higher levels of glass forming elements are segregated in the intercrystalline regions.

2) Intercrystalline layers of different thicknesses are observed. (Thick in the case of material 101 and thin in the case of material 121).

3) Although material 121 appeared by electron microscopy to be glass free, the A.E.S. evidence suggests a high concentration of glass formers in the grain boundary regions.

6. PROPERTIES OF THE 101-121 SERIES

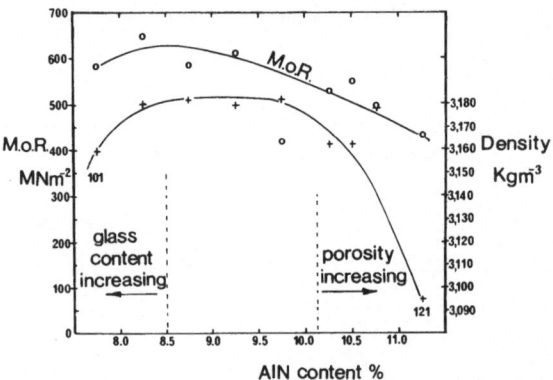

Figure 11. Variation in Density and Modulus of Rupture (M.o.R.) with A1N content. [9].

Figure 11 shows the range of strength and density obtained for the 101-121 range. Material 101 exhibits a low density due to its glass content which may well be influencing strength. At the other end of the range, density is falling off due to the lack of glass hindering the final densification process. Pores, observed optically as 10-20 micron holes in 121, become strength controlling.

The creep relationship first shown at Stoke [9] can now be compared with the evidence provided by electron microscopy (Figure 12). Across this range of composition there is a twofold increase in grain size. In 101 glass is impeding the growth of the β' grains and these, through growing freely in the liquid, are often hexagonal whereas grains in 121 are polygonal. Both the decrease in glass and the increase in grain size would be expected to contribute towards the improved creep resistance observed.

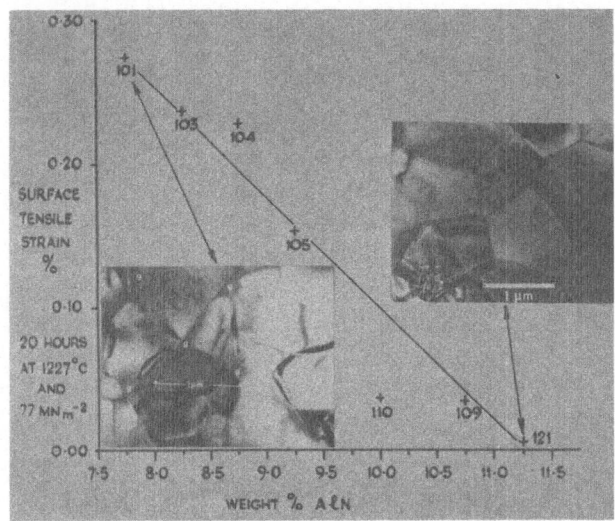

Figure 12. Effect of AlN content on Creep compared with electron
micrographs [9] .

7. CONCLUSIONS - Formation

1) MgO increases the rate of formation of a.silicate liquid
during the hot-pressing of $Si_3N_4/AlN/SiO_2$.
2) The early formation of the reactive silicate liquid enhances
the rate of dissolution of AlN.
3) MgO does not increase the rate of reaction of α and β Si_3N_4
to form β' sialon.

- Properties

1) Glass as well as pores appear to influence the room temperature properties.
2) The variations of glass content and the creep properties of
sialons appear to follow the expected relationship.

REFERENCES
1. R J Lumby and R F Coe, Proc. Brit. Ceram Soc. 15, 91, 1970.
2. R F Coe, R J Lumby and M F Pawson, "Special Ceramics 5"
361-375.
3. D W Richerson, Bull. Amer. Cer. Soc. 52 No. 7, July 1973
pp 560-562.
4. F Lange, J Amer. Ceram. Soc. 57 No. 2, pp 84-87 1973.
5. Ram Kossowsky, J Mat. Sci. 8 1973, pp 1603-1615.
6. S Wild, P Grieveson, K H Jack, and M J Latimer, "Special
Ceramics" 5 pp 377-382.
7. P Drew and M H Lewis, J Mat. Sci. 9 261, 1974.

404

8. W J Arrol, "Ceramics for High Performance Applications"
   Brook Hill 1974.
9. R J Lumby, B North, and A J Taylor, "Special Ceramics 6"
   pp 283-296.
10. M H Lewis, B D Powell, P Drew, R J Lumby, B North and
    A J Taylor, J Mat. Sci. to be published.
11. K H Jack, J Mat. Sci. $\underline{11}$ 1135 (1976).

ACKNOWLEDGEMENTS

The authors would like to thank the Directors of Lucas Industries
for permission to publish this paper. In addition, the help and
valuable discussions with Dr M H Lewis of the University of
Warwick are gratefully acknowledged.

DISCUSSION (Cannon, Lange, Moulson)

Jack: I think Dr. Lumby would agree that the reaction sequence
depends entirely on the reactants used and these can be varied
and still produce the same final composition. In the specific
example he gives, the mixture is of AlN, SiO_2 and Si_3N_4; the AlN
and SiO_2 react first to produce a high z-value sialon which then
reacts with the Si_3N_4 to give the equilibrium solid solution. If
he used a mix of Si_3N_4, Al_2O_3 and AlN, or Si_3N_4, Si_2N_2O and AlN,
the reaction sequence would be quite different. Can he say anything about the best mixtures and optimum conditions to produce
a β'-sialon?

Lumby: The reaction sequence does indeed depend on the combination of constituents, but preparing the same final composition
from different combinations of constituents is often found to be
very difficult.

STATUS REPORT ON DENSIFICATION OF β-Si_3N_4 SOLID SOLUTIONS
CONTAINING AlN:Al_2O_3 DURING CHEMICAL REACTION

L. J. Gauckler, S. Boskovic* and G. Petzow

Max-Planck-Institut fur Metallforschung
Institut fur Werkstoffwissenschaften
7 Stuttgart-80, W. Germany

T. Y. Tien

Materials and Metallurgical Engineering
The University of Michigan
Ann Arbor, Michigan, U.S.A.

The densification of β-Si_3N_4 solid solutions in the system
Si,Al/N,O was investigated. Different chemical mixtures giving
the same final composition were used as starting materials. A
nitride containing liquid was found during sintering. The liquid
became glass at lower temperatures.

1. INTRODUCTION

It is the intention of this investigation to develop a
fabrication method to produce dense silicon nitride ceramics and
to understand the densification mechanism. Because of the phase
equilibria in the system Si,Al/N,O (1) was well established, some
of β-Si_3N_4 solid solution compositions were chosen for this study.
The phase diagram in the System Si,Al/N,O is given in Fig. 1.

Compositions $Si_{6-x}Al_xO_xN_{8-x}$ can be produced by mixing either
Si_3N_4, AlN and Al_2O_3 powders, or Si_3N_4, AlN and SiO_2 powders. For
one particular composition, where x = 4, the starting materials
can be SiO_2 and AlN only. Different reaction mechanisms would be
expected for different starting materials, and hence different
sintering kinetics.

---

*Humbolt Fellow

406

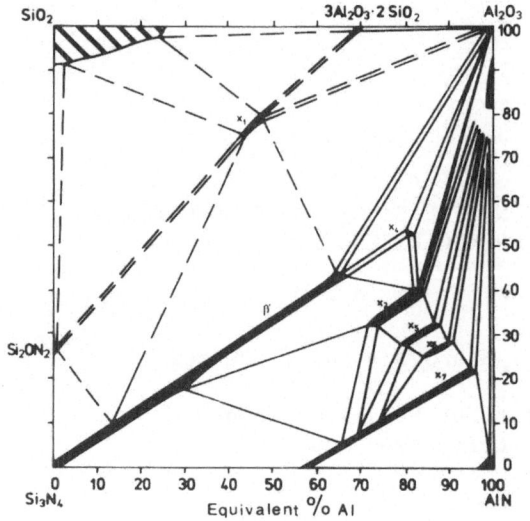

Fig. 1: Phase diagram of the Si,Al/N,O system at 1750°C after Gauckler et al.

2. EXPERIMENTAL RESULTS AND DISCUSSION

Appropriate amounts of powders were weighed and mixed in a hard metal mill under alcohol for three hours. The mixtures were dried and then compacted under an isostatic pressure of 650 MN/m^2. In the early stage of this investigation, the compacts were heated in a carbon resistance furnace under flowing nitrogen. During firing, the solid lost weight and the compact became a porous skeleton of AlN and other AlN-polytype phases. In the later stages, it was found that weight loss can be prevented by placing the compact in loose powders of the same composition in a covered alumina crucible during firing. The results reported in this paper were all obtained in this manner. The experimental arrangements are given in Fig. 2.

2.1 Si$_3$N$_4$, AlN and Al$_2$O$_3$ as starting materials.

Three compositions in the Series β-Si$_{6-x}$Al$_x$O$_x$N$_{8-x}$ were prepared, where x being 0.77, 2.18 and 4.0. These compositions are designated as β-10, β-30 and β-60, respectively. Specimens were prepared and sintered as described before. The heating time used was one hour at different temperatures. Green densities of these specimens were about 65% and the firing shrinkage at different temperatures are plotted in Fig. 3. Weight loss after firing are given in Fig. 4.

Pure Si$_3$N$_4$ composition did not densify as shown in Fig. 3.

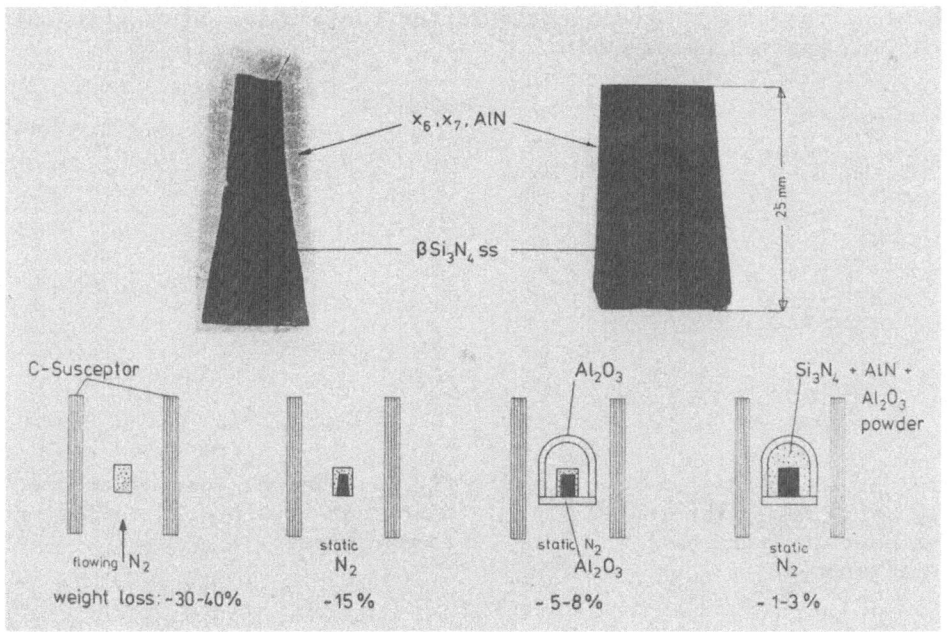

Fig. 2: Experimental arrangement of the results of sintering
β-Si₃N₄ solid solution from mixtures of Si₃N₄, AlN and Al₂O₃ at
1800°C for one hour.
Top left figure: Sintered in nitrogen atmosphere. The center
black part was single phase β-Si₃N₄ solid solution and skin was
identified to be a mixture of X_6; X_7 and AlN (see phase diagram in
Fig. 1). This result indicated that SiO was lost during firing.
Top right figure: The specimen was packed in loose powder of the
same composition during firing.
Bottom figures show the experimental arrangement during firing.
The extreme left was in flowing nitrogen. The second from left
was fired in nitrogen without flowing and the third was fired in
a covered alumina crucible. The extreme right shows the arrangement in which the specimen was packed in loose powder of the same
composition. Weight loss after one hour of heating at 1800°C are
shown at the bottom of the figure for β-30 material.

High aluminum containing compositions sintered to a higher density
than the low aluminum containing compositions. Phase changes
after sintering were analyzed for the β-30 compositions. The
results are given in Fig. 5. The volume percents of different
phases were computed from the measured x-ray diffraction peak
intensities. As shown in Fig. 3, densification started at 1600°C.
Longer sintering time has been used, however, no further shrinkage
was observed. From these data, one can suggest that the driving

408

force for densification of the β-Si$_3$N$_4$ solid solutions is mainly chemical reaction. Densification after the completion of chemical reaction was not observed.

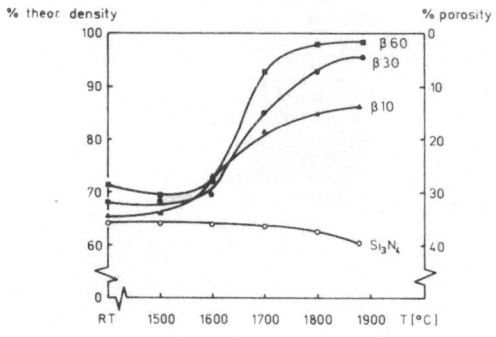

Fig. 3: Density of Si$_{6-x}$Al$_x$O$_x$N$_{8-x}$ sintered for one hour at different temperatures.

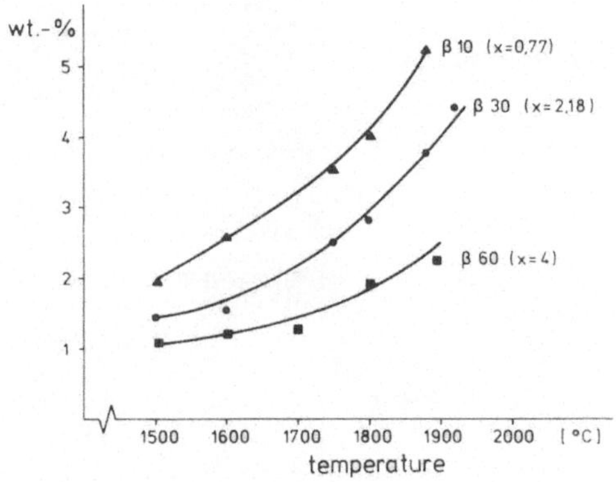

Fig. 4: Weight loss after one hour heat treatment at different temperatures.

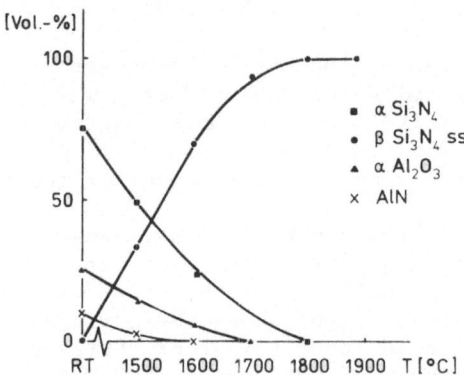

Fig. 5: Amount of phase present after heating for one hour at different temperatures for the composition β-30 material from the starting mixtures of α-Si$_3$N$_4$, AlN and α-Al$_2$O$_3$.

2.2 SiO$_2$ and AlN mixture as starting materials.

Quartz form of the SiO$_2$ and AlN powders were used for these experiments. From the phase diagram (Fig. 1), the composition at the intersect of the line connecting SiO2 and AlN and the β-Si$_3$N$_4$ solid solutions is β-60, i.e. x = 4.0. The same experimental procedure was followed as described before. The percent shrinkage after one hour of heating at different temperatures are given in Fig. 6. An increase in volume was observed for the sample heated at 1600°C for one hour. The amount of phases present after one hour of heat treatment at different temperatures were measured and the results are given in Fig. 7.

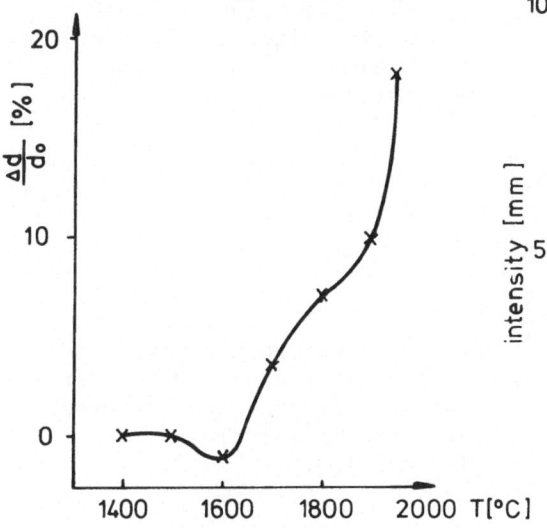

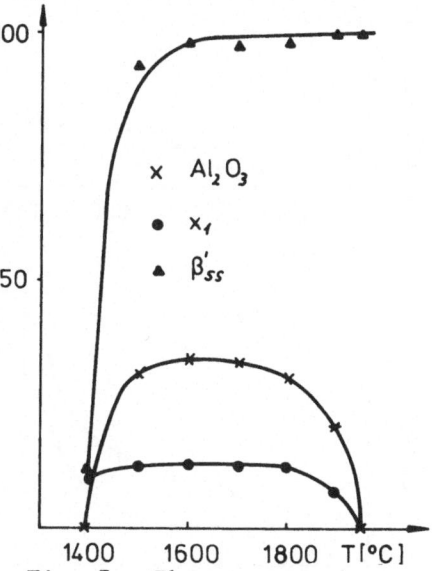

Fig. 6: Firing shrinkage of β-60 composition after one hour at different temperatures.

Fig. 7: Phases present after one hour heat

Crystalline SiO_2 was not detected after one hour of heat treatment at temperatures as low as 1500°C. It appears that $3 SiO_2 + 4 AlN \rightarrow Si_3N_4 + 2 Al_2O_3$ took place at the beginning of the reaction. X_1 and Al_2O_3 are intermediate products. In the later stage of the reaction, X_1, Al_2O_3 and AlN reacted to give β-Si_3N_4 solid solutions. The observed AlN diffraction peak intensities in the mixtures are lower than expected. Weight loss observed for these specimens are shown in Fig. 8.

The lattice parameters of the β-phase changed to a higher value for specimens heated at higher temperatures. The lattice parameters of specimens heated at different temperatures are given in Fig. 9. The data indicated that aluminum entered the β-solid solution lattice at high temperatures.

The sintering behavior of these specimens may be discussed in two separate temperature regions. At temperatures between 1400 and 1600°C, reaction in the solid state predominate. The slight volume increase can be accounted for by the chemical reaction. At

temperatures above 1600°C, liquid formed. The increase in weight
loss at 1700°C indicated partial melting.

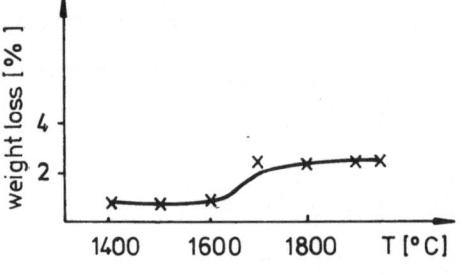

Fig. 8: Weight loss of β-60
composition after one hour of
heat treatment at different
temperatures.

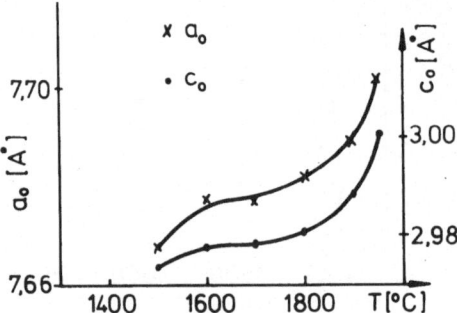

Fig. 9: Lattice parameters of
β-phase after one hour heat treatment at different temperatures.
The starting materials are
SiO_2 and AlN.

There should be a disappearance of crystalline phases at this
temperature. However, as shown in Fig. 7, no such discontinuity
was observed. There are still questions remaining to be answered.
At the present time, work is being continued to complete the rate
studies at different temperatures. Hopefully, better understanding
can be obtained in the future.

3. GLASS FORMATION IN THE SYSTEM Si,Al/N,O

A mixture of SiO_2 and AlN giving the composition β-60 were
heat treated at 1950°C for various lengths of time. The linear
shrinkage versus heating time are plotted in Fig. 10. Relative
intensities of the x-ray diffraction lines of the β-phase are
plotted in Fig. 11. It is clearly shown that the amount of β-
phase increased with time. It should be noted that there is no
other crystalline phase present. The relative amounts of β-phase
estimated from the intensity data were about 65, 75, 85, and 95
percent for 2, 30, 60 and 180 minutes samples, respectively.

The lattice parameters of the β-phase changed for specimens

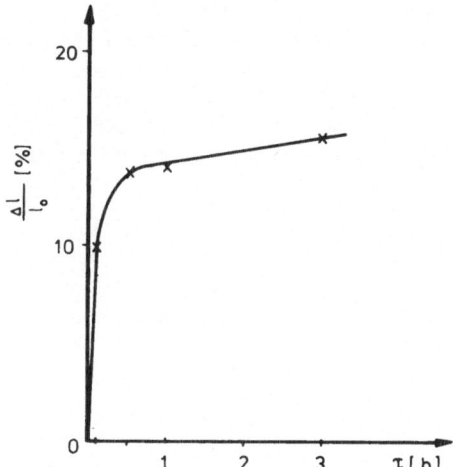

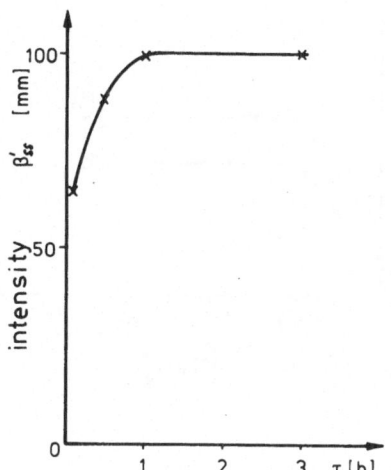

Fig. 10: Firing shrinkage of β-60 compacts sintered at 1950°C for different length of time.

Fig. 11: Relative intensities of β-phase x-ray peaks for specimens sintered at 1950°C for different length of time.

heated at 1950°C for different lengths of time. The lattice parameters of the β-phase in these specimens are given in Fig. 12. The x-ray diffraction traces of these specimens are given in Fig. 13. Note the higher background of the base line at low diffraction angles. This indicates the existence of an amorphous phase. Microstructures of these specimens are shown in Fig. 14. Quantitative metallographic analysis confirmed the amount of crystalline phase present as obtained by the x-ray analysis. From these data, one can estimate the composition of the glass. For instance, the 2 minute heat treated sample has a lattice parameter corresponding to $x = 2.18$. Using the lever rule, the composition of the glass is very close to that of X_4 on the phase diagram as shown in Fig. 1.

As shown in Fig. 10, densification was observed after the reaction was completed. One may suggest that solution-reprecipitation occurred during sintering. The characteristics of the curve at earlier time suggested a rearrangement of the solid particles in the presence of liquid.

Grain growth was observed in these samples as shown in Fig. 14. The crystals in these mixtures are $\beta\text{-Si}_3\text{N}_4$ solid solution, therefore, one can conclude that nitrogen species diffused through the liquid, hence the glass should be a nitride containing glass. As shown in Fig. 12, the lattice parameter of

412

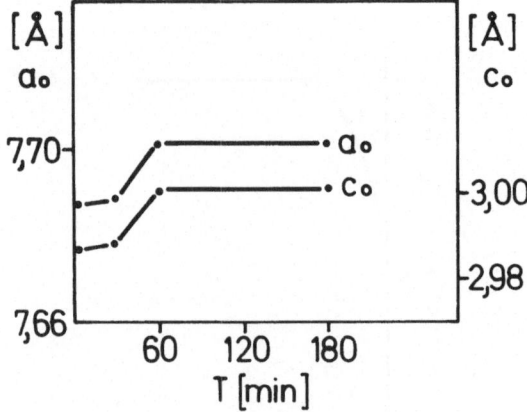

Fig. 12: Unit cell dimensions of the $\beta-Si_3N_4$ solid solution formed during sintering of $\beta-60$ composition at 1950°C for different length of time.

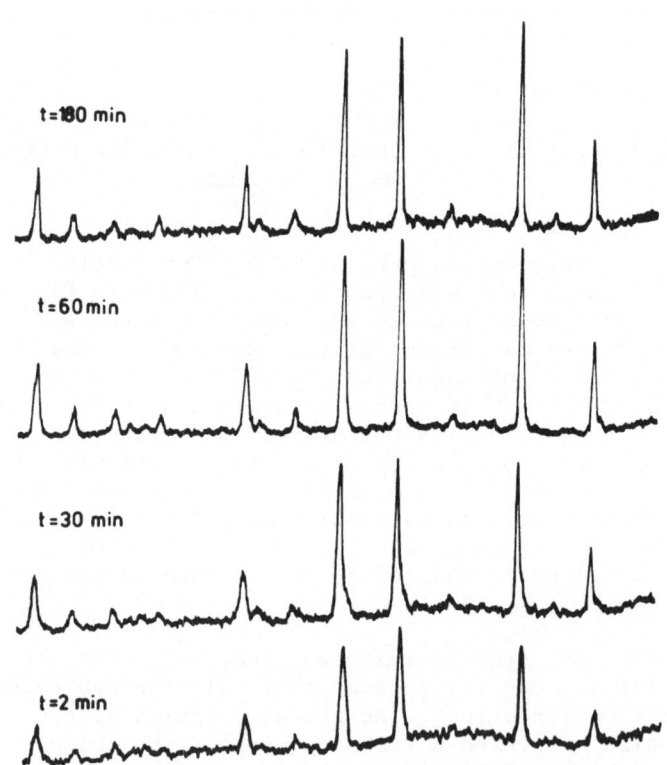

Fig. 13: X-ray diffraction patterns of $\beta-60$ composition fired at 1950°C for different length of time.

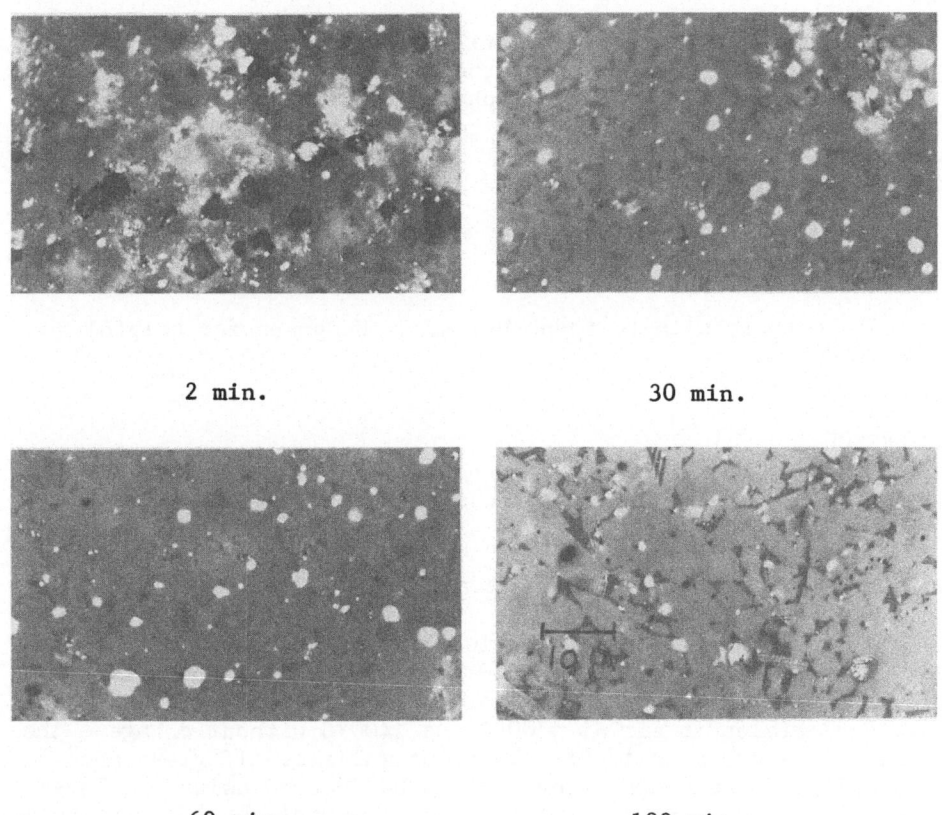

2 min.   30 min.

60 min.   180 min.

Fig. 14: Photomicrograph of specimens sintered at 1950°C for different lengths of time.

the β-phase changed with time. These data indicated the homogenization of the solid solutions containing different amounts of aluminum occurred.

Glass containing specimens (1950°C for 2 minutes) were heat treated at 1300°C for 8 hours. The x-ray patterns of these heat treated samples revealed the presence of the crystalline phases mullite, Al_2O_3, X_4, X_2 and an unknown phase. The intensity of β-phase was greatly reduced after heat treatment. These results indicate that the isothermal section at 1760°C is different from 1300°C.

4. CONCLUSION

It appears that there are still discrepancies in the Si,Al/N,O phase diagram. The location of the unknown phase in the diagram might be in the region of X_1, β-phase and Si_2ON_2. It is possible that there are two glass forming regions in this system, one being located at the Si rich side of X_4 and the other close to the composition of X_1.

ACKNOWLEDGEMENT

The authors wish to thank Dr. W. J. Huppmann for helpful discussion.

REFERENCE

1. L. J. Gauckler, H. L. Lukas and G. Petzow, J. Am Ceram. Soc. 58, 346 (1975).

DISCUSSION (Cannon, Lange, Moulson)

Jack: Your observations are quite remarkable; you find that a N-glass and β'-sialon react at 1900°C to give a β'-sialon with higher z value, yet at 1300°C there is no β'-sialon! We have heated β'-sialon in air at 1300°C and 1400°C without change - the surface develops a protective layer of mullite. If glass and crystalline phase compositions are to be deduced using the lever rule, it is important to come to some agreement about our respective behaviour diagrams and the phase designations. Your X_4 is our 8H, and X_2 is 15R, and both are in different positions on our two diagrams.

Thompson: Dr. Gauckler and Professor Tien's results on the sintering of β'-sialon are most interesting. Taken at face value they suggest that the minimum solidus temperature along the Si_3N_4-Al_3O_3N join occurs at some temperature below 1950°C and at 1950°C, the x=4 composition lies in between the solidus and liquidus curves. If this is indeed the case, the preparation of β'-sialon by melting is very attractive. The devitrification of the β'-sialon glass into mullite, X_4, X_2, and alumina at 1300°C occurs because the extent of β' solid solution at this temperature is less than the x=2 composition. It is important to confirm that this behaviour is characteristic of both crystalline and vitreous β'-sialon because, if so, the use of high x materials could be seriously restricted. It is also important to know at what x value β'-sialon melts rather than decomposes. If this occurs below the x=2 composition, then β'-sialons should be capable of being made from a melt and at the same time should not break down into other crystalline phases on further heat treatment.

Section G

MICROSTRUCTURE

GRAIN BOUNDARY ENGINEERING AND CONTROL IN NITROGEN CERAMICS

R. Nathan Katz and G.E. Gazza

Army Materials and Mechanics Research Center
Watertown, MA   02172 (USA)

ABSTRACT

   The deliberate selection and control of composition, structure, and processes occurring in the grain boundary both during fabrication and post fabrication treatment, with the aim of specific property modification, may properly be termed "grain boundary engineering". A review of the early developments in "grain boundary engineering" for ionic ceramics will be presented. The development of silicon nitride as a high performance engineering ceramic will be reviewed with emphasis on the central role of grain boundary engineering. In this review, various techniques of grain boundary control will be highlighted on the basis of the current state-of-the-art of grain boundary engineering in nitrogen ceramics. Needs for the development of improved characterization techniques for the grain boundary will be discussed. Opportunities for additional improvements in nitrogen ceramics will be considered.

GRAIN BOUNDARY ENGINEERING

   The deliberate selection and control of composition, structure, and processes occurring in the grain boundary, both during fabrication and post fabrication treatment, aimed at specific property modification may properly be termed grain boundary engineering (GBE). The concept of GBE in high performance, ceramics can be traced to Coble's development of Lucalox [1]. In this case starting powder, atmosphere and grain growth inhibitors were deliberately chosen to control grain growth, and pore migration during sintering. The result was a fully dense, translucent, cold pressed and sintered alumina body. This represented a major technical breakthrough in the area of ceramic fabrication, which lead directly to new applications for Al_2O_3.

A more recent example of grain boundary engineering is the development of "high strength" polycrystalline KCl windows with very low levels of optical absorbtion. Here the problem was to obtain strengths in excess of that available in single crystals and yet not degrade the optical properties by grain boundary associated pores, impurities and related effects. The latter condition precluded traditional sintering or hot pressing. The solution to the problem has been the deliberate addition of grain boundaries into single crystals, thereby producing optically "clean" grain boundaries. The creation of these optically "clean" grain boundaries involves press forging and a subsequent recrystal- lization anneal, [2-4] a process illustrated in Figure 1. Recent work [5] has demonstrated that press forging a single crystal of a KCl-KBr alloy significantly increased strength with very little effect on the absorption coefficient. The strength in one case for a press forged polycrystal was 5800 psi compared to 1400 psi for the single crystal, whereas, the absorption coefficient was only 3.4×10^{-3} cm^{-1} for the polycrystalline material compared to 3.14×10^{-3} cm^{-1} for the single crystal (at 10.6μm).

As the above examples illustrate the concept of GBE provides a useful perspective from which a program to improve materials properties (and, hence, materials performance) may be viewed. This is especially true in the case of nitrogen ceramics, and silicon nitride in particular. The production of fully dense silicon nitride requires the use of densification aids which have an impact on the processing, the room temperature and the high temperature properties of the material. Thus the development of high density, high strength, high temperature silicon nitride will be reviewed from the perspective of grain boundary engineering. While tracing these developments, various techniques and strategies of grain boundary control and modification will be highlighted. These may serve to assist in the further development of silicon nitride, or other similar covalent ceramics.

PRESS FORGING OF SINGLE CRYSTALS

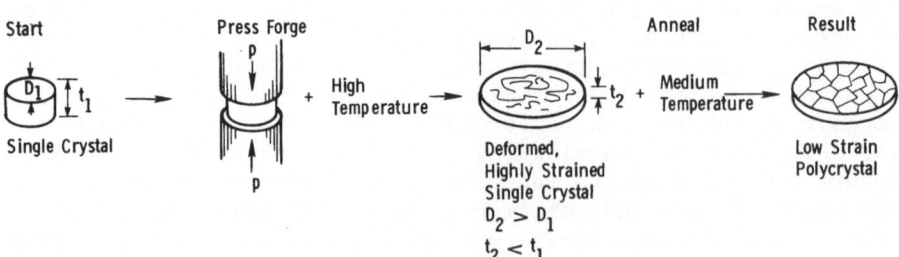

Figure 1   Press Forging of Single Crystals

BACKGROUND

The importance of grain boundaries to sintering and their influence on resultant material properties has been well recognized and documented in the literature for ionic compounds [1,6,7,8]. Early work on producing theoretically dense and, at times, optically transparent ionic ceramics demonstrated that control and/or manipulation of grain boundary chemistry was essential to produce desired results. Emphasis on the quality and character of starting powder reduced undesirable residual cation and anion impurities which often concentrated at boundaries. As finer particle size powders were produced to increase surface energy as a driving force for sintering, greater difficulty was encountered in eliminating residual impurity anions remaining from the powder precursors. Often, dopants were added to the primary material to be sintered to restrict boundary mobility, provide a high vacancy flux and prevent discontinuous grain growth. Grain boundary mobility was restricted by the presence of a dispersed second phase or by the formation of near boundary solid solutions. Ceramic systems considered as examples of such experimental studies are Al_2O_3 with MgO dopant [1,9], Y_2O_3 with ThO_2 dopant [10], and ThO_2 with either CaO or Y_2O_3 dopants [11].

Further work on producing fully dense ceramics were conducted with MgO using LiF additive [12, 13] and $MgAl_2O_4$ using a combined Li_2O-SiO_2 addition [14] or CaO addition [15]. Generally, where a liquid phase was formed at the grain boundaries, densification by the solution-precipitation mechanism was predominant. However, when a fugitive type addition was used, further heat treatment subsequent to densification might be required to eliminate residual additive and produce optimum properties of higher strength or optical transparency.

Controlling boundary mobility and chemistry by the mechanisms cited was principally successful with ionic solids, i.e., oxides, where sufficiently high diffusion coefficients could be obtained at temperatures where the materials were stable and where sufficient information was available on phase equilibria. In addition some direct characterization of the grain boundaries in ionic solids now exists, which confirms many of the assumptions used to control grain boundaries in these materials [6].

Sintering covalently bonded ceramics is more difficult because, in general, they require much higher sintering temperatures where problems with compositional instability exist. Also, at these high temperatures, exaggerated grain growth may occur to limit the densification process. Diffusion data is limited or non-existant, and only inferential characterization of the grain boundaries are available for most materials of interest. It is to this more difficult area of grain boundary engineering and control for covalently bonded ceramics, in particular silicon nitride, that we now direct our attention.

GBE AND THE OPTIMIZATION OF HOT PRESSED Si_3N_4

Since the late 1950's, considerable interest has grown in the development of Si_3N_4 and SiC materials which have high potential as component materials in various propulsion and power generation systems [16]. State-of-the-art high strength, hot pressed Si_3N_4 has resulted from a series of developments involving: empirical selection of densification aids, control of high temperature strength limiting reaction products in the grain boundary via a variety of strategies, and selection of new families of densification aids specifically chosen to yield grain boundary phases with improved properties. Thus, the development of hot pressed Si_3N_4 provides a continuing case study of grain boundary engineering. In contrast to materials improvements in many other ceramic systems, essentially all of the improvements in hot pressed Si_3N_4 as an engineering ceramic have resulted from grain boundary phase manipulations. These developments will now be reviewed.

MgO: Its Role and Importance

Since Si_3N_4 dissociates before significant sintering can occur, an additive was required which would permit densification at temperatures below which significant dissociation occurs. Early attempts to obtain a densification aid which would produce fully dense, high strength, hot pressed Si_3N_4 were successfully carried out by Deeley, et al [17] who demonstrated that MgO additions gave superior results over the many other additives considered. Although x-ray diffraction analysis of the hot pressed samples indicated only Si_3N_4 present, primarily beta phase, it was assumed that the MgO additive reacted with the surface silica on the Si_3N_4 particles to form a vitreous glass which produced a continuous bonding boundary phase. Further optimization of the system was carried out by Lumby and Coe [18], who showed the significance of using high alpha phase Si_3N_4 powder as starting material and the strength dependence on hot pressing time. This work implied the importance of the $\alpha \rightarrow \beta$ transformation in developing high strength material, as later rationalized by Lange [19]. Further attempts to define the role of MgO during densification were carried out by Evans and Sharp [20] who confirmed the existence of a glass phase at triple points by transmission electron microscopy. However, a similar uniformly distributed grain boundary phase was only inferred. Indirect evidence for a non-crystalline grain boundary phase of $MgSiO_3$ composition was obtained by Wild, et al [21] from annealing experiments on hot pressed Si_3N_4 containing 10% MgO. Visual evidence for glassy phases existing in high MgO content hot pressed Si_3N_4 by a transmission electron microscope study were also recently obtained by Drew and Lewis [22].

This study also confirmed the role of this glassy phase in the
$\alpha \rightarrow \beta$ transformation, thought necessary for the attainment of high
strength Si_3N_4.

In attempting to define the composition of the observed glassy
phase, studies with Auger spectroscopy and electron probe microanalysis were performed [23,24] and results suggested the
existence of various grain boundary glass compositions, i.e.,
$xSiO_2 \cdot yCaO \cdot zMgO$. These analyses also indicated that a proportion
of the smaller cations, (i.e., Mg, Al) have diffused into the
Si_3N_4 structure during hot pressing.

While hot pressed Si_3N_4 produced in this way had properties
which where encouraging for use in high temperature applications,
e.g., gas turbines, indepth studies of creep and high temperature
modulus of rupture (MOR) at ˜2200 F indicated that, in contrast to
reaction bonded Si_3N_4, hot pressed material exhibited a significant
fall off in properties, as shown in Figure 2. Much inferential
evidence from the results of Auger spectroscopy, TEM, creep
behavior, etc. pointed to the glassy grain boundary phase being
responsible for this behavior.

Studies [23,25] initiated to determine the influence of
residual impurities in the starting materials on the high temperature properties of the boundary phase have shown that the viscosity
of the boundary phase is purity dependent. In particular, the Ca
cation was found to be detrimental to high temperature properties
of Si_3N_4 which exhibited increased sub-critical crack growth [25]
and decreased creep resistance. [23] Evidence to date suggests
that this is due to the presence of the boundary phase which becomes viscous at high temperature and allows grain boundary
sliding.

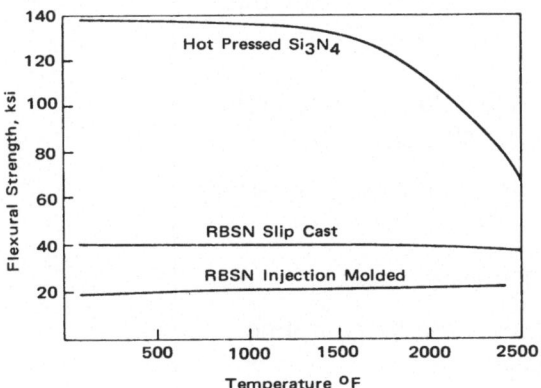

Figure 2 Flexural Strength of HPSN Compared to RBSN.

To summarize the above, the importance of MgO as an additive to hot pressed Si_3N_4 was that it permitted the attainment of full density and possibly provided the medium through which an $\alpha \rightarrow \beta$ solution-reprecipitation transformation could occur, with concomitant high RT strength. The principle drawback of the conventional use of MgO as an additive was the limitation it imposed on high temperature strength and creep behavior.

At this point the central problem faced by ceramists was how to find a way to create a more refractory grain boundary phase (GBP) or to eliminate it. In spite of the experimental difficulties in direct characterization of this grain boundary phase (namely that the small volume percent of this "amorphous" phase in a crystalline matrix is not amenable to x-ray diffraction methods and its small dimension rules out microprobe and similar direct techniques) and having to rely on inferential results, significant improvements in hot pressed Si_3N_4 have been made by applying the GBE strategies shown in Figure 3.

Increasing the Purity of the Grain Boundary Glass Phase

This approach was initially pursued by Kossowsky [23] and Lange [25] at Westinghouse and Richerson [26] at the Norton Company. These investigators elucidated the detrimental role of Ca, as well as other alkali and alkaline earth elements, on the high temperature strength and creep resistance of the magnesium silicate grain boundary phase. Efforts to produce higher purity Si_3N_4 powders were pursued in attempting to minimize the residual detrimental impurity content. This produced significant improvement in high temperature properties but the properties of the specific $MgO-Si_3N_4$ reaction products formed, i.e., $MgSiO_3$ or Mg-Si-O-N, then became

---

STRATEGIES TO INCREASE THE HIGH-TEMPERATURE
BEHAVIOR OF HOT-PRESSED Si_3N_4

1. Reduce Ca, Na, etc., impurities to make the grain boundary $MgSiO_3$ glass more refractory.

2. Develop a densification aid to yield a more refractory glass than $MgSiO_3$.

3. Develop a nonglass grain boundary.

4. Eliminate the grain boundary phase by promoting sintering via volume diffusion.

Figure 3   Strategies to Increase the High-Temperature Behavior of Hot-Pressed Si_3N_4

limiting on strength and creep resistance. It thus became apparent
that significant increases in the high temperature properties of
hot pressed Si_3N_4 would require other approaches.

Densification Aids other than MgO

Another approach for increasing high temperature properties
would be to induce the formation of more refractory reaction
products between the additive, and the SiO_2 layer on the Si_3N_4, to
form the boundary phase. Gazza, [27,28] explored this approach
and found that yttria was an effective additive for this purpose.
(Figure 4) It was reasoned that the probable formation of $xY_2O_3 \cdot$
$ySiO_2$ glasses and compounds would be more refractory than $xMgO \cdot$
$ySiO_2$ glasses. Further, that the large ionic radius of the Y
cation would tend to keep it in the boundary vicinity rather than
diffusing away into the Si_3N_4 grains. Also, it might be possible
to crystallize these "glassy" phases. At low Y_2O_3 additive levels,
<5 w/o, these glasses appear to form and some improvements in high
temperature properties were observed. However, it was found that
maximum strengths were obtained with 10-15 w/o Y_2O_3 (Figure 5),
more than required to completely react with the surface silica
on Si_3N_4 powder, and a more complex role of Y_2O_3 additions to Si_3N_4
was explained by work at Newcastle [29]. These studies have
shown that the refractory nature of the Si_3N_4-Y_2O_3 reaction
products are due to the formation of $xSi_3N_4 \cdot yY_2O_3 \cdot zSiO_2$ compounds
which may further react with Si_3N_4 to produce $Si_3N_4 \cdot Y_2O_3$ composition. In addition to producing high melting point refractory
phases, residual impurities such as Ca, Al, and Mg are accommodated
into solid solution by these compounds thus limiting their detrimental effect. Similar reaction products were reported by Tsuge
et al [30] in concurrent work. Although enhanced properties have
been demonstrated for the Si_3N_4-Y_2O_3 additive system, a problem
with anomalous strength degradation at approximately 1000 C has
been reported and further studied by several investigators [31,
32,33]. Lange [31] found that this behavior is related to the
lack of oxidation resistance of certain Si_3N_4-Y_2O_3 reaction phases
at 1000 C and severe cracking of specimens containing these phases
occurs. Brennan [32] is investigating Si_3N_4-15 w/o Y_2O_3 materials
using high and low purity Si_3N_4 starting powders, and thusfar,
finds the anomalous 1000 C behavior to be purity dependent. Tsuge,
Nishida and Komatsu [33] found that for their particular Si_3N_4-Y_2O_3
additive (~5%) system, strength problems at intermediate temperatures could be overcome by crystallizing the glass phase formed by
the additive reaction with Si_3N_4.

Other additives, such as CeO_2 [34,35] and ZrO_2 [36,37], have
also been studied to determine their effect on pressure sintering of
Si_3N_4 and potential for enhancing high temperature properties
of Si_3N_4.

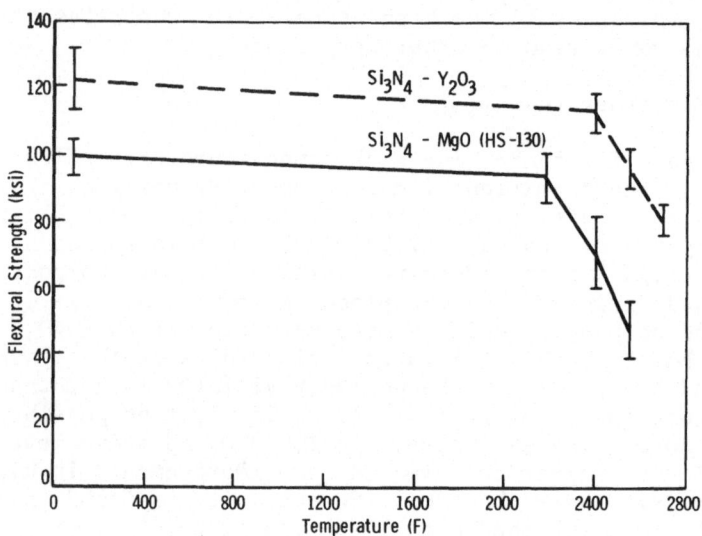

Figure 4   Strength Versus Temperature For Si_3N_4 + Y_2O_3 Versus Si_3N_4 + MgO

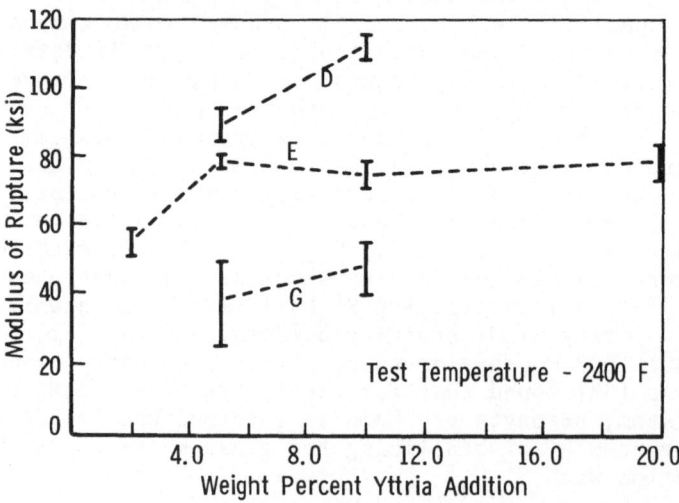

Figure 5   Effect of Yttria Addition Level on High-Temperature Strength of Hot-Pressed Silicon Nitride Formed From Various Powders

A study [38] on compounds formed by reaction of lanthanide compounds (La$_2$O$_3$, SM$_2$O$_3$, Dy$_2$O$_3$, Er$_2$O$_3$, and Yb$_2$O$_3$) with Si$_3$N$_4$ suggests that most will react to form oxynitrides which are isomorphous with silicates or oxyfluorides such as akermanite, cuspidine, or apatite compounds.

The use of the various additives mentioned to promote refractory grain boundary phases have been shown to react with Si$_3$N$_4$ in an analogous manner as "Sialons", phases produced in the silicon-aluminum-oxygen-nitrogen system, which were discovered independently in England [39] and Japan [40].

Direct Formation of Crystalline Grain Boundary Phases

The above section demonstrates the effectiveness of using additives which form a glassy GBP which eventually reacts with Si$_3$N$_4$ to form a crystalline GBP. The question naturally arises, "can one produce a crystalline GBP directly, without going through the glass transformation process?". The authors are aware of one study, by Wright and Niesz [41], which addresses this question. These authors used high β-Si$_3$N$_4$ powders, and removed much of the surface SiO$_2$ from them, by reaction to volatile SiO, under a N$_2$ partial pressure. A dual additive of AlN and Al$_2$O$_3$ which would produce a grain boundary "Sialon" was used. The results were interesting. Essentially full density was obtained, but the strengths were low, ca. 40 KSI (~280 MN/m^2). The crystalline nature of the grain boundary yielded an increasing as opposed to decreasing strength vs. temperature behavior, as well as a several order of magnitude reduction in creep rate compared to commercial Si$_3$N$_4$ (HS-130). Unfortunately, further work in this interesting area was not pursued.

Promotion of Volume Diffusion

Thusfar, additions used for enhancing the sinterability of Si$_3$N$_4$ have resulted in the formation of separate boundary phases which have controlled properties. A further approach to sintering, successfully used by Prochaska [42] with SiC, would be to use additions which promote greater degrees of volume diffusion, remove inhibiting species from the system by volatilization, form solid solutions and produce "clean" grain boundaries. While various teams are pursuing research on Si$_3$N$_4$ using this concept, no one has yet achieved the goal. The achievement, however, may result in differences in microstructural morphology and fracture mode of Si$_3$N$_4$ which may benefit some properties and be detrimental to others.

Additive Distribution

An important practical consideration in the implementation of any GBE concept for improved hot pressed or sintered Si$_3$N$_4$, is the technique of additive distribution. Generally, mechanical mixing of additive and Si_3N_4 is accomplished by ball milling. Problems associated with this technique are some pickup of impurity from the milling media and lack of optimum homogeneity in additive distribution. Although bulk compositions determined are near desired stoichiometry, localized areas within the product probably contain a compositional range. Such compositional variation might result in increased crack growth rates or inadvertant multiphase materials where separate phases are produced by reaction of Si_3N_4 with heterogeneously distributed additive under non-equilibrium processing conditions or by initial reactions which form volatile phases where some loss of component precludes reaching the desired end stoichiometry and equilibria. It is suspected that problems with synthesizing single phase "Sialons" can be traced to the latter problem. Additionally, if more than one additive is used, it may be preferable to pre-react or pre-mix additives by techniques promoting high degrees of compositional homogeneity before mixing with the Si_3N_4 powder.

Further studies show the need for optimum additive distribution as it has been reported [43] that the MgO/SiO_2 molar ratio in hot pressed Si_3N_4 is significant to maximize high temperature strength by producing possible chemistry changes in the grain boundary phase.

GRAIN BOUNDARY ENGINEERING IN REACTION BONDED SILICON NITRIDE

The concepts of GBE are not limited in applicability only to fully dense materials. Mangels [44] has used several of the concepts discussed above to improve the creep resistance of reaction bonded silicon nitride (RBSN). While one does not intentionally create a GBP in RBSN, as is done to densify hot pressed Si_3N_4, such a phase exists as a result of naturally occurring SiO_2 on the Si precursor powder and the Ca and Al in this Si powder. By choosing low Ca and Al precursor Si powder Mangles was able to reduce the creep of RBSN at 2200 F and 10 KSI. Further, by selecting nitriding schedules to increase the amount of edge to edge grain contact, as opposed to point to edge grain contact, this reduction in creep rate was made to exceed two orders of magnitude.

FUTURE RESEARCH OPPORTUNITIES

At this point in time it must be stated that GBE in nitrogen ceramics is more of an art than a science. The movement of an engineering art to an engineering science always creates research opportunities. Several of the more crucial of these are listed below:

a. Improved Techniques for Characterization of the Grain Boundary: At present research on improved silicon nitride and "Sialons" is limited by a lack of quantitative information on grain boundary phases, particularly amorphous phases.

b. Availability of Diffusion Data: At present there is practically no data on the diffusion of anything in Si_3N_4 or the various grain boundary reaction phases. The lack of single crystals has been a problem here, but the availability of CVD Si_3N_4 or Si_3N_4 in large fiber form may help overcome this problem.

c. Paucity of Phase Equilibria Data: Although much work is in progress, now, in the US, UK, and West Germany on phase equilibria in Si-M-O-N systems, much more will be required.

All of these areas are important to the further development of hot pressed, sintered, and reaction bonded nitrogen ceramics. It is also important that more input from actual systems evaluations of materials limitations on performance be fed into the basic research process, so that the limited research resources can be focused upon critical needs.

SUMMARY

Grain boundary engineering has proven to be a useful conceptual tool in advancing the technology of nitrogen ceramics. The tremendous improvements in high strength behavior of hot pressed silicon nitride achieved in the past five years (Figure 6) attest to this. Future research goals such as the attainment of pressure-less sintered, fully dense, high temperature capability silicon nitride and, "Sialons" will require even more application of GBE. In addition, emphasis on utilizing phase equilibria or behavior diagrams to produce tailored Si-M-O-N systems should benefit from information generated by previous GBE approaches.

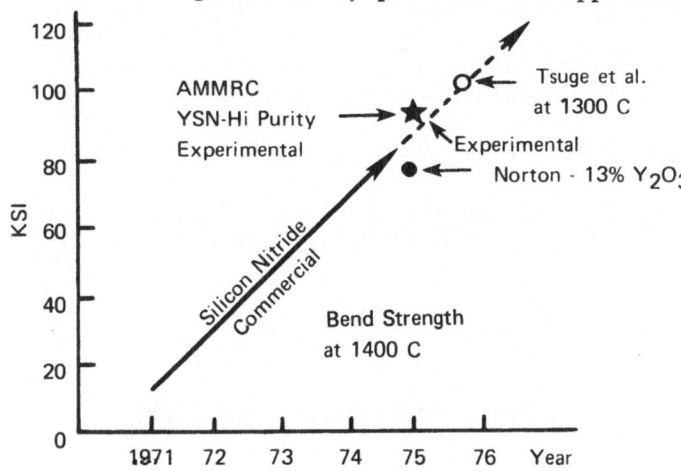

Figure 6 Increase in 4 pt MOR or Hot Pressed Si_3N_4 at 1400 C as a Function of Time Since the Inception of the ARPA "Brittle Materials Design" Program.

428

REFERENCES

1. Coble, R.L., Jour. App. Phys., 32, [5], (1961), pp. 787-99.

2. Spriggs, R.M., in Strengthening Mechanisms in Metals and Ceramics, Syracuse Univ. Press, NY, (1966), p. 181.

3. Rice, R.W., J. Amer. Ceram. Soc., 55, (1972), p. 90.

4. Rice, R.W., in Ultrafine Grain Ceramics, Syracuse Univ. Press, NY, (1970), p. 203.

5. Harrison, W.B., AFML-TR-75-109, July 1975, p. 210.

6. Kingery, W.D., Jour. Amer. Ceram. Soc., 57, [1] (Jan. 1974), pp. 1-8; [2] (Feb. 1974), pp. 74-83.

7. Burke, J.E., Jour. Amer. Ceram. Soc., 40, (3), (1957), pp. 80-85.

8. VanBueren, H.G. and Hornstra, J., in Reactivity of Solids, edited by J.H. deBoer, W.G. Burgers, E.W. Gorter, J.P.F. Huesse, and G.C.A. Schmit (Elsevier Pub. Co., 1961)

9. Jorgensen, P.J., and Westbrook, H.H., Jour. Amer. Ceram. Soc., 47, (1964), p. 332.

10. Jorgensen, P.J., and Anderson, R.C., Jour. Amer. Ceram. Soc., 50, (11), (1967), p. 553.

11. Anderson, R.C., U.S. Patent 3,574,645, Apr. 1971.

12. Rice, R.W., Bull. Amer. Ceram. Soc., 41, (1962), p. 271.

13. Atlas, L.M., Jour. Amer. Ceram. Soc., 40, (1957), p. 196.

14. Gatti, A., Mehan, R.L. and Noone, M.J., Naval Air Systems Comm. Rpt., Contract No. N00019-71-C-0126, Dec. 1971.

15. Bratton, R.J., Jour. Amer. Ceram. Soc., 54, (1971), p. 141.

16. Ceramics for High Performance Applications, ed. J.J. Burke, A.E. Gorum, and R.N. Katz, Brook Hill Pub., Chestnut Hill, MA (1974)

17. Deeley, G.G., Herbert, J.M. and Moore, N.C., Powder Met 8, (1961), p. 145.

18. Lumby, R.J. and Coe, R.F., Proc. Brit. Ceram. Soc., 15, (1970), p. 91.

19. Lange, F.F., Jour. Amer. Ceram. Soc., 56, (10), (1973), p. 518.

20. Evans, A.G. and Sharp, J.V., Jour. Mat. Sci., 6, (1971) p. 1292.

21. Wild, S., Grieveson, P., Jack, K.H., and Latimer, M., in Special Ceramics 5, ed. P. Popper, Brit. Ceram. Res. Assoc. (1972), p. 377.

22. Drew, P. and Lewis, M.H., Jour. Mat. Sci., 9, (1974), p. 261.

23. Kossowsky, R., Jour. Mat. Sci. 8, (1973), p. 1603.

24. Powell, B.D. and Drew, P., Jour. Mat. Sci., 9, (1974) p. 1867.

25. Lange, F.F. and Iskoe, J.L., Chapter 11, p. 223 ff in ref. 16.

26. Richerson, D.W., Amer. Ceram. Soc., Bull., 52, (7), (1973), p. 560.

27. Gazza, G.E., Jour. Amer. Ceram. Soc., 56, (12), (1973), p. 662.

28. Gazza, G.E., Amer. Ceram. Soc., Bull., 54, (9), (1975) p. 778.

29. Rae, A.W.J.M., Thompson, D.P., Pipkin, N.J. and Jack, K.H., in Special Ceramics 6, ed. P. Popper, Brit. Ceram. Res. Assoc., 1976.

30. Tsuge, A., Kudo, H. and Komeya, K., Jour. Amer. Ceram. Soc., 57, (6), (1974), p. 269.

31. Lange, F.F., Singhal, S.C., and Kuznicki, R.C., Westinghouse Elec. Corp. Tech. Rpt. 6, Contract No. N0014-74-C-0284, Apr. 1976.

32. Brennan, J.J., United Tech. Research Center Tech Rpt. R75-912081-2, Contract N62269-75-C-0137, Sept. 1975.

33. Tsuge, A., Nishida, K., and Komatsu, M., Jour. Amer. Ceram. Soc., 58, (7-8), (1975), p. 323.

34. Huseby, I.C., and Petzow, G., Powder Met. Int., 6, (1), (1974), p. 17.

35. Mazdiyasni, K.S. and Cooke, C.M., J. Amer. Ceram. Soc., 57, (12), (Dec. 1974), p. 536.

36. Rice, R.W., and McDonough, W.J., Amer. Ceram. Soc. Bull. 54, (8), (1975), p. 753.

430

37. Vasilos, T., AVCO Corp., Lowell, MA. personal communication.

38. Wills, R.R., Systems Research Laboratories, Inc. Quarterly Prog. Rpt. 3165-4, Sept. Nov. 1975, prepared for AFML, Processing and High Temp. Matls Br., Wright Patterson AFB, OH.

39. Oyama, Y. and Kamigaito, O., Japan Jour. App. Phys., 10, (1971), p. 1637.

40. Jack, K.H. and Wilson, W.I., Nature Phys. Sci. (London), 238, (1972), p. 28.

41. Wright, T.R. and Niesz, D., Battelle Columbus Laboratories Tech Rpt No. CR-134690, Oct. 11, 1974, NASA Contract NAS3-17766

42. Prochazka, S., Chapter 12, p. 239 ff in ref. 16.

43. Andersson, C.A., Lange, F.F. and Iskoe, J.L., Westinghouse Elec. Corp. Tech. Rpt. 3, Oct. 15, 1975, Contract No. N00014-74-C-0284, (ARPA)

44. Mangels, J., Chapter 9, p. 195 ff in ref. 16.

DISCUSSION

Brook: In view of the fact that 100 ppm MgO can be homogeneously distributed in Al_2O_3 by suitable methods, I wonder why inhomogeneities occur in Si_3N_4 with MgO additions at the few percent level. Do you see this as a thermodynamic problem involving wetting angles, or as a kinetic problem associated with inadequate initial mixing?

Katz: While one cannot rule out thermodynamic factors, I think that the kinetic ones dominate, especially since the liquid or glass phase may be going through a complex series of reactions to second and third phases, which might involve the main Si_3N_4 phase.

Cannon: As a result of studies on liquid phase sintering in CaF_2-NaF we find that the concentration of liquid phase can vary over wide limits within a particular piece. This is more than just a problem of incomplete mixing, but apparently results from segregation of the liquid to lower the total surface energy. The tendency to segregate is much stronger than we would expect for this system in which the contact angle is very low.

Clarke: I would like to interject a note of caution on interpreting the micrographs obtained in etching the grain boundaries in these materials. In etching of metals we know that the depth and width of the etched region depends on both the local composition and local crystallographic orientations. In addition the etchant is usually very selective to particular boundaries and not others.

Katz: If the phase is truly amorphous then etching should be
uniform. The fact that reaction-bonded silicon nitride, which has
very little glass at the grain boundaries and does not etch, tends
to support the coincidence of etching and presence of glass
phases.

Messier: Dr. Clarke questioned the relationship of the etched
microstructure in hot-pressed silicon nitride to the presence of
a grain boundary phase. It has been observed, however, that
etchants that will etch that material will not affect reaction
sintered silicon nitride, a material relatively free from oxide
impurities. It seems clear therefore that the observed etched
microstructure is consistent with the existence of a grain
boundary phase derived from the oxide sintering aid.

Jack: One method of removing grain boundary phase is similar to
the Battelle technique except that it produces a homogeneous β'-
sialon without removing silica. The silica is compensated by MgO
to produce Mg_2SiO_4 which subsequently reacts with Si_3N_4 and
appropriate additions of Al_2O_3 and AlN to give a single phase β'-
Mg sialon. In other words, balanced additions of MgO, Al_2O_3 and
AlN are made to Si_3N_4. The MgO first forms a liquid phase which
allows densification by liquid phase sintering, and then at
higher temperatures is incorporated into the structure to give a
"dilute" β'-sialon. The principles can be applied to pressureless
sintering as well as hot-pressing.

DIRECT OBSERVATION OF LATTICE PLANES AT GRAIN BOUNDARIES IN SILICON NITRIDE[*]

D. R. Clarke

Department of Materials Science and Engineering, and
Materials and Molecular Research Division, Lawrence
Berkeley Laboratory, University of California,
Berkeley, California 94720, USA

The rather dramatic decrease in the strength of hot pressed silicon nitride above about 1000°C has led investigators to postulate the existence of a glassy phase at the boundaries between the individual silicon nitride grains. The suggestion is that at these high temperatures the viscosity of the glassy phase rapidly decreases, flows and allows the grains of silicon nitride to slide past one another at unexpectedly low stresses.

Although recourse is frequently made to this explanation, the evidence is strictly circumstantial since no direct observation of the glassy phase between the boundaries has been made. Amongst the pertinent pieces of evidence suggestive of a glassy grain boundary phase are the following. The commonest sintering aid, magnesium oxide, used to hot press silicon nitride is known to form a magnesium calcium silicate with the silica invariably present on the surfaces of the silicon nitride powders used and the calcium impurities also present in the powder (1). Non-crystalline regions commensurate in size to the grain size have occasionally been observed by conventional electron microscopy, particularly at triple points (2,3). In addition internal friction measurements show that there is a viscous component present, which would be sufficient to form a glassy boundary phase perhaps 50-1000Å thick (4).

[*] This work has been supported by the National Science Foundation and by the Energy Research and Development Agency, through the Materials and Molecular Research Division of the Lawrence Berkeley Laboratory.

The purpose of this short contribution is to present, for the first time, direct observations of the crystal lattice planes up to, and on either side, of selected grain boundaries in silicon nitride. These observations show that there is no glassy phase at the grain boundaries at the particular boundaries investigated in a 5% MgO hot pressed silicon nitride. The technique used is that of lattice fringe imaging with the transmission electron microscope.

Lattice imaging by transmission electron microscopy enables the structure of a crystalline material to be studied directly at an atomic level. This approach has already proven to be extremely valuable in investigating phase transformations in metallic alloys (e.g. 5,6) and the structure of a variety of minerals (7).

In the following pages lattice images of a number of grain boundaries are presented, but before describing them in detail it is appropriate to mention how they were selected. Essentially this was done at random, since the position of the specimen was shifted under the beam (in the diffraction mode) until one, or two adjacent, grains were found to be oriented such that the

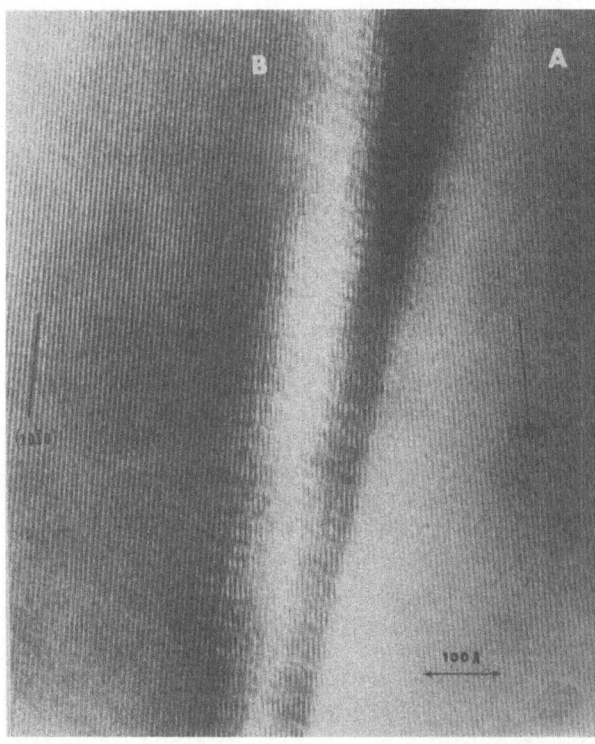

Fig. 1. A low angle (10.5°) twist boundary in silicon nitride. The (10$\bar{1}$0) prism planes on either side of the boundary have been imaged together with the grain boundary Moire fringes.

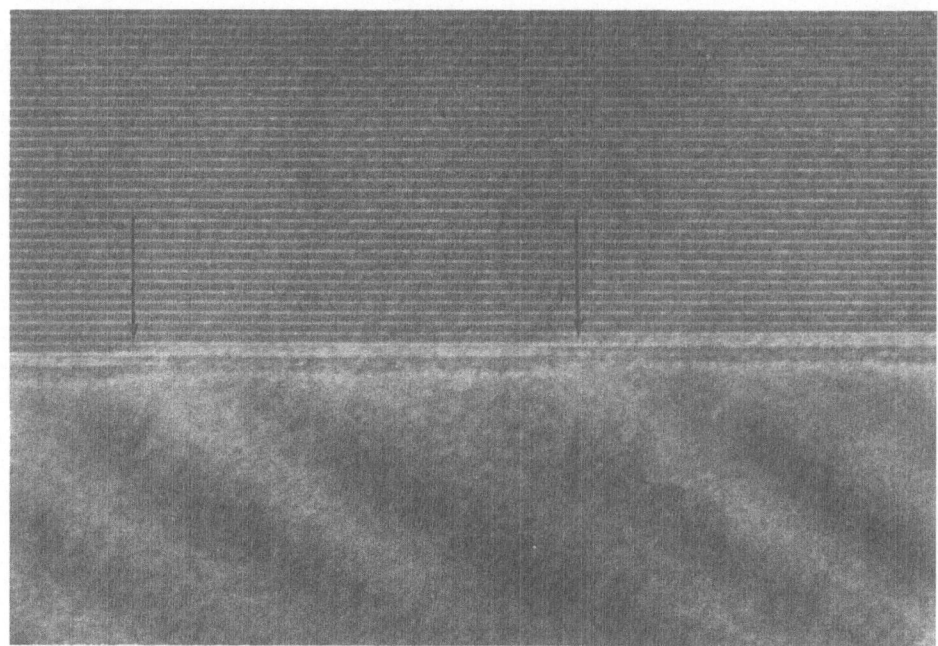

Fig. 2. (10$\bar{1}$0) lattice image showing unit cell high steps in the grain boundary of a β-silicon nitride grain.

lowest index planes were parallel to the electron beam. This method was chosen because it was not possible to tilt the specimens in the microscope inorder to bring a desired orientation into a strongly diffracting condition. The low index planes were chosen for imaging since the observing conditions are marginally less severe, so enhancing the probability of obtaining a lattice image.

Figure 1 is a lattice image of a low angle twist boundary in which the prism (10$\bar{1}$0) planes on either side can be seen. Whilst it was not possible to determine whether the grains are of α or β, the long, waisted appearance of whole B grain suggests that it is typically β silicon nitride. Grain A is also probably β silicon nitride since its fringe spacing is identical to that of grain B. In the boundary three sets of fringes can be discerned; the prism plane fringes extending from either grain and the coarser Moire fringes. Together these demonstrate that there is no amorphous region between the two grains. The characteristics of the Moire fringes serve to confirm this conclusion since their spacing and inclination are precisely those expected from the plane matching theory of crystalline grain boundaries (8).

Figure 2 is of a very small part of a boundary between a silicon nitride grain in which the (10$\bar{1}$0) prism planes can be clearly identified and an unidentified lower grain. Although the

actual planes cannot be distinguished in the area of the lower
grain the presence of the Moire pattern indicates that the grain
boundary must almost certainly be devoid of any amorphous phase.
The remarkable features of this micrograph, however, are the steps
in the grain boundary which are indicated by the black arrows.
These show that the grain boundary consists of a smooth crystal
plane (which we are looking down) with unit cell high steps to
accommodate the shape of the grain. They are also very suggestive
of a ledge mechanism for grain growth during sintering of the
material. In addition, the boundary appears to be formed by a
low index plane indicating that the boundary was determined by
a strongly anisotropic surface energy.

The stepped nature of the grain boundaries is also shown in
the next figure, where the boundary plane is inclined to the
(0001) planes in an α silicon nitride grain. (The irregular,
stepped profile is most clearly seen by sighting along the
boundary). In this figure the plane of the boundary is slightly
tilted with respect to the electron beam, giving rise to the
bright region at the boundary.

The full page micrograph, figure 4, is of a high angle
boundary between two α silicon nitride grains, looking almost
straight down the boundary plane. The fringes can be followed
right up to the boundary from either side demonstrating to a high
degree of confidence the absence of any intergranular amorphous

Fig. 3. Irregularly
stepped grain boundary
at a α silicon nitride
grain revealed from
the (0001) lattice
image.

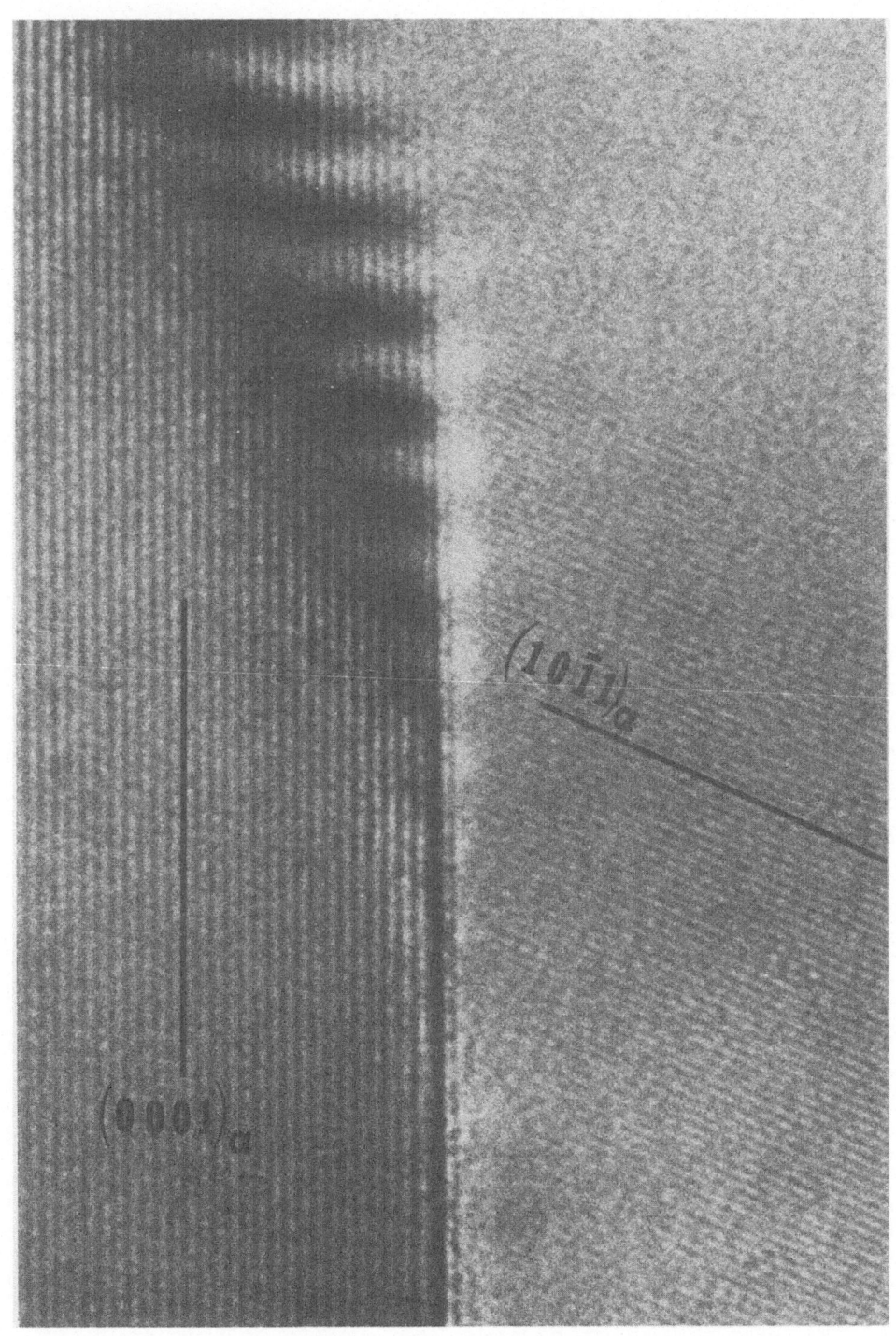

438

layer. The details of the Moire pattern once again confirm this interpretation. A striking feature of the micrograph is the smoothness and straightness of the boundary plane, again suggesting a very anisotropic surface energy. In this micrograph, as well as in the previous three, it is noticeable that the fringe spacing does not measureably alter in the vicinity of the grain boundaries suggesting that the unit cell and the composition do not change appreciably right up to the atomic planes forming the boundary.

A number of important conclusions concerning an amorphous inter-granular phase can be drawn from these observations. Firstly, there is no such phase at the particular grain boundaries studied, to a resolution of one unit cell. Secondly, the amorphous material detected by indirect methods must be inhomogeneously distributed throughout the material. Taken with the clear evidence from Auger analysis that fracture surfaces contain a glassy layer, this conclusion suggests that silicon nitride fractures intergranularly between only those grains which are separated by an amorphous region. A failure model in which the material is taken to be a composite of elastically deformable grains glued together in a softer viscous matrix would thus be most appropriate. Thirdly, the fact that the fringe spacings do not alter in the vicinity of the boundaries is striking confirmation that if silicon nitride is capable of taking into solid solution those elements not forming a glassy phase it does so by forming a uniform solid solution.

I am grateful to Dr. R. N. Katz of the Army Materials and Mechanics Research Center for the supply of the silicon nitride material.

REFERENCES

1. B. D. Powell and P. Drew, Journal of Materials Science, 9, 1867, 1974.
2. R. Kossowsky, Journal of Materials Science, 8, 1603, 1973.
3. K. Nuttall and D. P. Thompson, Journal of Materials Science, 9, 850, 1974.
4. D. R. Mosher, R. Raj and R. Kossowsky, Journal of Materials Science, 11, 49, 1976.
5. R. Sinclair, K. Schneider and G. Thomas, Acta. Met., 23, 873, 1975.
6. R. Gronsky, M. Okada, R. Sinclair, 33rd Annual Proc. EMSA, 1975.
7. Electron Microscopy in Mineralogy, edited H. R. Wenk, Springer-Verlag, 1976.
8. P. H. Pumphrey, Scripta Met., 6, 107, 1972.

DISCUSSION

Brook: I suspect that hot-pressed silicon nitride contains boundaries of at least two types, namely those between the crystallites of the silicon nitride particles and those between the particles themselves; of these, the first are probably clean whereas the second, from diverse kinetic measurements, are most probably associated with a second phase. Has anyone examined the structure of the original powder grains prior to hot pressing? If the grains are composed of a small number of crystallites, it would be difficult to draw conclusions about the 'average' boundary from particular examples in a hot-pressed microstructure.

Clarke: Not that I know of, and certainly not at this atomic level of resolution. In this particular material we know that the grain size measured in the hot-pressed microstructure is close to that of the starting powder size.

Jack: Your results are most impressive, but are the conditions for lattice imaging so rigorous that they automatically select only those regions which do not have a glassy interphase? i.e. you search for regions that give you the imaging conditions for two crystals and reject the rest. The α-particles are indeed agglomerates and so the two types of grain boundaries suggested by Professor Brook seem likely.

Cannon: In order to get reasonable intensities and satisfy lattice imaging conditions in two grains at a grain boundary, does one effectively select special boundaries which involve tilt or twist about low index axes? If so, such boundaries may have a particularly good atomic fit and low boundary energy. Such boundaries may not be wet by a glassy phase and not be typical of most high angle boundaries.

Clarke: The only selection process is to look at those boundaries for which there is one plane in each grain that lies parallel to the electron beam. This does not imply that there is any specific orientation relationship between the two grains. My answer to Professor Brook's question answers the second point satisfactorily. However, I am sure that is right about these special crystallographic boundaries.

Morgan: What is the potential of observing the structure and behaviour of grain boundaries at high temperature?

Clarke: The potential is enormous and not only in this particular field of materials. No-one has yet succeeded in carrying out direct lattice imaging at temperatures above room temperature. This is purely a problem of specimen stability and lens stability – the thermal stresses are so inhomogeneous and varying. Instrumental developments are however in progress so one day it may be possible.

Lange: Did you see any crystalline phase associated with a Mg compound?

Clarke: No.

MICROSTRUCTURAL ASPECTS OF DEFORMATION AND OXIDATION OF MAGNESIA-DOPED SILICON NITRIDE

N. J. Tighe

Institute for Materials Research, National Bureau of Standards, Washington, D. C. 20234

ABSTRACT. The microstructural changes that occurred in magnesia-doped silicon nitride as a result of slow crack growth, plastic deformation and oxidation were studied by transmission electron microscopy. Specimens which exhibited slow crack growth showed extensive crack branching along the fracture path and ahead of the primary crack tip. These primary and secondary cracks followed intergranular paths. In samples which were deformed by bending at 1400°C, dislocation arrays were found as well as intergranular cracks and voids. Silicon nitride oxidized during heating in air at 1400°C and enstatite and cristobalite were present in the oxide layer. At lower oxidation temperatures, crystalline and amorphous silica formed a semiprotective layer on the silicon nitride surfaces.

1. Introduction

Silicon nitride compacts have heterogeneous microstructures because the grains vary in morphology and in phase. Grain sizes within some compacts range from a few tenths to several micrometers in diameter. The sintered grain phases are α-silicon nitride, β-silicon nitride and silicon carbide, and there are large inclusions of silicon oxy-nitride and tungsten silicide. Tne tungsten particles are picked up during the grinding of the powder prior to compaction.

It is necessary to use electron microscopy to study the morphology and structure of individual grains in fine-grained silicon nitride compacts. This method requires special techniques that are not available in all laboratories, and microstructural

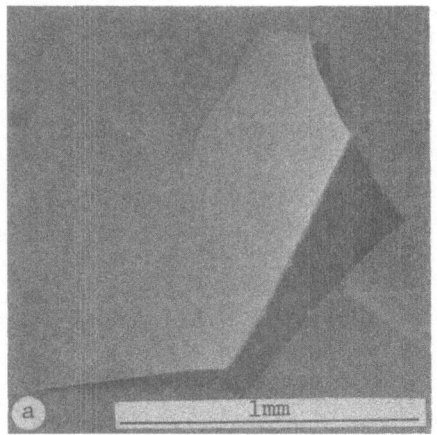 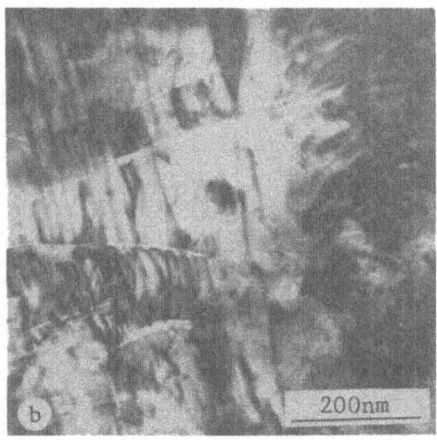

Fig. 1 Grinding damage (a) scanning electron micrograph of a thin foil specimen which curled-up as a result of strain on one ground side. (b) TEM showing sub-surface damage.

analyses of as-received and of tested specimens have previously been limited in scope [1-3]. Without such analyses it will become increasingly difficult to tailor silicon nitride materials to the desired applications in turbine engines. In the present paper we identify some microstructural features characteristic of deformation, of fracture and of oxidation in specimens made from magnesia-doped compositions.

2. Materials and Procedures

Samples were made from hot-pressed silicon nitride (HPSN) billets which had a hot pressing additive of MgO or of an MgO precursor. Spectrographic analysis by the manufacturer showed typical impurities as: Al 0.2 w/o, Ca 0.05 w/o, Fe 0.40 - 0.64 w/o Mg 0.70 - 0.74 w/o, W 2.6 - 2.7 w/o.

Microstructural examination was carried out on specimens from: (a) 4-pt. bend bars tested at 1400°C; (b) double torsion specimens [4] tested at 1400°C; (c) bulk and thin foil samples oxidized by heating in air at 1000°C, 1200°C and 1400°C; and (d) samples ground with 400 mesh SiC abrasives. Thin foils for transmission electron microscopy (TEM) were prepared by the ion sputtering technique [5]. In making foils to show the grinding damage it was necessary to sputter away ∿1/2 - 1 μm from the ground surface to keep the foil from curling-up as shown in Fig. 1b. Double torsion specimens were sectioned parallel to the tension surfaces and 3 mm disks for thin foils were cut out along the crack length.

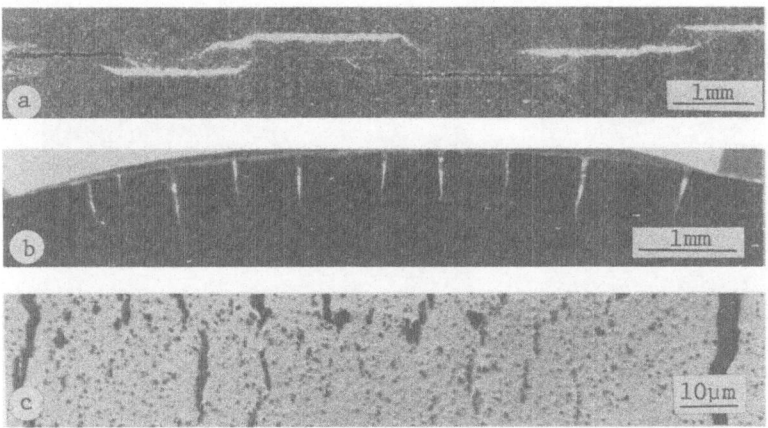

Fig. 2 Light micrographs of cracks in specimens deformed at 1400°C (a) tension surface of double-torsion specimen, (b) tension side of 4-pt. bend bar, (c) area between edge cracks in (b).

3. Results

3.1 Surface Damage

Test specimens usually are ground to produce uniformly flat surfaces. Cracks and chips which occur during the grinding process can penetrate several tens of micrometers. Such surface cracks become the strength limiting flaws of fracture mechanics theories.

The damage found near a ground surface is shown in Fig. 1b. The dislocation density is high and the strain contrast in the image reduced the resolution. The micrograph shows, however, the dislocations and cracks that constitute some of the near surface grinding damage.

3.2 Slow Crack Growth

At temperatures above 1200°C and at low strain rates magnesia-doped silicon nitride exhibits slow crack growth [4, 6] and samples deform plastically at relatively low stress.

Specimens which exhibited slow crack growth during crack propagation experiments of 1400°C were examined by light and electron microscopy. Cracks which had formed in the tension side of specimens were filled partially with oxide; and, therefore were easily visible. The segmented primary crack shown in Fig. 2a demonstrates the characteristic fracture pattern of slow crack growth in HPSN.

444

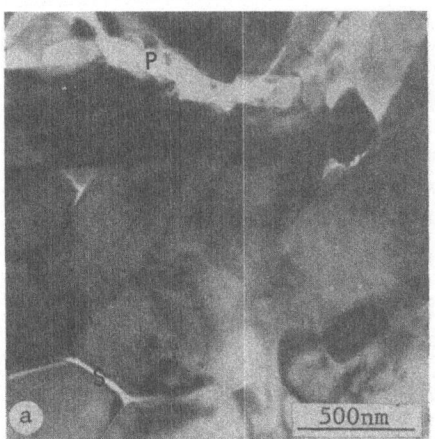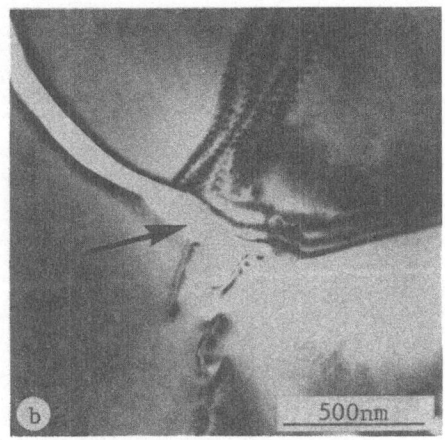

Fig. 3 Electron micrographs of cracks produced in double-torsion specimens tested at 1400°C, (a) primary (P) and secondary cracks (S), (b) dislocation arrays at the tip of a secondary crack.

Cracks were produced along the tension side of a 4-pt. bend bar which was tested at 1400°C. When the surface oxide was polished away the deep cracks shown in Fig. 2b were seen along the specimen edge; however, near the center of the bar cracks penetrated only a few tenths of a mm. Short crack segments which occurred between the deep edge cracks are shown in Fig. 2c. These observations suggest that the crack growth was influenced by edge effects and by surface oxidation.

In double-torsion specimens, electron microscopy showed that cracks occurred along grain boundaries. The primary crack and the secondary cracks were oriented approximately parallel to each other and were separated by one or more grains. An example of this configuration is shown in Fig. 3a. Figure 3b shows a crack tip which is several grain diameters away from the primary crack. The crack is several grains long and dislocations are seen emanating from the crack tip into the two adjacent grains.

Because of severe cracking along the tension surface of bend specimens thin foil specimens were made from regions several micrometers beneath the oxide layer. These samples showed primarily the microstructure similar to that seen in double-torsion specimens, namely intergranular fracture, and dislocation arrays near grain boundaries.

Grains on the compression side of the bent bars had extensive arrays of bubbles, (B), dislocation arrays, (P) associated with

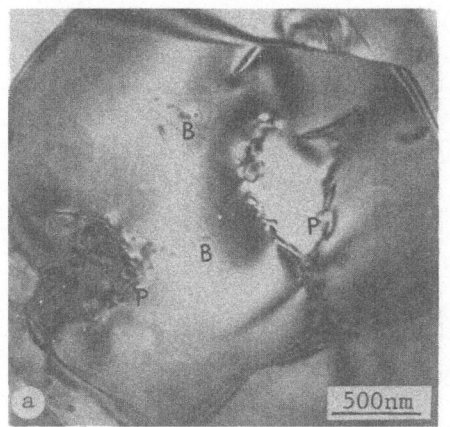 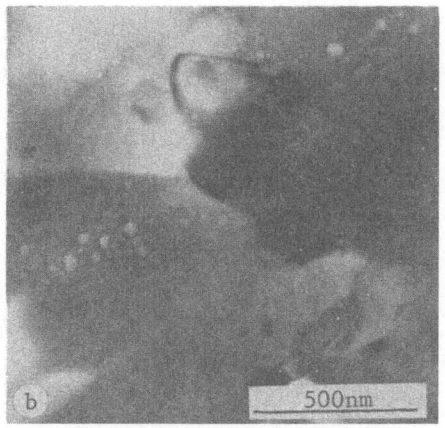

Fig. 4 Compression side of a bent sample showing (a) Bubbles, and dislocation arrays within a grain, (b) Extensive array of voids.

grain boundaries and in sub-grain boundaries. The pictures in Fig. 4a, b show typical areas in these samples.

3.3 Oxidation

Silicon nitride oxidizes readily in air above 1000°C, however the effects of the oxide on mechanical properties has not been established. The thickness and the structure of the oxide layer are related to the temperature and time of heating. The x-ray diffraction patterns obtained from oxidized specimen surfaces showed that β-cristobalite was present in samples oxidized at 1000°C, 1200°C, 1400°C and that enstatite ($MgSiO_3$) and possibly pigeonite (Mg_2Fe_2Ca) [Si_2O_6] were present in samples heated at 1400°C. When the oxides were examined by electron microscopy, two amorphous phases were found in addition to these crystalline phases.

A cross section of the oxide formed on a specimen heated 96 hrs at 1400°C is shown in Fig. 5a. The oxide is polycrystalline and polyphase and has many pores and cracks. This specimen was heated cyclically and the oxide appeared to have melted during this heat treatment.

Electron microscopy on the sample in Fig. 5a showed that the oxide nearest the Si_3N_4 matrix was Si_2N_2O. This phase was distinguished also by light microscopy, and, from the light microscopy examination it appeared that the oxide formed nodules which grew to cover the surface. Large grains of silicon oxy-nitride were found in the thin sections, and these grains were heavily faulted as shown in Fig. 5b.

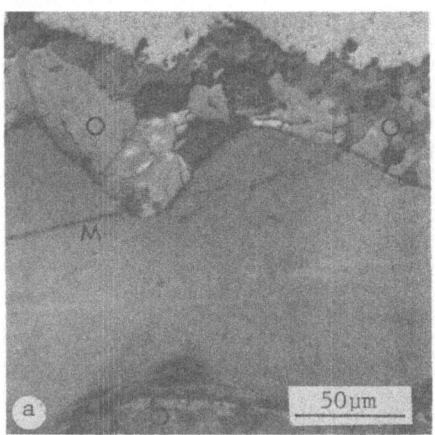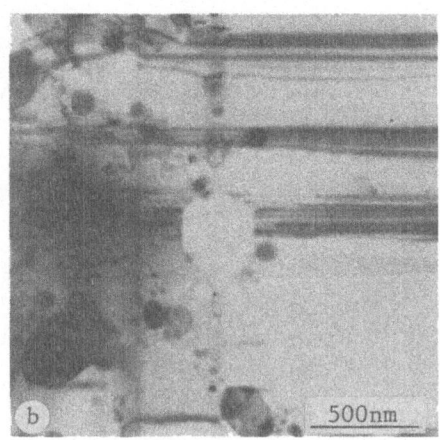

Fig. 5   Oxide produced at 1400°C, (a) light micrograph showing
oxide (O) on both surfaces of Si_3N_4 specimen (M), (b) Si_2N_2O
grain.

    In the layer adjacent to the silicon oxynitride, enstatite and
cristobalite were found. The electron micrograph in Fig. 6a shows
an enstatite grain with its characteristic faulting and a partially
vitrified cristobalite grain.

    Cristobalite was found in all the oxidized specimens. In thin
foils which were oxidized at 1000°C and at 1200°C, β-cristobalite
occurred as grains within a glassy matrix. However, under the influence of the electron beam, the β-cristobalite decomposed first to
α-cristobalite an amorphous phase. The grain shape did not
change during this vitrification. The vitrification occurred in
less than two minutes and Fig. 6b shows both transformed and nontrans-
formed grains. The diffraction pattern taken from the vitrified
β-cristobalite was used to differentiate the amorphous silica phase
from the other amorphous phase. It is possible that the second
amorphous phase is a magnesium silicate glass. Cristobalite and
various pyroxene phases can occur by devitrification of a magnesium
silicate glass at 1400°C as well as during cooling to room temperature.

    Enstatite is one phase in the pyroxene group of calcium, magnesium
iron silicates which form an impressive array of mineral structures.
These structures are difficult to identify from electron diffraction
patterns alone. The Ca,Mg and Fe are the major impurities in the
magnesia-doped silicon nitride and their presence in an oxide layer
indicates that considerable diffusion occurred during oxidation.

4.   Discussion and Conclusions

4.1 Microstructural Factors Affecting Strength

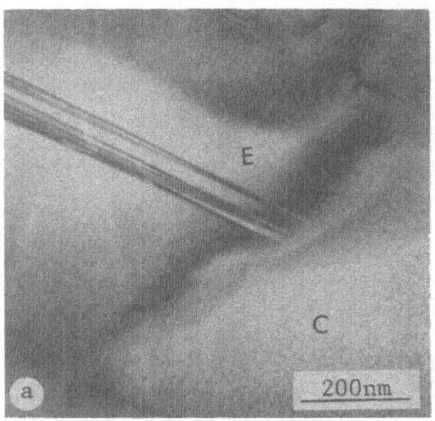 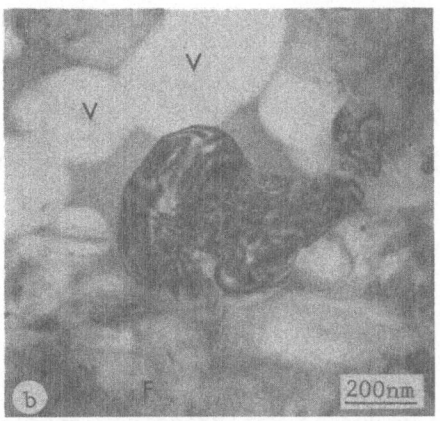

Fig. 6 Phases found in oxide produced by heating at 1400°C (a)
enstatite, MgSiO$_3$ (E) and β-cristobalite grains in the mid-oxide
layer. (b) β-cristobalite formed on a thin foil oxidized 1 hr at
1200°C; area shows faulted grains and some vitrified grains.

Other investigators attributed deformation of magnesia-doped
silicon nitride to grain boundary sliding from a glass binder
between grains [1, 6]. However this glass phase was not evident
in the samples discussed in this paper. Cracks did propagate
primarily along grain boundaries indicating that grain boundaries
presented easy fracture paths. It is possible that impurities
can segregate near grain boundaries and can weaken their bonding
without forming a distinct glass phase.

The electron microscopy results show that slow crack growth,
which occurs at low strain rates, is characterized by branching
of grain boundary cracks and by localized plastic deformation at
crack tips and along grain boundaries. Oxide formation during the
deformation may influence the crack growth.

The surface oxide can have an initial strengthening effect by
filling up and healing flaws produced during machining. However,
the oxide forms preferentially at fresh fracture interfaces and
penetrates deeply into the matrix during continued deformation.
This sub-surface oxide penetration can weaken the structure during
prolonged stressing at elevated temperatures. Changes in oxide
composition from silica to magnesium silicate decrease the viscosity
and the melting point and thus modify the oxide behavior at temperature.

Consideration of the microstructure suggests that the effects
of the oxide growth and of the structural inhomogenities which
contribute to uneven thermal stress distribution should be accounted
for in the development of deformation models for HPSN. It will
be useful to compare the microstructures of silicon nitride compacts

448

having different sintering additives and to examine specimens
after deformation in inert as well as in oxidizing atmospheres.

Acknowledgments

    This work was supported by the USAF under grant no.
F33615-76-F-6752. We thank Mr. L. Russell for performing mechanical
tests, and Mr. G. Garrett for printing the photographs.

References

1. R. Kossowsky, "The Microstructure of Hot-Pressed Silicon Nitride"
J. Matl. Sci. 8 1603-1615 (1973).

2. P. Drew and M. H. Lewis, "The Microstructure of Silicon Nitride
Ceramics During Hot-Pressing Transformation" J. Matl. Sci. 9 261-269
(1974).

3. N. J. Tighe, "Microstructure of Oxidized Silicon Nitride" Proc.
EMSA 420-471 (1974).

4. A. G. Evans and S. M. Wiederhorn, "Crack Propagation and Failure
Prediction in Silicon Nitride at Elevated Temperatures" J. Matl.
Sci. 9, 270-278 (1974).

5. N. J. Tighe, Experimental Techniques in Electron Microscopy in
Mineralogy, (ed.) H. R.-Wenk, Springer Verlag, Berlin 1976.

6. F. F. Lange, "High-Temperature Strength Behavior of Hot-Pressed
Si_3N_4: Evidence for Sub-Critical Crack Growth" J. Am. Ceram. Soc.
57, 84-89 (1974).

DISCUSSION

Lumby: Did you identify the large inclusions as tungsten
carbide? Did these inclusions anchor the grain boundaries?
What was the relationship between oxidation and surface damage?

Tighe: No; these inclusions were too thick to penetrate with
the electron beam.

    In other systems fresh fracture interfaces react with a gas
phase faster than the surface. Preferential reaction occurs in
MgO doped Si_3N_4 and a ground surface is more reactive than a
polished surface. When the surface is covered, the oxide grows
into the Si_3N_4 matrix.

Section H

MECHANICAL PROPERTIES

THE FRACTURE OF BRITTLE MATERIALS

A. G. Evans

Science Center, Rockwell International
Thousand Oaks, California   91360 U.S.A.

ABSTRACT

Structural design with brittle materials requires an inter-related series of tests and analyses to be performed on the candidate materials. These test requirements are discussed, with special emphasis devoted to three aspects of the design procedure; relations between microstructure and fracture toughness, statistical analyses of fracture, and projectile impact fracture.

1. INTRODUCTION

A comprehensive understanding of the mechanical properties of brittle materials is an essential prerequisite to the effective design of structural components, particularly those that are expected to sustain significant tensile stresses for long periods of time. The various elements of mechanical behavior that can be involved in the design process are summarized in Fig. 1. The essential details of many of these elements have been discussed in a recent review [1]. These details are not repeated here. Instead, recent developments are discussed and evaluated; especially the fracture properties that are expected to be important constituents in the design of structural components made from nitrides. These developments are in the following areas: (a) models for the relationship between microstructure and fracture

452

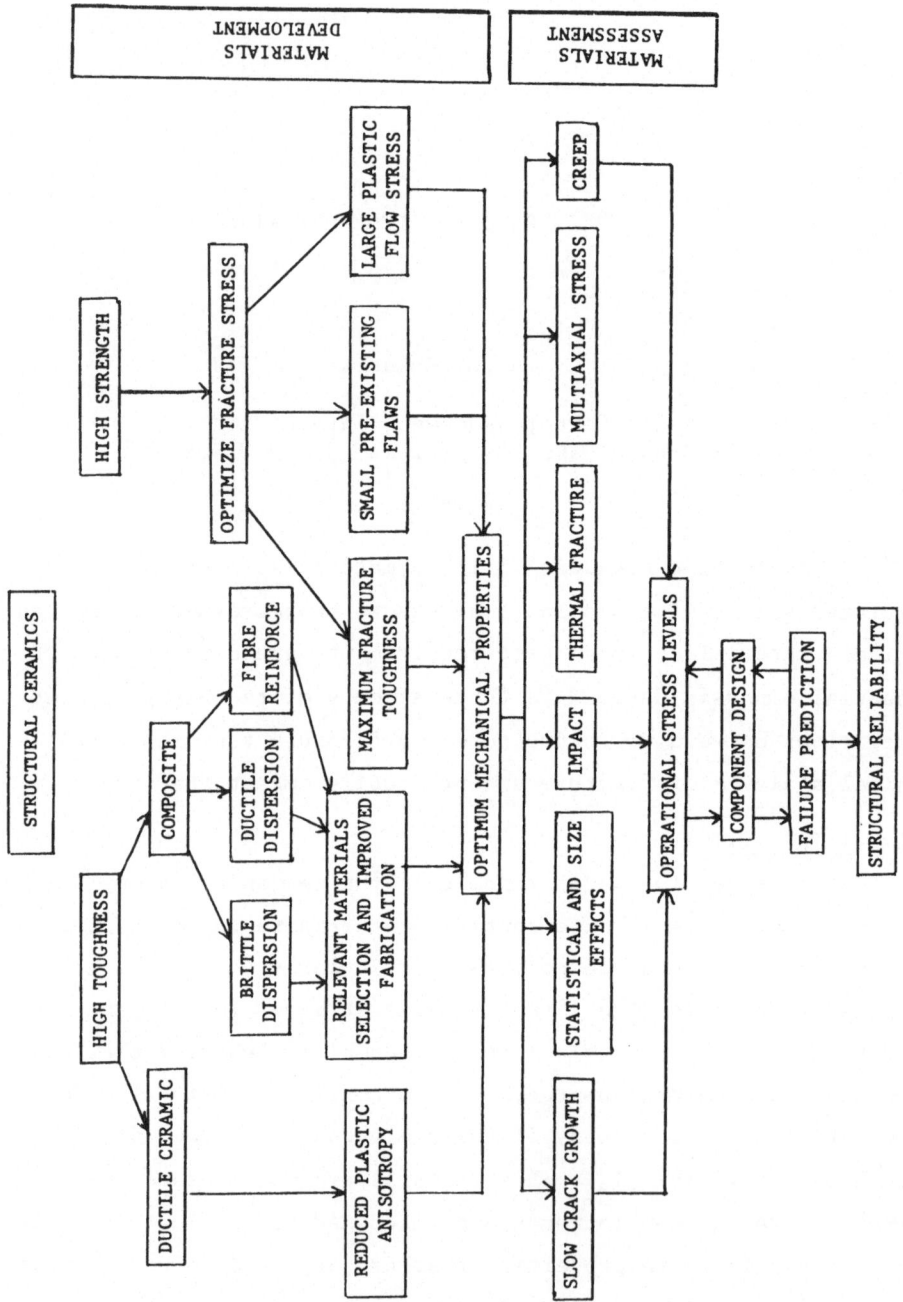

Fig. 1. A schematic of the sequence involved in the structural
design of brittle materials.

toughness, that might be utilized for materials development purposes; (b) statistical analyses of fracture from first principles, that form a framework for the prediction of size and stress state effects on the fracture strength; (c) analyses of projectile impact fracture, which indicate the material parameters that dominate this important mode of failure.

2. FRACTURE TOUGHNESS

Silicon nitride has an anisotropic (hexagonal) structure that can be used to advantage in the development of toughness. For the single phase material toughness might be developed in at least two ways, (i) the generation of microcracks within a 'process zone' around the primary crack, (ii) the 'pull-out' of elongated grains. For multiphase nitrides, additional mechanisms might be possible [2], but these are not considered in this paper. The toughening models to be described are based on stress field calculations, which enable variations in the stress intensity factor or strain energy release rate to be deduced [2]. (The alternate thermodynamic approach is not used because of the unknown contributions from the thermal and phonon energy terms [2].)

2.1 Microcracking

Radiographic techniques have been used to detect the incidence of secondary microcracking in high toughness silicon nitrides [3]. Microfracture should thus be given serious consideration as a mode of toughening in these materials. The microcracking of anisotropic polycrystals can be characterized by a distribution of microfracture strengths, S_m. A distribution function that seems to characterize the microfracture detected by acoustic emission studies [4] has a form suggested by extreme value statistics

$$\Phi = 1 - \exp\left[\left(-\frac{S_m}{S_o}\right)^m\right] \qquad (1)$$

where Φ is the microfracture probability, S_o is a scale parameter

and m is a shape parameter (c.f. the Weibull function). This
function can be used to obtain rough estimates of the distribution
of microcracks that occur around a stressed primary crack, by combining with the elastic principal tensile stress field relations
[2]

$$\sigma_{1,2} = \frac{K_I}{\sqrt{2\pi r}} \cos \theta/2 \ (1 \pm \sin \theta/2) \tag{2}$$

where K_I is the mode I stress intensity factor, r is the distance
from the crack tip and θ is the orientation relative to the crack
plane. The average densities of microcracks as a function of distance from the crack tip, deduced from eqns (1) and (2) are plotted
in Fig. 2 for three values of m (2,6,12). These distributions are
only approximate, because the presence of microcracks modifies the
stress intensification by reducing the effective modulus of the material in the microcrack zone [2]. A fully quantitative determination of the microcrack densities (and hence, the modified stress
field), would thus require a stress solution for a crack penetrating a zone of continuously varying modulus; this would be used
iteratively to obtain a self-consistent result for the stress
field and microcrack distribution. Such a stress solution is not
available and, in consequence, a rigorous analysis of the crack
tip stress field in a microcracked material cannot yet be performed. However, a solution for a crack at the interface of a
low modulus inclusion [5] does at least indicate that the microfracture should reduce the crack opening displacement and thus,
tend to reduce the stresses near the crack (Fig. 3).

The effect of microfracture on the fracture toughness can
be deduced from the stress field, if a criterion for primary crack
extension is postulated. Since the microcracks that form at a
small distance from the primary crack tip, i.e. less than the
microcrack diameter $(2a_m)$, will tend to magnify the local stress
and thus link with the primary crack [6], one plausible crack extension criterion would require that the density of microcracks

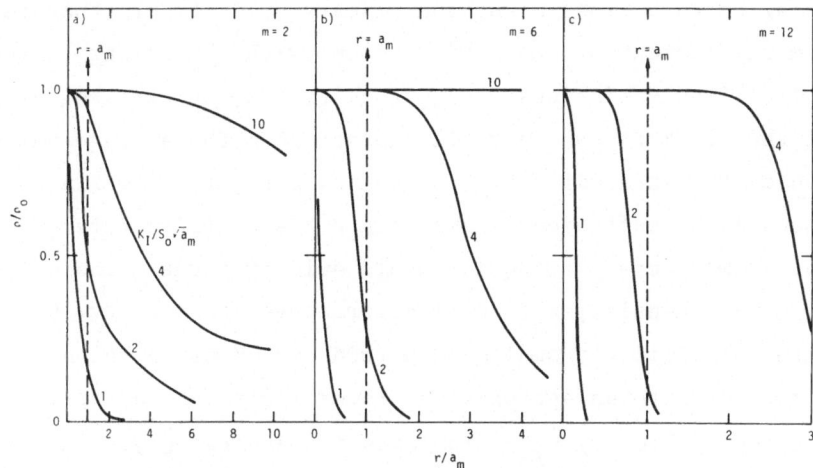

Fig. 2. Approximate microcrack densities as a function of
distance from the primary crack tip.

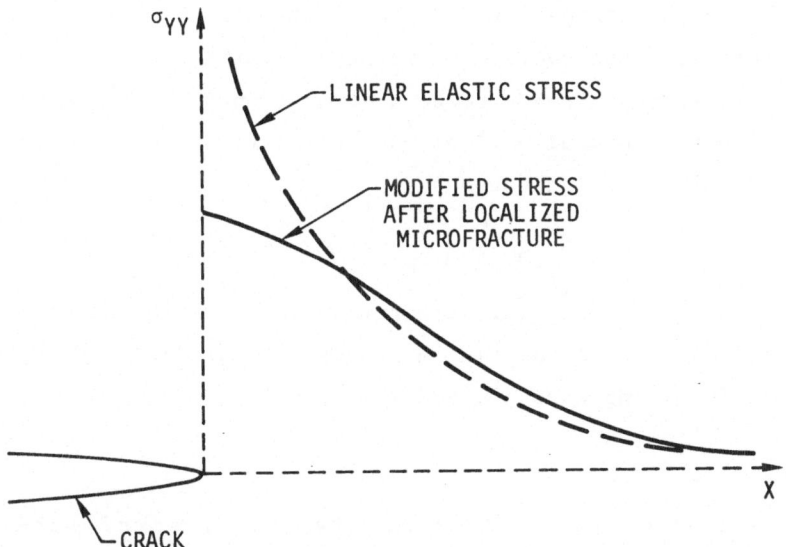

Fig. 3. A schematic showing the variations in stress around a
crack tip in a material containing a subsidiary
microcrack zone.

within an inner zone, $r \lesssim 2a_m$, exceed a critical value, ρ_c. Further, local perturbations along the primary crack front (that result from the linkage of small microcracks with the primary crack) lead to localized variations in K_I, such that K_I at trailing portions of the crack is substantially larger than the average value [7]. Hence, the intensification caused by the first few linked microcracks might sufficiently accelerate the microfracture process that primary crack extension would quickly ensue. The critical microcrack density, ρ_c, (or microfracture probability, Φ_c) might thus be relatively small and herein, we choose a value for Φ_c of 0.5 to compute an approximate* lower limit for the critical stress intensity factor, K_{Ic}, from eqns (1) and (2) [2],

$$K_{Ic} \gtrsim \xi(m) \, S_o \, \sqrt{a_m} \qquad (3)$$

where ξ is a relatively weak function of m (such that $\xi \approx 2$ for $4 < m < 30$). Qualitatively, eqn (3) suggests that the toughness can be enhanced by increasing the scale parameter S_o (or equivalently, increasing the average microfracture strength, $\overline{S}$). The microfracture strength tends to be a strong function of the primary microstructural dimension; one calculation, for a thermally anisotropic material, indicates that [2]

$$\overline{S} \approx K_s \, \sqrt{\frac{3}{a^*} \left(\frac{a^*}{a_m} - 1 \right)} \qquad (4)$$

where a^* is the critical microstructural dimension for spontaneous microfracture and K_s is the single crystal (or grain boundary) toughness. Combining eqns (3) and (4) gives

---

*An exact evaluation of the relation between K_{Ic}, S_o and m requires not only a more rigorous stress field analysis (see above) but also a detailed calculation of microcrack, primary crack interactions, to deduce Φ_c. Two-dimensional solutions for the interactions have been obtained by McClintock and co-workers [8,9] for idealized microstructures, but these need to be extended to three-dimensions before the present problem can be solved.

$$\frac{K_{Ic}}{K_s} = \xi \sqrt{3 \left(1 - \frac{a_m}{a^*}\right)} \tag{5}$$

This relation anticipates that the fracture toughness of systems that microcrack should tend to increase as the average size of the microstructure decreases, provided that a corresponding decrease in K_s does not occur. The use of fine scale microstructures would thus appear to offer the greatest potential for maximizing the microfracture induced toughness. This tendency is not fully realized, however, in alumina systems where the toughness tends to exhibit a maximum at intermediate microstructural sizes [10]. There are several possible reasons for this discrepancy; (a) K_s in these systems tends to diminish at fine grain sizes [2] (e.g. for reasons associated with the fabrication), or (b) parameters other than those considered in the present model might be important [11], or (c) microcracking may not be the primary source of toughness at all grain sizes (e.g. grain pull-out could become important at intermediate grain sizes - see below).

The role of the shape factor, m, in the control of toughness appears from eqn (3) to be minimal; but its effect might be accentuated when the stress field modifications caused by microfracture are included.[‡]

The present analysis of microfracture induced toughness, although it provides a reasonable physical appreciation of the origin of toughening, is not quantitative and guidelines for the microstructural design of high toughness materials can not yet be confidently established. All that the analysis really indicates is that anisotropy (thermal expansion or elastic) is necessary and that the size of the microstructure and its variability are important considerations. However, since further theoretical

---

[‡]The shape factor dominates the spread in microfracture strengths such that, by decreasing m, the zone of microfracture is increased and the stress intensification near the crack tip is reduced.

analysis will be hampered for some time by the lack of a solution for a crack penetrating a zone of variable modulus, it may be more expedient to establish microstructural design principles by empirical studies, on model systems suggested by the above analysis (e.g. the zirconia, alumina system [12]).

2.2 Pull-Out

An alternate mechanism involved in the development of toughness in silicon nitride systems is grain pull-out. The mechanisms, initially proposed by Lange [13], is illustrated in Fig. 4. The toughening derives from the closure tractions behind the primary crack front, whenever shear stresses (τ_i) at the elongated grain interfaces are large enough to resist pull-out, but small enough to prevent grain fracture. (At elevated temperatures, the shear resistance could derive from the viscosity of a second phase at the boundary; while at room temperature, the resistance could be frictional, if normal localized stresses are generated at the interface by thermal expansion and/or elastic anisotropy.)

The toughening can be evaluated if the compressive stress (σ_c) in the pull-out zone (of length L) caused by the interfacial shear resistance is regarded as the constraint in a Barenblatt cohesive zone [14]. Then the toughness is given by [2,14,15];

$$K_{Ic} = K_m + \sqrt{\frac{2}{\pi}} \int_{a-L}^{a} \frac{\sigma_c}{\sqrt{a-x}} \, dx \tag{6}$$

where K_m is the intrinsic toughness of the matrix. The force, F, exerted by each grain during pull-out is simply

$$F = 2\pi\tau_i \, r_f \, (\ell_f - u) \tag{7}$$

where r_f is the elongated grain radius (Fig. 4), ℓ_f is the grain half-length and u is the crack opening. Since the area of influence, A_ρ, of each grain is given by;

$$A_\rho \approx (2r_f + d_s)^2 \tag{8}$$

where d_s is the separation between elongated grains (Fig. 4), the equivalent compressive stress exerted by each elongated grain becomes:

$$\sigma_c = \frac{2\pi \, \tau_i \, r_f}{(2r_f + d_s)^2} \, (\ell_f/2 - u) \tag{9}$$

The crack opening in the cohesive zone increases with distance, $r = a - x$, from the crack tip [2];

$$u = \frac{K_I}{2E} \sqrt{\frac{r}{2\pi}} \, R\left(\frac{a-x}{L}\right) \tag{10}$$

where R is a function in the range 0 to ~ 1 that increases as $(a-x)/L$ increases (such that $R \approx 1$ at $a-x = L$). Hence, the zone length L at the critical condition is related to the grain length, ℓ_f, by;

$$L = 2\pi \left(\frac{2E\ell_f}{K_{Ic}}\right)^2 \tag{11}$$

Inserting eqn (10), with $K_I = K_{Ic}$, and eqn (11) into eqn (9) to obtain σ_c at the critical condition and then using σ_c to evaluate K_{Ic} (from eqn 6) yields a complex cubic equation. A simplified form of this expression (obtained by letting u be constant, $\ell_f/4$, within the ligament zone), which enables the primary variables affecting K_{Ic} to be more clearly delineated, is

$$2K_{Ic} = K_m + \sqrt{K_m^2 + 8\pi\ell_f^2 E \, r_f \, \tau_i / (2r_f + d_s)^2} \tag{12}$$

The important features of eqn (12) to note are that K_{Ic} increases as the interface shear strength increases, as the grain separation decreases (i.e. as the number of elongated grains increases) and as the grain dimensions r_f and ℓ_f increase. However, it is important to recognize that there must be an upper limit to the toughness that can be obtained by applying the appropriate microstructural modifications; this limit coincides with the incidence of grain fracture which occurs when;

$$2\tau_i \left(\frac{\ell_f}{r_f}\right) \gtrsim \sigma_f \tag{13}$$

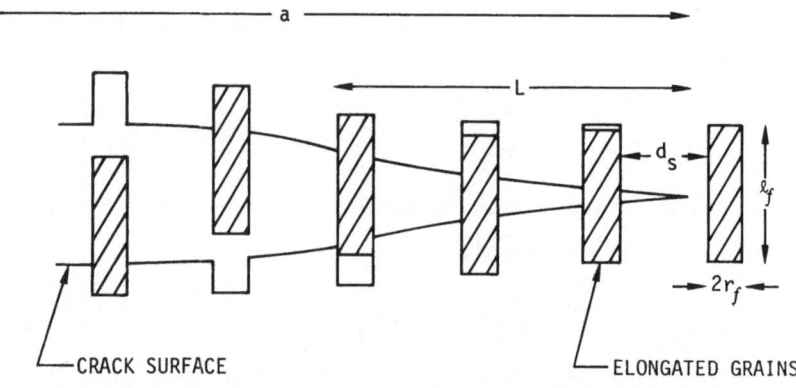

Fig. 4. A schematic of the grain pull-out mechanism.

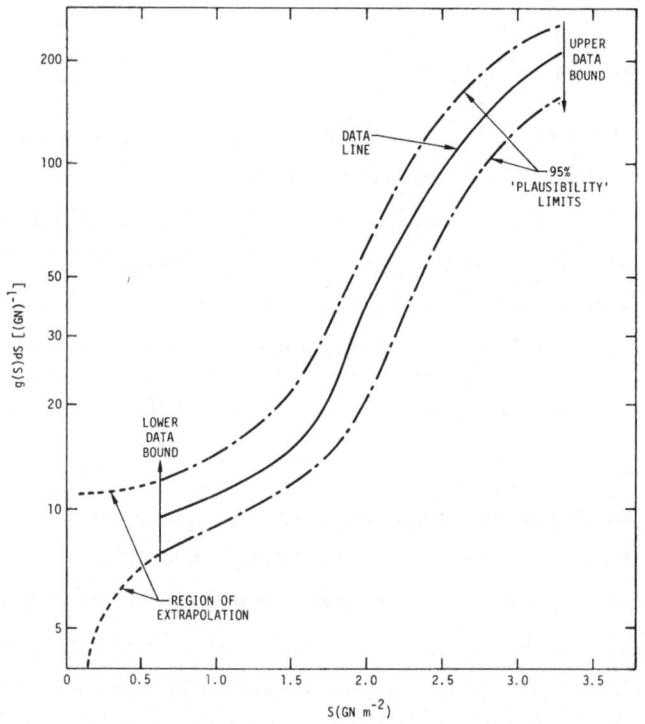

Fig. 5. A uniaxial strength distribution function for optical
fibres, showing the median and 95% 'plausibility'
limits. Also shown are approximate confidence limits
for data extrapolation.

where σ_f is the grain fracture stress. When this condition arises, the second term in eqn (12) essentially reduces to zero because grain fracture is likely to occur in close proximity to the fracture plane. However, the first term increases because the elongated grain fracture resistance must now be included. A quantitative computation of the effect of elongated grain fracture on K_m is not yet possible, because the calculation requires that the three-dimensional 'crack bowing' effects [7] be taken into account. However, if it is _empirically_ established that this effect on K_m is small, the requirements for optimum toughness in a system that exhibits grain pull-out are then given by eqn (12). These are, a large grain fracture stress on the plane orthogonal to the direction of elongation, and a large volume fraction of elongated grains; the other parameters, the interface shear strength and the grain dimensions should then be adjusted to the maximum composite value compatible with eqn (13), i.e. the interface shear strength should have the maximum possible value consistent with the grain dimensions.

3. STATISTICS OF FRACTURE

3.1 Characterization of Fracture Statistics

(a) Uniaxial Stress

The most fundamental, but general, theories of the statistics of fracture [1,16] have defined a flaw strength (flaw extension stress) distribution, g(S)dS, that represents the number of flaws per unit volume with a strength between S and S+dS. Then, when fracture occurs by the direct extension of pre-existing flaws the probability, 1-ϕ, that a small volume element, δV, will survive when subjected to a uniaxial tensile stress, S, is given by;

$$1 - \phi(S) = 1 - \delta V \int_0^S g(S)\, dS \tag{14}$$

The survival probability 1-Φ of body, V, is thereafter given by the product of the survival probabilities of each element, provided that neighboring elements are non-interacting, i.e. that the strength of an element is not influenced by the presence of its neighbors. Then, if the number of elements is large and the stress in the body is uniform [1,16];

$$1 - \Phi(S) \equiv \Pi \ (1-\phi(S)) = \exp \left[-V \int_0^S g(S) \, dS \right] \qquad (15)$$

When the stress is non-uniform the product must, in the general case, be evaluated numerically. However, for a uniaxial stress that varies in one-dimension such that

$$S = S_m (1 - z/\ell)^k \qquad (16)$$

where S_m is the peak stress, ℓ is a length dimension, z is the distance from the plane of maximum stress and k is a constant, the product can be evaluated analytically [16]

$$1 - \Phi(S) = \exp \left[-b_1 b_2 \int_0^\ell \int_0^{S_m (1-z/\ell)^k} g(S) \, dS \, dz \right] \qquad (17)$$

where b_1 and b_2 are the dimensions of the uniformly stressed planes. Changing the order of integration reduces eqn (17) to

$$\Phi(S_m) = 1 - \exp \left[-b_1 b_2 \ell \int_0^{S_m} (1 - (S/S_m)^{1/k}) \, g(S) \, ds \right] \qquad (18)$$

where $\Phi(S_m)$ is now the measured failure probability, as character-ized by the peak stress in the body. Differentiating eqn (18) twice and rearranging enables $g(S_m)$ to be expressed in terms of the failure probability [16];

$$g(S_m) = \frac{k \ S_m}{b_1 b_2 \ell} \left[G''(S_m) + \frac{(1+1/k)}{S_m} G'(S_m) \right] \qquad (19)$$

where $G = -\ell n(1-\Phi(S_m))$. The equivalent cumulative probability $g^*(S_m)$ is given, of course, by $g^*(S_m) = \int_0^{S_m} g(S_m) dS_m$. Hence, we

now appreciate that the characteristic statistical distribution
of strengths for a given material can, in principle, be deduced
from <u>measured</u> fracture probabilities, i.e. no <u>a priori</u> assumptions
about the nature of the distribution function (such as the power
function proposed by Weibull [17]) are needed. When surface flaws
are the only source of fracture, eqn (16) can be used to describe
the stress distributions for a wide range of test geometries (e.g.
tensile, bending, Hertzian [16]) and eqn (19) may be used to
evaluate $g(S_m)$ - which is then the <u>areal</u> distribution function -
by omitting b_2. However, for a volumetric distribution of fracture initiating defects, eqn (19) can only be applied to the tensile and pure bending geometries. But, an alternate tractable
function that can be used to analyze, for example, three or four
point bend test data, has the form [18];

$$S = S_m (1 - z/\ell) (1 - y/h) \tag{20}$$

where h is a specimen dimension and y is the distance from S_m.
The equivalent relation between $g(S_m)$ and $\Phi(S_m)$ that obtains for
this function is [18]

$$g(S_m) = \frac{1}{b\ell h} [G''' S_m^2 + 5G'' S_m + 4G'] \tag{21}$$

One difficulty involved in the derivation of $g(S_m)$
from fracture strength, probability data is the uncertainty in
the flaw origin, i.e. surface or interior. A knowledge of the
common flaw origin is needed to select the pertinent $g(S_m)$ function, e.g. eqn (19) or (21) when bend test data are being interpreted. But, equally as important, fracture mechanics indicates
that flaws at the surface propagate unstably at a lower stress
(by up to 40%) than internal flaws of the same size [1,15]. This
suggests that surface elements will generally be weaker than internal elements, even when the large flaws are randomly distributed throughout the volume of the body. This is a problem because
$g(S_m)$ obtained from eqn (21) would exhibit a significant stress
state dependence. These difficulties can be largely overcome if

fracture origins (surface or interior) are determined for each
sample. Then, the effective strength, S_ρ, can be deduced from
the fracture stress using

$$S_\rho = \beta S_m \qquad (22)$$

where β is unity for internal flaws and ~ 1.4 for surface flaws.
This approach should yield a unique $g(S_\rho)$ for a given batch of
material, independent of the test geometry.

One final step involved in the characterization of
the strength distribution function is to determine confidence
limits. Since the distributions are rarely normal in character,
the confidence limits must be determined numerically. A technique
for this purpose has recently been proposed [18]. The technique
entails repeated computer sampling from the test data, regarded
as the strength population. In detail, the cumulative probability
for the variate is plotted, using order statistics, and then plausible values for the variate are assigned at probabilities of zero
and unity. Thereafter, a test sample, i_t, is set up and a random
number, x, between 0 and 1 is selected. A value for the variate
corresponding to this number is then read from the original plot,
and a sample number is associated with this variate by taking the
next integer above $X(i_t+1)$. Continuing in this fashion a new
cumulative distribution of the variate is obtained that represents
the 'plausibility' limit, $1 - 1/i_t$.

An example of a strength distribution derived from
test data is summarized in Fig. 5. The data are tensile data
obtained on optical fibres [19] and $g(S_m)$ is derived from the data
by using eqn (19) and setting k equal to zero [18]. In this case,
the distribution is monotonic, indicating a single flaw population
within the measured stress range. Frequently, however, multi-
modal distributions are obtained [16]: clearly, a single function
(such as the Weibull function) could never afford an effective
description of the fracture statistics in these instances.

(b) Multiaxial Stresses

The strength distributions under multiaxial stress conditions cannot be obtained by simply evaluating failure probabilities in orthogonal directions from uniaxial data – an approach that has been frequently utilized in the literature – because this does not recognize that the flaws are unique entities. Rather, the relative strengths of bodies subjected to uniaxial and multiaxial stress states must be determined from the crack propagation criteria that pertain under uniaxial and multiaxial conditions [20]. A unique criterion for multiaxial fracture has not yet been agreed upon and hence, the statistical analysis of multiaxial fracture is incomplete. But, a fundamental approach to the multiaxial fracture statistics problem can be outlined, and herein one of the more plausible crack extension criteria – a critical strain energy release rate, G_c (see section 1) – is used for illustration purposes.

Consider a through crack of length 2a oriented at an angle θ to one of the principal tensile stress axes (σ_1). The mode I and mode II (plane stress) strain energy release rates for this crack, for a general state of biaxial tension, are [15]

$$G_I = \frac{\pi a}{4E} \left[(\sigma_1 + \sigma_2)^2 + (\sigma_1 - \sigma_2)^2 \cos^2 2\theta + 2(\sigma_1^2 - \sigma_2^2)\cos 2\theta \right] \qquad (23a)$$

$$G_{II} = \frac{\pi a}{4E} (\sigma_1 - \sigma_2)^2 \sin^2 2\theta \qquad (23b)$$

Now applying the relation [15];

$$G = G_I + G_{II} \qquad (24)$$

enables the critical crack extension stress, $(\sigma_1)_c$, to be defined as;

$$(\sigma_1)_c^2 = \frac{2G_c}{\pi a} \left[1 + \left(\frac{\sigma_2}{\sigma_1}\right)^2 + \cos 2\theta - \left(\frac{\sigma_2}{\sigma_1}\right)^2 \cos 2\theta \right]^{-1} \qquad (25)$$

This relation forms the basis for computing the biaxial fracture statistics of a body containing arrays of through cracks (Fig. 6) subjected to any stress ratio (σ_1/σ_2). An equivalent, but more complex, function can be derived for three-dimensional (e.g. penny) cracks under triaxial tension. Inspection of eqn (25) indicates that there is no dependence of the critical flaw extension stress on the flaw orientation under the equi-biaxial condition $(\sigma_1/\sigma_2 = 1)$. The fracture statistics obtained for this condition thus provide the most fundamental information about the distribution of flaw strengths (or flaw sizes).

Consider a cumulative biaxial strength distribution (Fig. 7) determined from biaxial flexure data [1] and analyzed as described above for the uniaxial data. The strength distribution in uniaxial tension, for example, can be estimated from this distribution by examining the strengths in angular elements, $\Delta\theta$, of a cylinder of unit volume (Fig. 6). For a random array of flaws, there is an equal distribution of flaws between elements, such that the number of flaws (in a given size range) located in each element, $\Delta\theta$, is given by;

$$\Delta\Theta = N_B \frac{\Delta\theta}{2\pi} \tag{26}$$

where N_B is the total number of flaws in that size range. The uniaxial strengths, S_u, of the flaws in such an element are related to the biaxial strengths, S_B, by;

$$S_u = S_B (1 + \cos2\theta_\eta)^{-\frac{1}{2}} \tag{27}$$

where θ_η is the orientation of the element mid-plane shown in Fig. 6. The strength distribution in each angular element can be derived directly from eqns (26) and (27) and the biaxial distribution, as shown schematically in Fig. 7. The final uniaxial strength distribution can then be obtained by summing the number of flaws in separate strength ranges over all elements, as indicated in Fig. 7 (remembering that there are four equivalent elements in the cylindrical geometry). This procedure is easily programmed on a small computer.

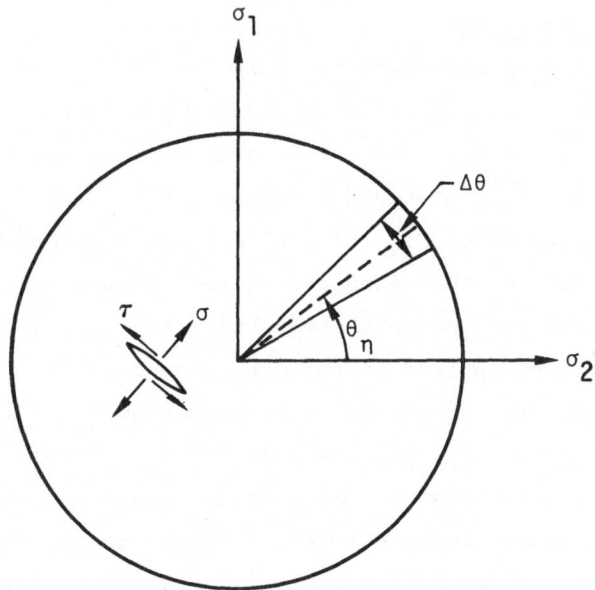

Fig. 6. A cylindrical body used for biaxial strength determinations on a body containing through cracks; $\Delta\theta$ is the
element to be considered.

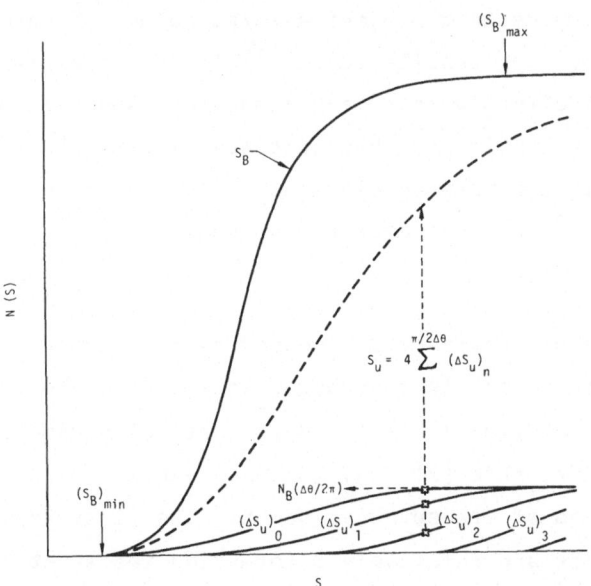

Fig. 7. A schematic of a cumulative biaxial strength distribution,
and the distribution of uniaxial strengths in each
element, $\Delta\theta$ (see Fig. 6) derived from it. Also shown is
the final uniaxial distribution obtained by summation.

3.2 Failure Prediction

Once strength distributions have been derived from test data, the problem can be inverted to predict fracture probabilities for any stress state. For simple stress states, eqn (17), or an equivalent, can be used and integrated numerically to obtain Φ. For more complex stress states finite element techniques are needed. One feature of the calculations that should be re-emphasized concerns the use of the weighting factor β (eqn 22) which should again be applied to surface elements.

The prediction of component strengths should only be applied to situations for which the peak tensile stresses lie well within the stress range of the test data used to characterize $g(S)$. Extrapolation to lower stresses quickly expands the confidence limits on $g(S)$ to unacceptably large values. This is a very important point that is not always appreciated. However, in order to obtain data at relatively low stress levels, either a very large number of tests must be performed, or the tests must be conducted on samples that subject a large volume of material to the peak stress, e.g. tensile tests. This is a constraint that can limit the effectiveness of stress state and size predictions. Also, it is emphasized that statistical strength predictions only apply if the flaw population remains invariant from sample to sample [1] (or component to component).

4. IMPACT DAMAGE

Recent studies of the fracture induced by the impact of projectiles on brittle materials have begun to indicate the relationships between the material (target and projectile) parameters and the extent of the fracture [21,22]. It has been established that at least two regimes of fracture can occur. The first arises when the projectiles are relatively deformation resistant (e.g. glass, WC, Al_2O_3, SiC), while the second damage mode develops when the projectiles are deformable (e.g. Nylon, water). The damage modes observed in these regimes, indicated in Fig. 8 are

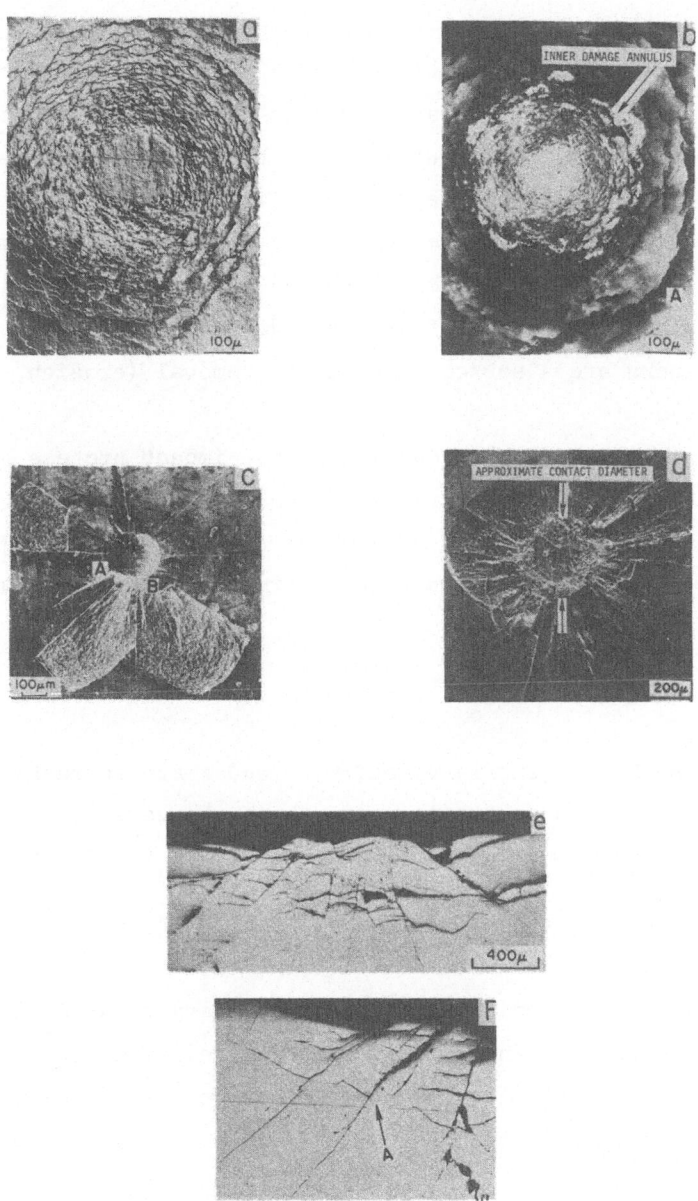

Fig. 8. Micrographs indicating the nature of the damage caused
by projectile impacts (a,b) by incompressible projectiles
(c,d,e,f) by compressible projectiles.

referred to as the 'plastic response' and 'elastic response' regimes by analogy with quasi-static indentation damage [23].

The damage in the 'plastic response' regime consists of (i) radial cracks that extend radially from the contact zone on planes roughly orthogonal to the surface, and thus develop into essentially semi-circular surface cracks, (ii) lateral cracks that initiate at the plastic zone and extend on planes approximately parallel (but slightly inclined) to the surface. The radial cracks are a prime source of strength degradation, whereas the lateral cracks are a source of material removal (erosion, abrasive wear, etc.).

An evaluation of the mechanics of the impact process have enabled the lengths of the radial cracks, C_r, and the depth, h, of the lateral cracks to be related to the target and projectile parameters (Fig. 9). In functional form these relations are: for radial cracks,

$$\frac{C_r}{v_o} \approx F_1 \left(\frac{\Gamma}{z}\right) F_2 \ (R_P) \tag{28}$$

where v_o is the projectile velocity, Γ and z are respectively the fracture energy and acoustic impedance of the target, R_P is the projectile radius and F_1 and F_2 are functions; for lateral cracks,

$$\frac{h}{R_P} = F_3 \left(v_o \sqrt{\frac{\rho_P}{H^*}}\right) \tag{29}$$

where H^* is the 'dynamic' hardness of the target and ρ_P is the projectile density. The details of these relationships are presented in Figs. 9 and 10. The curves can be used to directly predict the post-impact strength, σ_d, using

$$\sigma_d \approx \frac{K_{Ic}}{2} \sqrt{\frac{\pi}{C_r}} \tag{30}$$

and to estimate the material removal rate per impact, $\hat{v}$, using

$$\hat{v} \approx \pi \ C_r^{\ 2} \ h \tag{31}$$

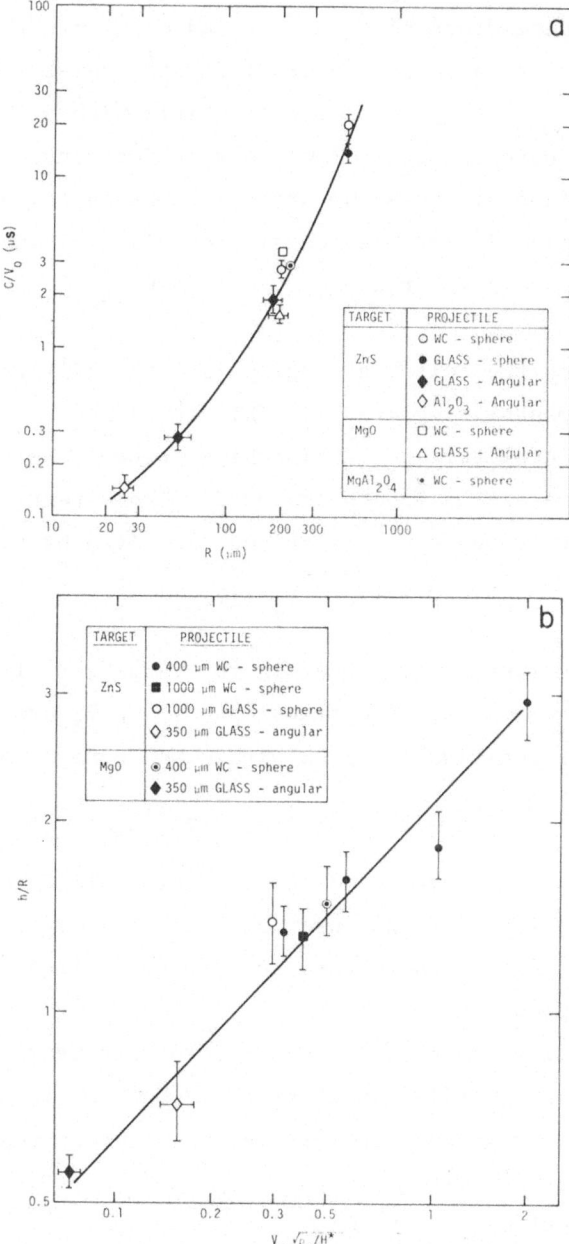

Fig. 9. Quantitative relations for the damage created by the
impact of projectiles, (a) the radial crack extension,
(b) the damage penetration as manifested by the lateral
crack depth.

The important aspects of damage relations that should be noted are the strong dependence of C_r on Γ/z and of h on H^*. These relations indicate that a good overall damage resistance in this regime can be realized by selecting materials with a large fracture energy and hardness and a low acoustic impedance. The optimum combination of these properties, in presently available materials, obtains in hot pressed silicon nitride, emphasizing the potential of this material for use in aggressive projectile environments.

The damage in the 'elastic' regime is quite different, and consists of segmental ring cracks. This damage has not yet been quantified, but qualitative relationships between the damage and the target parameters have been proposed. The extent, Δr, of the zone of fracture around the contact zone is given by;

$$\Delta r = F_4 \ (1/c_t) \ F_5 \ (1/S_c) \tag{32}$$

where c_t is the elastic wave speed in the target, S_c is the 'strength' of the near surface flaws and F_4 and F_5 are functions. The length (C) of individual cracks within the damage zone is;

$$C = F_5 \ (1/c_t) \ F_6 \ (K_{Ic}) \ F_7 \ (1/S_c) \tag{33}$$

In both cases a large wave speed (high modulus, low density) is important, as well as a large 'near surface' flaw strength and a high toughness. Again, therefore, hot pressed silicon nitride should exhibit good damage tolerance.

Further studies are clearly required to further elucidate the vagaries of the projectile impact problem, but the initial studies provide preliminary guidelines for designing impact resistant nitride ceramics.

References

1. A. G. Evans and T. G. Langdon, Progress in Materials Science, 19 (1976).

2. A. G. Evans, A. H. Heuer and D. Porter, Fourth International Fracture Conference, to be published.

3. C. C. Wu, S. W. Freiman, R. W. Rice and J. J. Mechdsky, Bull. Am. Ceram. Soc., 55 (1976) 395 (abstract only).

4. A. G. Evans and M. Linzer, Annual Review in Materials Science, in press.

5. F. Erdogan and G. Gupta, Intl. Jnl. Frac. 11 (1975) 13.

6. A. G. Evans and E. Charles, Acta Met., in press.

7. A. G. Evans and L. J. Graham, Acta Met., 23 (1975) 1303.

8. F. A. McClintock and F. Zaverl, Intl. Jnl. Frac., to be published.

9. F. A. McClintock and H. J. Mayson, ASME Applied Mechanics Conf. (June 1976), in press.

10. P. L. Gutshall and G. R. Gross, Eng. Frac. Mech., 1 (1969) 463.

11. R. W. Rice and S. W. Freiman, Bull. Am. Ceram. Soc., 55 (1976) 395 (abstract only).

12. N. E. Claussen and J. Steeb, Jnl. Amer. Ceram. Soc., to be published.

13. F. F. Lange, Jnl. Amer. Ceram. Soc., 57 (1974) 84.

14. G. I. Barenblatt, Adv. Appl. Mech., 7 (1962) 55.

15. B. R. Lawn and T. R. Wilshaw, Fracture of Brittle Solids (Cambridge Univ. Press) Cambridge, 1975.

16. J. R. Matthews, F. A. McClintock and W. J. Shack, Jnl. Amer. Ceram. Soc., 59 (1976) 304.

17. W. Weibull, Jnl. Appl. Mech., 18 (1951) 293.

18. A. G. Evans and F. A. McClintock, ARPA MRC Report (1976), to be published.

19. C. R. Kurkjian, R. V. Albarino, J. T. Krause, H. N. Vazirani, F. V. DiMarcello, S. Tarza and H. Schonhorn, Appl. Phys. Lett., 28 (1976) 588.

20. S. Batdorf and J. G. Crose, Jnl. Appl. Mech. 41 (1974).

21. A. G. Evans and T. R. Wilshaw, Jnl. Mater. Sci., in press.

22. A. G. Evans, M. E. Gulden and M. Rosenblatt, to be published.

DISCUSSION (Boch, Venables)

Räuchle: Assuming a certain distribution of volume flaws, for example a hyperbolic distribution which corresponds to a simple Weibull statistic, then machining may give a different failure distribution. Fracture mechanics show that these surface flaws are more stress sensitive, and predominantly surface flaws are seen in a bend test. A turbine rotor, for example, would now be dependent on the m-value of Weibull statistics whether surface or volume flaws dominate. Only for low m-values (m < 8) would volume flaws contribute to the failure probability.

Evans: It is true that in bend tests, many of the fractures originate from surface flaws; these can be a mixture of machining flaws and exposed 'volume' flaws (inclusion, pores, large grains). The method described in this paper might enable stress ranges wherein either the machining flaws or the 'volume' flaws dominate the strength distribution characteristic to be identified; especially if tensile test data are also obtained, whereas single valued Weibull-type functions can never have this flexibility. It is important, of course, to be able to distinguish the stress ranges, and the separate strength distribution function for different flaw types, if size effects are to be predicted with good confidence.

I do not agree that it can be assumed, a priori, that surface flaws are controlling when m is large and that volume flaws are controlling when m is small. In fact, in both cases there is likely to be a mixture of surface and volume flaws controlling fracture, with the mix depending on the character (size distribution function) of each flaw type. However, it is certainly possible that in certain instances, complete surface (or volume) flaw control is likely to be observed over a <u>limited</u> range of stress.

Davidge: For ionically bonded ceramics the effective surface energy can be estimated by summing the various contributions from the thermodynamic value, a plastic deformation term, a surface roughness term, etc. Reasonable agreement is obtained between experiment and theory without invoking a phonon/photon term as you suggested. Is this term negligible for these materials but of greater significance for covalent material?

Evans: It is certainly possible to infer from comparisons of measured fracture energies with the 'known' contribution to the fracture energy from the 'surface' energy, plastic deformation, etc., that the phonon/photon term is very small for ionics but substantial for covalent materials. It is possible, in principle, to verify this result from first principles by computer simulation of bond rupture in these materials (the origin of the force, displacement characteristic). This type of calculation has not yet been performed, although McMillan has presented some initial work on several ionic systems.

Krauth: The impact resistance $(\frac{\Gamma}{z})$ of silicon nitride was predicted to be good because of a high Γ and a low z. How do values compare with those for alumina and silicon carbide?

Evans: The ratio of Γ/z for hot-pressed silicon nitride is substantially larger than for all known aluminas and silicon carbides, because Γ is usually larger and $z(=\sqrt{E\rho}\,)$ is smaller.

Clarke: The presence of the acoustic impedance in the impact deformation suggests that reflections of acoustic modes might feed energy back into the impact stress wave to cause cracking in the body of the material when they constructively overlap as grain boundaries. Is this phenomenon of any importance in these fine-grained ceramics?

Evans: The acoustic impedance enters the expression for the impact damage because it is the prime factor affecting the amplitude of the shock wave generated during initial contact. The observed deformation and fracture in fine-grained polycrystalline ceramics is relatively isotropic in character, and does not indicate that the shock wave interacts sufficiently with the grain boundaries to produce localised cracking by constructive interference.

Lange: What test differentiates between volume and surface flaw distributions?

Evans: It is only possible to be unequivocal about the distribution function (surface or volume) that should be used to characterise the fracture statistics if a flaw origin on each test sample is identified by subsequent microscopy. It should be noted, however, that it is often possible to infer from the strength data (by trying both the surface and volume function), the ranges of stress over which fracture in surface or volume is controlled.

Davidge: Mose people use Weibull statistics to analyse the strength data of ceramics, which although an empirical approach, does seem to work for many materials. Have you compared your more general approach to the Weibull method for particular ceramics, and if so, with what results?

Evans: The general approach often indicates multi-modal distributions, showing that a single valued function (such as the Weibull function) cannot really apply. However it is still possible in many cases to obtain a reasonably good fit to the data with a single valued function over the range of the data. Several difficulties then arise of course which may, or may not, be important. For example, no confidence can be applied to data extrapolation; whereas limited extrapolation can be effected when the general method is utilised. Also, problems that might arise from effects such as transitions from surface flow control to volume flow control cannot be easily overcome when best fit single valued functions are used.

MONTE CARLO TECHNIQUE FOR ESTIMATING UNCERTAINTIES IN LIFETIME PREDICTIONS*

J.E. Ritter, Jr. and D.C. Coyne

Mechanical Engineering Department, University of
Massachusetts, Amherst, Mass. 01002, U.S.A.

ABSTRACT

The uncertainty in lifetime predictions based on static and
dynamic fatigue strength data is estimated by a Monte Carlo
computer simulation technique and compared to that obtained from
the statistical law of propagation of errors. The results show
that the Monte Carlo technique is preferred, since it makes no
assumptions regarding linearity of the lifetime equations. This
estimated uncertainty in the lifetime predictions can be used to
place confidence limits on failure predictions. Although the
confidence limits may extend over several orders of magnitude, it
will be shown that much of the uncertainty in the predictions can
be eliminated by decreasing the allowable stress in service or by
increasing the required proof stress by a small increment.

---

*This work was supported by the National Science Foundation.

PROPERTIES OF POLYPHASE CERAMICS

R.G. Cooke

School of Materials Science, University of Bath,
Bath, U.K.

ABSTRACT

   Nearly all ceramic materials of engineering interest can be
regarded as composites of more than one phase in a general sense.
The problem of the prediction of the properties of the overall
material from a knowledge of the behaviour of the individual
component phases is receiving considerable theoretical attention.
It is not the purpose of this discussion to review in any detail
the progress that has been made, whilst noticing that considerable advances have been made in predicting properties such as
elastic moduli, thermal expansion and conductivity, dielectric
constants and so on. Rather, it is hoped to examine some of the
difficulties that arise and particularly suggest possible
approaches to problems such as fracture and electrical breakdown
where 'weakest link' as opposed to 'average' features dominate.

   One particular aspect of the composite problem is that in
the cases considered so far, the different components have
usually been assumed to be made of phases of widely differing
properties and particular geometrical arrangements (e.g. fibre
reinforced systems). It might therefore be useful to consider
also the problem of how different two such 'phases' (in the
broadest possible sense, i.e. including porosity, cracked regions
and small scale inhomogeneities) have to be before phenomena such
as fracture will be modified significantly. An experimental
programme is being started at Bath to investigate just this
situation using model systems.

ASSURING MECHANICAL RELIABILITY OF CERAMIC MATERIALS*

J. E. Ritter, Jr.

Mechanical Engineering Dept., University of Massachusetts
Amherst, MA   01002

ABSTRACT. Fracture mechanics provides the background for making
long-term failure predictions to assure the mechanical reliability
of ceramics. The crack growth parameters necessary for making these
failure calculations can be obtained directly from crack velocity
measurements or indirectly from static and dynamic fatigue strength
experiments. These failure calculations are best understood by
expressing the predictions in terms of a design diagram. From a
design diagram the probability of failure in service can be obtained
as well as the proof test stress necessary to insure a minimum lifetime in service.

1. INTRODUCTION

Today, ceramic materials are being used in structural applications
where tension stresses must be accepted and where a highly reliable
and safe structural element must be achieved. Unfortunately, ceramics exhibit delayed failure (commonly known as static fatigue) and
a wide variability in fracture strength that complicates designing
for their mechanical reliability. This paper will review the fracture mechanics foundation for making failure predictions for ceramics based on fatigue strength data and will show how these predictions can be incorporated into a design diagram. Design diagrams
are most useful to the engineer in deciding if a proof test is
necessary and, if necessary, in determining the proof test stress
required to assure a given lifetime.

---

* This work was supported by the National Science Foundation.

2. FUNDAMENTALS

Wiederhorn and Evans [1-3] have provided a sound, fundamental background for making failure predictions and for the determination of proof test stresses and the establishment of proof test conditions. Their analysis is based on the reasonable assumption that failure of glass occurs mainly from stress-dependent growth of pre-existing flaws to dimensions critical for spontaneous crack propagation. By assuming a power functional relationship between crack velocity during the period of subcritical crack growth and the applied stress intensity factor, they derived that the time to failure (t_f) under a constant applied tensile stress (σ_a) is:

$$t_f = [\frac{2}{A \, Y^2 \, (N-2) \, K_{IC}^{N-2}}] \, S_i^{N-2} \, \sigma_a^{-N} \tag{1}$$

where K_{IC}=critical stress intensity factor, S_i=fracture strength in an inert environment, Y=geometric constant, A and N=constants. For a given glass and test environment, the term in brackets in Eq. (1) is constant, hence,

$$\log t_f = \log B + (N-2) \log S_i - N \log \sigma_a \tag{2}$$

where B=constant. By using the average value of S_i, it can be seen from Eq. (2) that a log-log plot of t_f as a function of σ_a gives a straight line from which the constants B and N can be determined.

Wiederhorn and Evans [1-3] also derived that the fracture strength (S) at a constant stressing rate $(\dot{\sigma})$ is:

$$S^{N+1} = B \, (N+1) \, S_i^{N-2} \, \dot{\sigma} \tag{3}$$

Using Eq. (3), B and N can be determined from a log-log plot of S as a function of $\dot{\sigma}$ and using the average S_i.

The probability of failure for a given t_f and σ_a (or a given S and $\dot{\sigma}$) can be obtained from Eq. (2) (or Eq. (3)) by expressing the inert strength (S_i) in terms of its failure probability distribution. By so doing, we assume that the sample with the shortest fatigue life has the lowest inert strength and that the origin of fracture is the same for both fatigue and inert failures. As will be shown, this statistical approach often leads to low design values for the allowable stress in service if low failure probabilities are required. Fortunately, the proof test method removes this disadvantage of the statistical approach to fatigue and assures that every specimen surviving the proof test will have a minimum service life since weaker specimens are eliminated. Procedures of proof testing and precautions that must be followed during proof testing are discussed fully elsewhere [3]. For our purposes we will only note that for effective proof testing the proof test should be conducted in a relatively inert environment and that moderately rapid unloading rates should be used. Also, the proof test must duplicate in the component the actual state of stress expected in service; and, after proof testing, the components must not incur damage that will negate the value of the proof test. In summary, if these precautions are followed, proof testing can be a practical method for assuring the mechanical reliability of ceramics [3].

From a fracture mechanics viewpoint, the value of proof testing is that it characterizes the largest effective flaw possible in the tested components since any larger flaw would have caused failure during the proof test. The minimum time to failure after proof testing is the time it takes for this maximum flaw to grow to critical dimensions for spontaneous fracture and is given by [1-3]:

$$t_{min} = B \, \sigma_p^{N-2} \, \sigma_a^{-N} \tag{4}$$

where σ_p=proof test stress. By comparing Eq. (4) with Eq. (2) it is seen that σ_p simply represents the minimum inert strength of the components after proof testing.

In summary, Eq. (2) and (4) give lifetime predictions before and after proof testing and are dependent on the inert strength distribution and the fracture parameters B and N. These parameters must be measured under the service conditions and can be obtained from fracture mechanics measurements [2] or from either of two types of fatigue strength experiments. In one, the time to failure under constant stress is determined as a function of applied stress and Eq. (2) is used to calculate B and N. This type experiment is commonly termed static fatigue [4]. In the other, fracture strength is measured as a function of stressing rate and Eq. (3) is used to calculate the two parameters. This type of experiment is known as dynamic fatigue [4]. Being able to base failure predictions solely on strength data is most significant since it cannot always be assumed that data from large, preformed cracks (as present in fracture mechanics samples) are relevant to the propagation of microscopic cracks present in structural ceramics.

3. DESIGN DIAGRAMS

The failure prediction techniques discussed above are most easily understood in terms of a design diagram. A design diagram for a ceramic in a given environment gives the probability of failure for a given lifetime and applied stress and also the proof stress necessary to insure a minimum lifetime at the applied stress in service. To illustrate the construction and use of these diagrams, dynamic

484

fatigue data [5] for hot pressed silicon nitride in air at 1400°C was analyzed.

A least square analysis of the data [5] gives:

$$\log S \ (MN/m^2) = 2.38 + 0.082 \log \dot{\sigma} \ (MN/m^2 \cdot s) \tag{5}$$

Using the average inert strength as measure at room temperature, 509.86 MN/m^2, the constants in Eq. (3) can be determined to be log B=3.1 and N=11.2. The inert strength distribution of these samples can be reasonably assumed to be given by the Weibull relationship:

$$\ln \ln \frac{1}{1-F} = 8.0 \ln S_i/533.76 \tag{6}$$

where F=failure probability. Fig. 1 gives the design diagram based on this data where the curves for failure probability are obtained by coupling Eq. (6) with Eq. (2) and the proof test ratio (σ_p/σ_a) from Eq. (4).

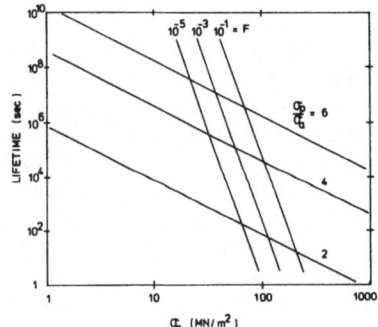

Fig. 1.   Design diagram for hot pressed silicon nitride based on dynamic fatigue data in air at 1400°C [5].

Design diagrams are most useful to the designer in deciding if a proof test is necessary and, if necessary, in determining the proof test required to insure a given lifetime. For example, from Fig. 1 a failure probability of 0.1 is predicted for a lifetime of 10^{5}s under an applied stress of 100 MN/m^2. To assure no failures under these same service requirements, a proof test ratio of 4.2 would be necessary.

It should be realized that the failure probability curves in Fig. 1 are based on laboratory specimens and have not been corrected

for the fact that these samples will not have the same inert strength
as actual components. This correction factor could be predicted
from Weibull statistics [3] or experimentally determined.

It is important to note that fracture mechanics data [6] for
this same hot pressed silicon nitride at 1400°C give quite different
fracture parameters (log B=6.0 and N=5.5) than those based on the
dynamic fatigue data above. These parameters obtained from fracture
mechanics experiments yield different failure predictions which em-
phasizes the importance of experimentally demonstrating agreement
between fracture mechanics and strength data. For the case at hand,
this disagreement between the two sets of data is probably related
to the fact that the microscopic flaws present on the strength
samples interact differently with the microstructure than the large,
preformed cracks present on fracture mechanics samples.

Finally, it should be realized that there is an uncertainty in
the failure predictions connected with the design diagrams due to
the experimental scatter associated with determining the fracture
parameters and inert strength. Statistical techniques have been
derived to estimate this uncertainty in the failure predictions [7]
and the analysis of typical fatigue strength data show that this
uncertainty is quite sensitive to the experimental error in the
fracture parameters. Fortunately, much of this uncertainty in the
failure predictions can be eliminated either by increasing the required proof stress or by decreasing the allowable stress in service
by a small increment [7].

4. EFFECTIVENESS OF PROOF TESTING

The minimum lifetime predictions after proof testing are only valid
when no crack growth occurs on unloading during the proof test.
Crack growth during loading and at the proof stress presents no
problem since we can still characterize the maximum flaw size. However, if flaw growth occurs on unloading, then the maximum flaw size
cannot be defined and no assurance of a minimum service life can be
given. Before proof testing can be used as a means for assuring a
minimum lifetime, the effectiveness of the proof test in truncating
the inert strength at σ_p should be experimentally determined.

Evans and Wiederhorn [1] have shown that the inert strength
after proof testing (S_i'), considering crack growth during loading
but not unloading, is:

$$\left(\frac{S_i'}{S_i}\right)^{N_p-2} = 1 - \left(\frac{\sigma_p^*}{S_i}\right)^{N_p-2} \left[1 - \left(\frac{\sigma_p}{\sigma_p^*}\right)^{N_p-2}\right] \tag{7}$$

where N_p is the crack propagation parameter appropriate for the proof test conditions and σ_p^* is the equivalent proof stress under inert conditions (i.e., where no crack growth occurs during proof).

Using the initial Weibull inert strength distribution, σ_p^* is determined from:

$$\ln \ln \left(\frac{1}{1-F_p} \right) = m \ln \left(\frac{\sigma_p^*}{S_o} \right) \tag{8}$$

where F_p=failure probability during proof testing and m and S_o= Weibull inert strength constants. The failure probability after proof testing (F_a) is related to the initial failure probability before proof testing by:

$$F_a = \frac{F - F_p}{1 - F_p} \tag{9}$$

Eq. (7) can now be rewritten to give:

$$\left(\frac{S_i'}{S_o}\right)^{N_p-2} = \left(\ln \frac{1}{1-F_a} + \ln \frac{1}{1-F_p}\right)^{\frac{N_p-2}{m}} - \left(\ln \frac{1}{1-F_p}\right)^{\frac{N_p-2}{m}} + \left(\frac{\sigma_p}{S_o}\right)^{N_p-2} \tag{10}$$

Thus, from Eq. (10) it is seen that the inert strength distribution after proof testing depends upon the initial inert strength distribution parameters m and S_o, the crack propagation parameter N, and the failure probability during proof F_p. It is also significant to note that the S_i' distribution is truncated at σ_p and that the inert strength after proof testing will be greater than the initial inert strength at all levels of probability if $m < N_p-2$.

REFERENCES

1. A. G. Evans and S. M. Wiederhorn, Int. J. Fract. Mech., 10, 379 (1974).
2. S. M. Wiederhorn, in Fracture Mechanics of Ceramics, Vol. 2, ed. by R. C. Bradt, D. P. Hasselman, and F. F. Lange, Plenum Press, New York, 1974.
3. S. M. Wiederhorn, in Ceramics for High Performance Applications, ed. by J. J. Burke, A. E. Gorum, and R. N. Katz, Brook Hill Publishing Co., Chestnut Hill, 1974.
4. J. E. Ritter, Jr. and J. A. Meisel, to be published in J. Am. Ceram. Soc., 1976.
5. F. F. Lange, J. Am. Ceram. Soc., 57, 84 (1974).
6. A. G. Evans, et al., Metall. Trans., 6A, 707 (1975).
7. D. F. Jacobs and J. E. Ritter, Jr., to be published J. Am. Ceram. Soc., 1976.

DISCUSSION (Boch, Venables)

Katz: In the expression for t_f it is rather disconcerting that K_{IC} is in the denominator. Could you explain this apparent anomaly?

Ritter: To a good approximation it can be shown that the lifetime of a ceramic material does not depend directly upon K_{IC}. The lifetime can be derived from the following fundamental relations:

$$K_I = Y \sigma_a \sqrt{a} \quad \text{and} \quad V = \frac{da}{dt} = A K_1^N$$

to be

$$t_f = \frac{2}{\sigma_a^2 Y^2} \int_{K_{I_i}}^{K_{IC}} (K_I / A K_I^N) dK_I = \left[\frac{2}{(N-2)A \sigma_a^2 Y^2} \right] \left[K_{I_i}^{2-N} - K_{IC}^{2-N} \right]$$

Since $K_{IC} \gg K_{I_i}$ for structural ceramics, then to good approximation:

$$t_f = \left[\frac{2}{(N-2)A \sigma_a^2 Y^2} \right] K_{I_i}^{2-N}$$

Thus, from this relationship, it is seen that t_f does not depend directly on K_{IC}. However, there is some evidence that a relationship between A and K_{IC} exists where higher K_{IC} corresponds to lower A. If this is true, then higher K_{IC} values will result in longer lifetimes. It should be noted that the t_f equation given in my paper can be derived from that above by substituting in the equality $K_{I_i} = K_{IC}(\sigma_a / S_i)$.

Popper: I have an uneasy feeling that in this field unjustified extrapolations are being made. You quoted 6 weeks lifetime - in practice it would have to be more like 6-20 years and I do not see how you can safely make long term predictions from short term experiments.

Ritter: There are always assumptions involved in any extrapolation of experimental data. Based on both experimental static fatigue data and theoretical reasons, there is no reason to believe that the basic mechanism of slow crack growth changes during the lifetime of a sample; thus, one would expect the extrapolations to be valid. However, in some systems it is possible that there is a fatigue limit below which no fatigue occurs. For these systems the extrapolation of data to stress levels below the fatigue limit will be most conservative. In summary, in order to minimise the extrapolation of experimental data, static fatigue tests should be carried out to times that are as long as it is economically feasible.

APPLICATION OF PROOF TESTS TO SILICON NITRIDE COMPONENTS

N.J. Tighe

Inorganic Materials Division, National Bureau of
Standards, Washington, D.C., 23234, U.S.A.

ABSTRACT

In the ceramic turbine programme, the high cost of turbine
components and the disastrous effect of component failure require
development of both destructive and non-destructive tests for
detecting and rejecting defective components. Proof tests, in
which a load is applied to break weak components, will ensure that
the survivors will have a probability of failure that is acceptable
for the design requirements. Thus, the development of a reliable
proof test programme requires testing a statistically significant
number of specimens. Proof tests have been applied to assure the
reliability of glass components and are being developed for polycrystalline ceramic components, such as silicon nitride. The
initial proof tests are being done with 4-point bend specimens of
hot-pressed and reaction-bonded silicon nitride. The tests are
conducted at room temperature and at the maximum useful temperature
of the materials.

DISCUSSION (Boch, Venables)

Diness: Have you paid attention to anisotropy of microstructure
in the way in which you cut and finish the hundreds of specimens
from each billet of hot pressed Si_3N_4? If so, do you observe
differences in toughness and strength and in probability of failure
as a function of microstructure?

If there do exist "strong and whole" directions in the
material, then proof test (and other mechanical data) must be
characterised with reference to this anisotropy.

Information already in the literature (Lange) indicates that hot pressed Si_3N_4 is anisotropic in microstructure. Work by Bonsal (to be reported at the Berkeley Ceramic Microstructure Conference, August 1976) claims very large (about 20% or more) differences in fracture energies for specimens with major axes parallel or normal to pressing direction. Such work, if correct, needs to be kept in mind in specifying the character of the material, its mechanical properties, and in the manner it is used in components relative to microstructural features. Alternatively, work may be needed on minimisation of microstructural anisotropy in order to avoid its undesirable effects.

Tighe: We have not looked at the effects of billet anisotropy on the strength of silicon nitride. This was done earlier by Dr. Lange. Our proof test programme is directed towards evaluating the proof test and the initial testing will be done with the specimens axis aligned normal to the hot pressing direction.

Lange: Using bar specimens to obtain statistical strength data introduces one problem, edge cracks due to cutting the specimen. Such cracks are, of course, found at edges of components and their distribution should be separated from surface and volume cracks; thus, in calculating the failure probability of a component, edge flaw distributions should be used for edge volumes, surface flaw distributions for surfaces, volume elements, and volume flaw distributions for interior volume elements.

Godfrey: Going to the biaxial flexure test (NBS disc) to avoid edge problems introduces the problem of grinding/polishing damage transverse to the applied tensile test. Polishing to a high quality specular finish of RBSN gave a 7% reduction in the biaxial flexure failure stress (compared with as-fired surfaces). The effect of edge damage in HPSN could be investigated by comparing square and circular section bend bars, as the Nottingham University group did for RBSN (Sivill, Ph.D. Thesis); with specimens shaped before nitridation, there was no difference in the unit volume strength (Weibull and friction corrected).

DENSE Si_3N_4: INTERRELATION BETWEEN PHASE EQUILIBRIA, MICROSTRUCTURE AND MECHANICAL PROPERTIES

F. F. Lange

Westinghouse Research Laboratories
Pittsburgh, PA 15235 U.S.A.

ABSTRACT. The mechanical properties of dense Si_3N_4 will be reviewed from the viewpoint of related fabrication parameters and compositional content. It will be shown that the microstructure of dense Si_3N_4 is governed by the phase relations of the constituents in the starting powder and that these phase relations, in turn, govern the mechanical behavior.

1. INTRODUCTION

Powder routes are used to fabricate dense Si_3N_4. Deeley [1] was the first to discover that pore-free bodies could be fabricated by hot pressing Si_3N_4 powder containing a densification aid. Many metal oxides, e.g., MgO, BeO, Al_2O_3, Y_2O_3, CeO_2, etc., are known densification aids [1,2,3,4]. Although hot pressing currently results in a superior product, the feasibility of pressureless sintering, with incorporated densification aids, has been demonstrated [5,6,7].

Densification of Si_3N_4 powder with the aid of a metal oxide is generally attributed to the presence of a liquid at high temperatures according to the general reaction:

$$Si_3N_4 + SiO_2 + \text{metal oxide} + (\text{impurities}) \xrightarrow{\text{Temp.}} Si_3N_4 + \text{liquid}, \quad (1.1)$$

where SiO_2 is unavoidably present on the surface of each Si_3N_4 partical and/or incorporated in the powder in the form of Si_2N_2O. Above the solidus temperature, all four constituents of the LHS of (1.1) will react to form a liquid. Neglecting the possible mass losses due to volatilization [5], the composition of the liquid and the equilibrium fraction of solid Si_3N_4 will depend on 1) the composition of the starting powder (i.e., the LHS of Eq. (1.1)), 2) the phase equilibria of the composite system, and 3) the densification temperature. The role of the liquid is to 'pull together' the Si_3N_4 particles and to allow the solution-repreciptation of Si_3N_4, both of which result in the disappearance of voids and thus, densification.

Upon cooling from the densification temperature, the liquid solidifies according to the general reaction:

$$Si_3N_4 + liquid \xrightarrow{cooling} Si_3N_4 + secondary\ phases, \qquad (1.2)$$

illustrating that Si_3N_4 is a polycrystalline, polyphase material.

The number, chemistry and content of the secondary phases depend on the composition of the starting powder and the phase relations in the given composite system. In addition, the crystal structure of Si_3N_4 can be altered by the concurrent substitution of certain metal cations for silicon and oxygen for nitrogen [8,9]. As experience has shown, the secondary phases and/or the solid solution alloying of Si_3N_4, can significantly influence all properties.

The following text will be split into three interrelated sections: First, fabrication will be reviewed in terms of starting powders, phase relations, the $\alpha \rightarrow \beta$ phase transformation and resulting microstructures. Second, the theoretical effects of second phases on the low and high temperature mechanical properties will be discussed. And, third, the observed relations between microstructure and mechanical properties will be discussed in terms of the starting Si_3N_4 powder, effect of inclusions size, effect of impurities and compositional effects in two Si_3N_4-SiO_2-MO systems.

2. FABRICATION

2.1 Starting Powders

Silicon nitride exists as two hexagonal crystal structures, viz. α-Si_3N_4 (P31C) [10] and β-Si_3N_4 (P6$_3$/m) [11]. At first, the α and β structures were thought to be, respectively, low and high temperature polymorphs; hence, their designation [12]. Then, based on a structure refinement with x-ray powder pattern data, Jack and co-workers [13] suggested that ~1.5 w/o oxygen was incorporated into the α structure, which lead to the conclusion that it was an oxynitride with the approximate formula of $Si_{11.5}N_{15}O_{0.5}$. Previous workers (Marchand, et al. [11]) and later workers (Kohatsu and McCauley [14] and Kato, et al. [15]) used single crystal x-ray data to refine the α structure. These three groups concluded thet the α structure consisted of only Si and N atoms. Numerous workers [7,16,17,18] have also shown that $\alpha-Si_3N_4$ powder can have much lower oxygen contents than that hypothesized by Jack, e.g., Kato et al. [15] report on oxygen content of 0.05 w/o. Thus, most current workers believe that both α and β structures can be pure Si_3N_4. Although several suggestions have been made,* the thermodynamic interrelation between the two structures is still unknown.

The oxygen content of different Si_3N_4 powders ($\alpha/\beta > 8$) has been reported to range between 0.5 - 3.0 w/o, depending on the manufacturer and the particular batch [7]. Assuming that all the oxygen is in the form of SiO_2, the SiO_2 molar fraction can range between 0.02 to 0.12, with commercial Si_3N_4 powders in the higher end of this range. Thus, SiO_2 is an important constituent in the starting powder.

The major cation impurities in Si_3N_4 powder are Fe, Ca and Al, reflecting the impurities in the raw Si used to manufacture Si_3N_4 and a possible metal addition to promote the nitriding of Si powder [20]. The content of the major impurities can range between 10-5000 ppm (e.g., the range for Ca), depending on the purity of the raw Si prior to nitriding. A microprobe analysis of various Si powders by the author has indicated that the three major impurities can exist as oxides, silicates, and with the exception of Al, as silicides [21]. The effect of impurities on the mechanical properties will be reviewed in a latter section.

The conventional method of preparing Si_3N_4 powder (containing SiO_2 and impurities) for densification consists of mixing in the MO densification aid and reducing the particle size of the composite powders by wet ball milling. Milling can be carried out in containers which will not add additional contaminates except volatile hydrocarbons. The commonly used WC milling media is known to contaminate the powders. Much of the WC contamination can be removed by subsequent classification [22]. Dry alcohols are used during milling to minimize the hydrolization of Si_3N_4 to SiO_2. Measurements before and after milling have shown a negligible change in oxygen content.

Thus, the constituents of the starting 'Si_3N_4' powder are Si_3N_4 ($\alpha/\beta > 8$), metal oxide, and various impurities (Fe, Ca, Al, WC and hydrocarbons).

---

* Observations indicate that the α structure is formed by a vapor reaction which lead Blegan [19] to suggest that α is metastable at all temperatures.

2.2 Phase Relations

Phase equilibrium studies of various Si_3N_4-oxide systems are only recent and in many cases, SiO_2 was an unrecognized component. As pointed out by Gauckler et al. [8], one can not apriori consider any Si_3N_4-SiO_2-MO system an ordinary ternary system,* but if the reciprocal reaction $Si_3N_4 + 6/x\ M_2O_x = 3SiO_2 + 4/x\ M_3N_x$ (where x = the valence state of the metal ion) can be assumed the system can be represented by a square diagram with Si_3N_4, $3SiO_2$, $6/x\ M_2O_x$ and $4/x\ M_3N_x$ located at the four appropriate corners. The composition at each point within the diagram can be represented by an equivalent cation concentration (e.g., $[M^{+x}]$), and an equivalent anion concentration (e.g., $[O^{-2}]$). Such phase representation has been reported to illustrate the tie lines observed at 1750°C for the Si_3N_4-Al_2O_3-SiO_2-AlN [8,23] and Si_3N_4-BeO-SiO_2-Be_3N_2 [9] systems, and to illustrate the Si_3N_4-MO solid solutions that exists within these systems at a cation/anion ratio of 3/4.

The two important phase diagrams illustrated in Figs. 1 and 2, viz the Si_3N_4-MgO-SiO_2 [24] and Si_3N_4-Y_2O_3-SiO_2[25] systems, are represented in the conventional and more convenient manner. For these two systems, a tie line exists between all or most of the Si_3N_4-MO join. For all practical reasons, this ordinary representation will suffice. The tie lines in both diagrams were obtained by the author (phase identification by x-ray diffraction analysis) for compositions hot pressed at temperatures between 1500-1750°C.° High purity, low oxygen (0.7 w/o) α-Si_3N_4 powder (α/β = 9) was used to obtain these data.

As obviously shown in either diagram, the secondary phases change from one compatibility triangle to another. Not as obvious, but pertinent to material control, the type and amount of each secondary phase can be easily changed at low MO contents when different batches of Si_3N_4 are used which contain different SiO_2 contents. The effect of compositional changes on mechanical properties in these two systems will be described in a later section.

2.3 The $\alpha \rightarrow \beta$ Phase Transformation

Investigations have shown that an $\alpha \rightarrow \beta$ phase transformation occurs during fabrication [26,27,28]. It has also been shown that this transformation occurs both during and after densification,

---

* That is, a true binary may not exist along the Si_3N_4-MO join.
° Each diagram represents an isothermal section without defined regions of liquid. A constant temperature could not be employed since some compositions would exceed their liquidus temperature and destroy the hot-press dies.

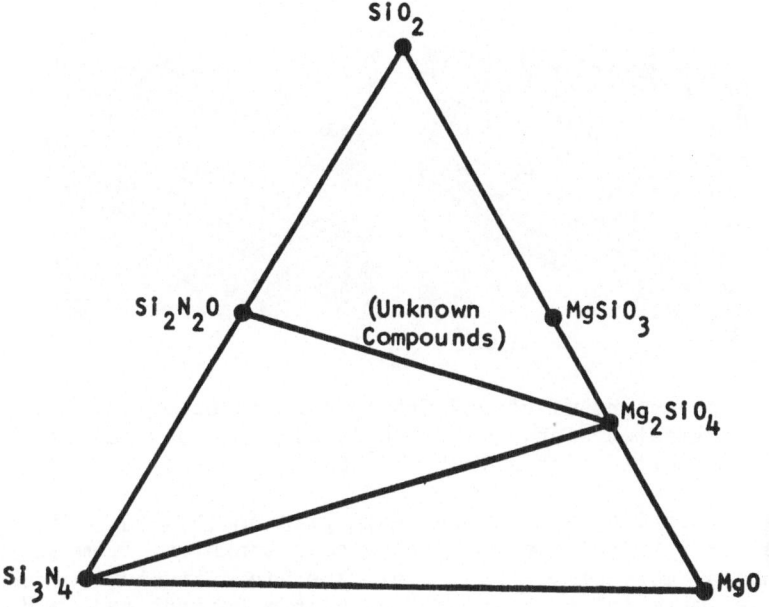

Fig. 1.   Tie lines observed in the Si_3N_4–SiO_2–MgO system between 1000°C and 1750°C.

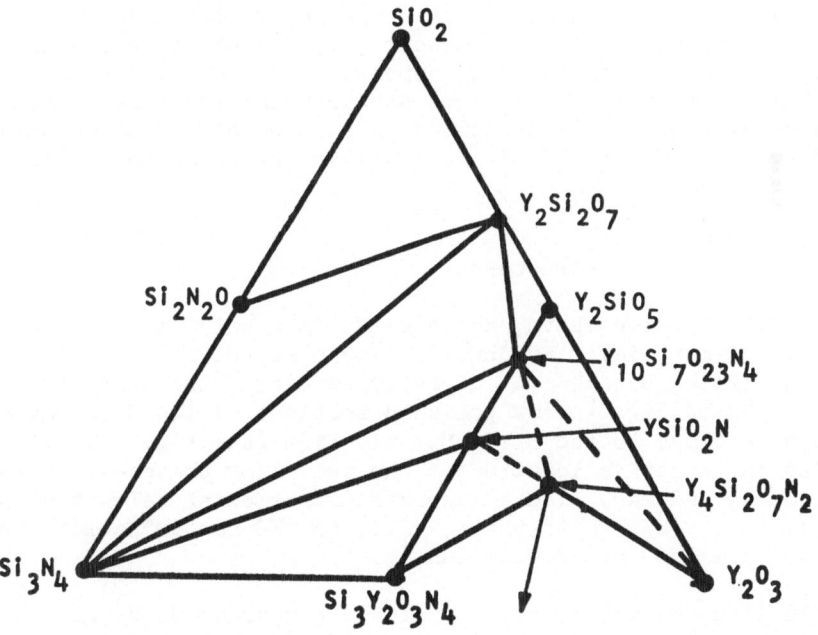

Fig. 2.   Tie lines observed in the Si_3N_4–SiO_2–Y_2O_3 system between 1500°C and 1750°C.

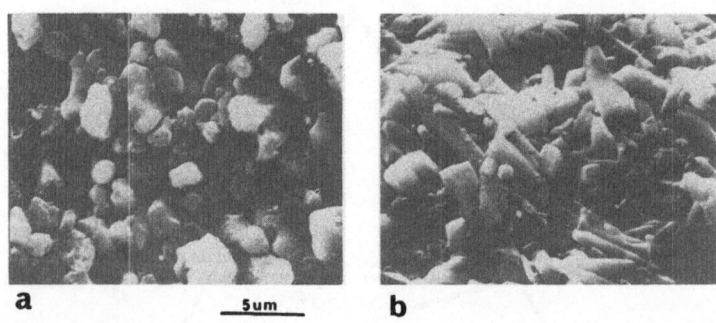

Fig. 3. Micrographs of etched fracture surfaces of (a) material
fabricated with β (α/β ∿ 0.1) Si₃N₄ powder and (b) material
fabricated with α (α/β = 9) Si₃N₄ powder.

i.e., densification is the more rapid process [27,28]. Simultaneous
with the α → β transformation, the grains transform from an equiaxed
to a fibrous, morphology [28], i.e., the microstructure of the fully
transformed material with an initial high α content (α/β > 8),
can be described as resembling a log and twig tangle. Observations
(e.g., Fig. 3) have suggested that the aspect ratio (R) of the
fibrous grains are related to the α/β ratio, i.e., R = 1 + α/β and
that the diameter of the fibers is similar to the size of the β
particles within the initial powder [28]. That is, grain growth
during and after densification appears to be restricted to growth
parallel to the fiber axis. The strength and fracture toughness
of material fabricated with either high α or high β phase powders
are discussed in terms of their resulting grain morphologies in a
later section.

2.4 Resulting Microstructures

     As illustrated above, dense Si₃N₄ is a polyphase, poly-
crystalline material. Evidence for this is not only obtained by
x-ray analysis, but also by directly examining the microstructure.
Micrographs in Fig. 4 show polished sections of two Si₃N₄ materials
fabricated close to the Si₃N₄-MgO tie line (see Fig. 1), illustrating the relative location of the two major phases—in this
case, Si₃N₄ and MgO. These microstructures exist below the solidus
temperature. (The white phase, viz., the WC contamination, should
be neglected for this discussion.)

     As illustrated, groups and single grains of Si₃N₄ (lighter
phase) are surrounded by MgO (darker phase); this is clearly
evident for material containing the larger MgO content (Fig. 4a).
As the MgO content decreases (Fig. 4b), the average separation

Fig. 4. Unetched micrographs of material close to the Si_3N_4-MgO
tie line: a) 70 mole % MgO, b) 40 mole % MgO.

distance between the Si_3N_4 grains (or grain groups) decreases as expected by volume fraction considerations. At lower MgO contents (not shown), the MgO phase is difficult to observe,* but observers using electron optics have detected noncrystalline second phases between the grains in such materials [30,31,32].

Above the solidus temperature, these same microstructures will, after reaching equilibrium conditions, look different. The MgO will react with Si_3N_4 to form a liquid and residual Si_3N_4, i.e., the amount of second phase between the Si_3N_4 grains (or grain groups) will increase as the temperature increases above the solidus temperature.

Both the effect of solid second phases below the solidus and the presence of liquid second phases above the solidus will be discussed in later sections.

Other microstructural features important for strength considerations are laminar inclusions of second phases resulting from density variations in pre-compacted powders [33], voids formed by rapid sintering, incomplete sintering, etc., and inclusions accidentally incorporated prior to densification. Only the last feature will be discussed later.

3. THEORETICAL EFFECT OF SECOND PHASES ON MECHANICAL PROPERTIES

3.1 Secondary Phases as Inclusions

One question that arises when observing a microstructure containing secondary phases is: "What is their effect on strength?" This question arises because residual stresses are associated with secondary phases and second-phase inclusions are occasionally located at fracture origins.

The details describing the residual stresses associated with spherical, second-phase inclusions are described elsewhere for

---

* Three factors complicate this observation. First, close to the Si_3N_4 corner of the phase diagram, the amount of Mg_2SiO_4 in the material increases due to the fixed amount of SiO_2 in the starting Si_3N_4 powder. Second, as shown on the micrographs, the volume fraction (VF) of MgO in Si_3N_4 is much smaller than the mole fraction (MF), due to the relatively low molecular weight of MgO, viz., VF (MgO) = 11.2 MF(MgO)/[44 MF(Si_3N_4) + 11.2 MF(MgO)]. Third, if all Si_3N_4 grains were equally spaced by the MgO phase, the separation distance $S_o \sim VF/3(1-VF)$ D; when the grain size (D) ~ 1 μm, and VF = 0.1; $S_o \stackrel{\sim}{-} 400$ Å [29].

the case of differential thermal contraction upon cooling from a
fabrication temperature [34]. Other causes for stresses arising
around and within second phase inclusions are phase transformation
and, when under stress, differential elastic properties. In
effect, all inclusions are stress risers and probable locations
of cracks, i.e., inclusions are crack precursors.

The effect of second phase inclusions, as found in particulate
composites, on strength, fracture toughness and elastic properties
has been previously reviewed by the author [35]. The subject
enlarged upon here is the dependence of inclusion size on its
effectiveness as a crack precursor.

The formation of cracks adjacent (or within) inclusions
during the cooling of a given particulate-matrix combination have
been observed to depend on the size of the inclusions, viz., cracks
are observed to be associated with large inclusions and not with
small inclusions [36,37]. This size effect was explained after
analyzing the thermodynamics of crack extension and arrest in the
highly localized stress field associated with an inclusion [38].

The analysis for crack extension was based on the fact that
the two energy terms that change during cracking, viz., the strain
energy within and around the inclusion and the energy associated
with the surface formed by the crack, have different functional
dependences on the particle size (D). That is, the strain energy
term depends on the volume under stress (D^3) and the surface
energy term is proportional to the surface of the particle (D^2).
By determining change in these two terms with changing crack length
and by invoking Griffith's condition for fracture, the extension
of a small pre-existing crack was found to be dependent on both
the maximum stress (σ_m) at the inclusion-matrix interface and on
the inclusion size (D), i.e., crack extension would only occur
when

$$\sigma_m{}^2 D \geq \text{constant}, \qquad (3.1)$$

where the factors included within the constant are material
properties and dimensionaless quantities related to the pre-existing
crack [38]. Expressed in another manner, the analysis shows that
for a given maximum stress (σ_m), a critical particle size exists
(D_c) below which cracking will not occur:

$$D_c = \frac{\text{constant}}{\sigma_m{}^2} . \qquad (3.2)$$

Although this analysis is not rigorous for the geometries
of the polyphase regions within the Si_3N_4 considered here, it
suggests that the size of the second phase regions (and the Si_3N_4
grains themselves) as well as their thermal and elastic properties

(which determine the residual stresses) is an important factor in determining the basic strength of these polyphase materials.

Strength measurements of different Si_3N_4-SiC composite materials [39] and SiC-Al_2O_3 composite materials [40] indicate that no strength degradation is observed when the size of the secondary phase regions is approximately the same as the grain size of the matrix. These observations along with the theoretical arguments outlined above suggest that when the second phase regions are smaller than the Si_3N_4 grains, they will not significantly degrade the basic strength of the material.

3.2 Mechanical Behavior at Elevated Temperatures

Host of measurements and observations have shown that the mechanical properties of hot-pressed Si_3N_4 begin to degrade at elevated temperatures. The contributions of dislocation motion and of solid-state diffusion related phenomena on degradation are uncertain, but overwhelming evidence strongly suggests that the presence of a liquid (or viscous) phase between the Si_3N_4 grains is a major contribution to degradation. The reaction of the secondary phases with Si_3N_4 above the solidus temperature and/or the rapid increase in fluidity of metastable silicate glasses are likely reasons for a liquid between Si_3N_4 grains (or grain clusters).

Two models have been hypothesized to suggest how a liquid may influence the mechanical behavior of a solid-liquid composition. In both models, the liquid is a minority phase and it is located between rigid grains of the solid.

The model proposed by Stocker and Ashby [41] suggests that ". . . the liquid phase enhances creep by providing regions, or paths of high diffusion conductance." Applied shear stresses are visualized to set up chemical potential gradients due to localized high stresses (e.g., at points where grains are in contact), causing the solution and reprecipitation of the solid which permits the stored strain energy to dissipate as work. In this case, the liquid phase provides channels of fast material transport. The strain rate function for this model is the same as described for the conventional diffusional flow mechanism except that the effective diffusivity must include the properties and volume content of the liquid. This model suggests no difference for deformation rates in tension and compression.

In the model proposed by Lange [29], diffusion (within the liquid or solid) is neglected, and the mechanics of non-elastic deformation is visualized as due to the separation and sliding of the grains. This model is discussed in the following paragraphs.

    Grain boundary sliding is first brought to mind when visualiz-
ing a stressed polycrystal with a viscous 'grain boundary phase.'
But, if the grains are considered to be relatively rigid, accommodation for grain boundary sliding must be eventually met by the
separation of grains. The question then becomes, which is the
rate controlling step, the grain boundary sliding or grain boundary
separation? Based on theories of liquid adhesives, it was shown
that at small strains (ε < .1) and for small volume fractions of
the liquid (VF $\leq$ 0.1), grain boundary separation was the rate
controlling step in the mechanics of deformation, and that the
volume fraction of the liquid phase was a dominate microstructural
feature. In addition, the growth of vapor bubbles rather than
flow of liquid from one part of the material to another was shown
to be required for boundary separation. When these important
results were combined with fracture mechanics concepts, it was
shown that the deformation rate should be significantly greater
in tension than in compression.

    It was also pointed out that deformation zones would arise
in the highly stressed region of a crack front and that such
materials will exhibit slow crack growth due to the separation
and sliding of grain pairs ahead of pre-existing cracks. Homogeneous deformation within the material, large deformations within
the zones associated with cracks and the slow growth of cracks
(which leads to a change in compliance) will all contribute to the
apparent non-elastic strain.

    It is likely that contributions toward non-elastic deformation
may be attributed to both models discussed above, but experimental
evidence shows that the deformation rate is significantly larger
in tension than in compression for both Si_3N_4 [42] and glass
ceramics [43] at elevated temperatures. These observations
strongly suggest that a mechanism similar to that of separation
and sliding appears to be dominate.

4. OBSERVED RELATIONS BETWEEN MICROSTRUCTURE AND MECHANICAL
    PROPERTIES

4.1 α and β Starting Powders; Strength Anisotropy

    All evidence has shown that α starting powder ($\alpha/\beta \geq 8$)
consistently leads to high strength hot-pressed Si_3N_4, whereas β
starting powder ($\alpha/\beta \leq .1$) only results in moderate strengths.
Lange [44] has shown that the critical stress intensity factor
(K_c) for materials fabricated with either α or β powders are
6.5 MN·m$^{-3/2}$ and 3.1 MN·m$^{-3/2}$, respectively. As expected from
these data, the strength of these materials also differed by a
factor of two. Concurrent microstructure observations lead to
the hypothesis that the relative high strength and fracture

toughness of material fabricated with α powder was due to the fiber-like grain structure of this material relative to the equiaxed grain structured development with the β powder (see Fig. 3). More recent strength and fracture toughness work by Iskoe and Lange [28] for materials where the α → β phase transformation was incomplete, neither substantiated nor validated the above hypothesis suggesting a need for further work in this area.

Recognition that the fracture toughness and strength of Si_3N_4 appeared to be due to its fibrous grain morphology suggested that the fibers may exhibit preferential orientation during hot pressing which, in turn, suggested that the strength of bulk Si_3N_4 may be anisotropic [44]. Subsequent x-ray diffraction experiments showed that the fiber axes (hexagonal, C axis of the β grains) were preferentially aligned perpendicular to the hot-pressing direction [44]. Subsequent flexural strength (and K_c measurements*) showed that specimens cut and stressed perpendicular to the hot-pressing direction were 20% stronger than specimens cut and stressed parallel to the hot-pressing direction [44]. Thus, the weak and strong directions of hot-pressed Si_3N_4 appear to be due to preferred orientation of the fiber-like Si_3N_4 grains.°

4.2 Large Inclusions

Large inclusions accidentally incorporated during powder processing prior to densification are crack precursors as discussed in the last section. The most commonly observed inclusions in hot-pressed Si_3N_4 were WC (or WSi_2), BN and large aggregates of unreacted Si_3N_4. The source of WC was previously discussed; large mill-ball chips are not uncommon. Large, hard Si_3N_4 aggregates are present in unclassified Si_3N_4 powder. These aggregates can be transferred through most all powder precessing procedures and end up as low density inclusions within the hot-pressed material. Boron nitride coatings on graphite dies are used by some manufactures to prevent a graphite-Si_3N_4 reaction during densification. Occasionally, BN aggregates become dislodged when Si_3N_4 powder is poured into the die to end up as low density BN inclusions within the hot-pressed material.

---

* K_c from specimen in which the crack phase is parallel and perpendicular to the hot-pressing direction were 7.2 $MN \cdot m^{-3/2}$ and 6.5 $MN \cdot m^{-3/2}$ for material fabricated at Westinghouse. The K_c for Norton material (perpendicular direction) ranges between 4.6 and 5.0 $MN \cdot m^{-3/2}$ [44,45].

° Other hot-pressed materials that develop an equiaxed grain morphology, e.g., SiC do not exhibit strength anisotropy [46].

The large WC and aggregated Si_3N_4 inclusions can be removed by classifying the powder just prior to densification.* Other means can be taken to eliminate BN inclusions. High density inclusions ($\geq$ 125 µm) can be observed within hot-pressed material by x-radiography [45,47]. Low density inclusions require sonic techniques and they are more difficult to detect [45,47].

4.3 High Temperature Behavior and Effects

4.3.1 <u>General Behavior</u>. The strength and creep resistance of hot-pressed Si_3N_4 degrades at high temperatures. The particular temperature where degradation is first observed depends on compositional effects described below and for the case of strength, the rate in which the specimen is stressed.

The strength dependence on stressing rate at elevated temperature was the first indication that slow crack growth was the cause of strength degradation [48]. One of many different chemical or microstructural mechanisms can be responsible for the growth of pre-existing cracks at velocities and stress intensities much lower than that required to cause catastrophic fracture. Although atmospheric effects have been reported [49], the apparent solid-liquid microstructure of Si_3N_4 appears to be the cause of slow crack growth at elevated temperatures [48,50,51].

Slow crack growth causes time dependent failure, the details of which can be found elsewhere [52,53]. Slow crack growth, and thus time dependent failure, can be characterized by crack velocity (V) vs stress intensity factor (K) measurements. Phenomenological V vs K curves for many ceramic materials have satisfied a simple $V = V_0 (K/K_0)^n$ relation [53]. Unfortunately, this relation may not be valid for commercial Si_3N_4 (in effect, n appears to decrease with decreasing velocity), preventing long-term failure prediction, but not influencing material comparisons when the data is obtained over the same velocity range [53].

4.3.2 <u>Effect of Impurities</u>. Impurities were the first compositional factors known to influence the high temperature mechanical properties of hot-pressed Si_3N_4 [48,54]. The effect of the major impurities on the strength and creep resistance of hot-pressed Si_3N_4 (approximate molar composition: Si_3N_4 = 0.79; MgO = 0.15; SiO_2 = 0.06) at 1400°C is shown in Fig. 5 [55]. To obtain these data, the impurities were added as oxides (or carbonates) to relatively pure α-Si_3N_4 powder prior to hot pressing. For the

---

* Classification appears to be the method used to increase the average strength and to decrease the scatter of strength for the Norton hot-pressed Si_3N_4 [22].

504

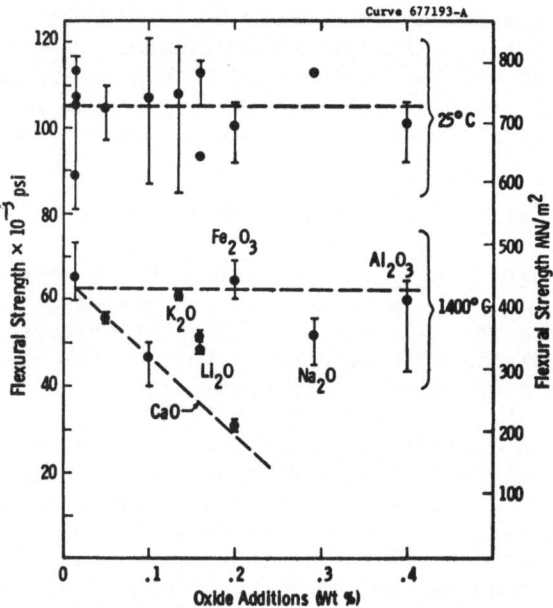

Fig. 5. Flexural strength for Si_3N_4 with incorporated oxide impurities.

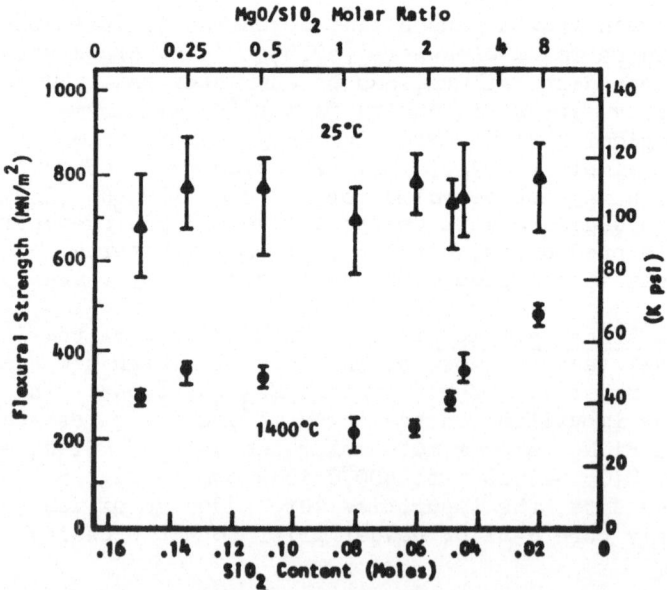

Fig. 6. Flexural strength of Si_3N_4 (0.833 mole fraction) as a function of the MgO/SiO_2 molar ratio.

case where MgO is the densification aid (i.e., material in the Si_3N_4-SiO_2-MgO system), CaO was the most detrimental impurity.* In other systems, other impurities may be important culprits.

4.3.3 <u>Compositional Effects: Si_3N_4-SiO_2-MgO System</u>. The effect of composition was first reported by Andersson, Lange and Iskoe [56] and more thoroughly studied by Lange [24]. As shown in Fig. 6 for compositions containing 0.833 mole fraction Si_3N_4, the strength at 1400°C (and the ratio of strengths at 25°C and 1400°C) are strongly related to the MgO/SiO_2 molar ratio. Exact behavior was also obtained for materials containing both 0.91 and 0.755 mole fractions of Si_3N_4.° The lowest strengths (at 1400°C) were observed for a MgO/SiO_2 molar ratio of 2:1, viz., along the Si_3N_4-Mg_2SiO_4 tie line.† Maximum strength is obtained when MgO/$SiO_2 \to \infty$, viz., along the Si_3N_4-MgO tie line. High strengths are also observed close to the Si_3N_4-Si_2N_2O tie line. Strengths at 25°C are relatively independent of composition.

It is obvious that the high temperature strength is strongly compositional dependent. Strength appears related to phase equilibria parameters, e.g., the solidus temperature+ but further phase equilibria work (viz., the determination of solidus temperatures) is required for better judgment.

4.3.4 <u>Compositional Effects: Si_3N_4-SiO_2-Y_2O_3 System</u>. Materials in the Si_3N_4-SiO_2-Y_2O_3 system (see Fig. 2) which contain one or more of the following secondary phases: $Si_3Y_2O_3N_4$, $YSiO_2N$, $Y_{10}Si_7O_{23}N_4$ and/or $Y_4Si_2O_7N_2$, exhibit unusual oxidation behavior and catastrophic strength degradation at elevated temperatures [25].

Detailed oxidation experiments have shown that $Si_3Y_2O_3N_4$ exhibits linear oxidation kinetics, i.e., the oxide surface scale does not protect the $Si_3Y_2O_3N_4$ [25]. Preliminary experiments indicate that $YSiO_2N$, $Y_{10}Si_7O_{23}N_4$ and $Y_4Si_2O_7N_2$ also exhibit linear oxidation kinetics [25]. All evidence shows that when these Y-Si-O-N compounds are present as secondary phase, they preferentially oxidize to cause cracks to develop in the polyphase Si_3N_4 material. Thus, the presence of these four compounds is unwanted.

---

* Na_2O and K_2O volatized during fabrication; $\sim 2/3$ of Li_2O disappeared during fabrication [55].
° Lower room temperature strengths ($\sim 15\%$) were obtained for the 0.755 mole fraction material.
† The non-ealstic portion of the stress-strain surve is also a maximum along the Si_3N_4-Mg_2SiO_4 tie line [24].
+ Fabrication work in this system suggests a solidus temperature of 1300°C within the Si_3N_4-Si_2N_2O-MgO system [24].

The four unwanted compounds can be precluded as secondary phases when material in this system is fabricated within the Si_3N_4-Si_2N_2O-$Y_2Si_2O_7$ compatibility triangle [25]. Due to the compatibility of SiO_2 with $Y_2Si_2O_7$ (the highest yittrium silicate),, materials within this compatibility triangle also exhibit the best oxidation resistance observed for hot-pressed Si_3N_4 materials.

5. CONCLUDING REMARKS

As illustrated above, both the constituents within the starting powder and the phase relations between these constituents govern the mechanical properties of dense Si_3N_4. Similar to mechanical properties, all other properties are governed by these same relations.

As first shown for the Si_3N_4-SiO_2-Y_2O_3 system, the oxidation behavior of dense Si_3N_4 depends on its composition [25]. Compositions that lie within the Si_3N_4-$Y_2Si_2O_7$-Y_2O_3 area will contain one or more of the unstable compounds, whereas compositions within the Si_3N_4-$Y_2Si_2O_7$-Si_2N_2O compatibility triangle will exhibit unusually good oxidation resistance apparently due to the compatibility of SiO_2 (the oxidation product of Si_3N_4) with $Y_2Si_2O_7$. Namely, from an oxidation viewpoint, the highest metal-silicate is a desirable second phase. Materials densified with the aid of MgO do not have this excellent resistance to oxidation apparently due to the preclusion of the highest silicate (Mg_2SiO_4) by the phase relations (see Fig. 1) [24].

Thermal properties are also governed by the types and amounts of the secondary phases [7]. It has been pointed out that the solid-solution of Al and O in the Si_3N_4 structure is the apparent cause for the lower thermal conductivity of SiAlON materials [57].

Conditions of phase equilibrium have been emphasized in the text. It should be remembered that nonequilibrium conditions may prevail due to the sluggish crystallization kinetics of silicates. Evidence for glassy secondary phases [31-32] and discussions concerning the effect of glassy grain boundary' phases on high temperature mechanical properties is found in the literature [29, 50,51,54,55]. Although this author does not dispute this viewpoint, it should also be remembered that a liquid will be present in these polyphase materials whenever the solidus temperature is exceeded. Thus, the determination of solidus and liquidus temperatures for pertinent compositional areas is essential.

In conclusion, it can be seen that the effect of different densification aids and impurities on fabrication, microstructure and properties should be approached through related phase equilibrium studies. Three guidelines can be suggested to achieve improved high temperature structural materials based on Si_3N_4. First, the solidus temperature should be as high as possible to prevent the formation of a liquid within the desired temperature range of application. Second, from an oxidation viewpoint, the secondary phase should be compatible with SiO_2. Third, to minimize thermal stresses (during, e.g., thermal transients), the thermal properties of the secondary phases should not significantly detract from those intrinsic to Si_3N_4. In practice, compromises will be necessary to achieve fabricability.

ACKNOWLEDGMENTS

Thanks are due Westinghouse colleagues who were essential participants in the work reviewed above. Special acknowledgment is due Dr. A. M. Diness, Scientific Officer, Office of Naval Research, for personal support of the effort which lead to this review.

REFERENCES

1. G.G. Deeley, Brit. Pat. No. 942,082, Nov. 20, 1963. G.G. Deeley, J.M. Herbert, and N.C. Moore, Powder Met. No. 8, 145, 1961.
2. G.E. Gazza, J. Am. Ceram. Soc., 56, 662, 1973.
3. I.C. Huseby and G. Petzow, Powder Met. Int., 6, 17, 1974.
4. K.S. Mazdiyasni and C.M. Cooke, J. Am. Ceram. Soc., 57, 536, 1974.
5. G.R. Terwilliger and F.F. Lange, J. Mat. Sci., 10, 1169, 1975.
6. K.H. Jack and W.I. Wilson, Nature Phys. Sci. (London) 238, 28, July 10, 1972.
7. F.F. Lange, "Task I: Fabrication, Microstructure and Selected Properties of SiAlON Compositions," Final Report NAVAIR N00019-73-C-0208, Feb. 26, 1974.
8. L.J. Gauckler, H.L. Lukas and G. Petzow, J. Am. Ceram. Soc., 53, 346, 1975.
9. I.C. Huseby, H.L. Lukas and G. Petzow, J. Am. Ceram. Soc., 58, 377, 1975.
10. R. Marchand, Y. Laurent, J. Lang and M. Th. LeBihan, Acta Crys. B25, 2157, 1969.
11. D. Hardie and K.H. Jack, Nature (London) 180, 332, 1957.
12. E.T. Kurkdogan, P.M. Bills and V.A. Tippett, J. Appl. Chem., 8, 296, 1958.
13. S. Wild, P. Grieveson and K.H. Jack, Special Ceramics, 5, pp. 385, Ed. by P. Popper, B.C.R.A., Stoke-on-Trent, 1972.
14. I. Kohatsu and J.W. McCauley, Mat. Res. Bull., 9, 917, 1974.
15. K. Kato, Z. Inque, K. Kijima, I. Kawada, H. Tanaka, and T. Yamane, J. Am. Ceram. Soc., 58, 90, 1975.
16. H.F. Priest, F.C. Burns, G.L. Priest, and E.C. Skaar, J. Am. Ceram. Soc., 56, 395, 1973.

17. A.J. Edwards, D.P. Elias, M.W. Lindley, A. Atkinson, and A.J. Moulson, J. Mat. Sci., 9, 516, 1974.
18. D. Campos-Loriz and F.L. Riley, J. Mat. Sci., 11, 195, 1976.
19. K. Blegen, Special Ceramics, 6, p. 223 (see Ref. 13).
20. T. Yamauchi and H. Suzuki, U.S. Patent 3,206,318, Sept. 14, 1965.
21. F. F. Lange, unpublished work.
22. H.R. Baumgartner and D.W. Richerson, Fracture Mechanics of Ceramics, 1, p. 367, Ed. by R.C. Bradt, D.P.H. Hasselman and F.F. Lange, Plenum, 1974.
23. K.H. Jack, J. Mat. Sci., 11, 1135, 1976.
24. F.F. Lange, "Phase Relations in the Si_3N_4-SiO_2-MgO System: Compositional Effects on Strength and Oxidation Resistance," ONR Tech. Rept. No. 9, Contract No. N00014-74-C-0284.
25. F.F. Lange, S.C. Singhal, and R.C. Kuznicki, "Phase Relations and Stability Studies in the Si_3N_4-SiO_2-Y_2O_3 Pseudo-Ternary System," Tech. Rept. No. 6, ONR N00014-74-C-0284, April 1, 1976.
26. S. Wild, P. Grieveson, K.H. Jack and M.J. Latimer, Special Ceramics 5, p. 377, (see Ref. 13).
27. R.J. Weston and T.G. Garruthers, Proc. Brit. Ceram. Soc. No. 22, 197, 1973.
28. J.L. Iskoe and F.F. Lange, "Development of Microstructure, Strength and Fracture Toughness of Hot-Pressed Si_3N_4," Tech. Rept. No. 7, (see Ref. 25).
29. F.F. Lange, Deformation of Ceramic Materials, p. 361, Ed. by R.C. Bradt and R.E. Tressler, Plenum, 1975.
30. J.V. Sharp and A.G. Evans, J. Mat. Sci., 6, 1292, 1971.
31. R. Kossowsky, J. Mat. Sci., 8, 1603, 1973.
32. P. Drew and M.H. Lewis, J. Mat. Sci., 9, 261, 1974.
33. F.F. Lange, J. Mat. Sci., 10, 314, 1975.
34. J. Selsing, J. Am. Ceram. Soc., 44, 419, 1961.
35. F.F. Lange, "Fracture of Brittle Matrix, Particulate Composites," Composite Materials, 5, pp. 1-44, Ed by L.J. Broutman, Academic, 1974.
36. D.B. Binns, Science of Ceramics, 1, p. 315, Ed by G.H. Stewart, Academic, 1972.
37. R.W. Davidge and T.J. Green, J. Mat. Sci. 3, 629, 1968.
38. F.F. Lange, Fracture Mechanics of Ceramics, 2, p. 599, (see Ref. 22).
39. F.F. Lange, J. Am. Ceram. Soc., 56, 445 (1973).
40. F.F. Lange, "Strong, High-Temperature Ceramics," Ann. Rev. Mat. Sci., 4, p. 364, 1974.
41. R.L. Stocker and M.F. Ashby, Rev. Geophysic and Space Phys., 11, 391 1973.
42. A.F. McLean, E.A. Fisher and R.J. Bratton, Brittle Materials Design, High Temperature Gas Turbine, AMMRC CTR 73-9, Interim Rept., p. 149, March 1973.
43. R. Morrell and K.H.G. Ashbee, J. Mat. Sci., 8, 1253, 1973.
44. F.F. Lange, J. Am. Ceram. Soc., 56, 518, 1973.

45. R.J. Bratton, C.A. Andersson, F.F. Lange and S.C. Sanday, "Ceramic Rotor Blade Development - 2nd Semi-Annual Tech. Rept.," EPRI Cont. No. RP 421-1, May 15, 1976.

46. A.F. McLean, E.A. Fisher and R.J. Bratton, Brittle Materials Design, High Temperature Gas Turbine, AMMRC CTR 73-32, Interim Rept., p. 154, September 1973.

47. R. Kossowsky, Ceramics for High Performance Applications, p. 665, Ed by J.J. Burke, A.E. Gorum and R.N. Katz, Brook Hill, 1974.

48. F.F. Lange, J. Am. Ceram. Soc., 57, 84, 1974.

49. D. J. Rowcliffe and P.A. Huber, Proc. Brit. Ceram. Soc., No. 25, 239, 1975.

50. F.F. Lange and J.L. Iskoe, p. 223, (see Ref. 48).

51. A.G. Evans, p. 373 (see Ref. 48).

52. S.M. Wiederhorn, Fracture Mechanics of Ceramics, 2, p. 613, (see Ref. 22).

53. A.G. Evans and S.M. Wiederhorn, J. Mat. Sci., 9, 270, 1974; A.G. Evans, L.R. Russell and D.W. Richerson, Met. Trans. 6A, 707, 1975.

54. D.W. Richerson, Bul. Am. Ceram. Soc. 52, 560, 1973.

55. J.L. Iskoe, F.F. Lange and E.S. Diaz, J. Mat. Sci., 11, 908, 1976.

56. C.A. Andersson, F.F. Lange and J.L. Iskoe, "Effect of MgO/SiO$_2$ Molar Ratio on High Temperature Strength of Hot-Pressed Si$_3$N$_4$," ONR Tech. Rept. No. 3, Contract No. N00014-74-C-0284, Oct. 15, 1975.

DISCUSSION (Räuchle, Ritter, Tighe)

Cannon: If the actual composition with respect to Na or Li is used instead of the nominal additive composition, what is the relative effect of Na and Li compared with Ca on degrading the high temperature properties?

Lange: Both Na and K were lost during hot-pressing; the residual amounts Na and K were 1% to 10% of the initial amounts. Approximately $\frac{2}{3}$ of the Li was lost. Since the strength of material with the small residual Na content was reduced, it can be expected that if all the Na were retained, it might have a stronger degradation effect relative to Ca. The same might be expected of K.

DEFORMATION AND MASS TRANSPORT IN SILICON NITRIDE

R.M. Cannon

Department of Materials Science and Engineering,
Massachusetts Institute of Technology,
Cambridge, Mass. 02139, U.S.A.

ABSTRACT

Densification, creep and creep rupture in dense silicon nitride are all significantly affected or controlled by a viscous grain boundary phase. However, there is not yet general agreement about the role of the liquid in contributing to deformation or mass transport, nor have the actual rate limiting steps been identified. Densification has been attributed to dissolution and diffusional transport of silicon nitride in the liquid phase, whereas creep and creep rupture have frequently been described in terms of a simplified model of sliding of rigid grains in a viscous boundary phase in which continual deformation involves cavitation rather than accommodation of the grain shape. Both mechanisms neglect deformation of the grains by slip although a modest crystallographic texture develops from hot pressing and a stronger texture can be obtained by hot forging silicon nitride.

The reported creep, densification, hot forging, grain growth and slow crack growth data are considered in order to develop a consistent view of the role of the liquid phase and the rate limiting steps for transport within the liquid. The results suggest that either the dissolution step or diffusion can be rate limiting depending upon the particular conditions, that an appreciable viscoelastic or recoverable strain may result from the rigid body motion of the grains in the liquid phase, and that the mean hydrostatic stress has an important effect on the relative modes of deformation.

DISCUSSION (Räuchle, Ritter, Tighe)

Greskovich: Both you and Dr. Lange have attempted to explain why creep rate is higher for Si_3N_4 tested in tension than in compression by mechanisms such as grain boundary sliding, separation of grains, diffusion in a liquid phase and/or a dissolution-precipitation process. We have found (1) that in SiC metastable-stable grain boundaries form in hot-pressed material, which upon annealing at elevated temperatures (but below the hot pressing temperature) in the absence of an applied stress, such grain boundaries become unstable and "spin up", thereby causing considerable grain growth and pore growth. Could it be that in hot pressed silicon nitride specimens tested in compression at high temperature most of the grain boundaries remain stable whereas many of these grain boundaries slowly become unstable when tested in tension, spin-up or disconnect and lead to void formation at triple points? A critical test of this idea would be to use annealed, hot-pressed specimens for both tension and compression tests so that one starts with a microstructure in nearly geometrical equilibrium. The creep mechanism may then be more easily interpreted in light of the models currently proposed.

Ref. (1): C. Greskovich and J.H. Rosolowski, "Sintering Covalent Solids", J. Amer. Ceram. Soc. **59** (7-8), 336-343 (1976).

Cannon: Most of the creep specimens have not been pre-annealed. I doubt that the silicon nitride boundaries in the presence of an amorphous second phase would be unstable in the manner you describe; however if they were simultaneously crystallizing or further reacting it could be possible they would cavitate more.

Lange: Grain growth during liquid phase sintering can be prevented when the contact (wetting) angle is approximately 0° and there is sufficient amount of liquid to prevent grain coalescence. Prevention of grain growth is, in fact, found in $SiC-Al_2O_3$ system if the Al_2O_3 content is $\geq 7^w/o$.

Cannon: The general experience in oxide ceramics is that a wetting liquid phase usually causes rapid grain growth relative to the situation where the impurities are in solid solution or solid second phases. If increasing the volume fraction liquid in SiC retards grain growth, it suggests that the coalescing may be diffusion controlled in SiC and so is slowed by the larger thickness of glass. Quantitative studies of coalescing in silicon nitride as a function of the volume fraction liquid phase would be useful in distinguishing the slow step between dissolution and diffusion.

Brook: Under the combined driving forces of hot-pressing and the
α-β phase transformation, the original α equiaxed particles have
been claimed to convert to a β-grain structure comprising inter-
linked columnar grains (F.F. Lange: this meeting). It seems likely
that such a structure would be relatively resistant to grain
growth because of impingement of grains with low energy surfaces.

Cannon: The fact that the growth is columnar, of course, means
that the growth rate is anisotropic and that dissolution/precip-
itation is difficult at least on some faces.   The impingement
argument is an intriguing one;   however, one is struck by the
surprising infrequency of secondary growth of even occasional
large grains in silicon nitride.

Clarke: How would the form of the internal friction curves be
different if the model of the material was of purely crystalline
grains with glassy material only at triple points - i.e. no glassy
material in the grain boundaries?

Cannon: If the glass were only at the triple points the peak for
boundary sliding would be moved to a much higher temperature or
the boundary viscosity would be higher without a glassy phase.
There would probably be a much reduced sliding peak at the low
temperature representative of the area fraction of boundary
covered by glass.   There may also be an additional peak due to
viscous flow within the glass.   The high temperature background
from diffusion of the matrix species would be moved to a much
higher temperature without the enhanced transport in the boundaries
by the glassy phase.

Morgan: I suspect that it is possible, because the lattice of
Si_3N_4 is a very open one, that the liquid phase may be relatively
denser than the solid. By Le Chatelier's principle there is an
extra driving force for dissolution at points of compression over
and above what really occurs.   Dissolution at points of compression could be greatly favoured over the normal dissolution (and
reprecipitation) at other points that might lead to grain growth.
Therefore creep is favoured more than grain growth.   Incidentally,
if this was so, there would be an increase of liquid phase with
pressure so that, under actual hot pressing conditions, there
would be slightly more liquid than would be apparent on returning
to normal pressure.

Cannon: The extra driving force at points of compression is
balanced by a comparable reduction in driving force at points of
tension.   Therefore there is only a second order modification in
the response to a shear stress or in the creep rate.   In densification there would be a small increase in densification due to this
effect, perhaps 5-15%.   The changes in volume fraction liquid and
solidus temperature would also be small because $\overline{PV}$ is very small
in comparison to ΔH_{fusion}.

ELASTIC AND ANELASTIC PROPERTIES OF NITROGEN CERAMICS

Jean-Claude Glandus and Philippe Boch

Centre de Recherches et d'Etudes Céramiques,
Université de Limoges, France.

1. ELASTIC AND ANELASTIC PROPERTIES

Ceramics are brittle materials: it is only at high temperatures that some plasticity develops. If for metals plastic properties are the most important, elastic properties are also in the foreground for ceramics. The stresses in a ceramic body under deformation agree well with elastic theory, since there is no plastic relaxation. A solid is elastic if deformations disappear when applied loads disappear: for the most part, ceramics exhibit a linear elastic behaviour, stresses and strains being proportional. At a point in a deformed medium, if the stress and the strain states are described by a second order tensor, σ_{ij} and ε_{kl}, then $\sigma_{ij} = C_{ijkl}\varepsilon_{kl}$ where the fourth order tensor C_{ijkl} defines the elastic constants. In the case of a triclinic crystal, there are 21 independent C_{ijkl}; in the case of higher symmetry, this number is reduced. For the nitrogen ceramics (1) there are 3 C_{ijkl} for the cubic structures BN, AlON, ZrN, HfN, TiN, TaN; 9 C_{ijkl} for the orthorhombic Si_2N_2O, Ge_2N_2O. For an isotropic medium, only two terms are needed: Young's modulus E and the shear modulus G. Glasses are isotropic on a microscopic scale; polycrystalline ceramics are isotropic on a macroscopic scale, the crystal grains having more or less random orientations. If preferential orientations occur, an anisotropic behaviour appears. Measurements of the elastic moduli along several directions of a sintered ceramic body are found to be a most sensitive procedure of discovering such texture. Various approximate calculations enable macroscopic moduli of agglomerates to be derived from the elastic constants of single crystals. Linear elasticity is a first requirement though real ceramics exhibit an anelastic behaviour, where the internal friction Q^{-1} describes energy losses during a cycle of periodic

deformations. Anelasticity arises from relaxation mechanisms
(dependent on frequency) or hysteresis mechanisms (dependent on
vibration amplitude).

2. EXPERIMENTAL PROCEDURES

Elastic and anelastic quantities can be measured by various
experimental procedures (2,3):

Static methods (tensile, twist, and in particular, for ceramics,
bending test) are used for measuring macroscopic elastic moduli,
but are inaccurate and not suitable to estimate Cijkl and internal friction.

Dynamical methods, at low frequencies (torsion pendulum $\simeq 1$ Hz)
and particularly at middle frequencies (vibrating beams, slabs or
spheres, from 100 Hz to 100 kHz) are amongst the best measurements
of E, G, and Q^{-1}, with great accuracy. These methods allow
measurements on small samples (spheres of diameter $\simeq 1$ mm).

Ultrasonic methods at high frequencies (MHz range) give results on
E, G, Q^{-1}, and on the numerous Cijkl of low symmetry crystals.
These high frequency procedures are very accurate. It is possible
to use small samples (cylinder of height $\simeq 1$ mm).

An important advantage of internal friction and elastic
moduli measurements, in comparison with other mechanical measurements, is that the tests are non-destructive.

3. INTERPRETATION OF THE RESULTS

Apart from the technological interest of knowing elastic
properties of materials in relation to microstructure, the elastic
constants Cijkl represent second derivatives of interatomic potentials with respect to deformations. From results on Cijkl it is
possible, in principle, to obtain information on cohesion forces
(intensity, direction). Often used to study metals, these applications of elastic moduli measurements are now being used on
ceramics. Thus, it is necessary to measure not only Cijkl but also
their derivatives with respect to two thermodynamical parameters,
temperature T and pressure p. Present measurements can be obtained
from $T \simeq 0K$ to $T \simeq 3300K$, with pressure to $p \simeq 100$ kbars. Third order
elastic constants can be evaluated, for studying non linear
effects due to anharmonicity of potentials. Internal friction gives
information about structural defects: solute diffusion, phase
transformations, grain boundary properties, etc.

4. NITROGEN CERAMICS

a) Nitrogen ceramics are very recent materials: few studies have
been devoted to their elastic and anelastic properties. At room
temperature for Si_3N_4 (4,5,6,7) $E \simeq 250,000$ N/mm^2, $G \simeq 120,000$ N/mm^2,
$\nu \simeq 0.2$, for samples with some porosity. These values are inferior

to those for Al_2O_3 ($E \simeq 360,000$ N/mm^2) but greater than for SiC ($E \simeq 190,000$ N/mm^2). For TaN, TiN, ZrN, $E \simeq 200,000$ N/mm^2 to $300,000$ N/mm^2 (8).

b) Temperature T is the main external parameter acting on E,G,Q^{-1}. Bibliographical data are essentially for Si_3N_4 (7): the values of the moduli and their variations depend on the microstructural state of materials, particularly porosity. In all cases the decrease of E or G is small as T increases. For the most dense sample, where the decrease of the elastic moduli is the most important, the diminution is about 10% only, between 20 and 1000°C. This fact confirms the suitability of Si_3N_4 to high temperature use, concurrently with SiC (5). In addition to the "normal" effect of temperature on elastic moduli, some anomalies appear, at given temperatures, especially "ΔM effects" caused by thermally activated relaxation peaks (2). Thus, creep due to vitreous phases at grain boundaries induces an internal friction peak, associated with a modulus loss. In Si_3N_4, measurements on a torsion pendulum at 1 Hz (9) show such a peak, the characteristics of which give information about the viscosity of the vitreous phase, activation energy to the involving mechanisms etc. This non-destructive technique constitutes a fine method for studying creep of ceramics.

c) Porosity P is the main internal parameter acting on E or G. As P increases E and G decrease, according to a relation that can be exponential, homographic, depending on P^2 or P^3 terms or linear. It is a linear decrease that is obtained to Si_3N_4(4,6,7). Scatter of the data points can be attributed to secondary effects such as pore shape, sintering conditions, anisotropy defects etc. It may be noted that E, extrapolated to P = 0, reaches the high value of 340,000 N/mm^2, comparable to Al_2O_3.

SAMPLES			$E(10^{10}$ Pa)	$G(10^{10}$ Pa)	ν
Si_2N_2O		Porosity			
+ MgO	11%	6.5%	19.74	8.33	0.184
	10.5%	7.5%	19.08	8.09	0.179
	10%	8%	17.10	7.05	0.213
	5%	16.5%	12.50	5.23	0.195
	2.5%	26.5%	7.04	2.96	0.189
SiAlON		Density			
40% 40% 20%		3.06	22.47	8.86	0.268
70% 20% 10%		2.53	16.43	6.55	0.254
Si_3N_4 Al_2O_3 AlN					

TABLE 1

518

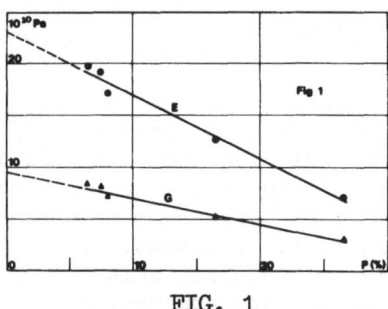

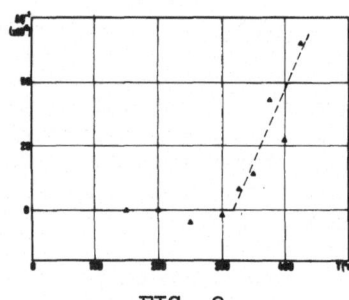

FIG. 1   FIG. 2

d) We have measured E and G for Si_2N_2O and other SiAlON compounds, prepared by the Chimie Minérale et Cinétique Hétérogène Lab. of Limoges (10). Studies have been performed using an ultrasonic apparatus, at 10 MHz, quartz transducers generating longitudinal or shear pulsed waves, where a phase comparison method leads to a relative accuracy of 10^{-4} on velocities. The results are in Table 1, and in Fig. 1, for samples $\phi \approx 20$ mm, $l \approx 12$ mm. The variations are quite linear. For Si_2N_2O, extrapolated values to P=0 are: $E \approx 230,000$ N/mm^2, $G \approx 100,000$ N/mm^2, $\nu \approx 0.2$.

5. CHARACTERIZATION OF THERMAL SHOCK DAMAGE

The damage caused by thermal shock is usually studied by destructive test, such as measure of the strength of several samples, before and after shocks (11). Such destructive tests offer two inconvenients: a lot of samples are needed, and it is impossible to study accumulative effects of successive shocks. On the other hand, non-destructive measurements of the elastic moduli and internal friction (12) do permit the setting out of a "damage curve" by using one sample only, and taking into account the cumulative damages. For Si_2N_2O, results are in Fig. 2. Internal friction of a quenched sample ($\phi \approx 14$ mm, $l \approx 10$ mm) is given versus quenching temperature. The "critical temperature" Tc of 325°C corresponds to a sudden increase of Q^{-1}. The subsequent increase of Q^{-1}, for T >Tc, is associated with the progression of microcracks.

6. REFERENCES

1) L.E. Toth. "Transition Metal Carbides and Nitrides", Academic Press, New York (1971).

2) A.S. Nowick, B.S. Berry. "Anelastic Relaxation in Crystalline Solids", Academic Press, New York (1972).

3) E. Schreiber, O.L. Anderson, N. Soga. "Elastic Constants and their Measurements", McGraw-Hill (1973).

4) A.G. Evans, R.W. Davidge. J. Mater. Sci. 5, 314 (1970).

5) "Ceramics for high temperature engineering", British Ceramic Society, Stoke-on-Trent (1973).

6) F.F. Lange. J. Am. Ceram. Soc., 56, 9, 445 (1973).

7) W.A. Fate. J. Appl. Phys., 46, 6, 2375 (1975).
8) G. Bentz, G. Provost. C. Urban Bulletin SFC 78, 5, 79, 1, 81, 23 (1968).
9) D.R. Messier, R. Kossowsky. J. Mater. Sci., 11, 49 (1976).
10) P. Goursat, P. Lhortholary, M. Billy. Rev. Int. Hautes Temp. et Réfract., 8, 149 (1971); and this meeting.
11) D.P.H. Hasselman. J. Am. Ceram. Soc., 53, 9, 490 (1970).
12) J. Boisson, F. Platon, P. Boch. Ceramurgia International 2, 74 (1976).

DISCUSSION

Katz: With regard to the non-destructive nature of internal friction tests, I would like to introduce a cautionary note. Several years ago Seaton and I did a series of internal friction tests on hot-pressed TiB$_2$, Al$_2$O$_3$ and Si$_3$N$_4$ using Forster's technique (1,2). We found that we could get acoustic fatigue effects with a degradation in room temperature strength of TiB$_2$ of about 35% after resonance. No acoustic fatigue in hot-pressed silicon nitride was observed. As a result of this work I feel that it is important for one to confirm with unresonated samples that the resonance technique used is indeed non-destructive.

1) Seaton, C.C. and Katz, R.N. J. Am. Ceram. Soc., May, 1973.
2) Seaton C.C. and Katz, R.N. Proc. 1972 IEEE Ultrasonics Symposium (referred to in Ref. 1).

Boch: We knew the references you indicate, and we always checked that the measurements were really non-destructive. In fact, this non-destructive character can be obtained, in our case, provided that two main conditions are obeyed: (a) to make the measurements in vacuo (to avoid contamination of microcracks), (b) to work at a very small amplitude of vibration. The ultrasonic apparatus we use operates at much lower amplitude than Forster's does. In addition, of these two conditions, the measurements are made during a short time only. Thus, there are not acoustic fatigue effects.

IMPROVING THE IMPACT PROPERTIES OF SILICON NITRIDE

John J. Brennan

United Technologies Research Center
East Hartford, Connecticut

ABSTRACT

The application of energy absorbing surface layers to Si_3N_4 and SiC is being studied under a current NASA contract. Among the layers studied, iron titanate and a silica-zircon material gave the most promising results. Iron titanate (Fe_2TiO_5) layers on Si_3N_4 resulted in Charpy impact strengths on the order of 2.5 joules at 1250°C and 1370°C compared to control values of 0.40 joules. Silica-zircon layers on Si_3N_4 gave RT Charpy impact values of 1.5 joules but only moderate (~50%) improvement at elevated temperatures. Ballistic impact results tend to agree with the Charpy impact observations. In addition, the severe impact and strength degradation suffered by commercial hot-pressed Si_3N_4 after high temperature oxidation was studied.

INTRODUCTION

Hot-pressed silicon nitride (Si_3N_4) is a leading candidate for use in advanced gas turbine engines. However, its relatively low impact strength may seriously limit its usefulness. This problem can be alleviated in two ways: (1) through the use of fiber reinforcement and (2) through the use of energy absorbing surface layers. Both of these approaches provide energy absorption modes not available in the monolithic material.

From work done under a U.S. Naval Air Systems Command contract, as reported elsewhere [1], hot-pressed Si_3N_4 containing 25 vol percent tantalum wire reinforcement has been found to increase the Charpy impact strength of Si_3N_4 by a factor of over 30 from RT to 1300°C. In addition, the threshold energy below which no damage occurs upon impact has been increased by a factor of 4-6, both in Charpy and ballistic impact up to 1300°C. More recently, the second approach to increasing impact strength has been employed under a NASA contract dealing with the toughening of Si_3N_4 and SiC through the use of energy absorbing surface layers.

ENERGY ABSORBING SURFACE LAYERS ON Si_3N_4

The objective of this work was to develop surface treatment methods that will consistently result in Si_3N_4 Charpy impact specimens (6.4 x 6.4 x 51 mm) having impact strengths greater than one joule at temperatures up to 1370°C. The basic Si3N4 material used in this study was Norton NC-132 hot-pressed Si_3N_4.

The rationale for the investigation of energy absorbing surface layers to improve the impact strength of Si_3N_4 is based primarily on the use of microcracked materials, the microcracking being caused either by thermal expansion anisotropy, thermal expansion differences between phases, or volume changes during phase transformations. It was theorized that the microcracks, upon impact, would extend causing extensive fracturing and crushing of the coating with energy being absorbed by the creation of surface area. Previous NASA work appeared to support this theory [2]. Three types of microcracked materials were initially chosen; partially stabilized zirconia, a silica-zircon mold material (70 wt % SiO_2-30 wt % $ZrSiO_4$), and magnesium titanate. Iron titanate was also evaluated. The titanates exhibited microcracking due to their large thermal expansion anisotropy while the silica-zircon and zirconia materials were microcracked due to phase changes (high cristobalite to low cristobalite for the SiO_2-zircon and tetragonal to monoclinic for the partially stabilized ZrO_2). Thin plates of these materials (6.4 x 12.8 x 1 mm thick) were cemented onto Si_3N_4 Charpy impact specimens using a silicon carbide based refractory cement (Carbofrax 3445). The coated samples, as well as Si_3N_4 controls, were subjected to instrumented Charpy impact tests, using a modified Physmet machine, at RT, 1250°C, and 1370°C.

The Charpy impact strength of NC-132 Si_3N_4 remained essentially constant at 0.40-0.45 joules from RT to 1370°C. The partially stabilized ZrO_2 and the magnesium titanate coated Si_3N_4 samples exhibited a moderate increase in Charpy impact to approximately 0.60 joules. The silica-zircon and iron titanate coated specimens, however, exhibited a significant increase in Charpy impact resistance with the silica-zircon performing better at RT (1.48 joules) and the iron titanate performing better at elevated temperatures (2.56 joules at 1250°C). The results of the latter two coated samples are given in Table I. The instrumented Charpy impact traces for the silica-zircon at RT and the iron titanate at elevated temperatures are very similar. The RT trace for silica-zircon is shown in Fig. 1. The double load peaks for the silica-zircon coated Si_3N_4 at RT and the Fe_2TiO_5 coated Si_3N_4 at 1350°C are characteristic of coated

Table I

Charpy Impact Tests of Coated and Uncoated Si_3N_4

Coating	Temp. (°C)	Impact Resistance (joules)	Maximum Load (kN)
None	RT	0.40	3.7
	1250	0.45	3.2
	1370	0.40	2.8
Silica-zircon	RT	1.48	3.6
	1250	1.14	3.3
	1370	0.66	3.0
Fe_2TiO_5	RT	0.69	3.9
	1250	2.56	3.1
	1370	2.14	3.1

LOAD = 178 lbs/div   TIME = 0.1 ms/div

ENERGY

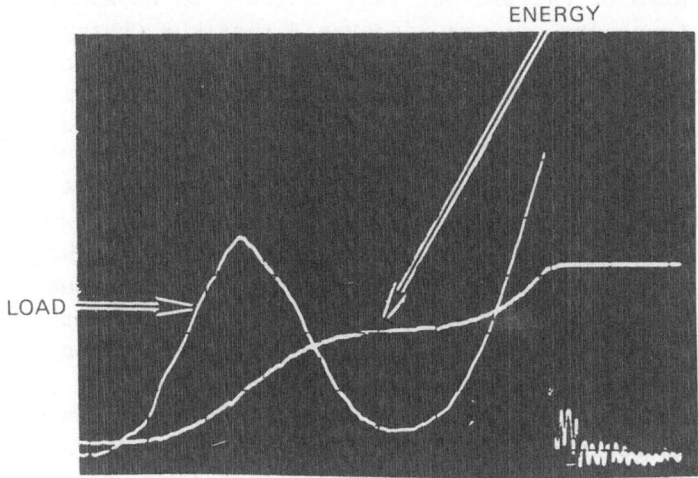

LOAD

Figure 1. RT Instrumented Charpy Impact Test of Cemented Silica-Zircon Layer on Si_3N_4.

samples giving high impact energies (>1 joule). Apparently, the
first load peak is due to the crushing of the cemented energy absorbing layer while the second load peak is due to the fracturing
of the Si_3N_4 substrate.

It is not evident at this time that the increase in impact
energy is due to the presence of microcracks in the cemented
layers, since the microcracked materials do not absorb energy in
a consistent manner at different temperatures and that a silica-
zircon layer with the SiO_2 entirely in the form of fused silica,
exhibited equally as high an impact energy as the standard
silica-zircon material. The presence of porosity (35%) in the
silica-zircon material and onset of plasticity in the iron titanate at elevated temperatures may be important factors in the
observed energy absorption. These factors are under study in an
ongoing program.

Although the energy absorbing surface layers studied can be
considered as somewhat model systems; in that they would not likely
withstand an advanced gas turbine environment, (especially as
coatings on rotating parts) the observed increases in impact
energy, which has been substantiated by ballistic impact tests,
argue for the continuation of study into this concept.

IMPACT ENERGY AND STRENGTH DEGRADATION OF OXIDIZED Si_3N_4

During the course of the present NASA contract on developing
Si_3N_4 of improved toughness, it was discovered that the Charpy
impact strength and maximum load to failure of Norton NC-132 hot-
pressed Si_3N_4 is drastically reduced after exposure to oxidizing
conditions at temperatures of 1100°C and above. It was initially
found that NC-132, when oxidized in air at 1350°C for 48 hrs,
suffered a drop in RT impact strength from 0.40 joules to 0.14
joules with a corresponding drop in maximum load to failure from
3.7 kN to 2.0 kN. In order to verify this drastic drop in maximum
load to failure, a slow bend test was run on similarly oxidized
samples with the resultant strength being 410 MN/m^2 compared to
910 MN/m^2 for unoxidized NC-132. It was then decided to run a
series of oxidation tests on NC-132 from 1000°C to 1350°C for
various times up to 75 hrs and measure the weight gain, RT Charpy
impact strength, and maximum load to failure. The Charpy impact
results are shown in Fig. 2. The maximum load to failure follows
the same trend as the curves in Fig. 2. It can be seen that

SPECIMEN SIZE = 6.4 x 6.4 x 5.0 (mm)

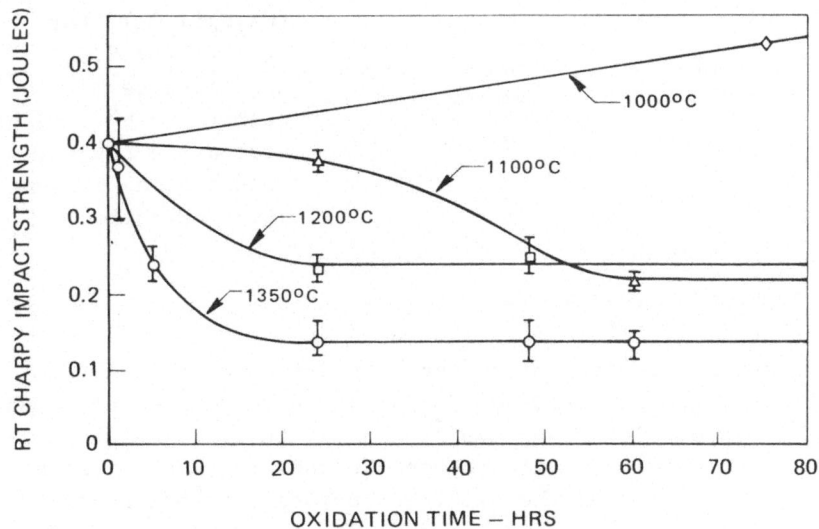

Figure 2. RT Charpy Impact Strength of Oxidized Si_3N_4.

temperatures as low as $1100°C$ can result in severe strength degradation and that at $1200°C$ or higher only a very few hours of
exposure to oxidizing conditions results in decreased impact
strength and load to failure. Somewhat surprisingly, exposure at
$1000°C$ appears to increase the strength of the material, at least
in impact.

     In contrast to the results obtained for NC-132 Si_3N_4, Si_3N_4
fabricated at UTRC with 15 wt % Y_2O_3 additive after a 60 hr oxidation at $1350°C$ exhibited a drop in impact strength of 22% and a
drop in maximum load of 13% compared to 68% and 48%, respectively,
for NC-132. These results are shown in Table II. This drastic
reduction in strength of NC-132 Si_3N_4 has also been reported by
the Westinghouse Corp. in the last ARPA Interim Report on Brittle
Materials Design, High Temperature Gas Turbine [3]. It appears
to be due to the formation of silicates on the oxidized surface
which forms voids or pits on the surface that act as crack initi-
ators. Further X-ray analysis of oxidized surfaces revealed that,
in addition to α-cristobalite approximately an equal amount of
enstatite ($MgSiO_3$) was present. Kiehle, et al [4] also found cristobalite and enstatite on oxidized ($1350°C$) surfaces of Norton

526

Table II

RT Charpy Impact Tests on Oxidized NC-132
Si_3N_4 and Si_3N_4 + 15% Y_2O_3

Material	Condition	Impact Energy (joules)	Maximum Load (kN)
NC-132 Si_3N_4	As ground	0.40	3.7
"	Oxidized 60 hrs-1350°C wt gain = 0.935 mg/cm²)	0.14	2.0
Si_3N_4 + 15% Y_2O_3	As ground	0.35	3.2
"	Oxidized 60 hrs-1350°C wt gain = 0.625 mg/cm²	0.27	2.8

HS-130 Si_3N_4 in addition to lesser amounts of akermanite ($Ca_2MgSi_2O_7$), forsterite (Mg_2SiO_4) and diopside ($CaMg(SiO_3)_2$). Kiehle also noted the surface pitting present after oxidation.

From scanning electron microscopy studies conducted at UTRC, the oxidized surface of NC-132 Si_3N_4 was found to consist of rough particles of $MgSiO_3$ with minor amounts of Mn, Ca, Fe, and Al present, the latter three primarily located between the $MgSiO_3$ grains. On occasion, large pits on the oxidized surface are noticed and form the fracture origin during impact. The fracture origin of one oxidized NC-132 Si_3N_4 sample is an exceptionally large surface pit (Fig. 3). In contrast to the oxidized surface appearance of NC-132 Si_3N_4, UTRC Si_3N_4 + 15% Y_2O_3 when oxidized 60 hrs at 1350°C has a surface consisting of large tabular crystals of yttrium

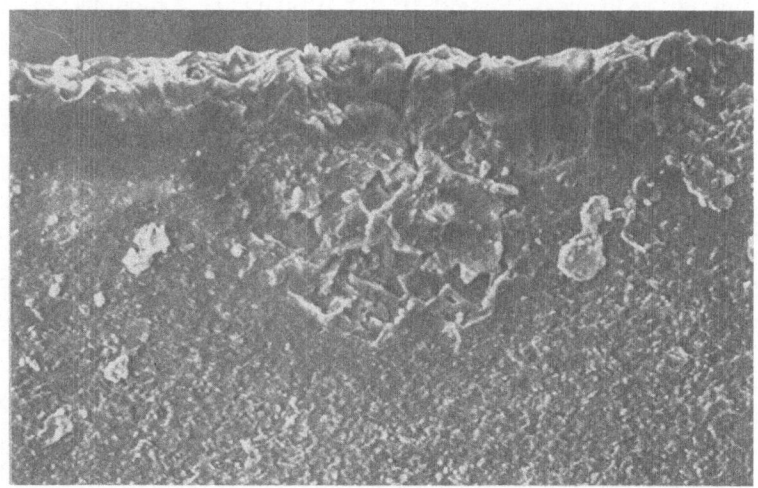

Figure 3.   Fracture Initiating Flaw on Surface of Oxidized NC-132 Si_3N_4 (500X).

silicate (Y_2SiO_5) with the underlying matrix being SiO_2 with a
minor amount of Al present. No Ca, Mn, or Fe was detected. Thus,
while the surface of oxidized Si_3N_4 + 15% Y_2O_3 is rougher in terms
of silicate grain size than NC-132 Si_3N_4, the lack of large surface pits leads to a much less severe drop in impact strength and
maximum load to failure.

From these oxidation results, it is apparent that Si_3N_4 hot-
pressed with MgO additives is severely limited in its role as a
high temperature gas turbine material in high stress applications
if indeed the MgO is responsible for the formation of the observed
surface flaws. However, it does appear that yttria additive Si_3N_4
materials are much more resistant to this type of degradation with
further study being necessary to determine the mechanisms
responsible.

REFERENCES

1. J. J. Brennan, "Increasing the Impact Strength of Si_3N_4
    Through Fibre Reinforcement", Special Ceramics 6, ed. by
    P. Popper, BCRA, June 1975, pp 123-134.

2. H. P. Kirchner and J. Seretsky, "Improving Impact Resistance
    of Ceramic Materials by Energy Absorbing Surface Layers",
    NASA CR 134644, March 1974.

3. "Brittle Materials Design, High Temperature Gas Turbine",
    AMMRC CTR 75-28, Interim Report #8, October 1975, p 133.

4. A. J. Kiehle, L. K. Heung, P. J. Gielisse, and T. J. Rockett,
    "Oxidation Behavior of Hot-Pressed Si_3N_4", J. Am. Cer. Soc.,
    Vol. 58, No. 1-2, January-February 1975, p 17.

DISCUSSION (Räuchle, Ritter, Tighe)

Davidge: I can see that the coating techniques you describe
could be useful in a 'one-shot' type of application where the
component is impacted just once. For longer term applications at
high temperatures I am sceptical about the use of oxide layers on
non-oxides. This will lead to complex chemical interactions and
almost certain degradation of mechanical properties. Furthermore,
I am not aware of any successfully developed glazes for carbides
and nitrides.

Lange: If you want to develop a glaze for the Si_3N_4 (or SiC)
systems, one must know the composition of the materials (i.e.
secondary phases) that are in equilibrium with the Si_3N_4 and use
that phase as a glaze. Such a glaze would be compatible with
Si_3N_4 and if you are in the right system, it should also be compatible with SiO_2, the oxidation product of Si_3N_4.

Greskovich: Is there an optimum porosity for the coating material
and an optimum thickness of the coating material which permits
high impact strength of Si_3N_4 and SiC?

Brennan: No experiments have been done to date to optimise these
parameters. Whether or not porosity and/or coating thickness
play a role in the improvement of the impact strength has still
to be studied.

Cannon: Do you expect porous layers to oxidise, filling with SiO_2
- and severely degrading the energy absorbing ability?

Brennan: If the porous layers are oxides to begin with, they
should not fill with SiO_2 unless they are so porous that they
have inter-connected porosity, allowing the Si_3N_4 to oxidise.
Then SiO_2 might fill the pores and change the impact characteristics.

Ritter: It would appear that the thermal shock behaviour of the
coated Si_3N_4 is of more importance than the impact strength. Do
you have any thermal shock results for the coated systems?

Brennan: We have found that thermal cycling of the coated samples
between RT and 1370°C quickly results in debonding of the
cemented coating. A cement with a stronger bond may help;
however, if the thermal expansion coefficient of the coating is
significantly different from that of the substrate then the system
will never withstand thermal shock or thermal fatigue. An energy
absorbing surface layer on Si_3N_4 or SiC that could withstand
thermal shock or fatigue must have a coefficient of thermal expansion near that of the substrate.

SOME PROPERTIES OF β-Si$_{6-x}$Al$_x$O$_x$N$_{8-x}$

L.J. Gauckler, S. Prietzel, G. Bodemer, G. Petzow

Max-Planck-Institut für Metallforschung,
Institut für Werkstoffwissenschaften,
7000 Stuttgart-80, W.-Germany

Because of the conflict data in the phase relationships in the system Si-Al-O-N a systematic study of the properties changes as a function of aluminum content in the β-silicon nitride solid solution was not possible. The composition of the β-Si$_3$N$_4$ solid solutions containing aluminum has been cleared and extent of solubility is also known at present (1). In this paper some of the results will be presented for hot-pressed compositions (2).

1. PHYSICAL PROPERTIES

1.1 Lattice parameters

The hexagonal lattice parameters were calculated from x-ray diffraction measurements of the β-Si$_3$N$_4$ (002), (430), (420), (330), and (610) reflections, using monochromated CuKα radiation. The lattice parameters increase with increasing aluminum and oxygen content as shown in Fig. 1. The value x gives the amount of replaced silicon atoms and ranges from x = 0 to x = 4.27 at temperatures of 1850°C, where 71% of the silicon is replaced by aluminum.

1.2 Density

The bulk densities were measured using Archimedes principle with the open pores sealed as well as being calculated from x-ray data. The density decreases with increasing Al-content as shown in Fig. 2. The density of other phases in the system Si$_3$N$_4$-AlN-Al$_2$O$_3$-SiO$_2$ was found to be 3.084 g/cm^3 for X$_2$ (15R) and 3.05 g/cm^3 for X$_1$ phase.

530

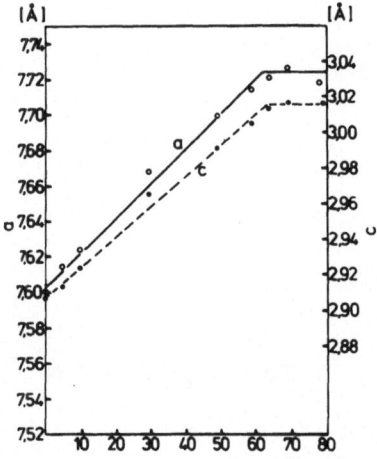

Equivalent-% Al

Fig. 1:

Lattice parameters of the
β-$Si_{6-x}Al_xO_xN_{8-x}$ at 1760°C.

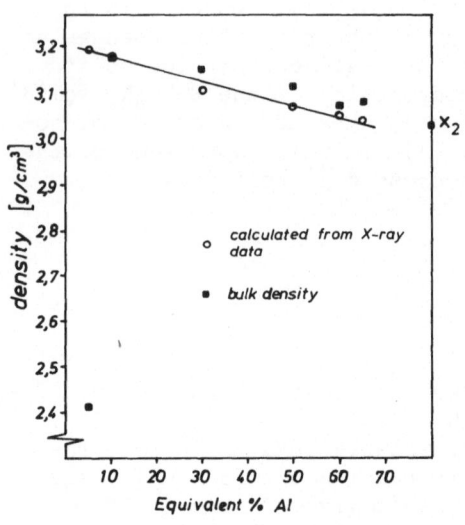

Fig. 2:

Density of β-$Si_{6-x}Al_xO_xN_{8-x}$.

2. MECHANICAL PROPERTIES

2.1 Young's Modulus

The Young's moduli of the β-Si_3N_4 solid solutions were measured by
dynamic and static methods. The dynamic measurements were carried
out on specimens 50 to 80 mm long, having a cross section of 3x5
mm^2. The resonance frequencies were in the range of 10 kHz. The
data from static measurements were obtained from the stress/strain

slope of bend strength experiments. The results are given in Fig. 3, where Y was plotted as a function of composition.

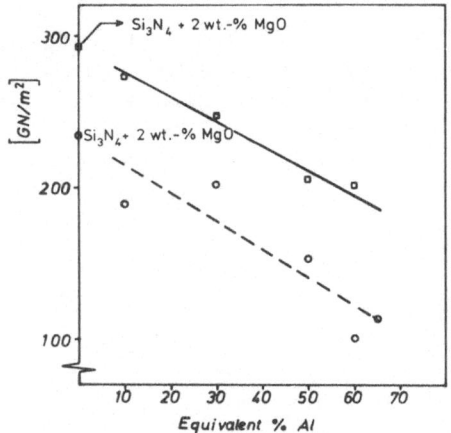

Fig. 3:

Young's modulus of $\beta-Si_{6-x}Al_xO_xN_{8-x}$ versus Al-concentration.

Young's modulus decreased linearly with increasing Al-concentration.

2.2. Hardness

The hardness measurements (Vickers) were carried out with a pyrami-. dal diamond indenter (angle between the planes = 136°) on polished surfaces under a load of 10kp. The measurements were carried out on 5 different specimens of each composition with 15 to 20 indentations. The hardness decreased with increasing Al-content as shown in Fig. 4.

The hardness measurements were very sensitive to the residual porosity of the specimen. This explains the low value obtained for the $\beta-5$ composition. The hardness of the X_2 material is comparable with the highest values obtained from the $\beta-Si_3N_4$ solid solutions.

2.3 Bend Strength

Hot pressed specimens were diamond cut to a cross section of 3x5 mm^2. The bend strength measurements were carried out at RT using a 28 mm span and a cross-head of 0.05 mm/min. At least five specimens of each composition were tested. The results are shown in Fig. 5.

The values obtained for the $\beta-Si_3N_4$ solid solutions were all lower than those of the MgO containing hot pressed specimens. Despite the large scattering, a slow decrease of the strength with increasing Al concentration is visible. The highest bend strength was observed on specimens containing Al_2O_3, AlN and MgO in small

532

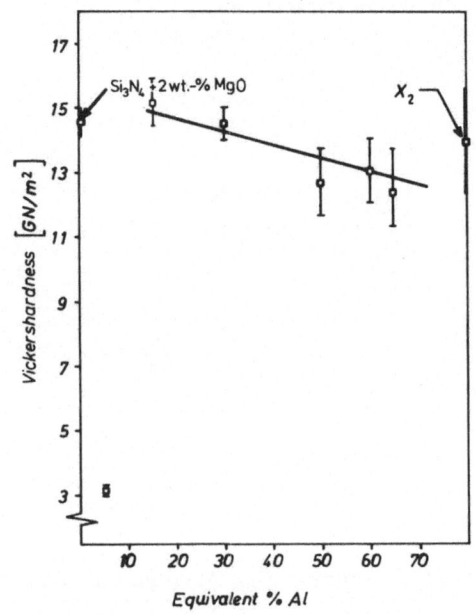

Fig. 4:

Hardness of β-$Si_{6-x}Al_xO_xN_{8-x}$, Si_3N_4 + 2 wt.-% MgO and X_2 phase. Bars are standard deviation.

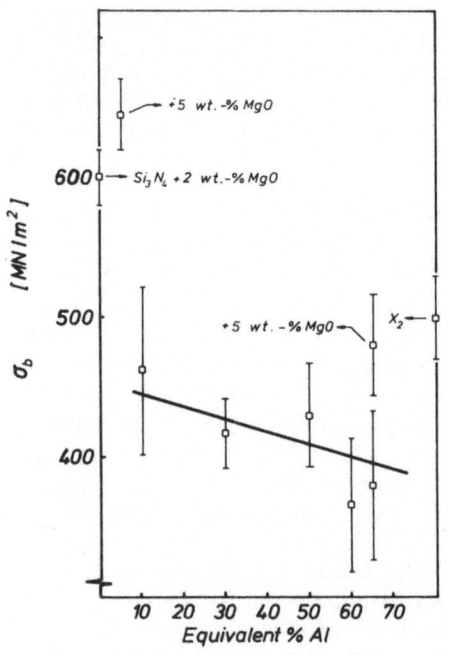

Fig. 5:

Three point bend strength of $Si_{6-x}Al_xO_xN_{8-x}$, X_2 (15R) and MgO containing materials.

amount. X-ray analysis, however, showed in these specimens small amounts of other phases beside β-Si$_3$N$_4$ solid solution.

Bend strengths versus temperature were measured on x = 1.5 composition.

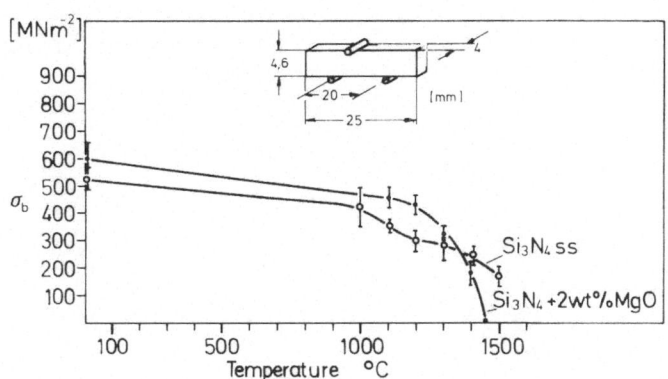

Fig. 6:

High temperature bend strength of β-Si$_{4.5}$Al$_{1.5}$O$_{1.5}$N$_{6.5}$ and MgO containing Si$_3$N$_4$.

Higher bend strength data have been observed for a two phase material (β + X$_1$), prepared from 10 wt.-% Al$_2$O$_3$ and 90 wt.-% β-Si$_3$N$_4$ (3).

2.4 Toughness

The toughness of the β-Si$_3$N$_4$ solid solutions is characterized by the critical stress intensity factor K$_{Ic}$. For the measurements a 2 mm notch was made in the specimens (30x3x5 mm^3) with a diamond saw (0.1 mm width). The specimens were tested under the same conditions as used for the bend strength specimens. The results are shown in Fig. 7.

For all β-Si$_3$N$_4$ solid solutions the toughness was remarkable lower than the values for the MgO containing Si$_3$N$_4$. The toughness of the X$_2$ (15R) phase being slightly higher than those of the β-Si$_3$N$_4$ solid solutions.

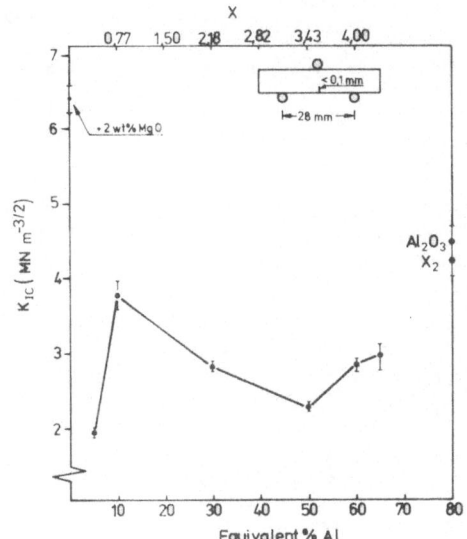

Fig. 7:

Toughness of β-Si$_3$N$_4$ solid solutions versus Al-concentration.

3. THERMAL PROPERTIES

3.1 Thermal Expansion

The thermal expansion was measured in a dilatometer against an Al$_2$O$_3$ reference. In the temperature range from 40 to 1200°C a linear expansion of the specimens was observed at a heating rate of 10°/min. The results are shown in Fig. 8.

An almost linear decrease of the thermal expansion coefficient with increasing Al-concentration was observed for the β-Si$_{6-x}$Al$_x$O$_x$N$_{8-x}$ solid solution. The value extrapolated to 0% Al is in good agreement with the average value of 3.4 10^{-6} °C^{-1}, obtained by high temperature x-ray method by Henderson and Taylor (4) for pure β-Si$_3$N$_4$. The value from Jack (5) (x-ray method) of sialon was in good agreement with ours, after recalculating his composition from RT x-ray data together with the new formula for the β-Si$_{6-x}$Al$_x$O$_x$N$_{8-x}$ solid solution. It was found to be 42.5 eq.-% Al (x = 3). From these data one would expect an improvement of the thermal shock resistance.

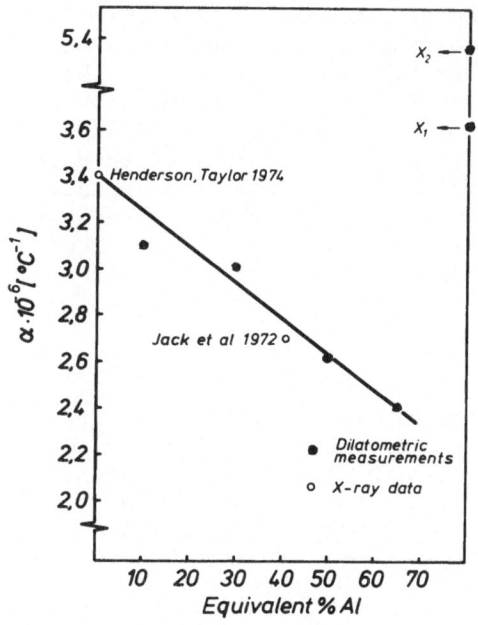

Fig. 8:

Coefficients of thermal expansion of β-Si$_3$N$_4$ solid solutions, X$_1$ and X$_2$ (15R) phase. (x-ray data ($^{\circ}$) from Henderson and Taylor (4) and Jack (5)).

3.2 Thermal Conductivity

Thermal conductivity was measured using cylindrical specimens ($\emptyset$ 20x35 mm^3) in a TCFCM comparative thermal conductivity instrument (Dynatech Corporation). For the β-Si$_3$N$_4$ solid solutions Al$_2$O$_3$ was used as the reference material, whereas iron was used in the case of the 5 wt.-% MgO containing Si$_3$N$_4$. The measurements were carried out in the temperature range 170 to 800°C. The results are shown in Fig. 9.

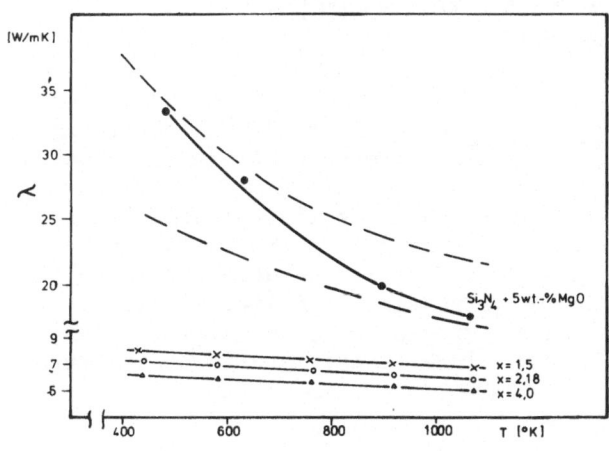

Fig. 9:

Thermal conductivity of β-Si$_{6-x}$Al$_x$O$_x$N$_{8-x}$ with x = 1.5, x = 2.18, x = 4 and 5 wt.-% MgO containing Si$_3$N$_4$.

The β-Si$_3$N$_4$ solid solutions showed a thermal conductivity three times lower than the MgO containing Si$_3$N$_4$. The dashed lines envelop the values obtained for other MgO containing Si$_3$N$_4$. Most of these data were measured on HS 130 from Norton company (6). The large scatter is probably due more to imperfection and inclusions than to different amounts of MgO. The large decrease of conductivity with increasing temperature is not observed in the case if the Si$_3$N$_4$ solid solutions. The latter showing a decrease with increasing Al concentration. The absolute accuracy of these measurements were ±10%, the accuracy of one set of data being ± 5%.

3.3 Thermal Shock Behaviour

The bend strength remaining after thermal shock is plotted versus the temperature difference of the quench in Fig. 10. The β-Si$_3$N$_4$ solid solutions and the MgO-hot-pressed Si$_3$N$_4$ exhibit a sudden strength drop at critical temperatures.

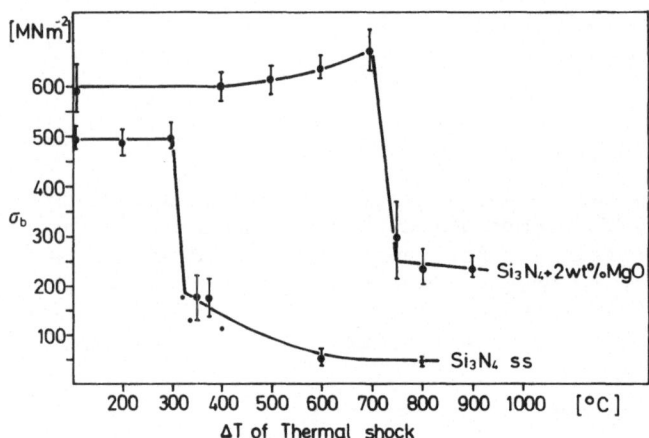

Fig. 10:

Bend strength of β-Si$_{4.5}$Al$_{1.5}$O$_{1.5}$N$_{6.5}$ and 2 wt.-% MgO containing hot-pressed Si$_3$N$_4$ versus temperature difference of thermal shock.

The critical temperature of the Si$_3$N$_4$ solid solution is T$_c$ = 320°C and that of the MgO containing T$_c$ = 750°C. This poor thermal shock behaviour of the β-Si$_3$N$_4$ solid solution, despite its lower coefficient of thermal expansion (see Fig. 8), is caused by the low thermal conductivity of this material (see Fig. 9) and the poor toughness of the single-phase material (see Fig.7).

SUMMARY

Hot pressed $Si_{6-x}Al_xO_xN_{8-x}$ showed comparable strength to MgO containing Si_3N_4.

The MgO containing Si_3N_4 showed higher toughness than the β-Si_3N_4 solid solutions.

Despite low coefficients of thermal expansion of β-Si_3N_4 solid solutions the thermal shock behaviour is poor compared to the MgO containing Si_3N_4. This is due to the very low thermal conductivity of the β-$Si_{6-x}Al_xO_xN_{8-x}$ solid solutions with Al-concentrations greater than x = 1.5.

ACKNOWLEDGEMENT

Thanks are due to the German Bundesministerium für Forschung und Technologie, BMFT, and to Dr. Wiedenmann from Bosch GmbH for the thermal conductivity measurements.

REFERENCES

1. L.J. Gauckler, H.L. Lukas and G. Petzow, J. Am. Ceram. Soc. 58 346 (1975).

2. L.J. Gauckler, "Gleichgewichtsuntersuchungen in den Systemen Si-Al-N-O und Si-Al-Be-N-O", Thesis, Univ. of Stuttgart, 1976.

3. N. Claussen and C.P. Lahmann, Powder Met. Int., Vol. 7, No. 3 (1975).

4. C.M.B. Henderson and D. Taylor, Trans. and J. Brit. Ceram. Soc. 74 (1975 49-53.

5. K.H. Jack, J.Mat. Sc. 11 (1976) 1135-1158.

6. "Engineering Property Data on Selected Ceramics", Vol. 1, MCIC Report, March 1976, Battelle, Columbus, Ohio, USA.

DISCUSSION (Räuchle, Ritter, Tighe)

Popper: Your results showing a poorer thermal shock resistance of a particular "Si-Al-O-N" material compared with "pure" Si_3N_4 are of considerable interest since the most important potential application of these materials is in hot engines. If this is due to a shorter phonon wavelength as a result of foreign ion substitution, then there is a fundamental limitation to the thermal shock resistance of the more complex material in spite of its lower thermal expansion.

Godfrey: Hasselman has obtained data for a range of sialons which show similar low thermal conductivity. These results are probably due not solely to the presence of glass, but also to phonon scattering by dissimilar ions in the lattice (similar to the effect of Cr_2O_3 dissolution on the thermal conductivity of Al_2O_3).

Clarke: Were the grain sizes for the Si_3N_4+MgO (2%) and the $Si_{4.5}Al_{1.5}O_{1.5}N_{6.5}$ commensurate? A large difference might account for differences observed in thermal conductivity.

Gauckler: The median grain sizes were in the range 3 to 5 μm.

Katz: According to Hasselman's models one would predict that the higher the strength preceding ΔTc, the lower the retained strength after ΔTc has been exceeded. Your materials seem to show the reverse effect.

Gauckler: The location of the damage generated by the shock is important. I would expect that in material having low thermal expansion, the cracks would be on the surface and internal cracks would be very limited. If this is the case, the strength decrease would not be too great.

Thompson: Our X-ray thermal expansion data for the β' sialons show the same trend but with a slightly smaller slope. Our values for X_2 phase (7.0 x 10^{-6}/°C //a; 4.0 x 10^{-6}/°C //c) are in good agreement with the dilatometric 5.4 x 10^{-6} value, but our X_1 value of ~ 6 x 10^{-6} is considerably larger than your value (~ 3.6 x 10^{-6}/°C).

SYNTHESIS, CHARACTERISTICS AND SINTERING OF β' (Si,Al,O,N) COMPOUND POWDERS[+]

J. Demit

Research Department, CERAVER, Tarbes, France.

ABSTRACT

Synthesis of pure β'-sialons from appropriate mixtures of γ-aluminium oxynitride, silicon nitride, and aluminium nitride powders was realized. Their properties were analysed and compared with the literature data. Samples were sintered, and flexural strengths and oxidation behaviour were measured.

[+]This work has been supported in part by D.R.M.E. (Direction des Recherches et des Moyens d'Essai) of France "Ministère de la Défense.

MECHANICAL PROPERTIES OF REACTION BONDED SILICON NITRIDE

R.W. Davidge

Materials Development Division, Building 552,
AERE Harwell, Didcot, Oxon, OX11 0RA.

ABSTRACT. Reaction bonded silicon nitride (RBSN) is one of the more important of the silicon-based ceramics for engineering applications. Although the level of the mechanical properties of RBSN is not particularly outstanding, the importance of RBSN lies in the extremely versatile range of green fabrication techniques that are available. The reaction bonding process does however lead to an open structure with usually > 15% porosity (typically 25%), and this has important consequences on the mechanical properties particularly at temperatures > 800°C where appreciable oxidation can occur. The microstructure of RBSN is generally a fine scale mixture of the α- and β-polymorphs and, in the absence of major amounts of deleterious impurities, the material is highly refractory in neutral atmospheres and retains strength to near the sublimation temperature. The properties of RBSN is discussed firstly from a materials science point of view in relation to microstructural features and secondly from an engineering viewpoint where fracture mechanics and statistical data relevant to engineering applications is presented.

1. INTRODUCTION

Reaction bonded silicon nitride (RBSN) is the longest-established of the new nitrogen-based engineering ceramics. Because of the relatively high porosity level, RBSN does not have outstanding mechanical properties compared with more recent denser materials. Nevertheless, RBSN possesses a number of important advantages not least of which is the very versatile range of fabrication techniques available. Many of the fabrication methods are based on the plastic forming routes where temporary binders are

used with the silicon powder to assist in the moulding of the
green article. There is very little dimensional change during
reaction bonding and thus machining operations can be kept to a
minimum. In addition, silicon nitride has a good refractoriness,
especially in inert atmosphere, and also a low coefficient of
thermal expansion, $\sim 3.10^{-6}$ K^{-1}, which leads to a good thermal
shock resistance.

The aims of this paper are firstly to review briefly the main
mechanical properties of RBSN from a materials science point of
view, including elastic properties, strength and creep behaviour,
as a function of temperature and including the effects of oxidising environments; and secondly, to present engineering design
data for RBSN with discussions of the time dependence of strength
and statistical variations in strength.

2.   MICROSTRUCTURE

The microstructure of any material is a key feature which has
a strong influence on the mechanical behaviour. Microstructure
represents the link between the fabrication conditions of the
material and its properties.

Porosity is a particularly significant feature that results
from the specific fabrication route for RBSN. This is generally
> 15% and typically 20-25%. Such high porosities are necessary
to allow the penetration of nitrogen to the centre of the article
being nitrided to give a high conversion of silicon to silicon
nitride; it is difficult to achieve lower values without introducing unacceptably long nitriding schedules. Porosity is thus
mainly open-type. The pore size has a wide distribution but the
majority of pore volume is concentrated in very fine pores,
typically 80% < 0.1 μm[1]. The maximum pore sizes observed are
in the range 20-30 μm, and although these are few in number they
clearly have a pronounced effect on the strength. Figure 1 shows
the distribution of the larger pores on a polished section of
RBSN, and pores < 20 μm are observed. The morphology of pores is
better observed by observations of fracture faces. Figures 2 and
3 show single large pores which were observed in materials nitrided predominantly below and above the melting point of silicon.
The biggest pores observed are smaller in the former case. These
often contain fine fibres of silicon nitride and are believed to
occur at the interstices between the original silicon particles.
For material nitrided predominantly above the melting point of
silicon, larger pores are observed. Here, the largest pores are
of similar dimension to the largest original silicon particles.
It is thought, therefore, that when some of the large silicon
particles melt the silicon flows away from its original site to

Fig. 1. Scanning electron micrograph of polished surface of RBSN[2].

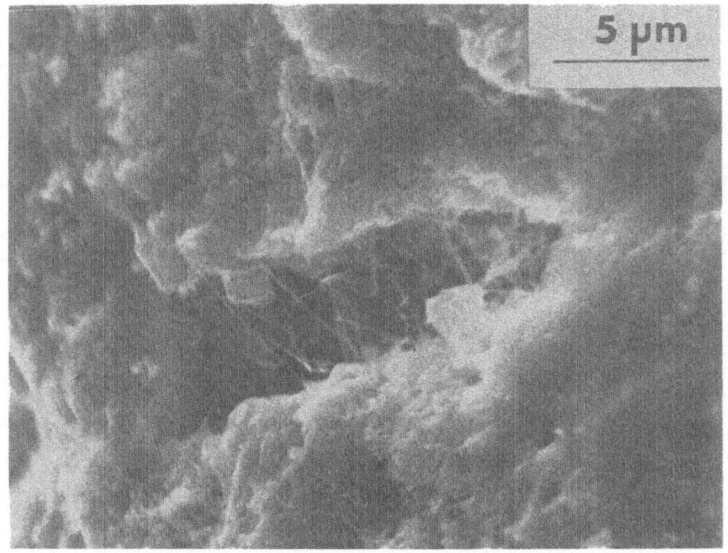

Fig. 2. Scanning electron fractograph of RBSN nitrided to completion at 1350°C showing whiskers in the larger pores of the original compact[3].

544

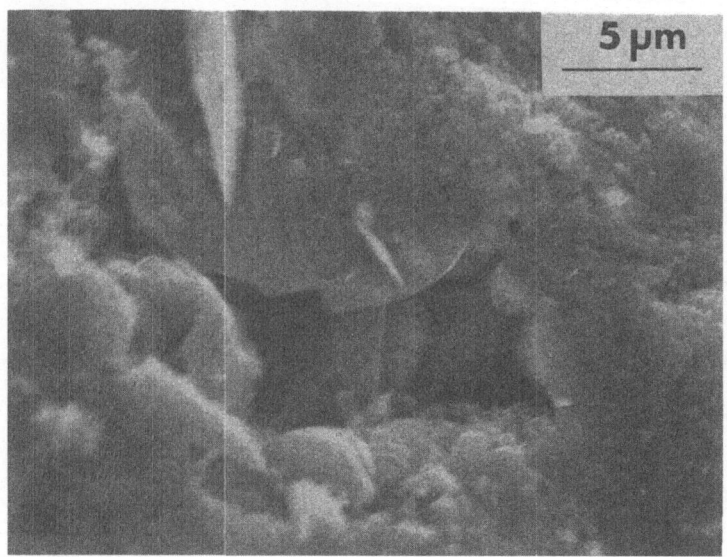

Fig. 3. Scanning electron fractograph of RBSN nitrided 5 h at 1350°C plus 24 h at 1450°C showing a pore created when a large silicon particle melts[3].

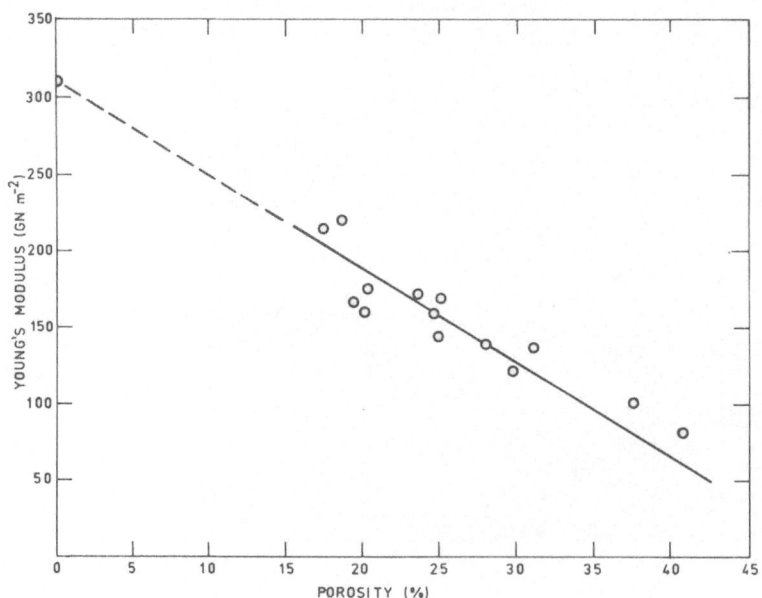

Fig. 4. Variation of Young's modulus of RBSN at ∿20°C as a function of porosity[3,5].

leave the observed pores.

Both the α- and β-forms of silicon nitride are generally detected in RBSN. The relevant amounts of these phases varies widely between different materials but generally the α-Si3N4 formation tends to be favoured by low nitriding temperatures and β-Si3N4 by high nitriding temperatures. Electron microscopy[4] of the material reveals that the grain size is very fine. Individual grains < 5 μm are observed but these are surrounded by a very fine grain sized material << 1 μm. Electron defraction reveals a tendency for the larger grains to be β-Si3N4 and the smaller grains α-Si3N4. Many pores contain fine fibres of - Si3N4.

RBSN often contains small amounts of unreacted silicon, typically 1%. Total metallic impurities are also commonly 1%, comprising mainly Fe,Al, and Ca. Oxygen and carbon are the commonest non-metallic impurities, present as -Si3N4, Si2ON2 and SiC.

3.   ELASTIC CONSTANTS

The variation in Young's modulus (E) with porosity (P) at ∿20°C is shown in Figure 4 for material taken from a wide variety of sources[3,5]. Assuming that the value[5] for fully dense material (E0) is 310 GN m^{-2}, the data show good agreement with the empirical relationship

$$E = E_o (1 - 2.0 P) \tag{1}$$

Young's modulus decreases slighly with increase in temperature[6] by ∿2% at 1000°C in material of density 2.51 Mg m^{-3}.

4.   STRENGTH

The strength of RBSN of a given density at ambient temperatures shows a much greater scatter than does Young's modulus. Data[5] for a variety of materials are given in Figure 5. This scatter is to be expected in that strength depends on the largest pore present whereas the elastic constants are dependent on all the pores present. Apart from the fact that strength increases with increase in density it is thus not possible to establish quantitative strength/density relationships of general significance.

Rather than seek empirical relationships of the type mentioned above, it is more satisfactory to apply fracture mechanics

techniques to gain a deeper understanding of the factors controlling strength. The fracture stress σ_f, is given quantitatively
by[7]

$$\sigma_f = \frac{1}{Y}\left(\frac{2E\gamma_i}{C}\right)^{\frac{1}{2}} \tag{2}$$

where C is the size of the largest inherent flaw, γ_i is the
effective surface energy for fracture initiation and Y is a geometrical constant. Values of E, γ_i and C are thus required for
each material to enable the stresses to extend the inherent flaws
to be evaluated, and compared with directly measured values.

Techniques for the measurement of the effective surface
energy of ceramics are now very well established and have been
reviewed recently[8]. A particularly simple specimen to use
is the three-point bend centre-notched bar. It is important
however to ensure that the radius at the tip of the notch is less
than a small critical value. Table I shows how the surface
energy decreases sharply as the notch root radius decreases and
only approaches the value obtained from sharp-cracked specimens
when the radius is < 20 μm. The need to use specimens with sharp
cracks is disputed by Barnby and Taylor[9] who state that provided
the notch root radius is < 120 μm then surface energy values are
obtained which are similar to those from specimens with sharp
cracks. Their conclusions are based on only two results for
sharp cracked specimens and the unusually high quoted values for
surface energy, 24-32 J m^{-2}, may in the absence of more detailed
evidence be an overestimate. The surface energy does not appear
to vary strongly with the α/β ratio for a given density of
material, but for two independent sets of data the surface energy
decreases with decrease in density. The surface energy is invariant with change in temperature from 20-1450oC.

The most obvious flaws in RBSN are pores. Substituting
values for the largest pore size into equation (2) with the measured surface energy values enables comparisons to be made between
calculated and observed fracture stresses. The data in Table I
show that very good agreement is obtained using this analysis
which thus gives confidence to the applicability of fracture
mechanics to RBSN.

To obtain material of the highest strength one thus needs to
maximise E and γ_i, and minimise C. Both E and γ_i increase with
increase in density, and C generally decreases. High density is
thus expected to be associated with high strength, as indicated
very roughly by the data in Figure 5. The proviso is that the
pore size can vary quite markedly with grain microstructure and
heat treatment for materials of the same nitrided density, as

Density (Mg m^{-3})	Surface Energy (J m^{-2})	Notch Radius (μm)	α/β Ratio	Fracture Stress Measured (MN m^{-2})	Fracture Stress Calculated (MN m^{-2})	Temperature (°C)	Reference
2.61	32 + 8	320	65:35			20	1
	34 + 6	160					
	29 + 4	80					
	9 + 3	20					
	9 + 2	sharp					
2.55	6 + 2	sharp	80:20	285 + 30	300 + 100	20	3
	6 + 2		25:75	220 + 30	240 + 80	20	
	6 + 2		60:40	275 + 30	290 + 100	20	
	6 + 2		60:40	235 + 30	250 + 80	1450	
2.13	4 + 1.5	sharp	60:40	100 + 20	120 + 40	20	3
2.61	8.7 + 2	sharp	65:35	292 + 15	320 + 70	20	1
	9.1 + 3		30:70	265 + 20	295 + 100		
2.63	10.1 + 3	sharp	60:40	247 + 30	260 + 80	20	1
	9.7 + 3		25:75	183 + 23	195 + 60		
2.53	6.4 + 1.5	sharp	70:30	165 + 15	150 + 35	20	1

Table I. Measured and calculated strengths for RBSN

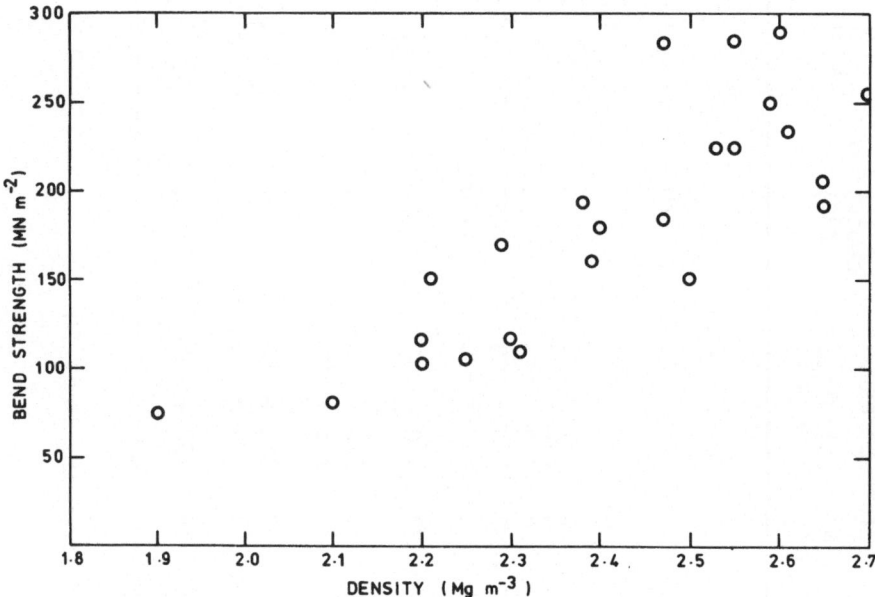

Fig. 5.   Strength vs density for RBSN at $\sim$20°C$^{(5)}$.

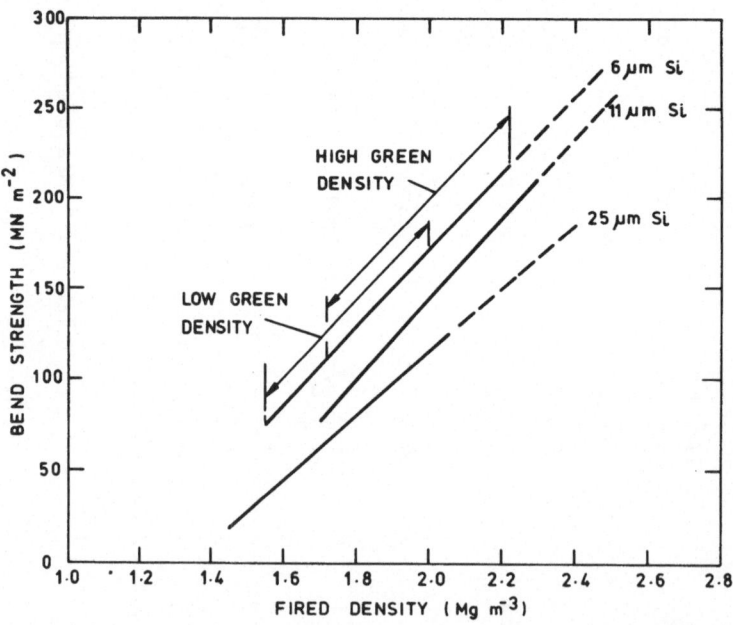

Fig. 6.   Strength vs density for part-nitrided RBSN at 20°C$^{(10,11)}$.

mentioned in Section 2. A decrease in the pore size and therefore the flaw size should in principal be possible by the use of silicon powders of finer particle size. However, the practical advantages may be limited in that fabrication with very fine silicon powder is rendered more difficult and the ultimate fired densities may be lower. It is also advantageous from a strength point of view to nitride predominantly below the melting point of silicon to produce smaller pores. This does of course lead to an economic penalty in that increased nitriding times are necessary.

Although strength is thus understood in a quantitative way from a fracture mechanics analysis, significant improvements to the best strengths currently available will be difficult unless fabrication techniques can be developed to give the desired microstructural characteristics associated with higher strengths.

An alternative, but at present empirical, approach to the estimation of the ultimate strength of RBSN has been taken by Jones and Lindley[10,11] who determined strength/density relationships for partly nitrided materials. Samples from three batches of silicon powder of mean grain size 6, 11 and 25 μm, were isostatically pressed at 31, 92 and 185 MN m^{-2}, and heated in argon for 5 h at 1175°C to form compacts of varying green density. These compacts were then fired in nitrogen at various temperatures in the range 1200-1350°C for 5 h. Although the resulting compacts had a very wide range of Si/Si$_3$N$_4$ ratios the results for each type of silicon powder gave a linear relationship between strength and the density of the part-nitrided compacts. Results are shown in Figure 6. For the upper line the range of strength and density values is indicated for specimens isostatically pressed at the lowest and highest pressures. For a given particle size therefore the maximum strength will be limited by both the initial green density and the increase in density that can be produced during nitriding. For example, the highest green density in these experiments for the finest particle size silicon was 1.39 Mg m^{-3}, with a maximum weight gain of 59%. The prediction from the data would be that a green density of 1.60 Mg m^{-3} and a weight gain of 62% (the maximum thought feasible, but lower than the 66.7% theoretical weight gain for complete conversion to silicon nitride) would give a strength of 300 MN m^{-2}. The requirements are thus the same as those stated above - to produce high density compacts from fine silicon powders: the solution however is again limited by the fabrication considerations.

5. STRENGTH VS TEMPERATURE & EFFECTS OF OXIDATION

The high temperature mechanical properties of RBSN are

dominated by the effects of oxidation which are of particular
significance for a porous material. This is of great importance
because many of the envisaged applications for RBSN involve
operation in oxidising environments. Oxidation is quite rapid
at temperatures as low as 1000°C; for example[2], 100 h in air
gives a 5% weight gain, equivalent to a silica content of 14%.
Although the rate of the oxidation reaction increases with increase in temperature the total amount of oxidation of a specimen
does not. Much greater weight gains are found on oxidation at
1000°C than at 1400°C. This apparent anomaly depends on the
relative fluidity of the oxidation product at the various temperatures. Figures 7 and 8 show surfaces of RBSN oxidised at
1000 and 1400°C. At 1000°C the silica is not very fluid and the
pores are becoming filled with silica. At 1400°C, however, the
silica is very fluid and after very short oxidation times the
surface of the specimen becomes covered by a continuous film.
This acts as a barrier to further internal oxidation and the oxidation effects become concentrated at the surface. Another
important difference between oxidation at these two temperatures
is that the oxidised surface at the higher temperature contains
a network of cracks after cooling to ambient temperatures. The
main oxidation product is crystobalite, and this has a large
thermal expansion mismatch with RBSN on cooling through temperatures ∿270°C where a phase transition from α- to β-crystobalite
takes place. Cracks have not been observed in material oxidised
at 1000°C, presumably because the much thinner layers of silica
can more readily accommodate the expansion mismatch.

Strength data as a function of temperature are shown in
Figure 9. When tested in argon, RBSN shows strengths which fall
only very slightly with increase in temperature; there is no
evidence of any plastic effects in these short-term tests even
at the highest temperature used, 1800°C. Although dislocations
have been observed by electron microscopy in material deformed at
high temperature, the deformation was under very high compressive
stresses[4]. Dislocation movement is thus not considered to be
an important factor in determining the high temperature strength
of RBSN.

When tests are conducted in air the strength at temperatures
> 1000°C is slightly enhanced above the values obtained in argon.
Very pronounced effects on strength are obtained by oxidising
specimens at 1400°C for 24 h and then cooling directly to the test
temperature. The results in Figure 9 show that the strength
increases markedly to ∿500 MN m^{-2} as the test temperature after
oxidation is reduced. Here the surface layer is believed to act
in the manner of a conventional glaze and the development of a
suitable glaze for RBSN could lead to a significant enhancement
of strength. However, once the specimens are cooled through the
crystobalite transformation temperature the strength drops to a

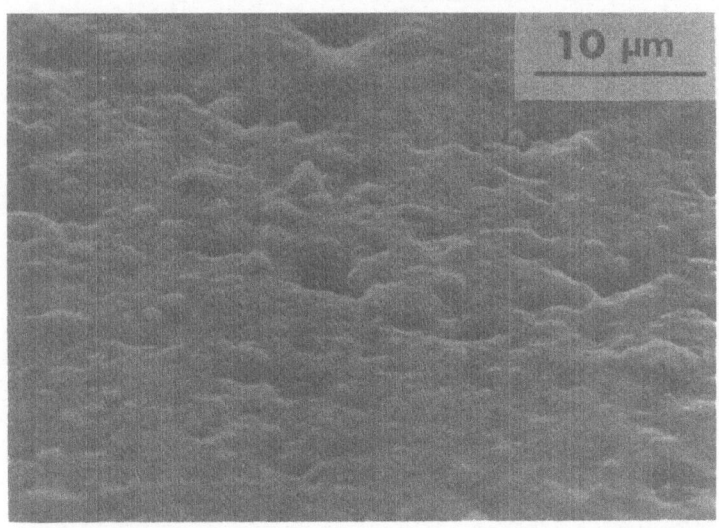

Fig. 7.   Scanning electron micrograph of surface of RBSN oxidised
          100 h at 1000°C.

Fig. 8.   Scanning electron micrograph of surface of RBSN oxidised
          24 h at 1400°C.

552

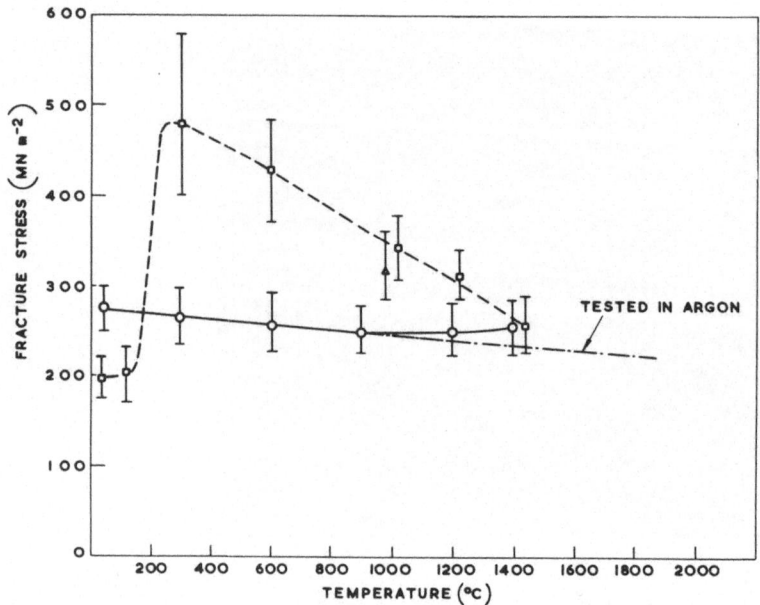

Fig. 9. Strength of RBSN (density 2.55 Mg m^{-3}) vs temperature[3].
0: air 20 min; □:24 h, 1400°C and cooled to test temperature;
Δ: 100 h, 1000°C.

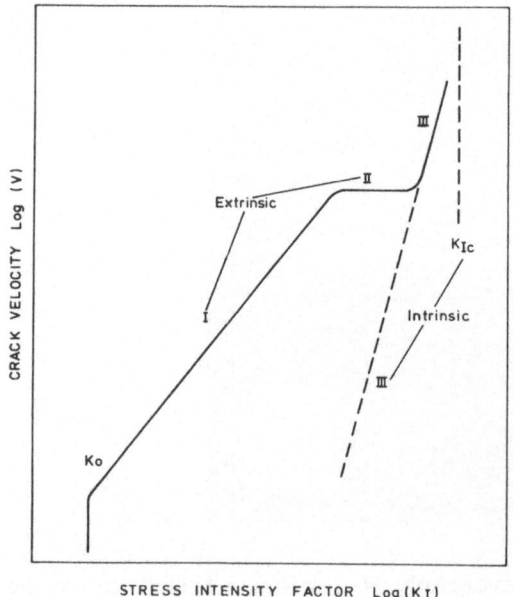

Fig. 10. Idealised K_I/V diagram for ceramics[15].

value lower than that for the original material. In this case
the specimen surface is highly crazed as shown in Figure 8.
Once having been cooled to ambient temperature the material then
exhibits low strength values if the temperature is again raised.
Oxidation at 1000°C produces quite different effects and strength
is enhanced by ~25%, irrespective of whether the material has
first been cooled to ambient temperature or not[2]. Oxidation
at 1000°C is thus beneficial and has been related to the
rounding of pores by oxidation.

6.   CREEP

Most of the available creep data for RBSN have been obtained
in bending deformation and in air at temperatures generally
≳1200°C. This makes detailed analysis difficult. Maximum observed creep strains are usually << 1%. The detailed mechanisms
of creep probably involve grain boundary sliding and the formation
of voids at grain triple points. This can also enhance sub-
critical crack growth and thus the maximum creep strain is likely
to be limited by delayed fracture.

Grathwohl and Thuemmler[12] compared creep rates for RBSN of
density 2.11-2.40 Mg m^{-3} at 1400°C when tested in air or in
vacuum. Results are given in Table II. The data show clearly
that the creep rate is much greater in air than in vacuum.

Stress (MN m^{-2})	Temperature (°C)	Strain in 300 min (10^{-4})	Atmosphere
50.0	1200	35	vac
20.7		75	air
41.1		130	air
35.4	1300	25	vac
50.0		55	vac
70.0		75	vac
20.6		105	air
40.8		215	air
50.9		305	air
66.7	1400	35	vac
93.0		50	vac
20.7		120	air
41.2		305	air

Table II.   Creep rates for RBSN, of density 2.11-2.40 Mg m^{-3}, at
12-1400°C in air and vacuum[12].

The strains quoted include both elastic and plastic deformation; when due allowance is made for the elastic component the plastic creep rate at 1300°C is about 30 times less in vacuum than in air. This difference is associated with a textural instability caused by internal oxidation of the material when tested in air. Similar experiments by Engel and Theummler[13] show that the creep rate at 1250°C under a stress of 40.8 MN m^{-2} varies quite markedly with material density. The creep strain in 300 min increases about threefold as the density varies from 2.33 to 2.05 Mg m^{-3}.

Some evidence for the effect of impurities on creep rate comes from the data of Din and Nicholson[14]. Three materials were tested in air at temperatures of 1200-1450°C at stresses from 70-140 MN m^{-2}. All the materials had similar aluminium and iron contents (0.5 and 0.7% respectively) but the calcium content varied from 0.04-0.10%. The calcium level had a marked influence on the creep rate, with the creep rate increasing with increase in calcium content. It was suggested that the calcium lowers the refractoriness of the glassy phases at grain boundaries.

7. ENGINEERING DESIGN DATA

In assessing the capability of a material for a particular engineering application, the generation of design data relative to the operating conditions is essential. The imposed conditions include temperature, environment, state of stress, and stress-time requirements. The problem is simply stated as the determination of the maximum stress capability under the operating conditions.

Two materials aspects are however also of major importance: firstly the relatively large statistical variations in the strength of brittle materials and secondly the time dependent degradation of strength.

The statistical variation in strength can be understood in terms of the range of flaw sizes in a piece of material. Empirical analyses are available and that of most widespread (but not universal) applicability is the weakest link model due to Weibull. For many ceramics, including RBSN, a plot of ln 1/P vs ln σ_f gives a straight line with a slope equal to the Weibull modulus m, where P is the survival probability. The general Weibull approach also enables data for a particular size or geometry to be estimated from data on specimens of another size or geometry.

The time dependent degradation of strength is due to sub-critical crack growth occurring under stress, sometimes assisted by environmental factors. Of crucial importance is the stress intensity factor/crack velocity (K_I/V) diagram for the material. This can be determined by a number of well-established techniques using specimens with large preformed cracks[8]. Figure 10 shows an idealised diagram in which there are a number of stages. Starting with the highest K_I values, K_{IC} represents the largest stress intensity factory that the material can withstand and where fracture is virtually instantaneous. There is then an intrinsic stage III (independent of environment) where there is a linear relationship between log K_I and log V, ($K_I^n \alpha$ V). This probably occurs in all ceramics and is controlled by various mechanisms ranging from plastic deformation induced crack growth at high temperature to interactions between the crack and lattice structure at low temperature. The extrinsic regions II and I are dependent on interaction between the ceramic and its environment, such as water vapour attack of oxides. In region II the rate of diffusion of the corrosive species to the crack tip is rate controlling and V is independent of K_I. In region I the rate of reaction between the corrosive species and the atoms under stress at the crack tip is controlling and again log K_I is proportional to log V.

For crack growth in regions I or III of the K_I/V diagram, which covers many cases of practical interest, it can be shown[16] that the ratio of strengths (σ) at two strain rates ($\dot{\varepsilon}$) is given by

$$\left(\frac{\sigma_1}{\sigma_2}\right)^{n+1} = \frac{\dot{\varepsilon}_1}{\dot{\varepsilon}_2} \tag{3}$$

(n is the slope of the K_I/V diagram). Further, the time to failure in a constant strain rate test ($\tau_{(\dot{\varepsilon})}$) is related to the time to failure under the maximum stress sustained in a constant stress ($\tau_{(\sigma)}$) by

$$\tau_{(\sigma)} = \frac{\tau_{(\dot{\varepsilon})}}{n+1} \tag{4}$$

It is thus possible to relate the data obtained in standard mechanical tests to that from the K_I/V diagram.

At ambient temperatures most oxides show an environment sensitive behaviour and the important part of K_I/V diagram relevant to delayed fracture is in stage I. There is little reported data available for nitrides and carbides; the sensitivity to the environment is not strong and only stage III behaviour tends to be observed with a relatively high n value, ~100. Limited K_I/V data has been reported for RBSN using the double torsion test[17]. A single linear stage (III) is observed at K_I values close to K_{IC}. Data was obtained for specimens tested in

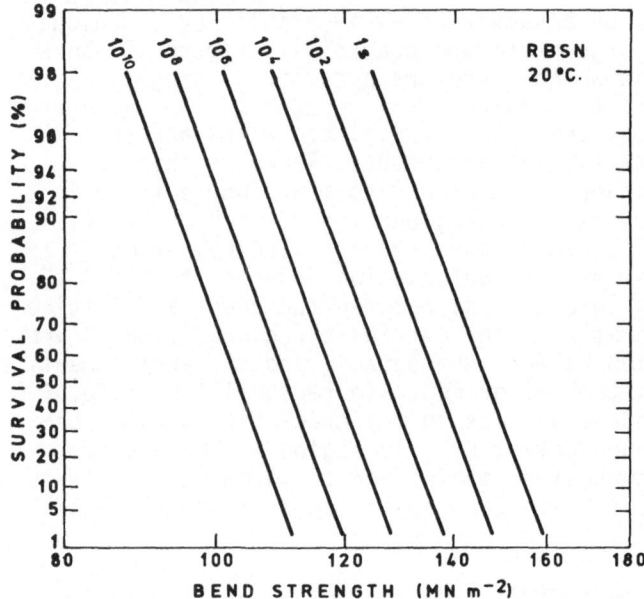

Fig. 11. SPT diagram for RBSN (density 2.30 Mg m^{-3}) at 20°C[18].

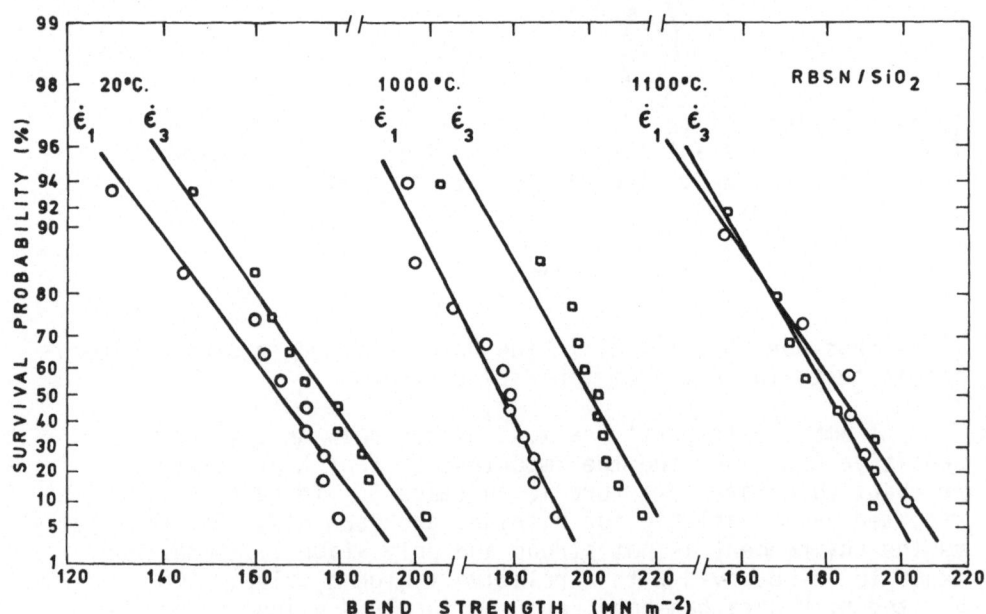

Fig. 12. Strain rate dependence of strength of RBSN (density 2.30 Mg m^{-3}) after oxidation for 100 h at 1000°C[18].

both water and in argon and the calculated values of n were not significantly different and ~ 60.

The statistical and time dependent properties can be combined to give a strength-probability-time (SPT) diagram[16]. For RBSN of density 2.30 Mg m^{-3}, SPT data has been obtained by studying the strain rate dependence of strength[18]. The SPT diagram for RBSN tested in bending at 20°C is given in Figure 11. The Weibull modulus is 22 and the n value, 70.

Data was not obtained for the as-nitrided material at high temperatures in air because of the problems associated with the changing nature of the material with time. Instead, specimens were oxidised for 100 h at 1000°C to produce a microstructure that would be relatively stable at high temperature. This treatment increased the strength at 20°C by $\sim 20\%$. The strain rate dependence of strength for this oxidised material is shown in Figure 12, where under the low strain rate ($\dot{\varepsilon}_1$) fracture occurred in 600 s, and under the high strain rate ($\dot{\varepsilon}_3$) in ~ 6 s. At 20°C the strain rate dependence is similar to that for unoxidised RBSN and n = 80; the Weibull modulus is however significantly reduced by oxidation by a factor of 2. At 1000°C the strain rate dependence is greater and the n value is 37; this effectively doubles the spacing of the lines on the SPT diagram. At 1100°C the strain rate dependence of strength of oxidised RBSN is insignificant which, in terms of the SPT diagram, means that all the lines coincide. This phenomenon has been ascribed to rounding of tips of cracks in the presence of a partly-plastic oxide phase[19]. Crack growth is still possible under these conditions but its effect in the short term is effectively cancelled by crack rounding. This should not however be taken to mean that strength will not degrade in long times; degradation is highly likely but tests have not yet been performed.

8.   CONCLUSIONS

(a)   RBSN is a very useful engineering ceramic that can be fabricated in a wide range of component geometries. Porosity is usually in the range 15-40%.

(b)   The important mechanical properties include a moderate strength with reasonable refractoriness, particularly in inert atmospheres, and a good thermal shock resistance which is associated with the low coefficient of thermal expansion.

(c)   Young's modulus decreases with increase in porosity by a proportional amount approximately twice the fractional porosity.

(d)   Strength decreases with increase in porosity. Practical strengths of ~ 300 MN m^{-2} are feasible. High strength is·generally associated with fine particle size silicon, high green density, and nitriding at temperatures below the melting point of silicon.

(e)   Small amounts of creep under tensile stresses are observed on testing at high temperatures in inert atmospheres.

(f)   Engineering design data is available for ambient temperatures; RBSN has a high Weibull modulus and a small time dependence of strength.

(g)   Oxidation of RBSN is of great significance for material used at high temperatures in oxidising environment. Oxidation at 1000°C produces an increase in strength, but oxidation at 1400°C produces a reduction. Creep rates are enhanced manyfold by oxidation effects at high temperatures. Oxidised RBSN exhibits effects due to crack blunting at 1100°C which leads to a short-term elimination of the time dependent effects.

(h)   Data for oxidised RBSN indicate that the development of a satisfactory glaze for RBSN could give practical strengths of ~ 500 MN m^{-2}.

REFERENCES

1.   B.J. Dalgleish and P.L. Pratt, Proc. Brit. Ceram. Soc., 25 (1975) 295.

2.   R.W. Davidge, A.G. Evans, D. Gilling and P.R. Wilyman, Special Ceram., 5 (1972) 329.

3.   A.G. Evans and R.W. Davidge, J. Mat. Sci., 5 (1970) 314.

4.   A.G. Evans and J.V. Sharp, J. Mat. Sci., 6 (1971) 1292.

5.   J.W. Edington, D.J. Rowcliffe, and J.L. Henshall, Powder Met. Int., 7 (1975) 82 and 136.

6.   D.E. Lloyd, Special Ceram., 4 (1968) 165.

7.   R.W. Davidge and A.G.Evans, Mat. Sci. Eng., 6 (1970) 281.

8.   A.G. Evans, 'Fracture Mechanics of Ceramics', Ed. R.C. Bradt et al, Plenum Press, New York (1974) 17.

9.   J.T. Barnby and R.A. Taylor, Special Ceram., 5 (1972) 311.

10. B.F. Jones and M.W. Lindley, J. Mat. Sci., 10 (1975) 967 and 11 (1976) 191.

11. Idem, Powder Met. Int., 8 (1976) 32.

12. G. Gratwohl and F. Thuemmler, Ber. Deut. Keram. Ges., 52 (1975) 268.

13. W. Engel and F. Thuemmler, Ber. Deut. Keram. Ges., 50 (1973) 204.

14. S.U. Din and P.S. Nicholson, J. Amer. Ceram. Soc., 58 (1975) 501.

15. R.W. Davidge, Ceram. Int., 1 (1975) 75.

16. R.W. Davidge, J.R. McLaren and G. Tappin, J. Mat. Sci., 8 (1973) 1699.

17. K.D. McHenry, T. Yonushonis and R.E. Tressler, J. Amer. Ceram. Soc., 59 (1976) 263.

18. R.W. Davidge, G. Tappin and J.R. McLaren, Powder Met. Int., 8(3) (1976).

19. J.R. McLaren and R.W. Davidge, Proc. Brit. Ceram. Soc., 25 (1975) 151.

DISCUSSION (Messier, Siebels)

Singhal: Chemical-vapour deposition of a thin layer of pure, dense Si_3N_4 on the surface of reaction-bonded silicon nitride may provide extremely good oxidation-resistance by preventing any internal oxidation. Have you tried such coatings?

Davidge: No; but provided that the bonding between the surface layer and the bulk material were good and the surface defects were effectively filled in, one might expect to get a strength increase of ~1.4 times - the relative value to propagate volume and surface flaws of the same size.

A STRUCTURAL MODEL EXPLAINING THE DEVELOPMENT OF STRENGTH IN
REACTION SINTERED SILICON NITRIDE

B.F. Jones, K.C. Pitman and M.W. Lindley

Admiralty Materials Laboratory, Poole, Dorset.

ABSTRACT.   A model is postulated describing the development of
the structure of reaction sintered silicon nitride. The model
involves the gradual elimination of the necks between silicon particles during the early stages of nitridation and the subsequent
development of a continuous silicon nitride network within the pore
space of the original silicon compact. Consequently the critical
defects change from the voids between particles in the green silicon
compact, to voids in the developing silicon nitride network which
contain the reacting silicon particles.

1. INTRODUCTION

Reaction sintered silicon nitride is manufactured by heating a
porous silicon body in a nitriding atmosphere. Attempts to describe this process have been incomplete [e.g. 1,2] because few
studies have been made of structures intermediate between the
original silicon compact and the final silicon nitride. We have
been studying the development of strength in reaction sintered
silicon nitride [3-5]. In this paper we describe a structural
model for the formation of reaction sintered silicon nitride and
discuss some of its implications.

2. THE STRENGTH/DENSITY AND MODULUS/DENSITY RELATIONSHIPS

The key to recent developments was the discovery of linear relationships between strength and nitrided density [3] and between Young's
modulus and nitrided density [6] for a particular silicon powder
compacted to various green densities and nitrided under static

DENSITY Mg m^{-3}	% CONVERSION TO SILICON NITRIDE		PROPORTIONS OF PHASES BY VOLUME (%)					
			SILICON NITRIDE		SILICON		POROSITY	
	lgd	hgd	lgd	hgd	lgd	hgd	lgd	hgd
1.80	33.6	18.8	26.1	15.9	41.9	55.8	32.0	28.3
2.40	94.9	75.1	73.7	63.5	3.2	17.1	23.1	19.4

TABLE I STRUCTURES OF 'lgd' AND 'hgd' COMPACTS NITRIDED TO EQUIVALENT DENSITIES

conditions [5] to various degrees of conversion. A typical strength/ density relationship for a powder with a median particle size of 13 μm is shown in Fig 1a [6]. Standard procedures [3] were used to produce compacts with green densities of 1.47 Mg m^{-3} (lgd) and 1.60 Mg m^{-3} (hgd). The important implication of the existence of this relationship and the modulus/density relationship is that whilst 'lgd' and 'hgd' materials nitrided to the same density will have the same strength and Young's modulus, their structure will be significantly different (Table I). These observations provide important clues concerning the development of the structure, as well as being of immediate value to the fabricator [7].

An important influence of nitrogen gas flow in reducing strength during reaction sintering has been demonstrated (Fig 1a) [5,6]. The linear relationships between strength and density and between Young's modulus and density were significantly different from those found for 'static' nitriding. Even very slow flow rates, undetectable in conventional practice, produce a similar effect and have important practical implications. Additionally strengths lower than the green strength have been measured for materials nitrided to very low weight gains (~5%) and the strength/density relationship is independent of whether as nitrided or lapped surfaces are tested [6]. These observations must all be explicable in terms of any structural model.

3. CRITICAL DEFECT SIZE

It is difficult to directly identify the critical defects in reaction sintered silicon nitride by conventional means [8], except in rare circumstances when large quantities of impurities are present in the original silicon powder. Several workers [9-11] suggest the critical defects are either original inter-silicon particle voids which have remained unfilled during nitriding, or large silicon particles or holes remaining when such particles have melted, whilst another [12] suggests impurity particles may cause the important flaws. There is no direct evidence in the literature linking impurities with established critical defects but it has been our experience that where impurities are the cause, the fracture origin and associated impurities are easily detected and good linear

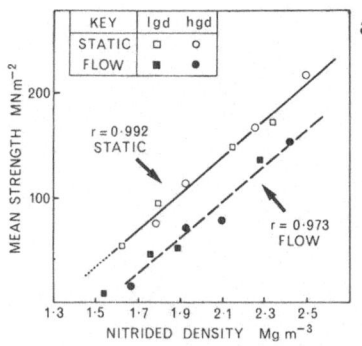 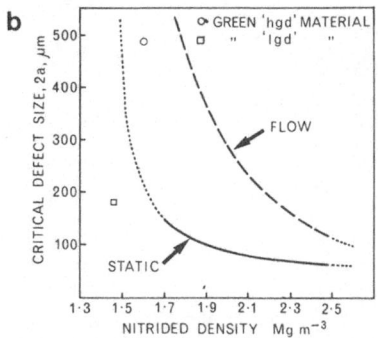

FIG.1 (a) Strength and (b) critical defect size versus nitrided density for compacts nitrided under STATIC and FLOW conditions.

strength/density relationships do not exist. For the present powder fracture origins could not be identified and good strength/density relationships exist (Fig 1a), therefore suggesting that in this case the critical defects are not impurity particles or impurity generated voids.

We have determined the critical defect size by a fracture mechanics method [6] and this has been a primary source of information on which the structural model is based. As in earlier work 'lgd' and 'hgd' compacts were nitrided to various degrees of conversion under either 'static' or 'flow' conditions. The elastic moduli of the compacts were determined by a resonant frequency method and the critical stress intensity factor, K_{IC}, by the double torsion method and the critical defect size calculated [6]. Fig 1b shows the result of this investigation and includes the critical defect size measured for 'lgd' and 'hgd' green silicon compacts. These data indicate that during the early stages of nitriding the critical defect size may increase, but decreases continuously in the density range 1.7–2.5 Mg m^{-3} always being greater for material produced in 'flowing' compared to 'static' nitrogen.

4. STRUCTURAL MODEL

A structural model has been evolved to explain the development of strength and Young's modulus of compacts made from this powder. The structure of the green compact is that of silicon particles joined by necks of silicon or silica and the critical defect size is determined by the size of voids between silicon particles. In Fig 2a ideal packing is used for illustration, but in practice, non-ideal packing occurs and the interparticle voids size will be a function of particle size and shape distribution, and compaction pressure. Machining of the compacts into test specimens may well result in the fracture of necks near the surface and this probably accounts for the larger defect sizes in 'hgd' compared to 'lgd' green compacts (Fig 1b).

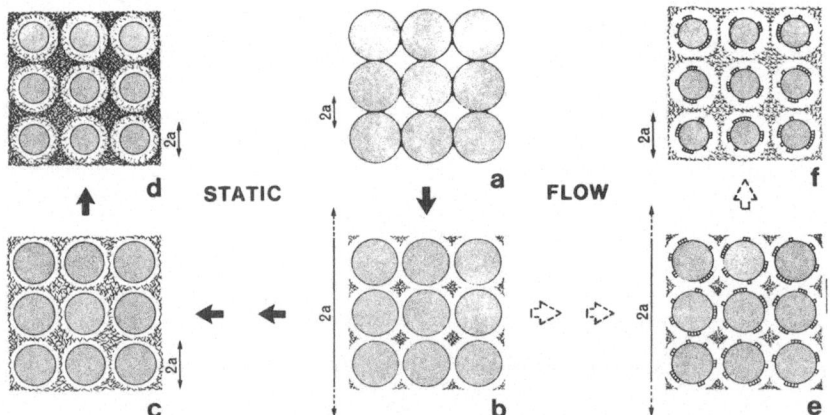

FIG. 2 Schematic model of development of structure of reaction sintered silicon nitride. (a) GREEN COMPACT – silicon particles bonded by silicon or silica necks: (b) EARLY STAGES, STATIC or FLOW (weight gain < 5%) – necks destroyed, nitride network not continuous: (c) STATIC (weight gain ∿15%) – nitride network continuous: (d) STATIC (weight gain∿25%) – nitride network thickening: (e) FLOW (weight gain∿15%) – nitride layer on particles, network not continuous: (f) FLOW (weight gain∿25%) – continuous nitride network just established.

During the early stages of nitriding the necks are eliminated (Fig 2b) and this is why green strengths have no effect on the strength/density relationship. The transition from a neck bonded structure to one bonded by a silicon nitride network (Fig 2c) is a gradual one occurring at different times in different parts of the compact. However it is clear that at an intermediate stage the critical defect size may be considerably larger than that of Fig 2c. This proved to be the case for 'lgd' material as shown in Fig 1b.

Thus the development of the strength and modulus of the partially converted compact requires the development of a continuous silicon nitride network linking the whisker-like matte which develops in the pore space of the original silicon compact. As the network develops and thickens and the silicon particles are consumed (Fig 2c and 2d) so the strength and modulus increase in a manner which proves to be linear with density. Because most commercial powders show no evidence that their strength is controlled by impurities this type of model probably applies for most powders.

5. DISCUSSION

To explain the existence of the strength/density and modulus/density relationships it must first be noted that at any particular density the surface energy for fracture initiation and critical defect size are also identical for 'lgd' and 'hgd' materials [6]. It is therefore necessary to consider the nature of the silicon nitride formed. In the early stages the silicon nitride forms as a whisker matte, and only those whiskers which "interlock" will contribute to mechanical strength. A proportion of the whiskers do not contribute to strength, i.e. they are "redundant", but as the matte

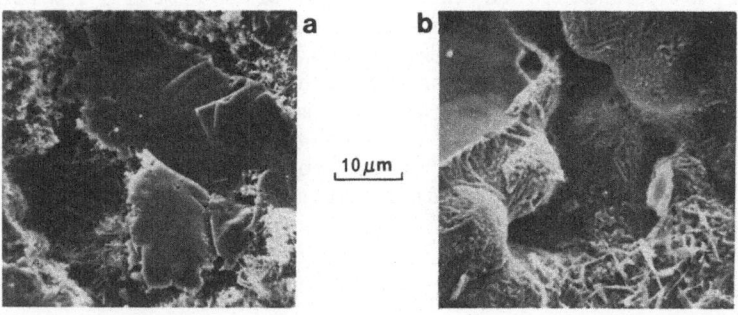

FIG.3 Partially nitrided silicon compacts (weight gain~15%) (a) STATIC, and (b) FLOW—note surface layers.

thickens the proportion of "redundant" whiskers decreases. The only difference between the green silicon compact structures of 'lgd' and 'hgd' is that larger interparticle voids must exist in the 'lgd' compact. Thus at a given density, although more silicon nitride will be present in 'lgd' material (Table I) it will possess a greater degree of "redundancy" compared to 'hgd' material because of the greater void space. It does not seem unreasonable to assume that there will be a direct relationship between the size of the voids in the green silicon compact and the proportion of silicon nitride which is "redundant" and that this results in the linear strength/density and modulus/density relationships being independent of green density.

The effect of nitrogen gas flow can be accounted for because a greater proportion of silicon nitride forms on the surface of the silicon particles [6] (Fig 3) and does not contribute to the continuous silicon nitride network. In effect this material can be added to the "redundant" portion of the network. It seems likely that such surface layers spall off as the reaction proceeds but they are probably not efficiently incorporated into the network. As a consequence of surface layer formation the development of a continuous network is delayed to higher degrees of conversion and since the surface layers are also "redundant" as far as mechanical properties are concerned, the strength is reduced (Figs 2e and 2f).

It is clear therefore that the strength of reaction sintered silicon nitride produced under specific nitriding conditions will be governed by the structure of the green silicon compact. The green compact structure will depend on the properties and compactability of the starting silicon powder. Whilst under conditions of ideal packing higher strengths are to be anticipated from fine powders since small closely spaced voids will be efficiently linked by the growing silicon nitride network, the fact that fine powders tend to be less compactable results in the existence of an optimum intermediate particle size for maximum strength. Thus a maximum effort must be applied to powder characterisation preparation and compaction.

It is also evident from this work that nitriding conditions

can have an important influence on the strength of reaction sintered
silicon nitride. Conditions which give rise to network rather
than surface layer formation (i.e. 'static' rather than 'flowing'
nitrogen) favour high strengths. Recently it has been shown [13]
that the use of flowing nitrogen/hydrogen gas mixtures can completely obviate the deleterious effect on strength observed for
flowing nitrogen. Other experimental conditions which might increase the proportion of silicon nitride forming via a vapour phase
reaction are being studied.

7. CONCLUSIONS

A model for the process by which reaction sintered silicon nitride
forms from a silicon compact has been formulated. All of the experimental observations concerning the development of mechanical properties of the material can be explained in terms of this model.
Based on the model predictions are made of the important steps
required to maximise the strength of the material.

ACKNOWLEDGEMENTS

This paper is published with the permission of the Ministry of Defence
(Procurement Executive) and of HMSO, holders of Crown Copyright.

REFERENCES

1. P. Popper and S.N. Ruddlesden, Trans.Brit.Ceram.Soc., 60,
    (1961), 603.
2. N.L. Parr and E.R.W. May, Proc.Brit.Ceram.Soc., 7, (1967), 81.
3. B.F. Jones and M.W. Lindley, J.Mater.Sci., 10, (1975), 967.
4.   idem   , Powder Met.Int., 8, (1976), 32.
5.   idem   , J.Mater.Sci., 11, (1976), 1288.
6. B.F. Jones, K.C. Pitman and M.W. Lindley, J.Mater.Sci.,
    to be published.
7. B.F. Jones and M.W. Lindley, Proc.Science of Ceramics 8 (1976)
    123.
8. D.J. Godfrey and K.C. Pitman, Proc.2nd Army Mater.Conf.
    Ceramics for High Performance Applications (Ed. J.J. Burke,
    A.E. Gorum and R.N. Katz) Brook Hill Publ.Co. 1974 p.425.
9. D.R. Messier and P.Wong,   ibid, p.181.
10. A.G. Evans and R.W. Davidge, J.Mater.Sci., 5, (1970), 314.
11. B.J. Dalgleish and P.L. Pratt, Proc.Brit.Ceram.Soc., 25,
    (1975), 295.
12. D.J. Godfrey, Proc.Brit.Ceram.Soc., 25, (1975), 325.
13. B.F. Jones and M.W. Lindley, J.Mater.Sci., to be published.

DISCUSSION (Messier, Siebels)

Jack: The difference between "static" and "flow" microstructures must depend upon differences in the compositions of the vapour phase and should approach to some extent the conditions where there is or is not a static gas boundary layer at the surface of each particle. Thus, there are two factors for an impurity such as O_2 or H_2O in the nitriding gas - a potential (e.g. p_{H_2} or p_{O_2}) which will be different for "static" and "flow" conditions, and a total quantity of impurity. Additions of H_2O, H_2 or O_2 to the gas should affect microstructure very markedly if Dr. Lindley's model is correct. It would also be interesting to see the effect of much higher flow rates.

Lindley: Under 'gas flow' conditions strength and microstructural changes are probably more a result of species removed from the effective reacting environment of the compact (probably SiO) rather than the introduction of impurity species. At complete conversion, compact weight gains are always less in flowing N_2 than static N_2 (about 1% relative difference) and the strength of material found under 'static' conditions exceeds that formed under flow conditions by some 25% over a wide range of oxygen impurity levels.

Katz: Do you think that choosing very narrow bimodal particle size distributions, such that the finer distribution would fit in the interstices of the coarser distribution, would yield stronger material? Your model would seem to suggest that this would be beneficial.

Lindley: The strength of reaction sintered silicon nitride under specific nitriding conditions will be governed by the structure of the green compact. Ways of increasing the green density should be studied; potential material improvements could be highly significant.

Moulson: Do you have any evidence that the linear strength/nitrided density relationships hold up to the highest density values?

Lindley: The evidence is conclusive that when the developing pore microstructure controls strength, the strength/nitrided density relationships are linear to the highest nitrided densities, corresponding to 100% silicon nitride formation. Departures from linearity are observed at the higher levels of conversion if liquid phases are produced as a result of silicon or eutectic melting.

STRENGTH-DENSITY-NITRIDING CYCLE RELATIONSHIPS FOR REACTION SINTERED Si_3N_4*

John A. Mangels

Turbine Development Department
Ford Motor Company
Dearborn, Michigan U.S.A.

ABSTRACT. The development of a nitriding process for reaction sintered Si_3N_4 of various densities was described with respect to nitriding cycle and atmosphere. The density-strength relationships (2.3-2.8 g/cc density range) for these various cycles was given. It was shown that in order to achieve the maximum strength density ratio, care must be taken to utilize the proper nitriding schedule. The microstructure of the various materials were discussed. It was shown that the pore size is the strength limiting factor in these materials. It was inferred, by microstructural examination, that the exothermic nature of the nitriding reaction was the major factor leading to the large porosity and therefore is the controlling factor in the development of high strength reaction sintered Si_3N_4.

INTRODUCTION

Reaction sintered silicon nitride has been shown to be a useful engineering material when applied to certain applications in a gas turbine engine. One major advantage is the ease with which the material can be fabricated into complex shapes, such as turbine stators or rotors. The reliability of these components could be increased if the strength could be improved. One obvious method of achieving this is by increasing the density. However, nitriding considerations limit the final density to only 85% of theoretical (2.7 g/cc). An effort must be made to achieve the maximum

---

* This work was supported in part by the Advanced Research Projects Agency under Contract Number DAAG-46-71-C-0162.

strength for any given density. This can be attained by proper microstructural development which, as will be shown, will be achieved by proper nitriding conditions.

EXPERIMENTAL PROCEDURE

The test samples were fabricated by injection molding silicon metal powder to yield green densities, ranging from 1.43 g/cc to 1.76 g/cc. The silicon powder had a maximum particle size of 30 μm with a mean particle size of 3-5 μm. The major impurities were 0.60% Fe, 0.2% Al and 0.02% Ca. All compositions yielding a density of 2.5 g/cc or greater had 2.5 weight percent Fe_2O_3 added as a nitriding aid.

Three basic nitriding schedules were employed as shown in Figure 1. The three step cycle is a modification of the two stage nitriding process proposed by Parr[1]. The multi-step cycle proposed by Messier[2] and the constant rate cycle proposed by Mangels[3] both employ maximum furnace temperatures of 1400°C, 20°C below the melting point of silicon, while the 3-step cycle increases the temperature above the melting point of silicon. All cycles utilized a 100% N_2 atmosphere or 1.0 to 4.0% H_2/Balance N_2 atmospheres[4] at furnace pressures of 3 psig.

The strength tests were performed using four point bending (95.2 mm x 190 mm top and bottom span) and test specimens (31.7 mm x 63.5 mm) with as nitrided surface conditions.

RESULTS

Table 1 shows the nitriding results in terms of percent

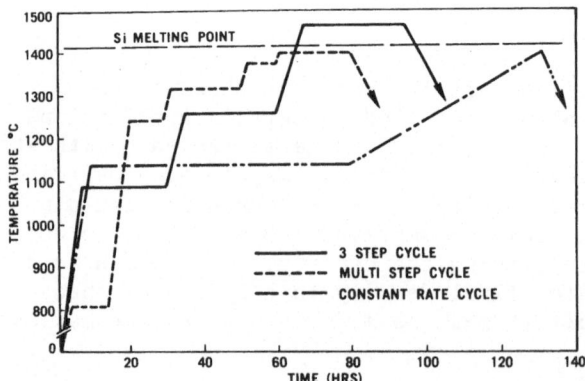

Figure 1   Time-Temperature Schedule for Nitriding Cycles

Table 1
Percent of Silicon Converted to Si_3N_4
For Various Nitriding Cycles and Density Levels

Nitriding Cycle	ATM	2.3 g/cc	2.55 g/cc	2.7 g/cc	2.8 g/cc
3 Step	100% N_2	97.5%	97.0%	98.5%	94.7%
	1%- 4% H_2/N_2	97.1	-	96.7	-
Multi-	100% N_2	97.0	99.0	98.1	95.9
Step	1%- 4% H_2/N_2	-	98.3	98.4	95.9
Constant	100% N_2	-	98.3	98.0	-
Rate	1%- 4% H_2/N_2	-	98.3	98.0	94.7

silicon converted to Si_3N_4. This was obtained by weight gain measurements, with compensations made for the amount of nitriding additive used. The data shows that the nitriding cycle or atmosphere does not effect the degree of nitriding. The 2.8 g/cc density level material has poor (94.7 to 95.9%) conversion when compared to the lower density materials, indicating that these cycles and atmospheres are not adequate to properly nitride such high density material.

The strength-density relationship for the various nitriding cycles and atmospheres are given in Figure 2. Each cycle results in a strength increase with density over a limited density range. The 3 step cycle yields strengths that increase to a maximum of about

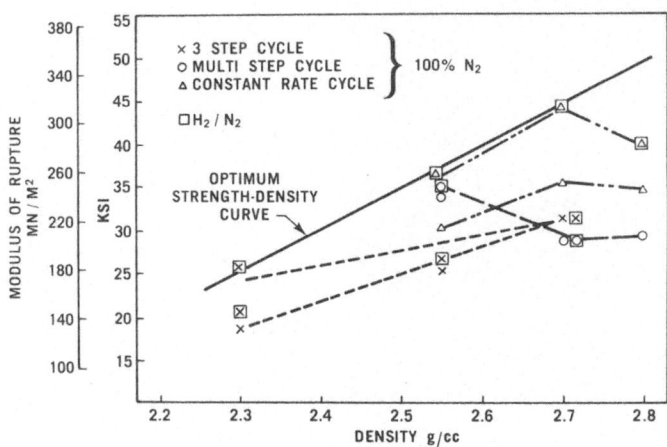

Figure 2   Strength-Density Relationships for Various
Nitriding Cycles and Atmospheres

200 MN/M^2 at a density of 2.7 g/cc. The multi-step cycle yields a
maximum strength of 250 MN/M^2 at the 2.55 g/cc level and then
decreases for higher densities. The constant rate cycle with a
H$_2$/N$_2$ atmosphere yields the highest strengths; 310 MN/M^2 at the
2.7 g/cc density level.

The strength-density behavior of these materials can be explained by examination of their microstructures. Figure 3 shows
the structure of the 2.7 g/cc material nitrided under various
conditions. The strength of the samples in Figure 3a through 3c
are between 190 and 250 MN/M^2 while Figure 3d had the highest
average strength of 310 MN/M^2. These materials have different pore
distributions even though the densities are the same. The low
strength microstructures show the largest pore sizes to be very
similar (25 μm), while the high strength structure has a much
finer and more uniform porosity. The large porosity is usually
associated with dense regions. The dense regions were probably
formed when a silicon particle melted and was dispersed into the
surrounding structure causing the large pores to form[5]. This
type of structure would be expected from the 3-step cycle, since
the nitriding temperature exceeds the melting point of silicon.
However, the multi-step and the constant rate cycles both exhibit
the same phenomena even though the maximum cycle temperature was
only 1400°C. From this information, it can be inferred that the
silicon temperature was higher than the furnace temperature due to

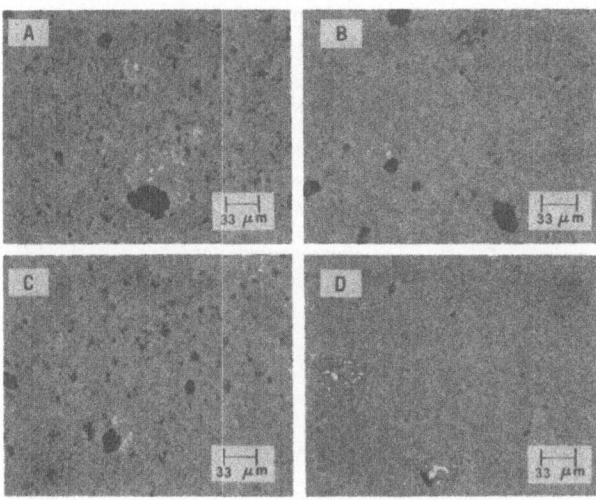

Figure 3   2.7 g/cc Material Processed Under Different Conditions.
            (A) 3-Step Cycle, 100% N$_2$, (B) Multi-Step Cycle,
            100% N$_2$, (C) Constant Rate Cycle, 100% N$_2$, (D) Constant
            Rate Cycle, 4% H$_2$/N$_2$.

the exothermic nature of the nitriding reaction.

Figure 4 shows the microstructure of materials of different densities but all having strengths between 25 and 30 Ksi. It is again obvious that the pore distributions are different for each density level; however, the largest individual pores are in the 25 to 30 μm range. These large pores again show evidence of having been formed by silicon "melt out" as previously described. The strength limiting factor appears to be the formation of the 25-30 μm pores, probably due mainly to the local nitriding temperatures exceeding the melting point of silicon.

DISCUSSION AND SUMMARY

The material strength increases with density for all cycles investigated. The strength, however, reaches a limiting value for each cycle, at which point a different cycle must be employed to further yield higher strength values. Examination of the microstructures shows different pore distributions for each density level and for each type of nitriding cycle. However, the materials exhibiting low strength all had large pores of approximately 25-30 μm. As shown with the 2.7 g/cc material, the pore distribution and pore size can be changed with processing.

It appears that the exothermic nature of the nitriding reaction is the principal factor affecting the strength of the material.

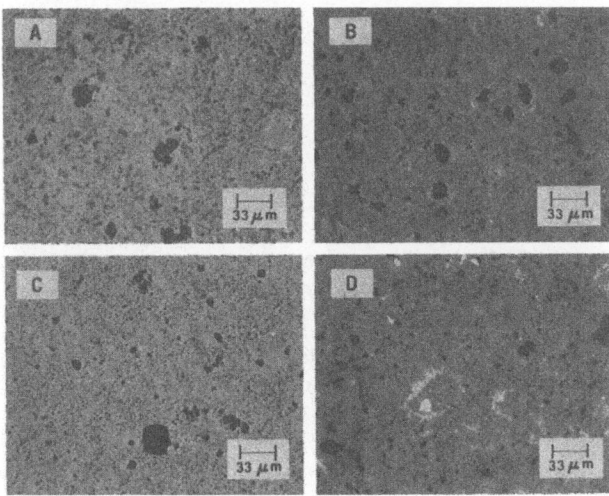

Figure 4   Different Density Materials With Strength Between 175 and 210 MN/M^2. (A) 2.3 g/cc, (B) 2.55 g/cc, (C) 2.7 g/cc, (D) 2.8 g/cc.

Mangels[3] has shown that for large furnace loads (10,000 grams of silicon) up to a 200°C difference was obtained between the control temperature of the furnace and the silicon part temperature.

The 3-step and multi-step cycles both have large temperature steps along with fast temperature ramp rates which would tend to accelerate the reaction and increase the effect of the exotherm. Conversely, the constant rate cycle slowly increases temperature, which should be helpful in controlling the exotherm. This does occur when a H_2/N_2 atmosphere is used, resulting in high strengths (310 MN/M^2) and uniform microstructure (Figure 3d). The true effect of the H_2/N_2 atmosphere in the process is unknown at present.

The constant rate cycle must still be modified to yield better control of the nitriding exotherm. Large furnace loads still cannot be reproducibly nitrided to yield high strength. Also, this cycle when applied to 2.8 g/cc material is not sufficient to nitride 2.8 g/cc density level material, or to produce a microstructure conducive to high strengths.

In order to realize the full strength potential of reaction sintered Si_3N_4, the nitriding exotherm must be understood and methods developed to completely control it during the nitriding process.

REFERENCES

1. Parr, N. L., et. al., "Structural Aspects of Si_3N_4", Powder Metallurgy, No. 8, 1961, p. 152.
2. Messier, D. R., Wong, P., "Kinetics of Nitridation of Si Powder Components", J. Amer. Ceramic Soc., Vol. 56, No. 9, (1973) p. 480.
3. Mangels, J. A., "Processing High Density Reaction Sintered Si_3N_4", Bulletin Amer. Ceramic Soc., Vol. 55, No. 4, (1976), p. 395.
4. Mangels, J. A., "Effect of H_2/N_2 Atmospheres on the Properties of Reaction-Sintered Si_3N_4", J. Amer. Ceramic Soc., Vol. 58, No. 7-8, (1975), p. 354.
5. Evans, A. G., Davidge, R. W., "The Strength and Oxidation of Reaction Sintered Silicon Nitride", J. Material Science, 5, 1970, p. 314.

GENERAL DISCUSSION (Messier, Siebels)

CRITICAL DEFECTS IN REACTION BONDED SILICON NITRIDE

Popper: If large particles (100 μm) of iron derived from the milling treatment are responsible for the large flaws in reaction sintered silicon nitride, then one should remove these by "magnetting" or acid treatment and then the small amounts necessary for accelerating nitridation added in a very finely divided form (e.g. as Fe_2O_3 or in solution).

Moulson: We have removed particulate iron from commercial silicon powder by magnetising a powder/water slurry - identified the particles as Fe and their typical size as 100 μm. These must lead to large (> 100 μm) inhomogeneities in the reaction bonded material.

Lange: Are the large pores formed by the melt-out of $FeSi_2$ liquid, instead of pure Si?

Moulson: All the work reviewed by Dr. Davidge is concerned with commercial silicon powder carrying impurity levels typically $1-1.5^W/o$, of which up to $\sim 0.8^W/o$ is Fe. We should therefore regard the Fe/Si eutectic temperature as critical, rather than 1412°C.

Davidge: The relative importance of Fe/Si eutectic melting and Si melting must depend on the relative sizes of the Fe and Si particles. Only if the Fe particles are of comparable size to the largest Si particles will this produce the largest defect. The production of clean silicon must be paid for, and probably slows down the nitriding reaction with additional economic penalty; these factors need to be balanced against the potential increase in strength that may be obtained.

Mangels: Examination of the microstructural development of the material shows no evidence of the pores being formed until the temperatures reached the 1400°C region, suggesting that the Fe-Si eutectic is not playing a part. The electron microprobe analyser shows no concentration of either iron or aluminium, and optical microscopy yields no evidence of $FeSi_2$ around the large pores (except in a few completely random cases). I conclude that the pores do act as critical flaws.

Godfrey: The role of large voids in RBSN strength control is debatable, as is the value (with commercial purity silicon) of nitriding above and below 1412°C. Iron/silicon eutectic melting produces holes; pure RBSN (99.95%) does not show these features. The strength of these materials is 200-220 MPa, as compared with 230 for commercial purity RBSN (25 μm silicon). If 99.95% purity RBSN is produced from 8 or 11 μm mean size powder strengths ~ 260 MPa are obtained; but 12.5 μm commercial purity RBSN gives a strength of 270 MPa, with melt holes. As pointed out by Godfrey and Lindley in 1973, the strength-controlling defects in RBSN are probably the large fields of very fine porosity, which appear as light regions in polarised light, not the discrete black holes caused by eutectic or silicon melting.

Davidge: I accept that in certain cases the critical defect may be a large region of very fine porosity but the consensus view expressed at the meeting appears to support my statement that in most cases the critical defects are the largest observed single pores.

Discussion contribution
CRITICAL DEFECTS IN REACTION SINTERED Si_3N_4

B.F. Jones and M.W. Lindley

Admiralty Materials Laboratory, Poole, Dorset, U.K.

A knowledge of the size of the critical defects in reaction sintered Si_3N_4 is important to both the fundamental understanding of the factors which control its strength and the engineering application of the material. The literature contains no positive direct identification of the critical defects in reaction sintered Si_3N_4. Excluding mechanical damage of surfaces, the following sources of defect have been suggested: (a) porosity developed from the inter-particle voids in the silicon compact, e.g. unfilled or partly filled holes (1-3); (b) 'inter-network' porosity replacing reacted silicon particles (4,5); (c) incompletely converted large silicon particles (e.g. 2); (d) holes formed by melting large silicon particles (e.g. 1,6); (e) impurity particles, or voids formed from impurities (7).

For (a)-(c) a linear correlation between strength and weight gain and strength and nitrided density should occur because inter-particle voids are progressively filled or the size of the remaining silicon particles reduces during reaction. For (d) the relative size of the silicon particles at the moment when melting occurs (1, 6) determines whether there is a departure from the linear strength/ density relationship to lower strengths.

Whether or not large impurity particles (e) or the melting of such impurity particles will influence strength depends on the relative size, composition and distribution of the impurity particles compared to the size of the defects arising from other mechanisms. There are many instances of reaction sintered Si_3N_4 materials with relatively high concentrations of impurity but where the impurity size and composition is such that 'strength degradation' does not occur (e.g. 2,6). Only one of six commercial silicon powders we have examined has not shown a well defined strength/

578

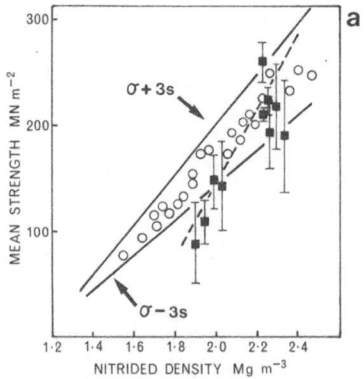

IRON   CALCIUM

Fig. 1: Impurity particles controlling strength (a) strength/
density relationship (■ ---, error bars ± one standard deviation);
note large strength scatter and lack of linearity of strength with
density compared to samples where strength controlled by inter-
network porosity (0 ____, strength lines ± 3 standard deviations);
(b) fracture mirror and identification of impurities in the
critical defect at its origin.

nitrided density relationship (Fig. 1a): the strengths of samples
prepared from this powder were very variable and were not controlled
by the normal 'inter-network' porosity (4,5) but by impurity particles (Fig. 1b). Although this particular atypical silicon powder
had an unusually high iron content (>1.0 wt% Fe), it was the large
size and composition of the impurities compared to the size of the
pores formed by mechanisms (a)-(d) that gave rise to fracture
mirrors enabling the critical defects to be identified. When the
more normal 'inter-network' porosity controls strength such fracture
mirrors are not evident.

The observation and measurement of critical defects from
polished microsections must be treated with considerable caution
since it can only be surmised that the observed microstructural
features may be critical defects: additionally, because of polishing artefacts (e.g. 2,8) and in some instances the lack of correlation between strength and observed pore size (e.g. 8), pore sizes
measured on polished sections may bear no resemblance to the effective defect size. Under circumstances where the identification
and measurement of critical defects is so difficult, an alternative
approach is the calculation of critical defect sizes from fracture
mechanics data (4.5).

REFERENCES

1. A.G. Evans and R.W. Davidge. J. Mat. Sci., 5 (1970), 314.
2. D.R. Messier and P. Wong. Proc. 2nd Army Mater. Conf.
    Ceramics for High Performance Applications, Brook Hill Publ.
    Co. (1974), 425.

3. B.F. Jones and M.W. Lindley. <u>J. Mat. Sci.</u>, <u>10</u> (1975), 967.
4. B.F. Jones, K.C. Pitman and M.W. Lindley. <u>J. Mat. Sci.</u>, to be published.
5. B.F. Jones, K.C. Pitman and M.W. Lindley. This volume.
6. J.A. Mangels. This volume.
7. D.J. Godfrey. <u>Proc. Brit. Ceram. Soc.</u>, <u>25</u> (1975), 325.
8. D.J. Godfrey and M.W. Lindley. <u>Proc. Brit. Ceram. Soc.</u>, <u>22</u> (1973), 229.

THE DEVELOPMENT OF MICROSTRUCTURE AND YOUNG'S MODULUS IN REACTION-BONDED SILICON NITRIDE

P. Longland and A.J. Moulson

Department of Ceramics, University of Leeds, Leeds, U.K.

ABSTRACT

In an attempt to increase understanding of how the mechanical strength of reaction-bonded silicon nitride develops during the nitriding reaction, sonic Young's Modulus has been measured at stages during the low pressure (50 torr) nitriding of pure silicon powder compacts, and the data compared with those predicted by the 3-phase composite model developed by Ishai and Cohen.

Some microstructural observations, made at early stages in the nitriding reaction, are presented.

INTRODUCTION

The first significant attempt to correlate the mechanical properties and microstructure of reaction-bonded silicon nitride was made by Evans and Davidge (1). Their study, and many subsequent studies, have been based on silicon powders contaminated with metallic impurities at approximately the $2^w/o$ level. It is now appreciated that chemical contaminants, in either the silicon or nitriding gas, do affect reaction kinetics and microstructure (2,3). A programme is currently under way at Leeds aimed at bringing the microstructure, and hence certain technologically important properties of reaction-bonded silicon nitride, under control. This paper, which describes an initial stage of the programme, is concerned with the development of microstructure and Young's modulus during the low pressure nitriding of pure (< 10 ppm metallic impurities) silicon. It has been demonstrated elsewhere (2) that high-purity silicon can be nitrided at low nitrogen pressures.

Room-temperature values of Young's modulus have been obtained for silicon powder compacts at stages of their conversion to

silicon nitride, and a comparison made with those calculated from
a simple three-phase composite model. Some observations have been
made of the microstructural changes which occur during the very
early stages of the nitridation of compacts of spheroidized, and
of unspheroidized, silicon particles. It was an original intention
that spheroidized silicon would facilitate observation and possibly
permit quantitative measurements to be made of the developing
nitride.

EXPERIMENTAL

Silicon powder was prepared from semiconductor grade ingots
(Monsanto Ltd., Ghent, Belgium) as described elsewhere (2). The
powder was classified into two size fractions (53-65 μm and <45 μm)
using bronze meshes, and then washed in 1M HCl followed by distilled
water, to remove the metallic impurities introduced during commin-
ution and classification.

The 53-65 μm particles were spheroidized by allowing them to
fall through the hot zone (1720°C) of a vertical alumina tube
furnace, in a hydrogen atmosphere. The spherical particles were
size-classified and washed as described above.

For the measurement of Young's modulus, powder compacts were
prepared from the silicon powder (<45 μm, mean 20 μm) by isostatic
pressing at a pressure of 140 MN m^{-2}. Bar samples (approx. 30 x 5
x 5 mm) were then cut from the powder compacts using a thin glass
slide.

The samples, contained in an alumina crucible carried on a
molybdenum suspension, were nitrided in a molybdenum-wound alumina
tube furnace. 'White-spot' nitrogen (<5 ppm oxygen) was passed
through a manganous oxide oxygen-trap and a vertical column of
phosphorous pentoxide before entering the furnace. The nitrogen
pressure in the apparatus could be maintained at fixed values in
the range 0-760 torr.

To carry out a nitridation 'run', the system was first evac-
uated to 10^{-3} torr for 30 minutes, then filled with nitrogen to the
required pressure; the sample was then lowered into the hot zone.
In all cases the reaction-conditions were 1370°C and 50 torr
nitrogen pressure.

Young's modulus was measured by a resonance method, the sample
being electrostatically driven (4). The frequency range, 10^3 to
2 x 10^5 Hz, could be covered allowing room-temperature moduli in
the range 10 - 300 GN m^{-2} to be measured on samples of length 30 mm.

MICROSTRUCTURAL OBSERVATIONS

Figs. 1(a) and (b) are micrographs which show the microstructural changes occurring during the initial stages of nitriding a
compact of silicon spheres. Neck formation between adjacent

583

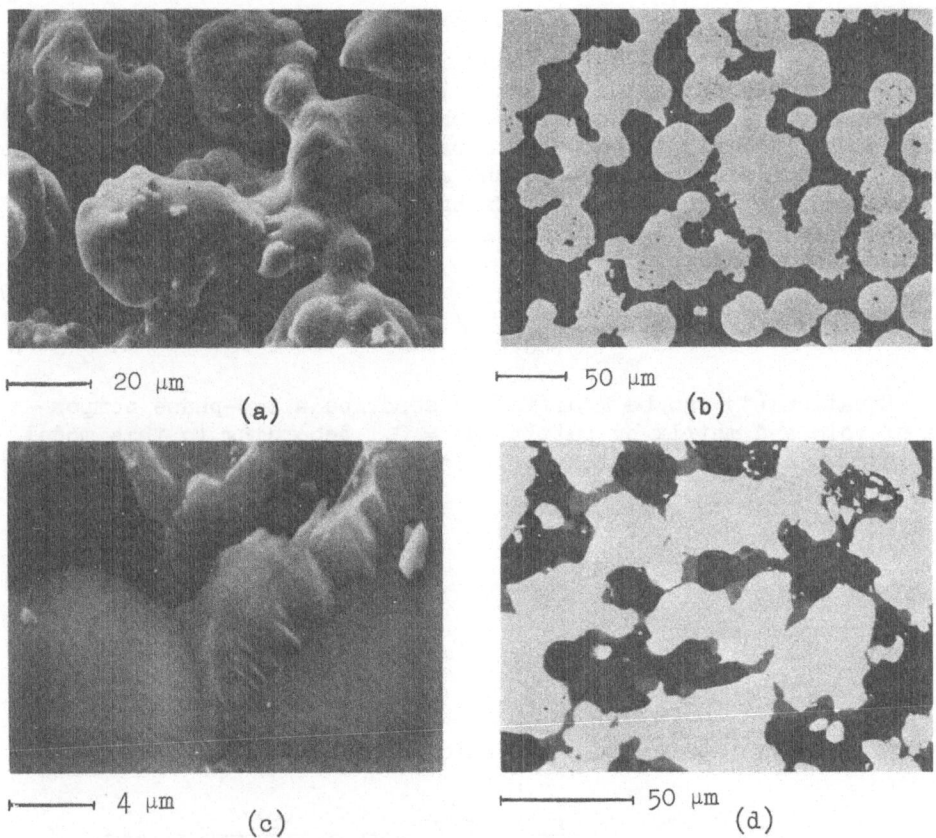

⊢——⊣ 20 μm (a)	⊢——⊣ 50 μm (b)
⊢——⊣ 4 μm (c)	⊢————⊣ 50 μm (d)

Figure 1: Microstructural changes occurring during early stages of nitriding.

(a) SEM of spheroidized particles
(b) Optical micrograph of spheroidized particles
(c) SEM of unspheroidized particles
(d) Optical micrograph of unspheroidized particles.

particles has occurred, apparently either by surface diffusion or vapour transport. There is evidence of some reaction product forming along what appears to be grain boundaries. Figs. 1(c) and (d) show the microstructure developed during the initial stages of nitridation of a compact of unspheroidized silicon particles. Reaction has proceeded to a greater extent than in the case of the spherical particles, but the formation of a continuous silicon network is still the dominant feature of the microstructure. Silicon nitride (light grey) is seen to have formed preferentially in the neck regions between the original particles.

THREE-PHASE COMPOSITE MODEL FOR DEVELOPING YOUNG'S MODULUS

The model is based on one described by Cohen and Ishai (5) in which a "system consisting of discrete dispersed particles can be represented by a single continuous phase confined within a homogeneous matrix". In the case of a partially nitrided silicon powder compact, the dispersed phase is the unreacted silicon; the matrix is the porous silicon nitride. Using the above model, Cohen and Ishai derived the following expression for the Young's modulus of a two-phase composite:

$$E_c = E_m \left[1 + \frac{c_f}{(\frac{m*}{m*-1}) - c_f^{\frac{1}{3}}} \right] \qquad \dots \text{(1)}$$

Equation (1) can be modified to describe a two-phase composite of void and matrix by putting $m* = 0$. According to this model the modulus of porous silicon nitride would be given by:

$$E_{PSN} = E_{SN} \left[1 - c_p *^{\frac{2}{3}} \right] \qquad \dots \text{(2)}$$

where
$$c_p* = \frac{c_p}{1-c_{Si}} \qquad \dots \text{(3)}$$

$$m* = m\left(\frac{1}{1-c_p*^{\frac{2}{3}}}\right) \qquad \dots \text{(4)}$$

Substitution of (2), (3) and (4) into (1) yields

$$E_{PSN-Si} = E_{SN} \left[1 - \left(\frac{c_p}{1-c_{Si}}\right)^{\frac{2}{3}} \right] \left[1 + \frac{c_{Si}}{\frac{m}{m-1 + \left(\frac{c_p}{1-c_{Si}}\right)^{\frac{2}{3}}} - c_{Si}^{\frac{1}{3}}} \right] \qquad \dots \text{(5)}$$

If ρ_g and ρ_n are the 'green' and 'nitrided' densities of a partially nitrided compact, then the following expressions can be obtained:

$$\left. \begin{array}{l} c_p = 1 - 0.281\,\rho_g - 0.148\,\rho_n \\ c_{Si} = 1.072\,\rho_g - 0.643\,\rho_n \end{array} \right\} \qquad \dots \text{(6)}$$

In deriving the above expressions it is assumed that the densities of silicon and silicon nitride are 2330 and 3200 kg m^{-3} respectively, and that the conversion of silicon to silicon nitride is accompanied by a volume expansion of 22% and a mass gain of 66.7%.

Substitution of (6) into (5) gives the relationship between Young's modulus of the composite, nitrided density, and green

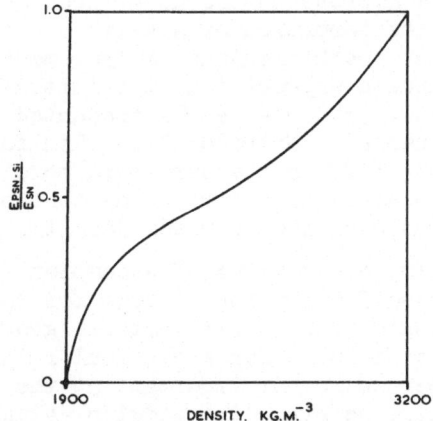

Figure 2: Computed Young's Modulus of 3-phase composite

density. Fig. 2 shows the computed relationship between Young's modulus and ρ_n for a green density (ρ_g) of 1900 kg m^{-3}, the green density which, theoretically, can be nitrided to a fully dense silicon nitride body.

Fig. 3 shows a comparison between theoretical and experimental data for a green compact density of 1500 kg m^{-3}.

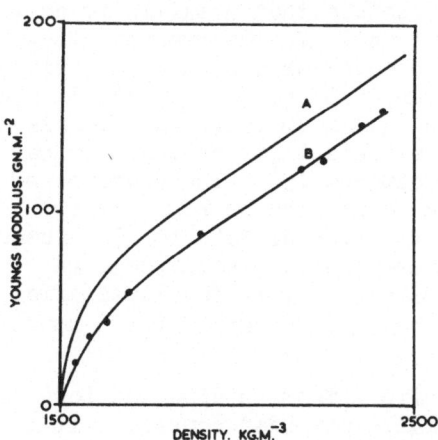

Figure 3: Comparison of computed (A) with experimental (B) Young's Modulus data for 3-phase composite.

DISCUSSION

The two salient features of the microstructure developed in the first stage of nitridation are the formation of a silicon network and the growth of nitride in the 'neck' regions. Measurements made on vacuum-sintered silicon compacts suggest that the initial steep rise in modulus to approximately 35 GNm^{-2} can be accounted for by the development of silicon 'necks'. Nitride which then forms preferentially in the 'neck' regions stiffens the structure more efficiently than the nitride formed later which gives rise to the nearly linear relationship between modulus and nitrided density.

It is surprising that such a simple modelling of a complex composite provides the remarkably good description it does for the observed rate of development of Young's modulus with nitride growth. The fact that the predicted values of modulus for a particular ρ_n are not in agreement with those observed is not important at the present stage of the study, especially so since the modulus values at later stages of the reaction are markedly sensitive to microstructural changes occurring early in the reaction. It is noteworthy that although the experimental data are linear with ρ_n over a wide range of values, they do not extrapolate to a value close to those quoted for the modulus of fully dense hot-pressed silicon nitride (6,7). This observation is consistent with the predictions of the model.

An interesting general feature of the microstructures is the absence of a 'whisker' phase. This inevitably brings into question the validity of the idea that the formation of a 'whisker' mat is an integral part of the reaction-bonding process (1,8). As mentioned previously, commercial purity silicon powders have been used extensively by other workers and it is possible that metallic impurities promote the formation of a 'whisker' mat. In the present study, 'whisker' formation has only been observed when aluminium contamination of the silicon spheres was known to have occurred. In some cases the silicon spheres picked up aluminium during spheroidization, presumably as a result of the instability of the alumina furnace tube under the hot-zone conditions (9). The presence of aluminium in some of the spheres was confirmed by EPMA and by etching a polished section in a 5% solution of NaOH for 5 minutes at 100°C when the aluminium-rich phase was completely removed (Fig. 4(a)). Fig. 4(b) shows the microstructure developed after the aluminium-contaminated spheres had been nitrided for 5 hours at 1370°C in nitrogen at 50 torr.

The study is being continued with attention directed towards refining the modelling of the partially nitrided compact. It is expected that an improved understanding of how the mechanical properties of the composite develop during nitriding, and how this development depends upon processing variables, will assist in the overall aim to optimise properties of the fully formed ceramic.

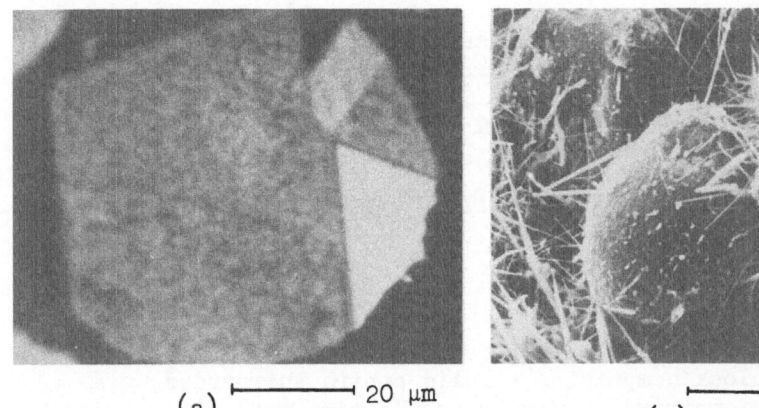

(a) ├──────┤ 20 µm

(b) ├──────┤ 40 µm

Figure 4: (a) Polished section optical micrograph showing
aluminium-contaminated silicon spheres after etching

(b) Nitride growth on aluminium-contaminated silicon
spheres.

ACKNOWLEDGMENT

The authors wish to thank the European Research Office of the
U.S. Army for financial support. The assistance of Dr. T. Dosdale
with the computation is gratefully acknowledged.

NOMENCLATURE

C_f = volume fraction of particulate filler.

C_p^* = volume fraction of pores in void-matrix composite.

C_p = volume fraction of pores in 3-phase composite.

C_{Si} = volume fraction of silicon in 3-phase composite.

E_c = Young's modulus; 2-phase composite.

E_m = Young's modulus; matrix.

E_{SN} = Young's modulus; silicon nitride.

E_{PSN} = Young's modulus; porous silicon nitride.

E_{PSN-Si} = Young's modulus; 3-phase composite.

m = modular ratio, silicon:silicon nitride.

m^* = modular ratio, filler:matrix of 2-phase composite.

REFERENCES

1. A.G. Evans and R.W. Davidge. J. Mat. Sci. 5, 314-325 (1970).
2. A. Atkinson, A.J. Moulson and E.W. Roberts. J. Am. Ceram. Soc.
   59, (7/8), 285 (1976).
3. J.A. Mangels. J. Am. Ceram. Soc. 58 (7/8), 353 (1975).

588

4. W.R. Davis. Trans. Brit. Ceram. Soc. 67 (11), 515-541 (1968).
5. L.J. Cohen and O. Ishai. J. Comp. Matls. 1, 390 (1967).
6. D.E. Lloyd. Special Ceramics 4, ed. P. Popper, 165 (1968).
7. F.F. Lange. J. Am. Ceram. Soc. 56 (10), 518 (1973).
8. D.S. Thompson and P.L. Pratt. Proc. Brit. Ceram. Soc. No. 6, 37-47 (1966).
9. D.W. Readey and G.C. Kuczynski. J. Am. Ceram. Soc. 49, 26-29 (1966).

DISCUSSION (Messier, Siebels)

Davidge: Your model seems rather oversimplified in that it uses just a single block comprising a solid centre surrounded by a porous region of matrix. Would it not be more realistic to use the model as a small building brick and build up a large structure from this?

Moulson: The fact is that Cohen and Ishai obtain remarkably good modelling of a porous two-phase composite using this simple approach. It certainly would seem likely that following Dr. Davidge's suggestion would lead to more satisfactory modelling, and this will be explored.

Lumby: Would you comment on the difference between whisker and laminar growth? Could this be not only impurity controlled but also, with special reference to the work of Grieveson, be kinetically controlled?

Moulson: In the system we are concerned with, pure silicon nitride at low N_2 pressures, we do not see whisker growth and we do not believe that this observation is anything to do with kinetics. We suspect that when we see whiskers the silicon has been contaminated with Al. Grieveson's system is very different from ours and therefore will almost certainly react differently.

Messier: Your observations regarding the lack of whisker formation when iron was placed on the surface of the silicon before nitriding are puzzling. We have done an experiment on the nitridation of a single crystal rod of silicon with iron wire placed on it and observed extensive whisker formation in the vicinity of the wire. It is possible that oxygen may also affect the whisker formation process and perhaps the difference is related to difference in oxygen levels in the two systems.

Moulson: It seems very likely that oxygen level is a variable which must be taken into account when considering growth morphology. It is possible, too, that the systems differ in other respects, for example cation (other than Fe) impurities arising from the Fe or from the hot walls of the reaction system.

HIGH TEMPERATURE, TIME DEPENDENT PHYSICAL PROPERTY CHARACTERIZATION OF REACTION SINTERED Si_3N_4*

John A. Mangels

Turbine Development Department
Ford Motor Company
Dearborn, Michigan U.S.A.

ABSTRACT - The high temperature properties of 2.7 g/cc injection molded reaction sintered Si_3N_4 was studied with respect to hot modulus of rupture, creep and stress rupture. The results show no degradation of strength with temperature, very small creep rates and no time dependent stress rupture failures. Qualitative tests were performed and no slow crack growth was observed in this material at temperatures up to 1400°C. Oxidation tests were performed on 2.55 and 2.7 g/cc Si_3N_4. The results show that the oxidation resistance of the material improves as the density is increased.

INTRODUCTION

The development and application of any material, metal or ceramic, requires precise knowledge of the physical properties. When developing a structural material to be used at high temperatures, as in a gas turbine, the understanding of the time dependent properties of the particular material is critical. This paper will look at various high temperature, time dependent properties of reaction sintered Si_3N_4 and will show that this is a viable material for use in high temperature structural applications.

EXPERIMENTAL PROCEDURE

The material used in this report was 2.7 g/cc injection molded

*This work was supported in part by the Advanced Research Projects Agency under Contract Number DAAG-46-71-C-0162.

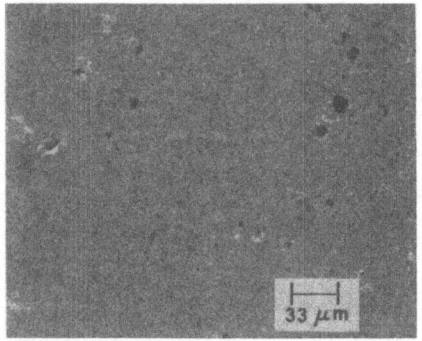

Figure 1. Typical microstructure of 2.7 g/cc Si_3N_4

Si_3N_4. The material was nitrided using a constant rate cycle and
a 4% H_2 - 96% N_2 static atmosphere[1]. The phase composition was
70% α Si_3N_4 and 30% β Si_3N_4. The major impurities were as follows:
1.4% Fe, 0.3% Al, 0.02% Ca, 1.0-1.5% O_2. The microstructure, as
shown in Figure 1, contained pores up to 10 μm. Some areas of
silicon "melt out"[1] were observed.

The strength testing was performed using four point bending
(9.5 mm x 19.0 mm fixture dimensions) and samples with a cross
section of 3.1 mm x 6.3 mm. Loading rates of 0.50 mm/min were
employed. The high temperature tests were performed using SiC
fixtures. The stress rupture testing utilized a similar test set
up with dead weight loading. The creep tests were also performed in
four point bending in accordance with the procedure of Nemeth[2].
The oxidation testing consisted of exposing the samples at various
temperatures in a static air atmosphere using an electrically
heated furnace for 200 hours, cooling the samples to room temperature, measuring any weight changes and then measuring the room
temperature strength.

RESULTS

The high temperature strength data is shown in Figure 2. The
short time hot strength decreases only slightly from a value of
260 MN/m^2 at room temperature to 230 MN/m^2 at 1400°C. The Weibull
modulus, M, is shown to increase slightly with temperature. Also
shown in Figure 2 are the stress rupture values obtained to date on
this material. These show 4 out of 16 samples breaking instantaneously on load, with three of these failures occuring at stresses
of 86% of the ultimate strength. All other tests were run at least
200 hours and in one case up to 800 hours without failure. In no
case was any time dependent failures observed for this material.

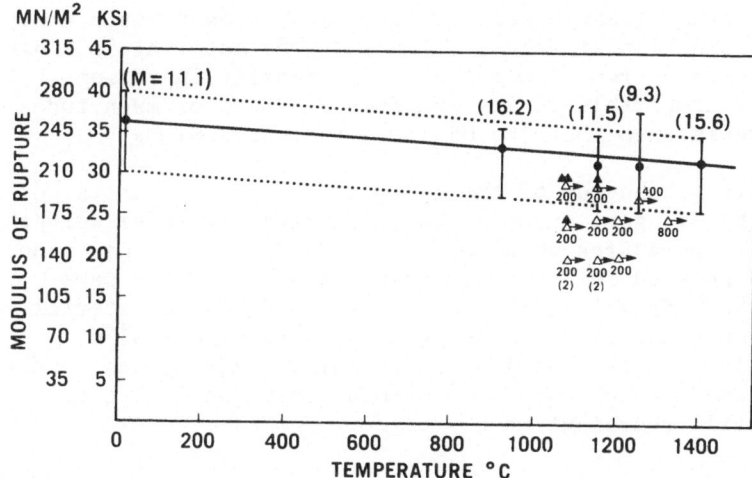

Figure 2.   High temperature strength properties of 2.7 g/cc
Si$_3$N$_4$. ♀ characteristic MOR and data range;
(   ) weibull modulus; ▲ stress rupture test,
instantaneous failure; △ stress rupture test,
test terminated with no failure after the number
of hours shown.

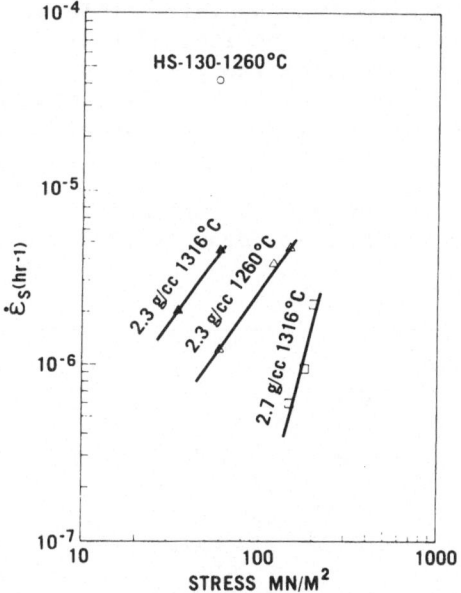

Figure 3.   Steady state creep rate ($\dot{\varepsilon}_S$) versus stress for
2.7 g/cc and 2.3 g/cc reaction sintered Si$_3$N$_4$[3]
and for hot pressed Si$_3$N$_4$ (HS-130)[4].

The creep data is expressed in Figure 3 as steady state creep rate ($\dot{\varepsilon}_s$) as a function of stress. The 2.7 g/cc material is shown to have greater creep resistance than lower density injection molded Si_3N_4 of comparable purity and over 2 orders of magnitude better creep resistance than the HS-130 hot pressed Si_3N_4.

The oxidation behavior of the 2.7 g/cc Si_3N_4 is illustrated in Table 1. The 2.7 g/cc material had significantly lower weight gains at both temperatures when compared to a 2.55 g/cc injection molded Si_3N_4 processed in a similar manner and having the same purity. More significant is the fact that the 2.7 g/cc material did not experience any strength degradation after the 1038°C exposure. However, after the 1260°C exposure, a 22% decrease in strength was noted, which is an improvement over the 2.55 g/cc material but still representing a problem.

DISCUSSION

Since the strength of the 2.7 g/cc reaction sintered Si_3N_4 does not degrade rapidly with temperature (as does hot pressed Si_3N_4[4] and many Sialon compositions[5]), it can be inferred that there is no glass forming at the grain boundaries of the material. The lack of glassy grain boundaries implies that the creep resistance of the material will be good, which was observed. It also implies that the stress rupture properties will exhibit little or no time dependency, which was also observed. The lack of time dependent strength failures also indicates that the phenomenon of slow crack growth is not a factor in this material. Preliminary tests show that no slow crack growth was observed at any temperature up to 1400°C[3,4].

The oxidation data can be interpreted with the assumption that internal oxidation is occuring in these low density (79% and 85%

Table 1

Results of Oxidation Tests in
Reaction Sintered Si_3N_4

Density	Test Temperature	Exposure Time	Δ Wt	Avg. MOR	MOR
2.55	-	-	-	245 MN/m^2	-
g/cc	1038°C	200 hours	+ 3.7%	195 MN/m^2	- 25%
	1260°C	200 hours	+ 2.0%	135 MN/m^2	- 46%
2.7	-	-	-	275 MN/m^2	-
g/cc	1038°C	200 hours	+0.75%	275 MN/m^2	0%
	1260°C	200 hours	+0.55%	215 MN/m^2	- 22%

of theoretical density) materials. An internal oxidation model would predict lower weight gains at higher test temperatures and lower weight gains for higher density material. Both of these predictions were observed. It is unsure what mechanism governs the strength degradation, however, it could be related to the formation of cristobolite as the major oxidation product and its large volume change at 250°C.

SUMMARY

The material properties investigated show that the 2.7 g/cc reaction sintered Si_3N_4 could be used advantageously as a structural material at high temperatures. It has good high temperature strength and oxidation resistance, it shows no time dependent failure mechanisms (within the temperature-stress-time regions studied) and it shows no evidence of slow crack growth up to 1400°C.

ACKNOWLEDGEMENTS

The author would like to thank Dr. J. Uy for performing the short time hot strength and Dr. K. R. Kinsman for performing the slow crack growth studies.

REFERENCES

1. Mangels, J. A., "Strength-Density-Nitriding Cycle Relationships for Reaction Sintered Si_3N_4", this volume.
2. Nemeth, J., Youdelis, W. V., Parr, J. G., "High Temperature Creep Behavior of Polycrystalline $SrZrO_2$", J. Amer. Ceram. Soc. 55, 125-129 (1972).
3. Mangels, J. A., "Development of a Creep Resistant Reaction Sintered Si_3N_4", in Ceramics for High Performance Applications, Ed. Burke, J. J., et. al., Metals and Information Center, Columbus, Ohio (1974).
4. McLean, A. F., Fisher, E. A., Bratton, R. J., "Brittle Materials Design, High Temperature Gas Turbine", AMMRC-CTR-74-59, Interim Report, September, 1974.
5. McLean, A. F., Baker, R. R., Bratton, R. J., Miller, D. G., "Brittle Materials Design, High Temperature Gas Turbine", AMMRC CTR-76-12, Interim Report, April, 1976.

594

DISCUSSION (Messier, Siebels)

Davidge: The data presented relating to delayed fracture data for RBSN at high temperature should be treated with caution in view of the small number of samples tested. The absence of delayed fracture for~ 10 specimens does not prove that time dependent effects are negligible. Our recent data (Powder Met. Int.$\underline{8}$ (3) 1976) for RBSN at $1000^{\circ}C$ shows that effective strengths are reduced by~7% for each decade increase of time under stress.

Mangels: This data was not intended to imply that there is no time dependency in the stress rupture properties of RBSN. However, from the data we have to date, no time dependent failures have been observed. Testing is still in progress.

Riley: Is any new information available on the effects of high temperature anneals on the strength of fully nitrided RBSN? Older work indicates that improvements in strength can be obtained at temperatures in the region of $1500^{\circ}C$.

Mangels: Our experience is that annealing gives weight and strength losses.

Section I

ELECTRICAL PROPERTIES

ELECTRICAL PROPERTIES OF SOME SIALONS

K.H. Jack

The Wolfson Research Group for High-Strength Materials,
Crystallography Laboratory, The University,
Newcastle upon Tyne, England.

ABSTRACT. Lithium-silicon nitride ($LiSi_2N_3$) and silicon oxynitride
($\square Si_2N_2O$) have similar crystal structures and show a limited
mutual solid solubility. It therefore seemed possible that silicon
oxynitride and its corresponding sialons might accommodate lithium
and perhaps other alkali cations to give solid electrolytes
analagous with β-alumina. The wide variety "AlN"-polytypes
observed in the magnesium, lithium and other sialon systems, and
the possibility of doping these as well as modifying the interatomic bonding, also suggest the possibility of new semi-conducting
materials.

    For these reasons, an examination of the electrical properties
of sialons has been started by Dr. J.S. Thorp and his colleagues
at Durham University. So far, only the electrical conductivities
of a few β'-sialons with and without lithium have been compared
with those of hot-pressed β-silicon nitride. The results are
consistent with the presence of a glassy phase in all the materials.

I. INTRODUCTION

Silicon oxynitride, Si_2N_2O, is potentially a good engineering
ceramic with excellent thermal shock and oxidation resistance. It
is built up of $SiON_3$ tetrahedron in such a way that parallel sheets
of silicon-nitrogen atoms are linked by Si-O-Si bonds; see Fig. 1.
Alumina can go into solution giving O'-sialons with the oxynitride
structure but with replacement of SiN by AlO and maintaining the
2M:3X atom ratio. The structure of lithium-silicon nitride, $LiSi_2N_3$,
is based on that of the wurtzite-type AlN but because the two
different metal atoms lithium and silicon are ordered the unit

598

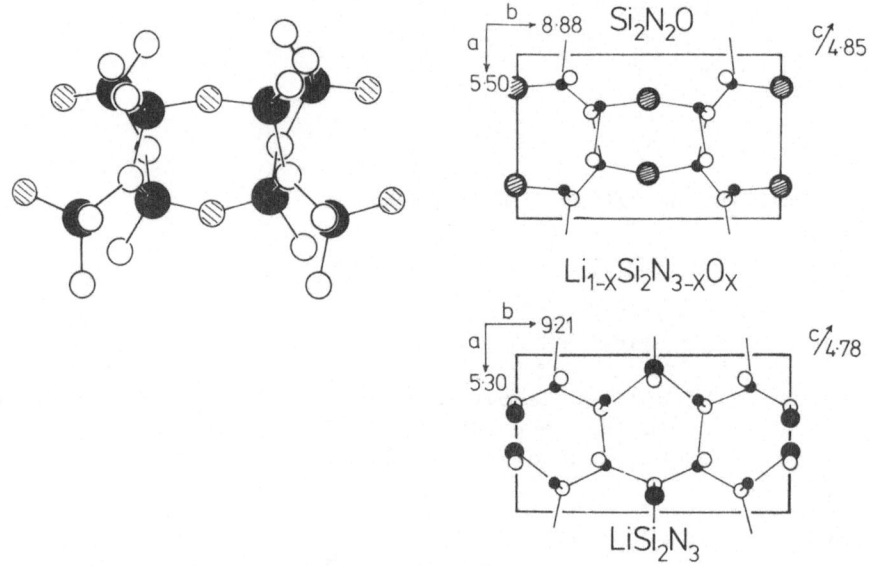

Fig. 1. The crystal structure of silicon oxynitride, Si_2N_2O.

Fig. 2. Comparison of the $LiSi_2N_3$ and Si_2N_2O structures.

cell is an orthorhombic superlattice of the hexagonal sub-cell and contains four formula weights.

Fig. 2 shows that Si_2N_2O and $LiSi_2N_3$ are essentially iso-structural. Indeed, if one nitrogen atom in $LiSi_2N_3$ is replaced by oxygen, charge compensation is maintained by removing the lithium:

$$Li^{1+} Si_2^{8+} N_3^{9-} \longrightarrow Si_2^{8+} N_2^{6-} O^{2-}$$

Unless very stringent precautions are made to exclude oxygen, preparations of $LiSi_2N_3$ always have some nitrogen substituted by oxygen and so are lithium deficient with compositions represented by $Li_{1-x} Si_2 N_{3-x} O_x$. The unit-cell dimensions of these defect oxynitrides extrapolate to the values for Si_2N_2O and even though there is apparently no continuous range of solid solution between $LiSi_2N_3$ and Si_2N_2O there is an appreciable mutual solubility. It therefore seemed possible for silicon oxynitride to accommodate lithium and perhaps more mobile cations on the sites between the rigid parallel planes of Si-N atoms and so provide a structure reminiscent of β-alumina.

$LiSi_2N_3$ is also similar in its structure to lithium meta-silicate, Li_2SiO_3, and can be derived from it by the simultaneous replacements:

$$Li^{1+} \longrightarrow Si^{4+} \qquad \text{and} \qquad O_3^{6-} \longrightarrow N_3^{9-}$$

It was thought that within the basic structure of silicon oxynitride a number of chemical variations might be possible:

(1) $\quad LiSi_2N_3 \longrightarrow Li_{1-x} Si_2 N_{3-x} O_x \longrightarrow Si_2N_2O$

(2) $\quad LiSi_2N_3 \longrightarrow Li_{1+y} Si_{2-y} N_{3-3y} \longrightarrow Li_2SiO_3$

(3) $\quad Si_2N_2O \longrightarrow Si_{2-z} Al_z N_{2-z} O_{1+z} \longrightarrow Al_2O_3$

and that these, with or without replacement of lithium by other cations, might provide stable, refractory, solid electrolytes.

It has already been said that lithium, magnesium, manganese and other metal-silicon nitrides and oxynitrides all have structures based on that of the wurtzite-type AlN. Defect AlN-polytypes are now known to exist widely in the Si-Al-O-N and M-Si-Al-O-N systems with compositions $M_m X_{m+1}$ and $M_{m+1} X_m$, and the possibility of doping such materials to produce ceramics with n- and p-type semi-conducting properties did not seem too remote.

For these reasons Dr. J.S. Thorp and his colleagues at Durham University have started to examine the electrical properties of some sialons prepared at Newcastle. None of the materials envisaged to have suitable conducting or semi-conducting properties have yet been obtained in a sufficiently well-characterized form and so in the meantime Dr. Thorp has examined a few β'-sialons, with and without lithium content, and has compared them with silicon nitride hot-pressed with and without MgO additions. Some results have recently been published (1).

2. ELECTRICAL CONDUCTIVITY IN β-Si_3N_4 AND β'-SIALON

Thorp & Sharif (1) measured d.c. and a.c. conductivities in the range 400-1000°C on the following hot-pressed β-Si_3N_4 and β'-sialon compacts:

Specimen	Composition	Phases
1	Si_3N_4 (washed with NaOH) hot-pressed at 1700°C for 1h	β-Si_3N_4 with glass impurity
2	Si_3N_4-5 wt% MgO hot-pressed at 1700°C for 1h	β-Si_3N_4 with Mg-Si-oxynitride glass impurity
3	z = 3.2 sialon hot-pressed at 1700°C for 1h	β'-sialon with trace amounts of X-phase and a glass
4	z = 4.0 sialon hot-pressed at 1700°C for 1h	β'-sialon with glass impurity

They also measured d.c. Hall and thermoelectric effects.

The variations of conductivities with temperature were similar for all specimens as were the variations of a.c. conductivity with frequency at different temperatures. Values of d.c. conductivity for the four specimens were:

Specimen	d.c. conductivity $(\Omega cm)^{-1}$		Activation energy eV	
	1000°C	400°C	1000-700°C	500 - 400°C
	$\times 10^{-6}$	$\times 10^{-11}$		
1	7	46	1.3	0.9
2	8	27	1.4	1.0
3	4	9	1.6	1.1
4	4	4	1.6	1.2

It is concluded that all the electrical properties reported are consistent with the presence of a glassy phase, and it seems probable that this determines the conductivity behaviour.

Limited measurements have been made on β'-sialons containing lithium and prepared by hot-pressing mixtures of α-Si_3N_4 with $LiAl_5O_8$. Although the d.c. conductivities are higher than for the β' materials without lithium, there are marked polarisation effects and the general conclusion is the same - the electrical properties are due to glass impurity.

3. FUTURE WORK

The results obtained so far do not contribute to an understanding of the electrical properties of sialons because the materials preparation was inadequate. An investigation of these compounds considered likely to show unusual electrical behaviour remains to be carried out when well-characterized specimens become available. The electrical properties of vitreous impurities are not without interest in their own right and will be studied now that bulk samples of nitrogen-glasses can be prepared.

REFERENCES

1.   J.S. Thorp & R.I. Sharif, J. Mater. Sci. 11 (1976) 1494.

DISCUSSION (Messier, Siebels)

Morgan: I feel that the comparison of the Si_2N_2O structure with $LiSi_2N_3$ is stretching the usefulness of crystal inter-comparison to the limits. The "anions" in $LiSi_2N_3$ are all tetrahedral, as for the wurtzite structure, while in Si_2N_2O all the nitrogens are trigonal planar (nearly) and the oxygen is only two-coordinate, as in many silicates. In fact, the sheets of six numbered rings in Si_2N_2O are very similar to those in $\alpha-Si_3N_4$ cross linked by O.

Jack: However imperfect the structural correspondence is between $LiSi_2N_3$ and Si_2N_2O, there is some mutual solubility and the lithium atoms or cations in $Li_xSi_2N_{2+x}O_{1-x}$ (where x is small) are associated with the bridging (O,N) sites located between parallel puckered sheets of Si-N atoms. The possibility of these lithium cations or other substituted alkali cations moving under a potential applied parallel to the Si-N sheets seems not unreasonable to me although, of course, it might appear so to others.

Popper: I do not think that the study of the electrical properties of nitrogen ceramics is justified unless some specific unusual property, e.g. ionic conductivity because of a vacancy structure, is expected. Otherwise we would not expect the electrical conductivity to be very different from that of similar oxides, i.e. they would possess a large energy gap between filled and conduction band and thus electronic and ionic contributions would be small and very impurity dependent. Professor Brook has reminded us that it has taken over 30 years to understand the conduction mechanism of Al_2O_3.

ELECTRICAL CONDUCTIVITY IN NITROGEN CERAMICS

H.M. Kizilyalli

Department of Physics, Middle East Technical
University, Ankara, Turkey.

ABSTRACT

Electrical conductivity studies have contributed greatly to
the growth of the ceramic industry. Many of the basic advances
in the characterization and fabrication of ceramic materials have
been motivated by the compositional and microstructural requirements imposed on them by electrical property applications. Conductivity studies may be also used to characterize the predominant
defects present and their charge states. One of the greatest
hurdles facing the inexperienced student in making meaningful
conductivity measurements is the avoidance of the many experimental
errors that can arise. Hence experimental methods for the measurement of conductivity (two, three, four probe, ac and dc methods)
will be reviewed and the importance of selecting a proper electrode
will be discussed.

DISCUSSION (Messier, Siebels)

Singhal: Have you obtained any electrical conductivity values on
AlN as a function of nitrogen pressure? Is it a pure ionic conductor? Have you come across any nitride which is purely ionic
conductor and can be used as electrolyte to measure nitrogen
activities in a way similar to ZrO_2-Y_2O_3 electrolyte?

Kizilyalli: No, but according to Francis and Worrall (1976), AlN
is electronic conductor. Calcium nitride may be ionic.

Popper: I do not think that in a strongly covalent-bonded
material such as AlN one would expect ionic conduction. The dielectric properties - particularly that of the dielectric losses -
are strongly influenced by the presence of unnitrided material

(silicon or silicides). It is for this reason that the use of Si_3N_4 as radomes has so far been unsuccessful in spite of its fabrication route being ideal. Similarly the porosity and the resulting absorption of moisture can affect the electrical conductivity greatly and very high temperatures are required to remove all traces of adsorbed water.

Lange: The dielectric loss in Si_3N_4 has been explained by the semiconductor people who have vapour deposited amorphous Si_3N_4 and found that free Si (normally termed 'non-stoichiometric Si_3N_4') causes losses.

Moulson: I believe one should be aware that very low (3) μ_e can be encountered in electron-hopping situations so it is not safe to simply assume that $\mu_e \gg \mu_i$.

There are many data relating to the electrical properties of amorphous Si_3N_4 in the literature, mainly in J. Electrochem. Soc. This is because of the interest which has been shown in the material as a possible passivation layer. The concensus of opinion is that amorphous Si_3N_4 is an electronic semiconductor having a band gap ≥ 10 eV.

Kizilyalli: One would, a priori, think the electrical conduction behaviour in amorphous and in crystalline forms of Si_3N_4 to be quite different, since the ionic and electronic conductivities have in recent years been shown to depend on defect structures of the compound.

Schaeffer: Electronic conductivity changes little when the material is transformed from the crystalline to the vitreous state; apparently this is due to the fact that the electronic defect structure is related to the short-range order which is unaffected by rearrangements of the lattice. Concerning ionic conductivity a different behaviour is to be expected, since activation energies and jump distances of the ions will depend on the long-range order.

Section J

OXIDATION AND CORROSION

OXIDATION OF SILICON NITRIDE AND RELATED MATERIALS*

S. C. Singhal

Westinghouse Research Laboratories
Pittsburgh, Pennsylvania, 15235, USA

ABSTRACT. The oxidation behavior of chemically vapor deposited (CVD), powdered and reaction-sintered forms of pure Si_3N_4 and of Si_3N_4-based hot-pressed structural materials between 1000 and 1500°C is reviewed. Powdered and CVD forms of pure Si_3N_4 oxidize at extremely small rates due to formation of coherent and protective surface SiO_2 layer. Reaction-sintered Si_3N_4 materials experience much greater magnitudes of oxidation because of oxidation in the pores until a continuous surface SiO_2 layer is formed. Rates of oxidation of Si_3N_4-based hot-pressed materials vary widely because different oxide additives in these materials cause formation of different secondary phases which greatly influence the rates of oxidation. It is shown through example of Y_2O_3-containing materials that it is possible to develop new, extremely oxidation-resistant Si_3N_4-based structural materials by carefully controlling the formation of secondary phases during processing.

1. INTRODUCTION

In the last decade, silicon nitride has attracted great interest because of its actual or potential application as refractory material in the ceramics industry for the production of tubes and crucibles [1], as diffusion mask and gate dielectric in semiconductor device technology [2], and as structural material in aggressive high temperature environments such as automotive and power-generating gas turbines, automotive thermal reactors and advanced combustion chambers [3]. For these varied applications, Si_3N_4 is made in many forms, important among which are

---

* Supported by the Defense Advanced Research Projects Agency.

chemically vapor deposited (CVD), powder, reaction-sintered and hot-pressed. The CVD, powder, and reaction-sintered forms are generally chemically-pure Si_3N_4, while the hot-pressed form of material is prepared by densifying Si_3N_4 powder with one or more of many oxides, e.g., magnesia, yttria, silica, alumina, etc. The oxidation data for these four different forms of Si_3N_4 are reviewed and compared in this paper. Before a discussion of their actual oxidation behavior, certain thermochemical aspects of oxidation of pure Si_3N_4 are presented.

2. THERMOCHEMICAL ASPECTS OF OXIDATION OF Si_3N_4

Similar to silicon and silicon carbide [4,5], Si_3N_4 can have two distinct types of oxidation behavior at high temperatures depending upon the ambient oxygen partial pressure. In "passive" oxidation at high oxygen partial pressures, a layer of silica (SiO_2) is formed on the surface according to the reaction:

$$Si_3N_4 (s) + 3 O_2 (g) \longrightarrow 3 SiO_2 (s) + 2 N_2 (g). \tag{1}$$

Based on equilibrium thermodynamic considerations, silicon oxy-nitride (Si_2ON_2) should also form as an intermediate phase during oxidation [6]. However, in most cases, any Si_2ON_2 formed is further oxidized almost completely to SiO_2 [7]. Formation of surface SiO_2 layer in this "passive" oxidation (characterized by an increase in weight of Si_3N_4) limits the rate of further oxi-dation, and thus provides excellent oxidation-resistance. During such oxidation, very high nitrogen gas pressures are produced at the Si_3N_4-SiO_2 interface according to reaction (1). For instance, using thermodynamic data from JANAF Tables [8], N_2 gas pressures of 7.5×10^{37} and 4.3×10^{23} atm at 1200 and 1800 K, respectively, are calculated for reaction (1) for 1 atm pressure oxygen. Evolu-tion of this N_2 gas from the interface causes formation of pores and fissures in the surface oxide.

As opposed to "passive" oxidation in highly oxidizing envi-ronments, severe "active" oxidation (characterized by a decrease in weight of Si_3N_4) can occur at low oxygen partial pressures according to the reaction:

$$2 Si_3N_4 (s) + 3 O_2 (g) \longrightarrow 6 SiO(g) + 4 N_2 (g). \tag{2}$$

In this "active" oxidation, material is continually lost from the surface due to the formation of silicon monoxide (SiO) gas. Any SiO_2 on the surface of Si_3N_4 (e.g., formed by pre-oxidation at higher oxygen partial pressure) could react with Si_3N_4 underneath to also form SiO(g) according to the reaction:

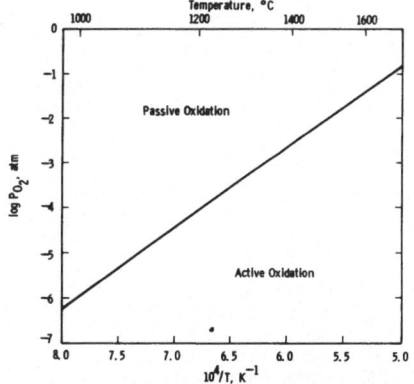

Fig. 1. Calculated regions for active and passive oxidation of Si_3N_4. After Singhal [6].

$$Si_3N_4(s) + 3\ SiO_2(s) \longrightarrow 6\ SiO(g) + 2\ N_2(g). \qquad (3)$$

In the event of all the surface SiO_2 being consumed according to this reaction, a bare Si_3N_4 surface is exposed causing further "active" oxidation according to reaction (2). The regions of "active" and "passive" oxidation for Si_3N_4 in terms of oxygen partial pressure and temperature, calculated [6] using Wagner's hypothesis [4] for the "active"-"passive" oxidation of Si, are shown in Fig. 1. The values of oxygen partial pressure required to prevent "active" oxidation would be also affected by the ambient nitrogen pressure. Slightly different oxygen partial pressure values, but essentially of the same order as those shown in Fig. 1, have been calculated by Jack [9] for preventing "active" oxidation in flowing nitrogen at 1 atm. Unfortunately, the "active" oxidation of Si_3N_4 has not yet been investigated by anyone to allow verification of these calculated values. In contrast, the "passive" oxidation of various forms of Si_3N_4 has been studied extensively, and this is reviewed and discussed in the following sections.

3. OXIDATION OF PURE Si_3N_4

3.1. Chemically Vapor Deposited Si_3N_4

Pure Si_3N_4 is extremely oxidation-resistant due to the formation at elevated temperatures of a protective SiO_2 layer on its surface. Galasso et al [10] commented that pyrolytic α-Si_3N_4 prepared by chemical vapor deposition (CVD) using SiF_4 and NH_3, did not oxidize in air below 1517°C. Greskovich et al [11] tried to thermogravimetrically measure the rates of oxidation of thin pieces of CVD-Si_3N_4 in air at 1410 and 1550°C. These investigators encountered difficulty in measuring the true specific

610

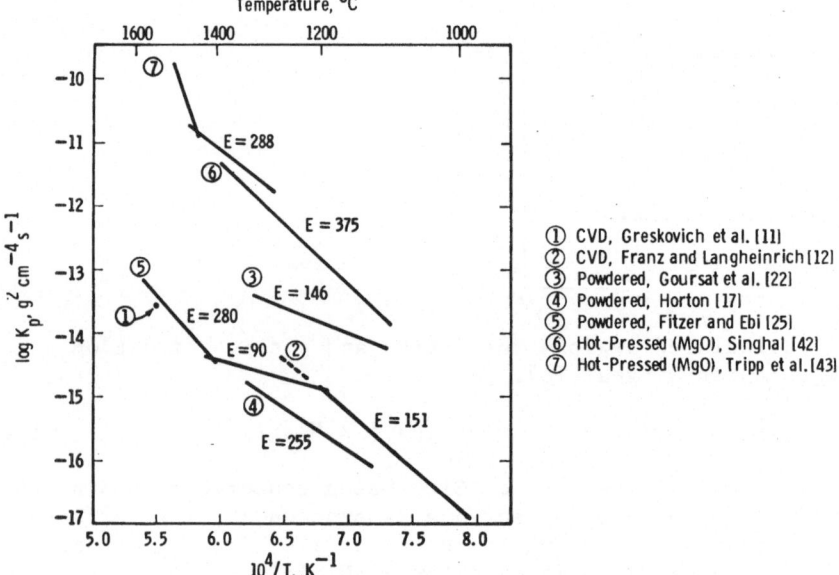

Fig. 2. Parabolic rate constants for oxidation of different forms of Si_3N_4. E indicates activation energy in kJ mol^{-1}.

surface area of the specimens due to extreme roughness of the surfaces. Nonetheless, they estimated that the oxidation rate at 1550°C should be less than 2.8×10^{-14} g^{2}cm^{-4}s^{-1} (filled circle in Fig. 2).

Franz and Langheinrich [12,13] studied the oxidation of amorphous Si_3N_4 films, formed on silicon slices by chemical vapor deposition using SiH_4 and NH_3, by interferometrically measuring the thickness of the SiO_2 layer formed in dry air at different temperatures. After a short initial period of a few minutes when a linear oxidation behavior [14] might be operative, the oxidation was observed to follow a parabolic relationship of the form

$$\chi^2 = B \cdot t$$

where χ is the oxide thickness at time t, and B is the temperature dependent rate constant. For the sake of comparison with other studies, their oxide thicknesses have been converted to corresponding weight gains (W) in accordance with reaction (1) under the assumption that oxide formed is theoretically dense SiO_2 with a specific gravity of 2.32. The oxidation behavior can now be expressed by the relationship

$$W^2 = Kp \cdot t$$

where Kp is the parabolic rate constant in $g^2cm^{-4}s^{-1}$. The parabolic rate constants* in dry oxygen at 1230 and 1260°C calculated by converting the oxide thickness data are shown in Fig. 2. These rates of oxidation are smaller than for the oxidation of silicon under similar conditions [12-15]. The lower rates of oxidation for Si_3N_4 may be due to formation of N_2 gas which can lower the equilibrium concentration of the oxidant in the oxide. Additionally, the lower rates could also be caused by consumation of part of the oxidant through formation of nitrogen oxides. Such formation of nitrogen oxide (NO) has been observed by Lin [16] in the mass spectrometric analysis of the gaseous species formed during oxidation of reaction-sintered Si_3N_4.

Franz and Langheinrich [12,13] found the rates of oxidation for CVD-Si_3N_4 in wet oxygen to be greater than in dry oxygen. Similar increases in oxidation rates in the presence of moisture have also been observed for silicon [14,15], powdered Si_3N_4 and SiC [17,18], and hot-pressed Si_3N_4 and SiC [19]. Such increases could be caused by different oxidant molecules with different diffusion coefficients through SiO_2 taking part in the oxidation process under dry and wet conditions.

In all of the above studies on oxidation of CVD-Si_3N_4 between 1200 and 1550°C, SiO_2 was identified as the only oxidation product. In contrast, Raider et al [20] found, through ESCA analysis of the oxide layers formed by oxidation of CVD-Si_3N_4 thin films in dry oxygen at 1070°C, that Si_2ON_2 was the major oxidation product unless oxidation was carried out for more than 1 hr. The formation of Si_2ON_2 is expected from equilibrium thermochemical considerations discussed earlier in Sec. 2. However, in most cases, particularly at higher temperatures, Si_2ON_2 oxidizes further almost completely to SiO_2 with the result that SiO_2 only is usually observed in oxide formed over Si_3N_4.

3.2. Powdered Si_3N_4

Oxidation of pure Si_3N_4 has also been investigated in powder form by many researchers. Krasotkina [21] observed that powdered Si_3N_4 oxidizes in air between 1100 and 1500°C mainly with the formation of SiO_2. Goursat et al [22,23] studied the oxidation of two different Si_3N_4 powders containing different amounts of α and β phases in oxygen at partial pressures between 2.2×10^{-1} and 2.1×10^{-2} atm in the temperature range 1100 to 1300°C. The

---

* The actual weight gains and hence parabolic rate constants may have been somewhat smaller due to the presence of pores in the oxide layer formed by release of the N_2 gas from the nitride/oxide interface.

powders were initially kept and brought up to the temperature in high vacuum to remove any pre-existing surface oxide. These investigators found that oxidation initially proceeds through a reaction-controlled stage characterized by an activation energy of 293 kJ mol^{-1}, which corresponds roughly to the Si-N bond energy. They also concluded that α-phase Si_3N_4 oxidizes faster than the β-phase in this initial linear region. Similar initial linear oxidation was also thought to occur by Franz and Langheinrich [12,13] in oxidation of CVD-Si_3N_4. However, in most other studies on oxidation of Si_3N_4, this initial linear region is not observed, presumably because the specimens already have a surface oxide layer which is not removed before starting the oxidation experiment. This pre-existing oxide increases in thickness during further oxidation through diffusion-controlled process resulting in a parabolic oxidation behavior. This parabolic oxidation region alone is observed in the oxidation of Si_3N_4 by most investigators.

Goursat et al also observed such a parabolic behavior after the initial linear region. In this parabolic region, the oxidation rates were equal for the two different Si_3N_4 powders, and independent of oxygen partial pressure. The investigators also postulated and confirmed no influence of N_2 pressure on oxidation rates. This is expected since, as discussed in Sec. 2, the N_2 pressures produced at the nitride/oxide interface are so high that small variations in ambient N_2 pressure in the range of a few atmospheres should have no effect on kinetics. The parabolic rate constants obtained by converting oxide thickness data of Goursat et al to corresponding weight gains in the same manner as described earlier are shown in Fig. 2, which give an activation energy of 146 kJ mol^{-1}. This value of activation energy is comparable to those obtained for the oxidation of Si and SiC, and also for the diffusion of oxygen through SiO_2 [24], indicating that oxidation of pure Si_3N_4 is controlled by diffusion of oxygen through surface SiO_2. Small differences in activation energy may be either from a slight influence of N_2 on the oxygen diffusion rate or from a very small effect of oxygen pressure on the rate.

Horton [17] and Fitzer and Ebi [25,26] also observed parabolic behavior for oxidation of powdered Si_3N_4 in oxygen at 1 atm pressure. The parabolic rate constants obtained by these investigators are also included in Fig. 2. Horton obtained an activation energy of 255 kJ mol^{-1} in dry oxygen in the temperature range 1065-1340°C, while Fitzer and Ebi observed three different oxidation regions in the range 1000-1600°C with three different activation energies. Between 1000 and 1200°C, amorphous SiO_2 was formed with an activation energy of 151 kJ mol^{-1} which is comparable to the value obtained by Goursat et al [22,23]. However, between 1200 and 1400°C, oxide scale consisted of cristobalite and oxidation proceeded with an activation energy of 90 kJ mol^{-1}. At 1400-1600°C, large increases in oxidation rates as well as activation energy were observed, apparently due to melting in surface SiO_2.

It is clear from the oxidation data on pure Si_3N_4 (CVD and powdered forms) summarized in Fig. 2 that there are wide variations in the rates and activation energies obtained by different investigators. The differences in the results on powdered Si_3N_4 might be due to the fact that Goursat et al took into account, as should be done, the progressive decrease in reactive area with advancing oxidation of Si_3N_4 powder, while Horton and probably Fitzer and Ebi also neglected to take this decrease in area into consideration in calculating the oxidation rates. However, differences in oxidation rates could also be due to differences in the structure of Si_3N_4 (amorphous, α-phase, or β-phase), its density and porosity, and also due to differences in experimental conditions, e.g., the amount of moisture in oxygen used, the amount of pre-existent surface oxide, and the method of bringing the specimen up to the oxidation temperature. Several factors might also cause different activation energies for parabolic oxidation; for example, a change from extrinsic to intrinsic controlled diffusion with increasing temperature in which case activation energy would depend on both the purity of the material and the temperature [27]. The crystal form of SiO_2 produced also depends both on temperature and on impurities which can affect crystal nucleation. Thus, even though parabolic kinetics, which has been observed by all investigators for oxidation of both CVD and powdered forms of pure Si_3N_4, suggests the oxidation of Si_3N_4 to be diffusion-controlled, different diffusion systems might be operative under different conditions.

3.3. Reaction-Sintered Si_3N_4

Although reaction-sintered Si_3N_4 is also chemically pure Si_3N_4 with no additives, it exhibits much different oxidation behavior than CVD and powdered forms because of the presence of large porosity. The oxidation behavior in air of many different reaction-sintered Si_3N_4 materials possessing different densities has been studied thermogravimetrically by many investigators [26, 28-32]. In these studies, the magnitude of oxidation has been observed to be quite different from material to material depending on the porosity, pore size, and pore size distribution. Even for the same material, the oxidation data usually have a large scatter and vary with the technique used to initiate the oxidation.

In spite of widely different results on different reaction-sintered materials, all materials have been observed to undergo much larger magnitudes of oxidation than CVD or powdered Si_3N_4, and to exhibit a similar oxidation pattern. The oxidation occurs

614

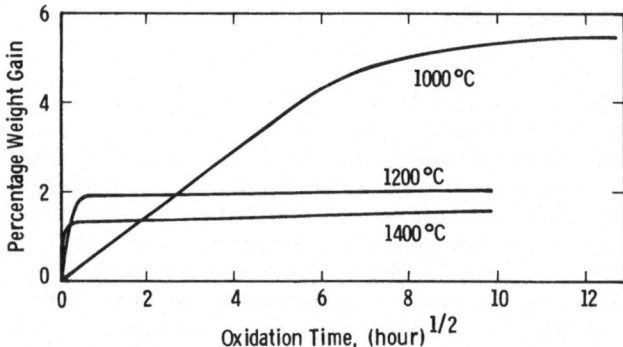

Fig. 3. Typical oxidation behavior of reaction-sintered Si_3N_4.
After Davidge et al [30].

in two stages, the relative importance of each being dependent on
time and temperature. In the first stage, oxidation occurs
internally in the pores until the pores are completely filled
with the oxide or a continuous dense surface SiO_2 layer which
limits the supply of oxygen to the interior is formed. Oxidation
in the second stage occurs on the external surface. In both
stages, oxidation follows parabolic behavior due to diffusion-
controlled process. But since the internal surface area is usually
much greater than the external surface area due to a large amount
of porosity in reaction-sintered materials, the rates of oxidation
in the first stage are usually much greater than in the second
stage.

At lower temperatures near 1000–1100°C, most of the oxidation
is usually internal since pores are not closed completely even
after extended periods of oxidation. This internal oxidation
causes a large increase in the volume of the material due to
formation of SiO_2. At higher temperatures above ∿1200°C, a transition occurs at very short times from the first stage to the
second stage because a continuous surface SiO_2 layer is formed
rapidly with only little internal oxidation. Higher the oxidation
temperature, sooner this transition occurs, allowing smaller
amounts of internal oxidation and hence lower total magnitude of
oxidation. This is illustrated in Fig. 3 by the oxidation data
for a reaction-sintered Si_3N_4 of 2.53 g cm^{-3} density [30], which
shows that the total magnitude of oxidation is much smaller at
1400°C than at either 1000 or 1200°C. The actual magnitude of
internal and external oxidation in a particular material depends,
in addition to time and temperature, on its density and pore
structure. Higher density materials experience, as expected,
smaller oxidation attack [31,32].

Both amorphous and crystalline SiO_2 have been reported to form in oxidation of reaction-sintered Si_3N_4, amorphous form being formed at generally lower temperatures [28-33]. Engel [34] and Thummler et al [35] observed formation of Si_2ON_2 also in the oxidation of a relatively low density reaction-sintered Si_3N_4 in air at 1400°C. The amount of Si_2ON_2 was found to be greater in the center of the specimens than in the surface. This is to be expected since the supply of oxygen to the interior to allow further oxidation of Si_2ON_2 to SiO_2 is restricted once a continuous surface oxide layer is formed.

Oxidation of reaction-sintered Si_3N_4 has been found to both increase and decrease its strength depending on the extent of internal and external oxidation, the nature of SiO_2 formed, the thermal cycling to which oxidized material is subjected, and the temperature at which strength is measured [29-33,36,37]. Internal oxidation in the first stage usually increases both the room and the high temperature strength, regardless of whether SiO_2 formed is amorphous or crystalline, due to rounding off of the internal pores and flaws. However, if a thick cristobalite layer is formed on the surface, the high temperature strength is increased but the room temperature strength is decreased due to cracking of cristo-balite on cooling below 250°C due to a phase transition [29,30, 33]. The internal oxidation of reaction-sintered Si_3N_4 enhances its creep rate also [34,35].

4. Si_3N_4-BASED HOT-PRESSED MATERIALS

Dense, high strength Si_3N_4-based materials are generally fabricated by hot-pressing Si_3N_4 powder with varying concentrations of an oxide additive, common additives being MgO, Y_2O_3, and Al_2O_3. These hot-pressed materials contain, in addition to one or both (α and β) Si_3N_4 phases, certain secondary phases formed during hot-pressing by reaction of the oxide additive with Si_3N_4 and with SiO_2 which is always present on the surface of Si_3N_4 powder particles. These secondary phases could also be formed or modified through reaction of the impurities (e.g., Ca, Fe, Al, Na, K, etc.) in the starting powder with Si_3N_4, SiO_2, and the oxide additive at the hot-pressing temperature. Because of the presence of these secondary phases, oxidation behavior of hot-pressed Si_3N_4-based materials is generally very different from that of additive-free, chemically pure Si_3N_4.

Depending on the chemical nature and concentration of the oxide additive and of the impurities in the starting Si_3N_4 powder, a number of different materials containing different secondary phases and possessing widely different properties are formed. These different hot-pressed materials, though usually designated only as Si_3N_4 (or sialons when Al_2O_3 is used), actually represent

a family of single-phase and multiphase materials, and the rates
of oxidation vary widely from material to material. Rates of
oxidation for a particular material depend on its chemical purity
and the nature and concentration of the oxide additive. Oxidation
of three different types of hot-pressed Si_3N_4-based materials,
fabricated using MgO, Y_2O_3, or Al_2O_3, has been discussed in a
recent review [38], and is only briefly described here.

4.1. Si_3N_4-MgO Materials

Oxidation of MgO-containing materials has been studied in
great detail since the only commercially available* hot-pressed
Si_3N_4 is prepared using ~1 wt % MgO. Magnesia in conjunction
with SiO_2 and certain impurities (e.g., calcium) forms mixed
silicates which concentrate in the grain boundaries as a glass
phase in these hot-pressed materials [39,40]. Kossowsky [40]
observed by Auger electron spectroscopy that Mg, Ca, Na, and K
elements concentrate in the grain boundaries and estimated the
composition of the grain boundary glass phase in the commercial
material to be approximately $CaO \cdot MgO \cdot 8 SiO_2$. Oxidation of MgO-
containing Si_3N_4 has been observed to start at these grain
boundaries and triple points [41]. Thus, the grain boundary phase
greatly influences the rates of oxidation of MgO-containing Si_3N_4
though the material still exhibits classical parabolic behavior
similar to that of additive-free pure Si_3N_4.

The parabolic rate constants for the commercial hot-pressed
Si_3N_4 in 1 atm pressure dry oxygen obtained by Singhal [42], and
in ~0.2 atm pressure dry oxygen obtained by Tripp et al [43] are
shown in Fig. 2 as a function of reciprocal temperature. Between
1100 and 1400°C, the rates of oxidation as well as the values of
activation energies from both investigations are comparable. The
small difference in data might be due to a difference in the
amount of moisture in the oxygen gas used by them since oxidation
rates are increased in the presence of moisture [19]. Above
1450°C, Tripp et al [43] observed a large increase in the oxidation
rate which was believed to be caused by melting in the oxide scale.

It is clear from the data summarized in Fig. 2 that the
oxidation rates for commercial hot-pressed Si_3N_4 are orders of
magnitude greater than for pure (CVD or powdered) Si_3N_4 at all
temperatures. The values of activation energy (288-375 kJ mol^{-1})
for oxidation of hot-pressed material are also higher than the
values obtained for powdered Si_3N_4 or for the diffusion of oxygen
through SiO_2 [24]. This suggests that the oxidation of MgO-

---

* Grade HS-130, Norton Company, Worcester, MA. Major
impurities Fe(0.6%), Al(0.1%), and Ca(0.07%).

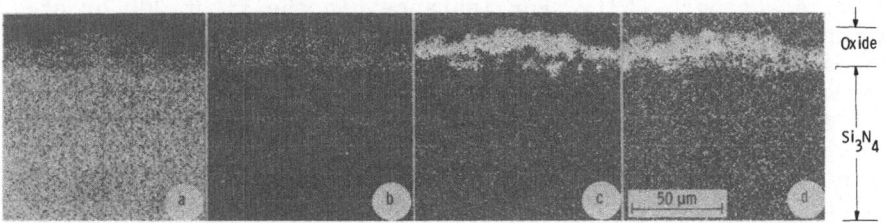

Fig. 4. Electron microprobe scans in the oxide layer formed over MgO–containing hot–pressed Si_3N_4 after 100 hrs of oxidation at 1370°C in 1 atm dry oxygen; (a) Si–Kα, (b) O–Kα, (c) Mg–Kα, and (d) Ca–Kα. After Singhal [38].

containing Si_3N_4 is controlled by a process other than oxygen diffusion through SiO_2. This is not surprising since the surface oxide formed on hot–pressed material at temperatures above 1000°C consists mostly of enstatite ($MgSiO_3$), though small amounts of cristobalite, Si_2ON_2, and glassy phases are also observed [26, 41–45]. In addition, impurity elements (Ca, Fe, Al, K, etc.) have been observed to concentrate in the surface oxide [42,43]. Such concentration of Mg and Ca in the oxide layer is illustrated in Fig. 4 by the electron microprobe scans in the surface oxide formed at 1370°C over commercial hot–pressed Si_3N_4. These impurity elements either form mixed silicates, e.g., akermanite ($Ca_2MgSi_2O_7$) and diopside ($CaMgSi_2O_6$), or dissolve in the glassy phases in the surface oxide depending upon the duration and temperature of oxidation [42,44]. Mixed silicates like Mg(Ca,Fe)· SiO_2 also have been detected in oxide films formed on other hot–pressed Si_3N_4 containing greater levels of impurities [46]. Thus, as discussed by Singhal [42], the high value of activation energy obtained for oxidation of MgO–containing hot–pressed Si_3N_4 might be due to diffusion of additive and impurity cations (mainly Mg^{2+} and Ca^{2+}) from the grain boundary glass phase to the surface through the oxide layer.

Because of the outward diffusion of additive and impurity elements, rates of oxidation for MgO–containing hot–pressed Si_3N_4 materials are greatly influenced by the amounts of these elements in a particular material. It has been observed that greater amounts of MgO (in the range 1 to 5 wt %) result in increased rates of oxidation [47,48]. Similarly, at a fixed MgO concentration, greater amounts of impurities, particularly calcium, in the material cause greater rate of oxidation attack [47]. Thus, rates of oxidation of different Si_3N_4 materials, even though hot–pressed with the same additive (MgO), can be significantly different depending upon the amounts of additive and certain impurities.

Extended oxidation of commercial hot–pressed Si_3N_4 above 1100°C greatly deteriorates both its room temperature and high–

temperature strength [49]. For instance, in the first 200 hr of
oxidation in air at 1205°C, the flexural strength at 1205°C
decreases drastically to ∿60% of its original strength, though it
then remains practically unchanged with further oxidation. Even
greater deterioration in strength has been observed after long-term
oxidation at higher temperatures. The reduction in flexural
strength is believed to be caused by surface degradation which
occurs during oxidation at elevated temperatures. It appears
that $MgSiO_3$ in the surface oxide reacts with Si_3N_4 to form sharp
microcavities on the surface which cause sharp reduction in the
strength of the material.

The corrosion behavior of MgO-containing hot-pressed Si_3N_4
materials in various combustion gases and simulated gas turbine
environments, with and without the addition of common turbine fuel
impurities, e.g., Na, S, and V, has also been studied [26,50-52].
In these high velocity environments, hot-pressed Si_3N_4 materials
generally experience a simultaneous corrosion-erosion attack
whereby a surface oxide layer is continually formed and removed.

4.2. Si_3N_4-Y_2O_3 Materials

Yttria has also been used in varying concentrations as the.
hot-pressing additive in making dense high-strength Si_3N_4 [53].
It is believed that Y_2O_3 forms a highly refractory crystalline
phase in the grain boundaries instead of a glass phase and thus
provides materials with better high temperature mechanical
properties than MgO-containing materials [54]. However, depending
upon the amounts of Y_2O_3 additive and SiO_2 which is normally
present on Si_3N_4 powder particles, many different secondary phases
can form in the hot-pressed material according to the phase
relationships [55] shown in Fig. 5.

Several experimental hot-pressed Si_3N_4 materials with
compositions lying in the phase fields labelled A, B, and C in
Fig. 5, and thus containing one or more of the $Si_3Y_2O_3N_4$, $YSiO_2N$,
and $Y_{10}Si_7O_{23}N_4$ phases, have been found to exhibit an unusual
oxidation behavior [55]. These materials oxidized catastrophically when heated in air at ∿1000°C but showed very little oxidation at higher temperatures near 1300-1400°C. After detailed
oxidation studies in air at 1000 to 1400°C on individually
prepared $Si_3Y_2O_3N_4$, $YSiO_2N$, and $Y_{10}Si_7O_{23}N_4$ phases (without any
Si_3N_4), it has been observed that all these three phases oxidize
linearly with time at an accelerated rate because the surface
oxide formed in each case is highly porous and non-protective.
Therefore, hot-pressed Si_3N_4 materials containing these phases
oxidize catastrophically too at ∿1000°C due to the accelerated
oxidation of the secondary phases. However, at higher temperatures,
Si_3N_4 phase in the multiphase materials initially oxidizes at a

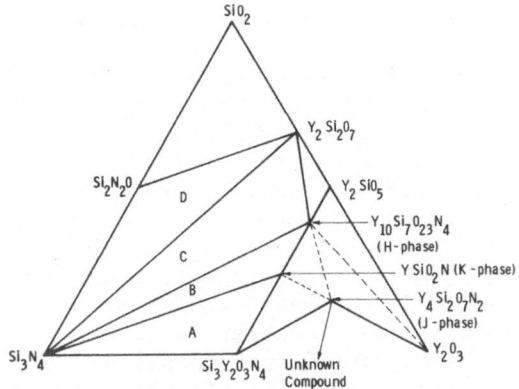

Fig. 5. Phase relationships in the Si_3N_4-SiO_2-Y_2O_3 system at $\sim$1600–1750°C. After Lange, Singhal, and Kuznicki [55].

much more rapid rate than the secondary phases and forms a protective surface SiO_2 layer limiting the supply of oxygen to the interior to allow any oxidation of the secondary phases. Thus, the same materials exhibit excellent oxidation resistance at higher temperatures in contrast to the catastrophic oxidation at $\sim$1000°C. This shows the great influence secondary phases can have in determining the oxidation behavior of a hot-pressed Si_3N_4-based material.

Silicon nitride materials hot-pressed with carefully controlled amounts of Y_2O_3 and SiO_2, which do not contain any of the above deleterious phases (e.g., materials with compositions in the phase field labelled D), follow parabolic oxidation behavior and exhibit extremely small rates of oxidation throughout the temperature range 1000 to 1400°C. For instance, one of the experimental hot-pressed Si_3N_4 material with composition in the phase field D oxidized at 1370°C in 1 atm pressure oxygen at a rate of only $\sim$2.6 x 10^{-14} $g^2cm^{-4}s^{-1}$. This rate of oxidation is smaller by a factor of about 200 compared to the oxidation rate of MgO-containing commercial hot-pressed Si_3N_4 and comparable to that for pure Si_3N_4. Preliminary experiments also indicate that long-term oxidation has much smaller effect on the flexural strength of this material than that on the MgO-containing commercial hot-pressed Si_3N_4. Thus, even though the detailed oxidation kinetics of Y_2O_3-containing Si_3N_4 materials still remains to be studied, these initial results indicate that Si_3N_4-based ceramic materials possessing extremely great oxidation-resistance comparable to that of pure Si_3N_4 can be prepared by careful selection of the densification additives and control of the secondary phases in the material.

4.3. Si_3N_4-Al_2O_3 Materials (Sialons)

Reaction of Si_3N_4 with different amounts of Al_2O_3 (up to ~60 wt %) results in the formation of silicon-aluminum oxynitride solid solutions which have come to be known as sialons. These materials have been shown to possess interesting physical, chemical, and mechanical properties for high temperature structural applications [56]. Early experiments [9,57] suggested that these materials may have better oxidation-resistance than MgO-containing hot-pressed Si_3N_4. The detailed oxidation behavior of several hot-pressed sialons has recently been studied by Singhal and Lange [58] and Schlichting and Gauckler [41]. Materials studied by Singhal and Lange varied in composition from 20 to 50 wt % Al_2O_3 and contained, in addition to β-sialon, an unidentified phase. The oxidation of these materials in 1 atm pressure dry oxygen at 1370°C followed parabolic behavior, and the rate constants for different materials are shown in Table I. These results show that the rates of oxidation decrease, as would be expected, with increasing amounts of Al_2O_3 in the material. Comparable rates of oxidation have also been obtained by Schlichting and Gauckler [41] for four single-phase β´-sialons containing from 6.1 to 57.4 wt % combined Al_2O_3 and AlN.

TABLE I

Parabolic Rate Constants for Oxidation of Different Sialon Materials in 1 atm Dry Oxygen at 1370°C [58].

Alumina Content wt %	Parabolic Rate Constant $g^2 cm^{-4} s^{-1}$
20	2.36×10^{-12}
30	1.67×10^{-12}
40	6.94×10^{-13}
50	2.22×10^{-13}
Commercial Si_3N_4	5.00×10^{-12}

The rates of oxidation for these sialons are smaller than for MgO-containing commercial hot-pressed Si_3N_4, though not as small as for Si_3N_4 material hot-pressed with controlled amounts of Y_2O_3 and SiO_2 described earlier in Sec. 4.2. The smaller rates of oxidation for sialons presumably result because a coherent and protective layer of predominantly mullite is formed on the surface.

5. CONCLUSIONS

After a short initial induction period of few minutes, Si_3N_4 and Si_3N_4-based materials generally follow a parabolic oxidation behavior in high oxygen potential environments at 1000 to 1500°C. Additive-free, pure Si_3N_4 oxidizes with extremely small rates and though there is large scatter between the data of various investigators, it appears that its oxidation is controlled by oxygen diffusion through surface SiO_2 similar to the oxidation of Si and SiC. But the presence of additives and significant amounts of certain impurities (particularly Ca, Na, and K) in the hot-pressed materials causes a large increase in their rates of oxidation compared to additive-free, pure Si_3N_4. The additive and impurity elements diffuse outwards during oxidation and concentrate in the surface oxide. The extent of oxidation for a particular Si_3N_4-based hot-pressed material depends on the chemical nature and concentration of the additive and impurity elements, and their diffusion through the surface oxide might be oxidation rate-controlling.

As exemplified by Y_2O_3 and Al_2O_3-containing Si_3N_4-based materials, different phases formed by reaction between Si_3N_4 and oxide additives play a very important role in determining the overall oxidation behavior of hot-pressed materials. This points to a need for carefully studying the phase relationships in relevant multicomponent systems and determining the oxidation behavior of individual phases which can form during hot-pressing in in a particular system. Based on such information, careful selection of additives and control of their concentrations can be made so as to avoid formation of deleterious phases during processing in producing new Si_3N_4-based structural ceramics with extremely high oxidation-resistance comparable to that of pure Si_3N_4.

REFERENCES

1. D. A. Oliver, Silicon Nitride: A Review of its Properties and Attractiveness for Use at Elevated Temperatures, in: High Temperatures in Aeronautics, ed. by C. Ferrari, Pergamon Press, New York and Oxford, 1964.
2. N. C. Tombs, H. A. R. Wegener, R. Newman, B. T. Kenney, and A. J. Coppola, Proc. IEEE (Corres.), vol. 54 (1966) p 88.
3. J. J. Burke, A. E. Gorum, and R. N. Katz, Ceramics for High Performance Applications, Brook Hill Publishing Co., Chestnut Hill, MA, 1974.
4. C. Wagner, J. Appl. Phys., vol. 29 (1958) p 1295.
5. J. E. Antill and J. B. Warburton, Corr. Sci., vol. 11 (1971) p 337.
6. S. C. Singhal, Ceramurgia Internat., vol. 2 (1976).

7. T. N. Zabruskova and I. Ya. Guzman, Trans. D. I. Mendeleev Inst. of Tech., vol. 18 (1971) p 127.
8. D. R. Stull and H. Prophet, JANAF Thermochemical Tables, U. S. National Bureau of Standards, Washington, DC, 1971.
9. K. H. Jack, The Production of High-Temperature High-Strength Nitrogen Ceramics, in: Ref. 3, 1974.
10. F. Galasso, U. Kuntz, and W. J. Croft, J. Amer. Ceram. Soc., vol. 55 (1972) p 431.

11. C. D. Greskovich, J. H. Rosolowski, and S. Prochazka, Ceramic Sintering, General Electric Co., Rpt. SRD-75-084, Advanced Research Projects Agency Contract N00014-74-C-0331, 1975.
12. I. Franz and W. Langheinrich, Solid State Electron., vol. 14 (1971) p 499.
13. I. Franz and W. Langheinrich, Formation of Silicon Dioxide From Silicon Nitride, in: Proc. Seventh Internat. Symp. Reactivity of Solids, ed. by J. S. Anderson, M. W. Roberts, and F. S. Stone, Chapman and Hall, London, 1972.
14. B. E. Deal and S. S. Grove, J. Appl. Phys., vol. 36 (1965) p 3770.
15. B. E. Deal, J. Electrochem. Soc., vol. 110 (1963) p 527.
16. S. Lin, J. Amer. Ceram. Soc., vol. 58 (1975) p 160.
17. R. M. Horton, J. Amer. Ceram. Soc., vol. 52 (1969) p 121.
18. P. J. Jorgensen, M. E. Wadsworth, and I. B. Cutler, J. Amer. Ceram. Soc., vol. 42 (1959) p 613.
19. S. C. Singhal, J. Amer. Ceram. Soc., vol. 59 (1976) p 81.
20. S. I. Raider, R. Flitsch, J. A. Aboaf, and W. A. Pliskin, J. Electrochem. Soc., vol. 123 (1976) p 560.
21. N. I. Krasotkina, Ogneupory, vol. 6 (1967) p 33.
22. P. Goursat, P. Lortholary, D. Tetard, and M. Billy, Silicon Nitride and Oxynitride Stability in Oxygen Atmosphere at High Temperatures, in: Proc. Seventh Internat. Sym. Reactivity of Solids, ed. by J. S. Anderson, M. W. Roberts, and F. S. Stone, Chapman and Hall, London, 1972.
23. D. Tetard, P. Lortholary, P. Goursat, and M. Billy, Rev. Int. Haut. Temp. Refrac., vol. 10 (1973) p 153.
24. K. Motzfeldt, Acta Chem. Scand., vol. 18 (1964) p 1596.
25. E. Fitzer and R. Ebi, Kinetic Studies on the Oxidation of Silicon Carbide, in: Silicon Carbide - 1973, ed. by R. C. Marshall, J. W. Faust, and C. E. Ryan, University of South Carolina Press, Columbia, SC, 1974.
26. J. Schlichting, Werkstoffe u. Korrosion, vol. 26 (1975) p 753.
27. P. Kofstad, High Temperature Oxidation of Metals, John Wiley and Sons, Inc., New York, 1966.
28. I. Ya. Guzman, Yu. N. Litvin, and G. V. Turchina, Ogneupory, No. 2 (1974) p 47.
29. A. G. Evans and R. W. Davidge, J. Mater. Sci., vol. 5 (1970) p 314.

30. R. W. Davidge, A. G. Evans, D. Gilling, and P. R. Wilyman, Oxidation of Reaction-Sintered Si_3N_4 and Effect on Strength, in: Special Ceramics, vol. 5, ed. by P. Popper, British Ceramic Research Assoc., Stoke-on-Trent, 1972.

31. A. F. McLean, E. A. Fisher, and R. J. Bratton, Brittle Materials Design, High Temperature Gas Turbine, Rpt. AMMRC CTR 73-9, Advanced Research Projects Agency Contract DAAG 46-71-C-0162, 1973.

32. A. F. McLean, E. A. Fisher, R. J. Bratton, and D. G. Miller, Brittle Materials Design, High Temperature Gas Turbine, Rpt. AMMRC CTR 75-8, Advanced Research Projects Agency Contract DAAG 46-71-C-0162, 1975.

33. J. V. Sharp, J. Mater. Sci., vol. 8 (1973) p 1755.

34. W. Engel, P. Porz, and F. Thummler, Ber. Dt. Keram. Ges., vol. 52 (1975) p 296.

35. F. Thummler, P. Porz, G. Grathwohl, and W. Engel, Kriechen und Oxidation von Reaktionsgesintertem Si_3N_4, in: Science of Ceramics, vol. 8, British Ceramic Research Assoc., Stoke-on-Trent, 1976.

36. D. J. Godfrey and K. C. Pitman, Some Mechanical Properties of Silicon Nitride Ceramics: Strength, Hardness, and Environmental Effects, in: Ref. 3, 1974.

37. M. E. Washburn and H. R. Baumgartner, High Temperature Properties of Reaction Bonded Silicon Nitride, in: Ref. 3, 1974.

38. S. C. Singhal, Oxidation of Silicon-Based Structural Ceramics, in: Properties of High Temperature Alloys, ed. by Z. A. Foroulis and F. S. Petit, The Electrochemical Soc., Inc., New York, 1976.

39. S. Hofmann and L. J. Gaukler, Powder Met. Internat., vol. 6 (1974) p 90.

40. R. Kossowsky, J. Mater. Sci., vol. 8 (1973) p 1603.

41. J. Schlichting and L. J. Gauckler, Oxidation of Some $\beta-Si_3N_4$ Materials, to be published.

42. S. C. Singhal, J. Mater. Sci., vol. 11 (1976) p 500.

43. W. C. Tripp, J. W. Hinze, et al, Internal Structure and Properties of Ceramics at High Temperatures, Wright Patterson Air Force Base, Aerospace Research Laboratories Rpt. ARL TR 75-0130, 1975.

44. A. J. Kiehle, L. K. Heung, P. J. Gielisse, and T. J. Rockett, J. Amer. Ceramic Soc., vol. 58 (1975) p 17.

45. N. Tighe, Microstructure of Oxidized Silicon Nitride, in: Proc. Electron Microscopy Soc. Amer., vol. 32, ed. by C. J. Arceneaux, Claitor's Publishing Div., Baton Rouge, LA, 1974.

46. R. Kossowsky, J. Amer. Ceram. Soc., vol. 56 (1973) p 531.

47. R. Kossowsky and S. C. Singhal, Effects of Grain Boundaries on the Properties of Hot-Pressed Si_3N_4 and SiC, in: Grain Boundaries in Engineering Materials, ed. by J. L. Walter, J. H. Westbrook, and D. A. Woodford, Claitor's Publishing Div., Baton Rouge, LA, 1975,

624

48. E. Gugel, A. F. Fickel, and H. Kessel, Powder Met. Internat., vol. 6 (1974) p 136.
49. S. C. Singhal and R. Kossowsky, Effect of Oxidation on the Flexural Strengths of Hot-Pressed Si_3N_4 and SiC, to be published.
50. W. A. Sanders and H. B. Probst, High Gas Velocity Burner Tests on Silicon Carbide and Silicon Nitride at 1200°C, in: Ref. 3, 1974.
51. S. C. Singhal, Oxidation and Corrosion-Erosion Behavior of Si_3N_4 and SiC, in: Ref. 3, 1974.
52. S. C. Singhal, Corrosion-Resistant Structural Ceramic Materials for Gas Turbines, in: Proc. 1974 Gas Turbine Materials in the Marine Environment Conference, ed. by J. W. Fairbanks and I. Machlin, Battelle's Metals and Ceramics Information Center, Columbus, OH, 1975.
53. G. E. Gazza, Amer. Ceram. Soc. Bull., vol. 54 (1975) p 778.
54. A. Tsuge, K. Nishida, and M. Komatsu, J. Amer. Ceram. Soc., vol. 58 (1975) p 324.
55. F. F. Lange, S. C. Singhal, and R. C. Kuznicki, Phase Relations and Stability Studies in the Si_3N_4-SiO_2-Y_2O_3 Pseudo-Ternary System, to be published.
56. K. H. Jack, J. Mater. Sci., vol. 11 (1976) p 1135.
57. W. J. Arrol, The Sialons - Properties and Fabrication, in: Ref. 3, 1974.
58. S. C. Singhal and F. F. Lange, Oxidation Behavior of Sialons, to be published.

DISCUSSION (Desmaison, Lindley)

Diness: What was the character of the CVD Si_3N_4 material which you gave oxidation data for? Was it crystalline or amorphous, and what was the impurity content? If the material was amorphous Si_3N_4, how relevant was this oxidation data to RBSN, HPSN and other powders? Does the possibility exist of HCl having been incorporated in the CVD Si_3N_4 and hence affecting diffusion in it?

Singhal: The data of Franz and Langheinrich are for amorphous Si_3N_4 films formed on silicon slices by chemical vapour deposition using SiH_4 and NH_3. No HCl is involved in the deposition process. The oxidation data on these CVD films are relevant from the point of view of understanding the intrinsic oxidation of Si_3N_4 in the absence of any impurities or additives.

Popper: Why is the oxidation rate of reaction-sintered silicon nitride greater than that of silicon nitride powder? Was the nitrided powder derived from high purity Si powder whilst the reaction sintered material was from commercial, i.e. impure Si powder?

Singhal: In making reaction-sintered silicon nitride materials, certain elements (e.g. iron) are intentionally added to Si powder to facilitate nitridation. These additives may be responsible for greater rates of oxidation for reaction-sintered material as compared to that for Si_3N_4 powder.

Jack: We have shown both by heat treatment of MgO hot-pressed silicon nitride and also by heat treatment of the synthesised Mg-Si-O-N glass that devitrification of this glass occurs at 1200-1400°C to give enstatite and Si_2N_2O. How much of this enstatite you observe is due to oxidation and how much to devitrification? Would the enstatite formed by devitrification act as nuclei for further formation by oxidation?

Singhal: The devitrification of the glass can give enstatite and Si_2N_2O in the interior of the material, but one would not expect these devitrification products to migrate to the surface, in the form of compounds. However, the enstatite formed by devitrification can indeed act as nuclei for further formation by oxidation. Some evidence for this has been observed by Schlichting.

Billy: In explaining the oxidation of silicon nitride, you speculated that molecular nitrogen may have diffused as bubbles to the surface. If so, the compactness of the surface layer would be greatly affected. How do you explain, in such a case, that the diffusion step may be still rate determining?

Singhal: The mechanism by which nitrogen is removed from the Si_3N_4/oxide interface is not clearly understood. A detailed oxidation investigation over a rather wide range of nitrogen partial pressures may give some clue as to what happens to this nitrogen. Mass spectrometric analysis of the gases formed during oxidation will also help in determining how nitrogen is removed from the interface.

Brennan: Have you identified the mechanism responsible for the large surface pits formed during oxidation of MgO-containing silicon nitride?

Singhal: Because of the heterogeneous nature of the material, it is diffficult to conclusively determine the one process responsible for the formation of pits on the surface during oxidation of commercial hot-pressed silicon nitride material. It is indeed possible that more than one process causes formation of such pits. In any case, the surface reaction between $MgSiO_3$ (the predominant oxide constituent) and Si_3N_4 appears to be the chief pit-forming process.

Lange: It should be pointed out that the Si_3N_4 materials fabricated in the $Si_3N_4-Y_2Si_2O_7-Si_2N_2O$ compatibility triangle do not pit as the $Si_3N_4-SiO_2-MgO$ materials even though these materials contain WC, WSi_2, Si_2Fe etc. This suggests that the accidentally incorporated second phases (WC, WSi_2, $FeSi_2$ etc.) are not necessarily the precursor of pitting during oxidation.

Davidge: In view of the very varied and often severe oxidation effects in silicon ceramics, with the concomitant degradation of mechanical properties, do you think that these materials have much future as engineering components for high temperature applications? Even if the oxidation rates are relatively low (e.g. CVD material), the environment will often introduce deleterious fluxing agents.

Singhal: The situation is not as bad as you seem to imply in your question. Firstly, the oxidative degradation of mechanical properties can be eliminated or at least drastically reduced in magnitude by developing materials with oxidation-resistance comparable to that of chemically pure silicon nitride. For example, the Si_3N_4 material hot-pressed with controlled amounts of Y_2O_3 and SiO_2 (in the phase field D in Figure 5) undergoes much smaller oxidative degradation in strength than the commercial material hot-pressed with MgO. As regards the fluxing agents from the environment, the effects of common impurities in the turbine fuel such as Na, S and V have been shown to be negligible at temperatures above about 900°C (see Ref. 52).

Morgan: In the past I have often been pessimistic about the rate of progress because I have felt that we are studying inferior materials; nevertheless, the manifold indications are that when we finally get, say, material with a total impurity content of 1% or less, then we will indeed see some dramatic improvements in strength, oxidation, hardness, creep etc. I am, therefore, optimistic that eventually we will succeed. I doubt, however, that the traditional ceramic powder technology can provide the answer and look instead to CVD and chemically prepared powders to provide the future direction.

OXIDATION AND HOTCORROSION BEHAVIOUR OF Si_3N_4 AND SiAlON

J. Schlichting

Institut für Chemische Technik, University
Karlsruhe, Germany

ABSTRACT. The oxidation kinetics of Si_3N_4 and SiAlON
corresponds to the known data of the permeability of
molecular oxygen through SiO_2-glass. As a consequence,
an extremely high oxidation resistance is observed at
temperatures up to 1600°C. In the case of reaction
sintered Si_3N_4 crystalline oxide layers are formed. Hot
pressed Si_3N_4 formes $MgO \cdot SiO_2$ layers with higher permeability of molecular oxygen. In the case of SiAlON partly
crystallized glasses are formed in which oxygen can only
penetrate through the glass-phase and not through the
mullite-crystals. The studies of hot corrosion with
Si_3N_4 and SiAlON at 1000-1200°C have shown an excellent
stability to compounds being present in the burnergas as
NaCl, V_2O_5 and H_2S, whilst alkali melts (Na_2SO_4 and
Na_2CO_3) do corrode the ceramics. It can be shown that
the livetime of Si_3N_4 and SiAlON is more than 10 times
longer than that of coated superalloys in a hot corrosion test.

1. INTRODUCTION

The use applicability of the both nitrogen ceramics
Si_3N_4 and SiAlON as well as of SiC and the silicides of
the IVth to VIth group of the periodic system as high
temperature materials is given by the formation of a
glassy or partly crystallized SiO_2 surface layer. In 1964
MOTZFELD (1) has shown that the activation energy of the
oxidation of silicon and SiC corresponds exactly to that
of the known diffusion data for the permeation of mole-

cular oxygen through a SiO_2-glass layer (2-4). These data were confirmed by FITZER, REINMUTH (5) for $MoSi_2$ and FITZER, EBI (6) for SiC and Si_3N_4.
In this work the oxidation as well as the hot corrosion behaviour of Si_3N_4 and SiAlON is described in detail. Hot pressed Si_3N_4 containing 5 w% MgO or 30 w% La_2O_3 is compared with reaction sintered Si_3N_4 and literature data for pure Si_3N_4 powder (6). Beside of this hot pressed SiAlON (β-$Si_{6-x}Al_xO_xN_{8-x}$) with 5, 10, 30 and 50 equivalent% Al was oxidized. These materials consist of β-Si_3N_4 with Al- and O-atoms on Si and N lattice sites. A solid solution exists in the system Si_3N_4-AlN-Al_2O_3-SiO_2 along a line of a constant metal : nonmetal ratio of 3:4 as described by GAUCKLER (7-9).

2. OXIDATION BEHAVIOUR OF Si_3N_4 and SiAlON

The oxidation experiments were carried out in air at temperatures between 1000 and 1600°C. For short time experiments up to 25 h a thermobalance (METTLER thermo-analyzer) for long time experiments up to 200 h a furnace was used. Fig. 1a shows the oxidation behaviour of diffe-rent Si_3N_4 at 1400°C. One can recognize low oxidation rates for reaction sintered Si_3N_4 materials. The scatter is caused by the different porosity of the used materials.

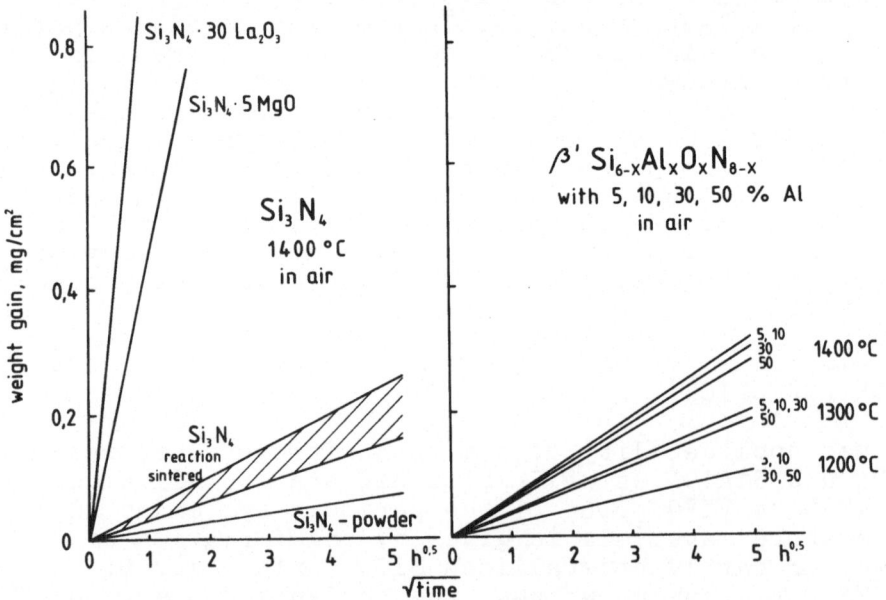

Fig.1: Oxidation behaviour of different Si_3N_4 and SiAlON in air

For comparison the slightly better oxidation behaviour
of Si_3N_4 powder as described by FITZER and EBI (6) is
shown. On the other side hot pressed Si_3N_4 with additional MgO or La_2O_3 oxidizes faster. FITZER and EBI (6)
have shown that on pure Si_3N_4 a crystalline SiO_2 layer
is formed during oxidation. The layer consists of glassy
SiO_2 with small crystals of cristobalit (up to 80 w% at
1500°C). Si_3N_4 specimens with MgO form $MgSiO_3$ layers and
materials containing La_2O_3 form La-Si-oxide layers with
a higher oxidation rate (10).
The oxidation of SiAlON is slower by one order of magnitude than that of hot pressed Si_3N_4 and in the same
order than that of reaction sintered Si_3N_4. Fig. 1b
shows this behaviour for different temperatures. It can
be recognized that the oxidation rate decreases slowly
with the Al-content. X-ray diffraction showed crystalline
phases on SiAlON too. Mullite and some traces of cristobalit were detected. Quantitative X-ray measurements
showed that the mullite concentration in the formed
glass layers is about 70 v/o (after 100 h oxidation
1400°C) for a SiAlON with 50 equivalent% Al and about
25 v/o for 5 % Al (10). In the early stage of oxidation
at 1400°C the formation of mullite depends on the Al-
concentration. Mullite could be detected after 2 h on
the materials with 5 % Al and after 5 min on the material
with 50 % Al. The formation of amorphous SiO_2 between
the mullite crystals was proofed by IR-spectroscopy on
isolated oxid layers of materials with lower Al content.
Fig. 2 compares in a ARRHENIUS plot the oxidation behaviour of the investigated materials compared with that
of SiC (6), Si_3N_4 powder (6) and $MoSi_2$ (5). For reaction

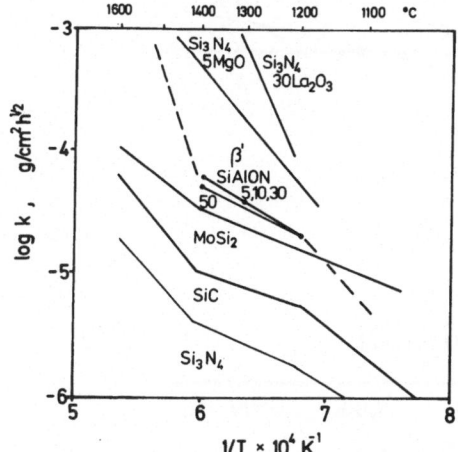

Fig.2: ARRHENIUS plot of log k versus 1/T for Si_3N_4,
       SiC, $MoSi_2$ and SiAlON

sintered Si₃N₄ material the curve will be displaced
slightly parallel. The mechanism of oxidation remains
the same. For the SiAlON material with 5, 10 and 30 % Al
an activation energy of 35 kcal/mol was found for the
temperature range between 1200 and 1400°C. For the material with 50 % Al the activation energy is slower. The
consequence is that all materials have the same mechanism. In all cases amorphous SiO_2 glass layers are
formed, which contain in the case of Si₃N₄ and SiC cristobalit, and in the case of SiAlON mullite. The oxidation of these materials is determined only by the oxygen
penetration through the ring structure of the SiO₄ network. The crystals diminish the area of the amorphous
glass and retard the oxidation. At the other hand MgO
and La_2O_3 additions form silicatic layers. The Si-O-rings
were widened by the Mg- or La-ions.

3. HOT CORROSION BEHAVIOUR OF Si₃N₄ and SiAlON

Hot corrosion tests were carried out with burnergas at
1100 - 1200°C. Fig. 3 shows the weight change under
cycling conditions (4 min 1200°C, 1 min 1000°C). Hot
pressed Si₃N₄ shows a parabolic weight gain with a high
oxidation rate. Reaction sintered Si₃N₄ has only a small
weight gain after 50-100 h when the pores of the material are closed. The same excellent resistance showed
SiAlON specimens with 10 and 50 w% Al (0,4 mg/cm² in
500 h for 10 % Al).

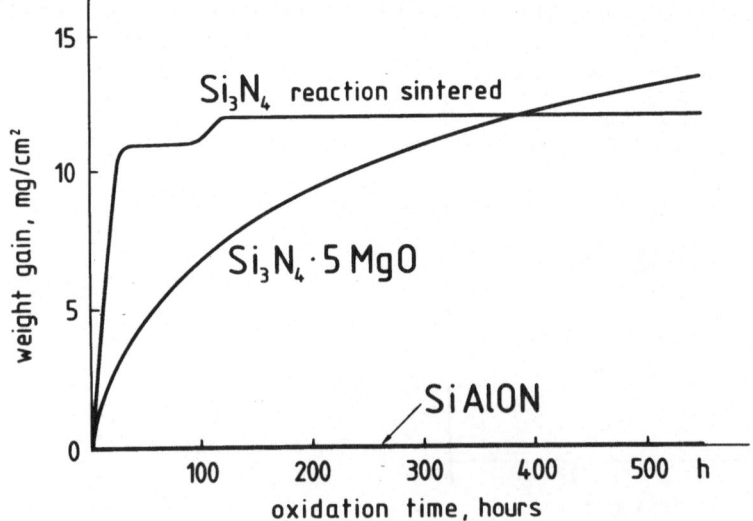

Fig.3: Oxidation of Si₃N₄ and SiAlON in burnergas,
1200°C

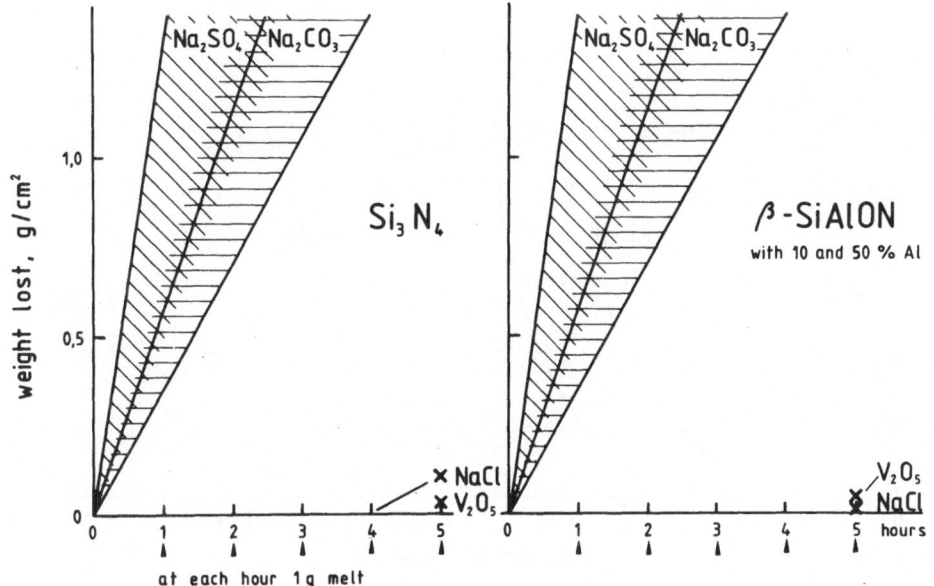

Fig.4: Solubility of Si₃N₄ (a) and SiAlON (b) in
different melts in burnergas of 1000-1100°C

The corrosion resistance of these materials in burnergas
with 10 % H₂S is excellent too. Both materials only
showed weight gains under 0,5 mg/cm² within 400 h at
1100°C. At the other hand chromaluminized superalloys
were destroyed within 50 h. X-ray diffraction of the
SiAlON-specimens showed the formation of crystalline
glass layers containing tridymit and mullite.
Si₃N₄ as well as SiAlON are soluble in alcaline melts
like Na₂SO₄ and Na₂CO₃ as shown in Fig. 4. The solubility in liquid NaCl and V₂O₅ is very poor (no solution
in NaCl within 300 h). With V₂O₅ melts V₂O₅-silicatic
layers with low penetration of V₂O₅ melt are formed at
the surface of the specimens.
In order to perform a more practicable hot corrosion
test Si₃N₄ and SiAlON were treated in burnergas under
cyclic conditions. During the cooling NaCl-, Na₂CO₃- or
Na₂SO₄-solutions (1 and 5 %) were sprayed on the specimens. For reaction sintered Si₃N₄ a lifetime of 50-100 h,
for hot pressed Si₃N₄ 150-200 h and for SiAlON 200-300 h
was found (Fig. 5).
During these experiments low melting Na₂O-SiO₂-glass
layers are formed at the surface. The solubility of
these ceramic materials is only a function of Na₂O content in these glasses (11). The lifetime of these
ceramics can be extended if the alkali content is
lowered.

632

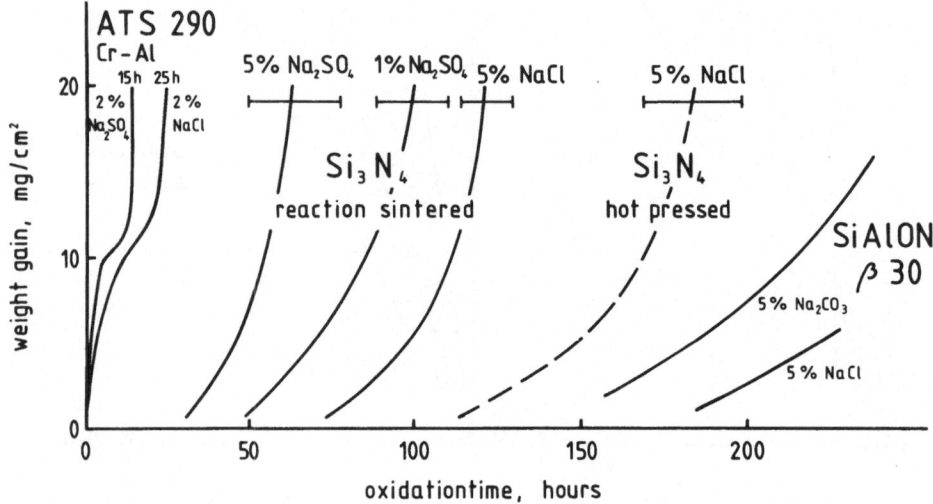

Fig.5: Corrosion of Si₃N₄, SiAlON and a Cr-aluminized
superalloy in burnergas under cycling conditions

For high temperature application new materials (the
nitrogen ceramics as well as SiC and metal silicides)
with high oxidation and hot corrosion resistance are
available. These materials can be used as bulk materials or as coating or composite materials.

REFERENCES

1. K. Motzfeld, Acta Chem. Skand. 18 (1964), 1596-1606
2. F.J. Norton, Nature 191 (1951), 859-867
3. E.W. Sucov, J. Am. Ceram. Soc. 46 (1963), 14-20
4. E.L. Williams, J. Am. Ceram. Soc. 48 (1965), 190-194
5. E. Fitzer and K. Reinmuth, in F. Benesovsky: Hoch-
   temperatur-Werkstoffe, Springer Verlag Wien, 1969
6. E. Fitzer and R. Ebi, Dechema 1972, Frankfurt
7. L.J. Gauckler, H.L. Lukas and G. Petzow, J. Am.
   Ceram. Soc. 58 (1975) 346-347
8. L.J. Gauckler, S. Prietzel, G. Petzow and J. Schlichting, Fall Meeting of the Am. Ceram. Soc. Sept. 1975
9. L.J. Gauckler, Dissertation University Stuttgart 1976
10. J. Schlichting and L.J. Gauckler, Powder Metallurgy
    International 8 (1976) in press
11. J. Schlichting, Werkstoffe und Korrosion 26 (1975)
    753-758

DISCUSSION (Desmaison, Lindley)

Godfrey: You gave data showing that Na_2SO_4 attacks Si_3N_4 strongly, but NaCl does not. Materials are often tested for hot corrosion resistance in sodium sulphate melts. Can you explain therefore why our 200 hour tests with 1% S fuel and sea-salt injections in our combustion rig at 750°C and 830°C gave no measurable attack or deterioration of strength with RBSN test specimens? The temperatures are around the maximum at which liquid would be present on turbine blades. Is the Na_2SO_4 test a good test?

Schlichting: It has been found that NaCl reacts in burner gas with S forming Na_2SO_4. This provides a good explanation for the corrosion of superalloys. These materials will react with NaCl as well as with Na_2SO_4. In the case of RBSN and NaCl, glass layers must be formed at the surface and which do not react with S. The Na_2SO_4 reaction described must therefore be at the surface of the material. In the case of RBSN, NaCl will be dissolved in the SiO_2-glass layer and will not be able to react with S to give Na_2SO_4. Thus there is no reaction. For the silicides (Si_3N_4, SiC, $MoSi_2$) results from the Na_2SO_4 test with melts are not comparable with those from NaCl and S in burner gas.

Krauth: Has the water vapour an influence on the oxygen diffusion through the SiO_2-glass layer in the case of 80% cristobalite?

Schaeffer: I think the answer is yes; the oxidation through SiO_2 glass proceeds faster than through cristobalite when speaking of volume diffusion. However, in the case of predominantly crystallised scales the increased oxidation is now due to a transport via grain boundaries.

Goursat: Have you studied the influence of oxygen partial pressure on the rate of oxidation?

Schlichting: The influence of oxygen partial pressure has been reported in the literature. Fitzer and Ebi have found that the oxidation rate decreases with decreasing oxygen partial pressure. In the case of CO and CO_2 as the reacting atmosphere the kinetics are controlled by the diffusion of nitrogen through the SiO_2-glass layer. For SiAlON all experiments were carried out in air.

Räuchle: We observed after a heat treatment of only 1 hour at 1300°C in air a room temperature strength decrease from about 220 N/mm^2 to only 100 N/mm^2. The reason for this observation was in all cases the formation of an oxide layer which cracked at the Cristobalite transformation point.

Schlichting: The decrease of strength is caused by the inner oxidation in the Si_3N_4. The oxide layers in the pores consist of cristobalite and SiO$_2$-glass. During cooling many bridges between the particles are broken at the cristobalite transformation point. This decrease in strength can be diminished by chemical vapour deposition of Si$_3$N$_4$ in the pores.

Godfrey: We too have found that the strength of RBSN is decreased after oxidation, as reported at the U.S. Hyannis Symposium. In a material with a strength of 230 MPa, 100 h at 1250°C can reduce room temperature strength to 135 MPa. Treatments have been developed which reduce this strength degradation, giving strengths at room temperature of ~ 200 MPa. Their effect on other properties now have to be determined.

PROBLEMS RAISED IN THE OXIDATION OF SINTERED CERAMICS :
THE CASE OF SILICON OXYNITRIDE *

M. Billy

Department of Ceramics, University of Limoges, France.

It can be expected that the microstructure of sintered ceramics,
and especially their porosity, must have a strong influence on
reactivity in modifying the reactive surface. This implies that
the oxidation tests neglecting the influence of the porosity,
such as those generally described, are suspected to give but a
little idea of the actual chemical behaviour of the material.

As far as nitrogen ceramics are concerned now, two other
features are to be considered. First, nitrides or oxynitrides are
covalent bonded compounds so that their sintering uses additives.
This leads to alterations in chemical properties ; the oxidation
behaviour, in particular, could be different for the sintered
material and the pure compound. Again, since nitrogen compounds
are known to be volatile, the onset of decomposition must not be
neglected in interpreting kinetics results at high temperature.

In this paper, we intend precisely to discuss the influence
of these factors (porosity, additives, thermal stability) on the
grounds of silicon oxynitride ceramics.

1. SILICON OXYNITRIDE CERAMICS

The properties of Si oxynitride are probably very similar to
those of silicon nitride. Both compounds in particular have about
the same stability and hot corrosion in oxygen results in the
formation of a same protective silica layer by a diffusionnel
process, so long as a thick scale is formed [1].

---

* This work has been supported in part by the CNRS : ERA n°539,
  "Nitrogen Ceramics".

636

The starting material was prepared using a pressureless method described elsewhere (2) which broadly consists in firing mixtures of silicon oxynitride powders with suitable additives (5 wt%) and in protecting the samples with a sheet of Si oxynitride itself in order to provide a stabilizing atmosphere around them. According to this method, it is possible to obain ceramics of a given porosity by choosing sintering parameters, such as firing temperature and initial grain size.

The influence of the additives on the reactivity has been studied from two sorts of specimens, one sintered with MgO, the other with BeO, corresponding to the formation of a liquid phase (forsterite) or not at the sintering temperature.

2. INFLUENCE OF POROSITY

The specimens to be oxidized are cylindrical pellets of about 400 mg which have been heated for 6 hours at 1500°C in an oxygen atmosphere, and subsequently weighted. Our results are reported in fig. 1 where the increase of weight Δm has been related to the degree of compaction τ.

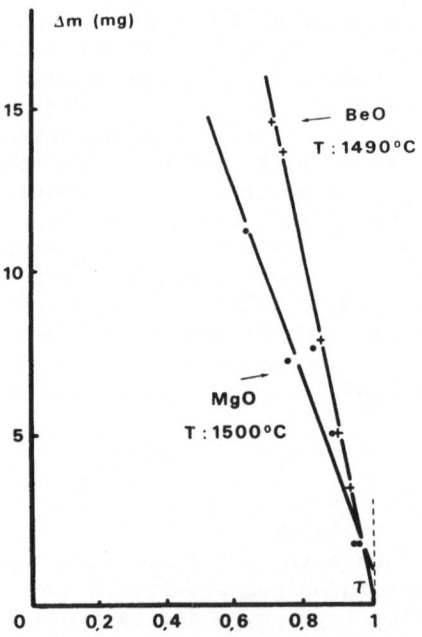

Fig. 1. Linear relationship between reactivity and porosity of silicon oxynitride ceramics oxidized for 6 hours at 1500°C.

One can see that a straight line relationship is obtained for each additive, MgO or BeO ; in both cases, the correlations are highly significant with a reliability of 5%. In other words, the reactivity of the sintered ceramics (which is proportional of course to increase of weight) decreases linearily with the degree of compaction, that is to say increases with porosity.

We can easily explain the influence of porosity in that the reactivity is directly proportional to the reactive surface S

$$\Delta m = KS = K(S_o + S_p),\tag{1}$$

where S is the total surface area including a constant term S_o, which depends on geometry of the sample, and S_p the surface of the open pores which varies with porosity.

Suppose now that there are N open spherical pores of a mean radius $\overline{R}$; their volume V_p corresponds to a fraction f of the total porosity $(1-\tau)$, so that we can write :

$$V_p = f(1-\tau) = \frac{4\pi}{3} \overline{R}^3 N = S_p \frac{\overline{R}}{3}.$$

By inserting the surface value for the pores S_p into (1), one derives the following expression

$$\Delta m = KS_o + 3 K \frac{f}{\overline{R}} (1-\tau),\tag{2}$$

which shows that the reactivity of a sintered specimen depends on the porosity as well as on the size of the open pores. This expression has some important consequences when testing the corrosion properties of sintered materials.

First, it is worth noting that the surface term S_o, which depends on the geometry of the specimens, is generally much smaller than S_p the term depending on the pores. This implies that, in general, the oxidation kinetics will not depend on the geometry of the specimens.

A more important conclusion, from a practical point of view, is that the reactivity, to be really significant, must be measured with zero porosity. Here, for example, silicon oxynitride sintered with BeO gives fully densified ceramics more resistant to oxidation than with MgO (cf. fig. 1). But a reverse conclusion would have been drawn from porous specimens ! This is because the reactivity depends on the pores size at a given porosity and because the BeO ceramics here tested had an average pores size smaller than for MgO.

3. INFLUENCE OF ADDITIVES

We have just seen that BeO silicon oxinitride ceramics have a better corrosion resistance at 1500°C than those sintered with MgO. Such a result does suggest the influence of additives.

638

An immediate explanation is to suppose the formation of different surface coatings from different additives. We found effectively that oxidation products contain not only silica but also forsterite, for MgO additive, and Be silicate for specimens sintered with BeO. Moreover, the silicate contents in the coating are of a higher concentration than those in the bulk. Thus, it can be concluded that a segregation of the additives takes place in the coating as the reaction proceeds.

Anyway, since the coatings are different in nature, for each additive, we can predict a different kinetic behaviour. It is also to be predicted that forsterite will be less protective than Be silicate, due to a lower melting point and to a higher plasticity. Our experimental results agree with these predictions.

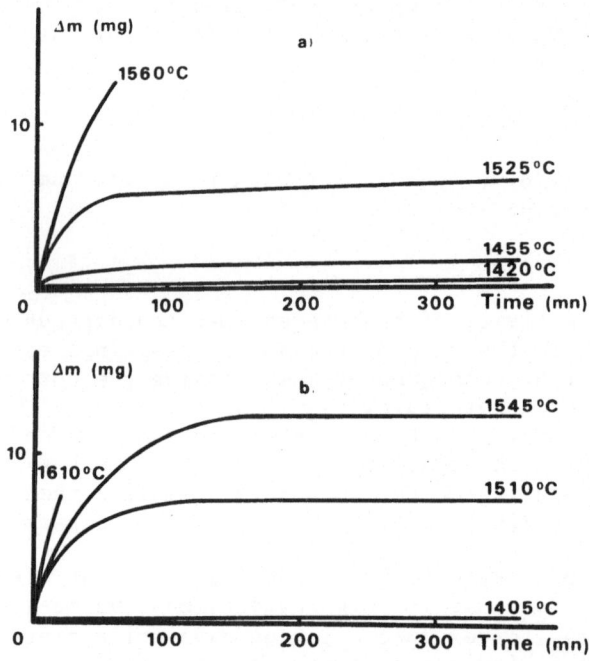

Fig. 2. Oxidation of silicon oxynitride ceramics sintered with BeO (5 wt%).

Fig. 2 is concerned with the oxidation of Si oxynitride ceramics sintered with BeO, the porosities of which were 10% (fig. 2,a) and 20% (fig. 2,b). The shape of the kinetics curves, expressing weight gain with time, is always asymptotic up to 1560°C, where oxidation becomes catastrophic because of the decomposition of Be silicate. The coating is then no more protective.

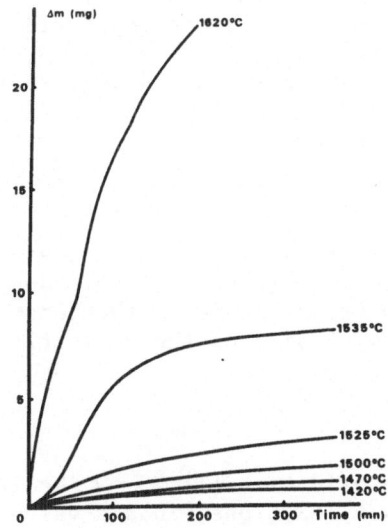

Fig. 3. Kinetic behaviour in oxygen (p = 32 torr) of silicon
        oxynitride ceramics sintered with MgO (porosity:4%).

Kinetic results about oxidation of ceramics sintered with
MgO have been reported in Fig. 3. One can see that the specimens
present a good oxidation resistance up to 1400°C : asymptotic
curves are still observed.

Above 1420°C, kinetic curves become parabolic, implying a
less protective character of the coating. This parabolic stage is
associated with an activation energy of 218 kcal/mol, which is
much higher than that observed for the parabolic oxidation of pure
Si_2N_2O (35 kcal/mol). The two phenomena are thus quite different.

Above the melting point of forsterite (i.e. 1550°C), the
coating is no longer protective so that the specimens are oxidized
very rapidly.

When now examining the coating by scanning microscopy, cracks
are always observed above 1400°C, especially in the case of MgO
additive. At higher temperature (1500°C), some craters are found,
on cracks in particular. The presence of these craters implies a
gaseous evolution through the oxidation product (N_2 and SiO have
been identified). In other words, we have to admit a decomposition of Si_2N_2O which interfers with the oxidation reaction above
1420°C. The presence of cracks or craters is not consistent with
a rate-determining diffusion step, though a parabolic law is
still observed.

The course of the reaction can be interpreted in the following way. Oxygen reacts with Si_2N_2O through the cracks and craters
which become closed with the newly formed oxidation product. The
oxidation rate is then proportional to the number of defects in

the coating, with v = kN. In another hand, cracks and craters being due to the internal gas pressure of the silicon oxynitride decomposition, their number will be directly related to the mechanical resistance of the coating, that is to say, N is a reciprocal fonction of the coating thickness of the form : N = k'/x. Hence, the rate expression for the oxidation :

$$v = \frac{kk'}{x} \ . \tag{3}$$

In such conditions, the overall kinetics will take into account not only the real oxidation reaction, by means of the constant k, but also the decomposition process by means of the constant k'. The activation energy will now consist of the activation energies of both phenomena :

$$E = E_o + E_D.$$

By using the known values for the oxidation (1) ($E_o = 43$ kcal/mol) and for the decomposition (3) ($E_D = 182$), one finds a value of 225 kcal/mol, very close to the experimental one (218).

3. CONCLUSIONS

The results here obtained from the oxidation of silicon oxynitride ceramics are of a real value in that they can be extended to other ceramic materials.

They show that a parabolic law with time does not correspond necessarily to a rate-determining diffusion.

The reactivity of sintered specimens varies both with porosity, as could be expected, and also with pores size. A linear relationship with porosity permits in a simple way to derive the reactivity for zero porosity for compact specimens, which is characteristic of the material.

Finally, the influence of additives on reactivity is very important, even when these are present in quantities as small as 5 wt%. The oxidation resistance of the pure compound is distinctly modified. Again, the refractive properties of the additive, as well as the mechanical properties of the coating, are predominant in predicting the chemical behaviour of sintered ceramics.

REFERENCES

1. P. Goursat, P. Lortholary, D. Tétard and M. Billy, Proc. of the 7th Int. Symp. on React. of Solids, Chapman and Hall, Bristol, 1972, 315.
2. M. Billy, P. Goursat, P. Lortholary, J.P. Mary and J. Mexmain, Bull. Soc. Fr. Ceram., 105, 1974, 1.
3. P. Lortholary and M. Billy, Bull. Soc. chim. Fr., 1975, 1057.

Section K

APPLICATIONS

APPLICATIONS OF NITROGEN CERAMICS--GAS TURBINES:
U.S. National Programs

R. Nathan Katz

Army Materials and Mechanics Research Center, Watertown, MA 02172

Before discussing current and potential applications for nitrogen ceramics, it is appropriate to consider the entire spectrum of ceramic materials and applications to gain a perspective from which the possibilities for nitrogen ceramics may be assessed. Figure 1, shows the various families of ceramic materials (of which nitrogen ceramics presently constitute a very small part), and the applications for ceramics. When looking at this list it is apparent that the unique properties of the various nitrogen ceramics will lead to increased or new applications in the areas of: refractories, abrasives, E-M windows and anti-reflection coatings, electronic devices, nuclear materials, bearings, military ceramics, heat engines, energy storage, and possibly bioceramics. Other participants at this meeting will be discussing many of these. I would like to limit my remarks to the application of nitrogen ceramics to heat engines and in particular gas turbines.

The high temperature strength, thermal shock and fatigue resistance, erosion/corrosion resistance, and other properties of many of the ceramics in the Si_3N_4 and "SiAlON" families of materials make them very attractive for possible application in Diesel, Stirling, and Gas Turbine Engines. Dr. Godfrey and Herr Siebels will discuss the applications of nitrogen ceramics to the diesel engine in some detail. In the Stirling Engine area opportunities include heater heads, pistons and displacers, however, relatively little work has been done to explore these opportunities. Gas Turbines, on the other hand, have been the object of many research, development and demonstration programs aimed at exploiting reaction bonded and hot pressed silicon nitride starting in the 1960's in the U.K. and the 70's in the U.S. and F.R.G. The driving force for the application of nitrogen (and other) ceramics to heat engines derive

644

from the benefits listed in Figure 2. The potential energy savings and use of abundant raw materials are particularly important to industrialized, modern societies.

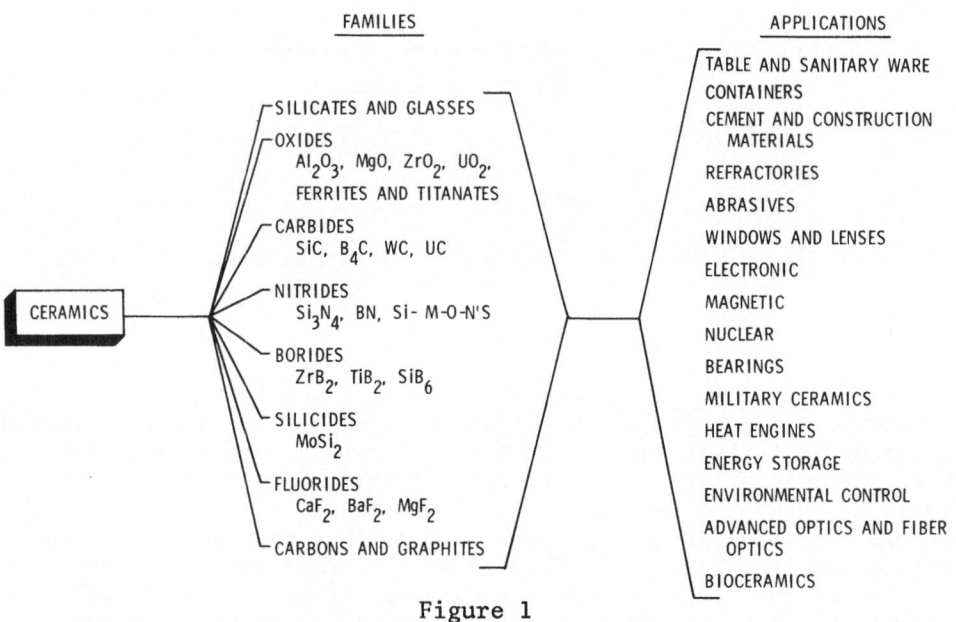

Figure 1

● INCREASED TURBINE INLET TEMPERATURE

    LOWER SPECIFIC FUEL CONSUMPTION
    HIGHER POWER DENSITY
    GREATER RANGE/PAYLOAD

● INCREASED RELIABILITY AND DECREASED MAINTENANCE

    ENHANCED EROSION - CORROSION RESISTANCE
    ENHANCED OVERTEMPERATURE CAPABILITY

● MULTIFUEL CAPABILITY

    LOGISTIC BENEFIT
    STRATEGIC BENEFIT - REDUCED DEPENDENCE ON
      FOREIGN FUELS

● REDUCTION IN STRATEGIC MATERIALS REQUIREMENTS

8700-0002   (Ni, Cr, Co, Cb)

Figure 2   Advantages of Ceramics
in Heat Engines

Based on favorable projections in the state-of-the-art for both materials and design capabilities, and the need for a systems approach, the United States Department of Defense's Advanced Research Projects Agency (ARPA) initiated a program, in 1971, to accelerate ongoing work in the area of gas turbine ceramics and awarded a contract to the Ford Motor Company, with Westinghouse Electric Company as subcontractor, to develop a design capability with brittle materials. This design capability was to be demonstrated via the successful application of ceramic hardware in hot flow path components to operate at 2500 F uncooled for 200 hours in a vehicular engine and 100 peaking cycles in an electrical power generating size turbine test rig. The major underlying goal of the ARPA program was to demonstrate and encourage the use of ceramics as engineering materials. This ARPA "Brittle Materials Design" Program because of its size and visibility rapidly became the focal point for turbine ceramic research in the U.S. (and abroad).

When the ARPA "Brittle Materials Design" Program was initiated, it was a debatable question whether brittle ceramics could survive the rigors of the gas turbine environment in any meaningful capacity. Today the viability of various forms of silicon nitride and silicon carbide as major static turbine components, such as combustors, stators, inlet nose cones and transitions, have been demonstrated in test rigs and engines, in the Ford portion of the ARPA Program, for 100 to 240+ hours, at 1930 F uncooled and in some cases have seen exposure at 2500 F or higher. In the summer of 1976 the first hot spin testing of ceramic rotors was initiated. Confirmation of the rotor attachment scheme, and the duodensity rotor concept, was obtained up to 80 percent speed, in a short time test of a partially bladed rotor in a modified engine at a turbine inlet temperature of up to 2650 F. Obviously, while a ceramic rotor has not yet been fully demonstrated, this first test of an aerodynamically loaded ceramic rotor in a modified engine is very encouraging.

During the past five years the dramatic change in the gas turbine engine manufacturing communities attitude towards ceramics can be chronicled by following the growing emphasis in ceramic components at the semi-annual ERDA Contractors Coordination Meetings in Ann Arbor, Michigan. Today the question is not can we use ceramics, but rather when will a meaningful demonstration of a ceramic rotor be made and can ceramic rotors be manufactured at acceptable costs?

Encouraged by the positive technical achievements of the past five years and the economic payoff of a potential national savings of ~$7 billion in imported petroleum annually, ARPA, AMMRC and the Office of Transportation Energy Conservation of the Energy Research and Development Administration (ERDA-TEC) entered into discussions in FY76 to transfer the civilian sector related technology from the ARPA program to ERDA-TEC. To effectuate this transfer ERDA will participate in funding of the ARPA "Brittle Materials Design" Program and in late FY77 or FY78 initiate a new program to be based on this technology. Execution of this plan will require close cooperation between ARPA, AMMRC and ERDA in FY77. The Army and other DoD exploitation of the ARPA program derived technology will continue as before. Figure 3 shows the ERDA and other major U.S. programs on turbine ceramics which the ARPA program has encouraged.

While the results to date on the ARPA and other programs indicate that there is every reason to be optimistic for ceramics in the gas turbine considerable additional work in the areas of design, materials development, fabrication, test and evaluation, further systems optimization and NDE and life prediction techniques are required. It is, therefore, very gratifying to see an ever broadening involvement by sponsoring agencies and industrial firms in this critical area, coupled with effective technology transfer. It is similarly gratifying to realize that nitrogen ceramics (so critical for ceramic turbine technology) are now a mature enough area of science to warrant a meeting of such scope as this NATO-Advanced Study Institute.

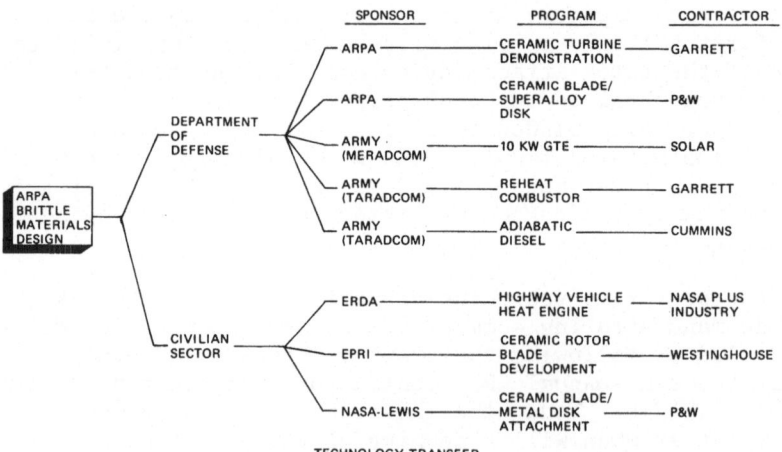

Technology Demonstration Programs Encouraged by The ARPA "Brittle Materials Design" Program

Figure 3

THE USE OF CERAMICS IN DIESEL ENGINES

D. J. Godfrey

Admiralty Materials Laboratory, Poole, Dorset, UK.

ABSTRACT. The properties of ceramics potentially useful for internal combustion engine applications are reviewed, and compared with those of cast iron and aluminium alloys. Potential benefits include improved efficiency due to the lower thermal conductivity of ceramics reducing heat losses, diminished thermal degradation of oil used to cool pistons, elimination of piston crown burning, better part-load operation, and reductions in size, cost and weight because cooling systems could be smaller. Successful utilisation of ceramics is described in a small petrol engine, and in two diesel engines with ratings of 690 and 1580 MPa brake mean effective pressure.

1. INTERNAL COMBUSTION ENGINE MATERIALS

The concentration on gas turbine application of the limited effort committed to investigations of the potential usage of engineering ceramics is probably the principal reason for the comparative neglect of their application as internal combustion engine components, such as pistons, valves, combustion chambers, cylinder head liners, cylinder liners, and supercharger turbines. Another reason is the deep and widespread doubt amongst engineers that "brittle ceramics" can be used in the severe conditions prevailing in internal combustion engines [1]. However, it is possible that their much greater thermal capability, low thermal conductivity, and small thermal expansion might prove very advantageous in engine technology, provided their use is feasible in such a highly-stressed situation.

For reasons of cost, fabricability and weight, cast aluminium-silicon alloy pistons have been used extensively, despite their

relatively-low melting point (ca. 577°C), poor elevated-temperature
mechanical properties, low hardness, and high thermal expansivity.
Ferrous materials have superior elevated-temperature properties,
but their greater density imposes larger inertial loadings on
other components. Relevant properties of materials are summarized
in Table 1.

Most mechanical properties of aluminium alloys are severely
degraded at modest elevated temperatures which makes successful
use above about 250°C difficult. The mechanical properties of
silicon nitride are unaffected up to about 1000°C or higher, and
use up to 1500°C in loaded situations is feasible. The softness
of aluminium alloys is not advantageous, and causes difficulties
in highly-loaded tribological situations, when wear can be excessive. The high hardness of ceramics is attractive, although their
wear behaviour may sometimes be affected detrimentally by their
lack of toughness (due to their lack of ductility). The thermal
capabilities of iron and steel are better than those of the aluminium alloys. However, the thermal expansivity of iron is high;
thermal cracking of iron components in diesel engines has been
experienced. Another consideration is the greater cyclic loading
imposed by the larger mass of iron or steel components, fatigue
failure of crankshafts being a serious possibility. Ceramics have
a density comparable with that of aluminium.

2. PRACTICAL EXPERIENCE WITH CERAMICS IN INTERNAL COMBUSTION ENGINES

The first experience with silicon nitride ceramics in an internal
combustion engine was gained at AML some years ago [1,2], with a
900 W (1.2 HP) Villiers 4-stroke air-cooled single cylinder petrol
engine. The engine was fitted with a 50 mm diameter redesigned
RBSN ceramic piston, which had two sets of triply-segmented RBSN
piston rings, and an RBSN solid surface-ground gudgeon pin. It has
been operated successfully for several years without any piston
or ring failure. On stripping the engine after the third period
of short evaluation runs, the RBSN gudgeon pin (which had the same
dimensions as the strong steel pin which it had replaced) was found
to have a central transverse fracture, although the engine had
functioned normally. This pin was subsequently replaced with a
pin of the strongest-available RBSN ceramic, and the engine has
operated satisfactorily ever since. Although remarkably successful, this exercise did not provoke any significant activity amongst
engineers, and some felt that the working conditions in the engine
must be of very moderate severity for ceramic components to have
survived. It was not possible to investigate heat flow, efficiency, or other engine operating parameters, and component survival
alone was demonstrated.

In 1973 it proved possible to interest the Royal Naval Engineering College in evaluating an RBSN piston in a diesel engine [3].

Lieutenant R.B. Stone RN designed and ran a ceramic piston which
was made at AML, and this was subsequently shown by Lieutenant
Commander B.D. Gibson RN and co-workers [4] to survive operation
in the engine even when run in the full-power fully-loaded condition. The engine was a Gardner single-cylinder 9 kW four-stroke
engine, having a compression ratio of 13:1, a bore of 108 mm and
a stroke of 152.4 mm, with a rating of about 690 kPa (100 psi)
brake mean effective pressure (BMEP). The engine was of pre-war
design, and more modern engines have over double this BMEP rating.
The piston was designed to minimize thermal stresses in the crown
region, by appropriate choice of a wall thickness of 15.2 mm and
a crown minimum thickness of 15.2 mm, replacing an aluminium alloy
piston with wall thickness and crown thicknesses of 24.13 and 10.16
mm respectively. Thermal stress calculations were carried out for
an estimated differential of 155°C with the ceramic piston, using
a computer programme embodying the approach of Fitzgeorge and
Pope [5]. The silicon nitride piston was made by turning and
boring a solid block of isostatically-pressed argon-sintered
silicon-powder compact, and reaction-bonding this to silicon nitride by heating in nitrogen at high temperatures. The resulting
silicon nitride component was lightly diamond-ground and lapped
where precise clearances were necessary in the gudgeon-pin hole.
The aluminium piston had a weight of 1.73 kg; the silicon nitride
piston design involved more material and weighed 2.12 kg. During
operation the ceramic piston would have absorbed about 20% of its
volume of lubricating oil (ca. 0.16 kg), since RBSN contains a
considerable amount of very fine (0.1-0.2 μm) porosity. The piston
design had generous thicknesses around the gudgeon pin hole, and
was not optimized for minimum mass.

The total running time recorded for the piston was 94 hours,
of which 9 hours was at zero developed torque, 31 hours were at
various part-load conditions up to 50% of full load, and 54 hours
fully-loaded developing a torque of 91 Nm. Operating experience
with the piston was very satisfactory. An attempt had been made
to monitor piston temperatures by fitting Templug metallographic
temperature steel sensor studs into special holes. These all detac-
hed during testing and one was found to have jammed between the
side of the piston near to the crown edge and the bore, although
the engine was still functioning satisfactorily. A very small
piece of RBSN had flaked away from the piston at the point of
contact, but this was the only damage or deterioration which was
observed. Measurement of the ovality of the gudgeon pin hole bore
with a Taylor Hobson Talyrond instrument showed a maximum ovality
of 10 μm; in view of the bearing load of 21 MPa this was considered
very satisfactory.

As a consequence of these investigations, a further exploration of the potential of RBSN as a piston material has recently
been made at Associated Engineering Developments Ltd, with an 80 mm
diameter RBSN piston in a single cylinder 2,000 rpm Petter diesel
engine [6]. Pistons have operated at full power for relatively

long periods, involving cycling through various operating regimes, in a fully-supercharged highly-rated (1580 kPa (230 psi) BMEP) diesel engine. Thermal stress calculations indicated that the maximum tensile stresses in the piston were about 100 MPa, in the hoop direction, around the piston crown, significantly below the tensile strength of RBSN [3], which is reported as 145 MPa [7] for a large volume without macroscopic defects.

3. DISCUSSION

Because ceramics such as silicon nitride have physical properties which differ markedly from cast aluminium alloy and iron materials, they merit serious consideration for applications in internal combustion engines. Whilst fabrication of silicon nitride pistons has so far been entirely accomplished by machining solid cylindrical compacts of isostatically-pressed and argon-sintered silicon powder, it is probable that a production engineering route comprising die moulding of a plastic silicon powder-polymeric binder mix would be economically more desirable, as is the practice with other RBSN components already.

Amongst the potential advantages of using RBSN in diesel engines, possibly the most obvious is the diminished energy loss which is attainable with lower-conductivity materials. The possibility of using the heat thus transferred to the exhaust by driving a supercharging turbine offers greater efficiency, as does the prospect of increased gas temperatures in the expansion part of the engine cycle. It appears that the proportion of the fuel energy lost as heat conducted through the piston is in the range 7-12% for many diesel engines. Assuming that 40% of this could be prevented by using a low-conductivity ceramic for the piston, this would transfer 2.8-4.8% of the fuel energy to the exhaust. A 25% conversion of this into work would give a gain in efficiency of 0.7-1.2%, a small but useful improvement. Further improvements could result from the thermal insulation of the cylinder head, prechamber combustor and exhaust duct, since approximately 10 to 25% of fuel energy is lost when high conductivity metals form the walls surrounding the hot gases. It seems reasonable to forecast a total increase in efficiency of 2%, representing an overall improvement of 5% with a 40% efficient modern engine.

The engineering advantages which might result from the use of ceramics can be summarized as follows:
(a) Greater efficiency because the heat lost to the engine structure and coolant could be reduced by making the piston of a low thermal conductivity material (15 instead of 100 W/mK), since a very considerable heat flux through the piston to the liner and lubricating oil exists in aluminium alloy piston diesel engines. This heat could be transferred to the exhaust, where it could be used to drive a supercharger fan or for intercooling.

(b) Greater efficiency because the low melting point of aluminium alloy and the deterioration of its properties above 200°C limit the temperatures which can be allowed in the combustion space walls; with ceramics, gas temperatures would only be limited by thermal stresses and by the capabilities of the exhaust valve and ducting.

(c) With appropriate design changes it might be possible to increase the efficiency of a diesel engine (typically just over 40%) to over 42%, a 5% improvement in efficiency. However, the development of high temperatures in ceramic components will require careful design, and control of thermal conductivity to prevent thermal stress problems.

(d) Piston crown "burning", which is due to a combination of melting, oxidation and gas erosion, could be eliminated by the use of ceramic materials.

(e) Reduction in the heat flux would reduce the degree of oil cooling necessary in the piston, decreasing the necessity for complex designs with forced circulation or 'cocktail-shaker' cooling. The deterioration of the lubricating oil should also be reduced.

(f) The use of low thermal conductivity material would allow significant reductions in the size of the cooling system; giving savings in its first cost and space requirements.

(g) Reductions in noise emission also might be achieved if the size of the cooling fan could be reduced. Another possible noise reduction might ensue if advantage were taken of the especially-low thermal expansion of certain ceramics (Si_3N_4 and some glass ceramics) to reduce the piston/liner clearance, and diminish the piston/liner slap.

(h) The increased combustion space wall temperatures possible with ceramics might improve efficiencies and decrease carbon deposition in engines which have a considerable amount of part-load or lightly-loaded running in their operational duty.

ACKNOWLEDGEMENTS

Grateful acknowledgement is made of helpful discussions with Mr Nottley and colleagues at MVEE, Professor Fessler and Dr Stanley at Nottingham University, Dr Smart and Dr Parker at AED Ltd, and Mr Panton, Ricardo Ltd.

REFERENCES

1. D.J. Godfrey, 'Ceramics for High-temperature Engineering', Proc.Brit.Ceramic.Soc., June 1973, No 22, 1.

2. D.J. Godfrey and E.R.W. May, 'The Resistance of Silicon Nitride Ceramics to Thermal Shock and Other Hostile Environments' Ceramics in Severe Environments, W.W. Kriegel and Hayne Palmour III (Eds.), New York, Plenum Press, 1971, pp 149-162.

652

3. D.J. Godfrey, 'Silicon Nitride Ceramics for Engineering Applications' SAE Paper 740238, 1974; Trans. SAE, 1974, 83, Sect.2, 1036.

4. B.D. Gibson and R.B. Stone, 'The Design and Performance of a Ceramic Piston' Private communication, in course of publication.

5. D. Fitzgeorge and J.A. Pope, 'An Investigation of the Factors Contributing to the Failure of Diesel Engine Pistons and Cylinder Covers', Trans. NE Coast Shipbuilders, February 1955, 71, Part 5.

6. R.F. Smart et al, In course of publication.

7. A.D. Sivill, 'Thermomechanical Stress Analysis of Silicon Nitride Components', PhD Thesis, University of Nottingham, October 1974.

TABLE 1

COMPARISON OF THE PROPERTIES OF METALS USED FOR PISTONS
WITH THOSE OF SILICON NITRIDE CERAMICS

| | Aluminium Alloy | | Grey Cast Iron | Nodular Cast Iron | Silicon Nitride | |
	LM13[*]	LM26[x] (SAE 332)	BS 1452	BS 2789	HPSN[ø]	RBSN[+]
Density Mg/m^3	2.70	2.74	7.3	7.2	3.2	2.2-2.75
Melting Point $^{\circ}C$	~570°	~550°	950-1230°	950-1230°	1750-1900° decomposition	
Thermal Expansivity $x10^{-6}/^{\circ}C$, linear, 20-1000°C	19	20	12	11	2.5	
Thermal Conductivity W/mK (at room temp.)	117	104	50	29	25-35	8-40
Specific Heat Capacity J/kg K	900	900	450	450	700	700
Strength M Pascal	190-280	250	200-300	370-725	800[÷]	230[÷]
Elongation %	~0.5	1.3	-	17-2	-	-
Young's Modulus G Pascal	71	80	103	166	300	175
Hardness	150-80 HB[v]	100 HB	180-260 HB	250 HB	1675-1950[k]	900-1350[k]

Analysis %:	Cu	Si	Mg	Ni	Fe	Mn	Ti	Zn	Sn	Pb
*LM13	0.7-1.5	10-12	0.8-1.5	0.7-1.5	1.0	0.5	0.2	0.5	0.1	0.1
xLM26	2.0-4.0	8.5-10.5	0.5-1.5	1.0	1.2	0.5	0.2	1.0	0.1	0.2

ø HPSN   Hot pressed silicon nitride   (fully dense)

+ RBSN   Reaction-bonded silicon nitride (incompletely densified)

÷ 3-point bend strength

v HB Brinell hardness

k Knoop hardness, 100 g load, kg/mm^2

ECONOMIC AND ENERGETIC CONSIDERATIONS FOR NITROGEN CERAMICS

R.W. Davidge

Materials Development Division, Building 552, A.E.R.E.
Harwell, Didcot,Oxon, OX11 ORA.

1.   INTRODUCTION

The development of a new material can result in two broad
types of market exploitation. Either the material can compete with
existing materials in existing products (which will almost certainly
require some modification) or, some unique property of the new material can result in the development of new products. The consideration of silicon-based ceramics for engine components is an
example of the former, where the main impetus is the belief that
improved engines of modified designs will be developed that are
more efficient than existing metallic engines. The successful
penetration of a new material into the market place probably requires both a technical and economic superiority compared with the
existing material. The economic factor includes not only the direct
replacement cost but additional advantages if the improved properties lead to an enhancement in the performance of the product, for
example by way of increased efficiency.

The main property advantage of silicon-based ceramics for
engine applications is the better refractoriness compared to metals,
which through higher engine operating temperatures, should lead to
greater engine efficiency. The relative lightness of ceramics is
also advantageous in moving components. The big disadvantage of ceramics lies in their brittleness and new design philosophies need development to allow the engineer to take advantage of the new materials.
These considerations are being considered in some detail in other
parts of this Conference and it is still too early to say whether or
not the ceramic engine will be technically feasible, although the
the signs are promising.

Technical feasibility is however an academic consideration

unless the economics are right. At present we can only consider
direct production costs. Cost is a simple concept but is affected
by a large number of factors, many of which are outside the control
of the materials engineer and often of a political nature. Some of
these external factors can produce a very rapid change in the situation, for example, the recent 'energy crisis', concern over toxicity
problems of particular materials, or worries about the supply of a
material from politically unstable parts of the world. Other
factors produce a more gradual change such as the steadily increasing desire to lessen the effects of technology on the environment.
All these effects however can be expressed in plain economic terms
although the short-term picture may be somewhat confused.

It is the purpose of this note to introduce a brief discussion of some of the external factors that are relevant to the introduction of silicon-based ceramics for gas turbine components. To
focus the discussion attention will be given[1] to the case of a
particular gas turbine component - a combustion chamber liner.
This is currently made from a nickel alloy but could be replaced by
reaction-bonded silicon nitride (RBSN). The factors considered include: the availability of raw materials; environmental effects;
the energy inventories for the two materials; a comparison between
raw materials and fabrication costs.

2.   MATERIALS AVAILABILITY

Table I shows the percentages by weight of relevant elements
in the earth's crust, and N in the atmosphere. Ni, Cr and Co can be
already classed as rare and super alloys are but a small outlet for
these elements.

Element	Crustal Abundance (wt. %)	Metal Concentration in Mined Ores (wt. %)
Ni	0.01	1
Cr	0.02	35
Co	0.003	1
Si	27	45
N	80% of earth's atmosphere	

Table I.   The availability of the principal alloying elements in Ni
alloy and of Si and N.

The figures for Si and N_2 require no comment, the reserves are
adequate for the calculable future compared with <100 years for Ni,
Cr and Co. In terms of resource availability there is no question

that silicon nitride is preferable to Ni alloy. Apart from the abundance of Si and N_2, these two elements are ubiquitous and in both strategic and balance of payments terms compare extremely favourably with the alloying elements.

3. ENVIRONMENTAL PROBLEMS

In the field of high temperature engineering neither of the two materials is likely to be used in large quantities and so disposal problems are minimal. However, Ni alloys are highly durable and recycling could be envisaged. Si_3N_4 could not be recycled in reaction sintering processes.

A much more serious aspect occurs in the winning of the constituents that go to make up these materials. Ni, Cr and Co are won by conventional ore mining processes and, for example, for every ton of Ni that is won 100 t of ore have to be processed, of which ∿99% is then dumped. However, because 20-40% voidage is introduced into the residue by the crushing operations not all the waste can be returned from whence it came, with the net effect that there is devastation of the landscape. A graphic example is provided by the Sudbury mine of the International Nickel Co. of Canada. This mining complex covers an area of 1800 km^2 and some 15 Mt of ore is hoisted annually. It is often suggested that improved mechanisation in the mining industry will allow ores containing of the order of 0.1 wt.% metal to be mined economically. This suggests the destruction of even greater terrain. In terms of conservation of land Si_3N_4 compares very favourably with Ni alloy.

4. ENERGY INVENTORY

Table II summarises recent estimates for the total energy requirements to win one tonne of each element.

Major Elements	kWh(t)/t of Element	Wt.% in Material	Contribution from Each Element kWh(t)/t Material
Ni	38,000	53.0	20,000
Cr	40,000	20.0	8,000
Co	24,000	18.0	4,000
Ti	154,000	2.0	3,000
Al	64,000	1.5	1,000
			36,000
Si	30,000	60	18,000
N	1,700	40	700
			∿ 19,000

Table II. Energy inventory for production of elements in nickel alloy and silicon nitride.

All the energies considered   have been converted into the equivalent coal energy or thermal energy, kWh(t). In order to convert electrical energy, kWh(e) into thermal energy an average power station efficiency of 30% has been assumed. The energy required for the raw materials in the nickel alloy is thus nearly twice that of those to produce silicon nitride.

The next factor to consider is the energy required to convert these raw materials into a useful form[1]. It is estimated that an extra 20% energy is required to convert the metal elements into nickel alloy tube. On the other hand to produce silicon nitride in the form of tubes requires about nine times more energy than to produce silicon and nitrogen. This is due to the relatively long high temperature nitriding treatment required. Finally, recalling that the density of nickel alloy is three to four times greater than that of RBSN, the energy to produce components of similar dimensions is about 1.25 times greater for RBSN than for nickel alloy.

Time, however, is likely to reverse this advantage for nickel alloy. A significant fraction of the energy content for nickel is in the mining stage and energy consumption here will almost certainly increase when poorer ores are mined. The energy content for RBSN is mainly in the nitriding stage and increased understanding of the nitriding process plus development of more efficient nitriding furnaces should lead to a reduction in energy consumption.

5.   COST COMPARISON

A detailed comparison is virtually impossible at this stage in that the well-established technologies for production of nickel alloys are competing with the rapidly developing ones for engineering ceramics. However, some broad generalizations can be made. Barnard[2] estimates that costs for nickel alloy turbine engine components in quantity production are £3.4/kg for raw materials and £13.2/kg for finished materials. In comparison, raw materials costs for silicon nitride including silicon, nitrogen plus other consumables such as temporary plastic binders, amount to £5-10/kg. Johnson[3] estimates that, for mass production of turbine components, the nitriding costs should be similar to the materials costs and the fabrication costs about twice this. RBSN is thus not likely to be competitive with nickel alloy on a weight basis but could be on a volume basis. Even so, higher costs for RBSN could be borne if its use resulted in significant increases in engine efficiency.

6.   CONCLUSIONS

This brief survey has shown that RBSN has advantages from a materials availability and environmental point of view compared with nickel alloys. The total energy requirements for production of components in these materials are slightly in favour of nickel alloys at present but this should reverse in the future. Production costs for RBSN are currently greater, for equivalent components, than nickel alloys but the major cost for RBSN is generated in the fabrication stages rather than the raw materials. The single most important factor therefore that could make RBSN more competitive with metal alloys is a significant reduction in the processing cost.

REFERENCES

1.   R.A.J. Sambell and R.W. Davidge, Atom, <u>215</u> (1974) 215.

2.   M.C.S. Barnard, Met. Mat. Tech., (1974) 62.

3.   R. Johnson, Advanced Materials Engineering Limited, private communication.

A CONSIDERATION OF SOME SILICON NITRIDE APPLICATIONS

E. Gugel

Annawerk Keramische Betriebe GmbH,
Rödental, Germany.

1. INTRODUCTION

Silicon nitride was described for the first time in 1857 as a thermally and chemically resistant material (1), but quite a long time elapsed until these properties were utilized for practical applications. The first patents appeared towards the end of the 19th century. Until 20 years ago little scientific work was done, as outlined in (2). At that time research and development work was started in Great Britain aiming for a practical utilization. Some activities followed in Germany. The world-wide intensive engagement with silicon nitride ceramics did not start earlier than 1970, when with the aid of government funds in several countries the development activities of gas turbine components based on silicon nitride ceramics grew remarkably. The enormous output of new research results demonstrates that this is still an expandable field, and one which is a long way from the end of development, especially when seen from the viewpoints of production and application.

It is worth mentioning that one of the first activities with silicon nitride led to an early application, to be the one with the greatest volume. This is that of silicon nitride as the bonding agent for silicon carbide refractories.

2. GENERAL CONSIDERATIONS

There is a great difference between making some specimens under simple reproduction conditions, and producing components of larger size, of complicated shapes, and in many pieces. There are the problems of mixing during body preparation, of the homogeneous mass flow during forming, the problems connected with exothermic heat exploration while nitriding and the surface influence on mechanical properties - just to stress some major problems. All these problems arise when we move from laboratory scale to technical production. In this field much more remains to be done. So the problem of application is not simply the fitting of the material, in the requested shape, with the appropriate properties; it is also that of the ceramic technology. Thus with today's state of the art it often happens that silicon nitride ceramics are far more variable in property than is desirable.

The potential field of application of silicon nitride ceramics for producer companies is very wide. However, typical and proved application areas have not yet been established. But there are, due to a combination of the chemical, mechanical and electrical properties, many possible applications. Silicon nitride is well known now in many industries by advertisements, research papers or other statements. Many people having material problems are asking for help with silicon nitride ceramics. In some cases help is not possible for technical reasons. In some other cases the material is much too expensive. And in those cases which seem justified from the technical and cost points of view, mutual development and innovation work is usually necessary, needing man-power, financial credit, and time.

Using a material such as silicon nitride will pay only when there are distinct advantages in life-time or in a more severe performance condition. The potential for gain is there, but it must be developed for any application. Several companies are already producing silicon nitride and they all need a sales volume to bring back at least some of the development expenses. Between these companies there is some competition of course, which might ruin the price. This would bring the danger that the companies themselves lose interest in continuing with such an interesting material, which without doubt could solve many technical problems.

3. HOT PRESSED SILICON NITRIDE (HS) AS A MATERIAL FOR BALL BEARINGS

Besides other potential, and partly proved applications, there is one which is of special interest: balls, rollers and races made of hot pressed silicon nitride. Because of their low specific gravity, high hardness and relatively low coefficient of friction, ceramics in general were considered attractive candidates for this application. In 1960 tests with glass ceramic balls showed, however, that their hardness and strength were too low (3). Some investigations with sintered and hot-pressed Al_2O_3, and with a SiC/Si material, did not lead to an advantageous performance relative to steel bearings (4,5). As was to be expected, the investigation of reaction sintered silicon nitride balls also failed (6,7). At the end of the 1960's HS came into consideration (7,8) and soon appeared to be the best candidate (9,10).

This application is all the more interesting because of the certain demand for improved engine design in the near future, especially for very high speed bearings such as in the gas turbine main shaft. The DN-value of 3×10^6 (bore diameter in mm times rpm) should be achieved (11).

Early experiments with HS balls or rollers were made by NASA in the USA (10). Very intensive studies followed by SKF (12,13), and by Federal Mogul together with Norton (14,15,16). Since 1974 balls, rollers and races (Fig. 1) have also been available from Annawerk-Ceranox in Germany (2). They are made of hot-pressed preshapes, diamond ground in several steps to final size and surface finish. The sphericity of balls up to 10 mm diameter is better than 0.25 µm, the maximum surface roughness lower than 0.2 µm.

3.1 Advantages of HS

3.1.1 Low Coefficient of Friction

The coefficient of friction of HS against HS and HS against steel under lubrication is about 0.1-0.15 (8,17,18) thus lying in the same range as steel against steel. Without lubrication the friction is still below 0.2, provided that the surface finish is

Fig. 1: 25 mm ceramic bearing: 10 mm rolls and races of hot pressed silicon nitride, cage of Nimonic.

of high quality. It is much lower than that of other ceramics (12) and of steel. This quality is essential for two reasons. Firstly a high temperature bearing may work under conditions where lubrication is not possible, and secondly if the lubrication system fails, the whole engine does not come to a catastrophic end.

3.1.2 Low Specific Gravity

The stress at the point contact, which is calculated as the so-called Hertzian stress, in a ball or roller bearing is considerable. It rises with increasing rotation and weight. The lifetime of a bearing depends on this stress. 3.2 HS has only 40% of the specific gravity of bearing steel, so a much higher rotation speed is allowed.

3.1.3 High Strength under Compression

To fulfil the function of a ball bearing the material should have a strength as close as possible to that of bearing steel. HS is the ceramic with the highest strength value found and as a brittle material the compressive strength is remarkably high. The limit for metallic bodies lies in their plastic deformation.

Fig. 2 shows some results of breaking strength of different 8 mm diameter balls. It shows that HS is much better than Al_2O_3 single or polycrystalline balls, but not as good as "hard metal". But this comparison is not a strictly correct one, since the "hard metal" can plastically deform to some degree and the stress at the beginning of the plastic deformation would be the correct value to take.

3.1.4 High Hardness and Abrasion Resistance

Silicon nitride is one of the hardest materials, exceeded only by diamond, cubic boron nitride, and boron carbide. Since the Young's modulus is relatively low, it is the toughest of all comparable ceramics and thus the abrasion resistance is very good too.

3.2 Performance Test

Some tests have been made especially in the United States and results have been published which are now highly interesting since HS ball bearings are becoming available.

3.2.1 Lubricability

The wetting of HS by hydrocarbon and ester-base lubricants is better than that of steel (12,13). This means that a HS bearing can be lubricated without any difficulties. The friction coefficient is then about the same as in lubricated steel ball bearings.

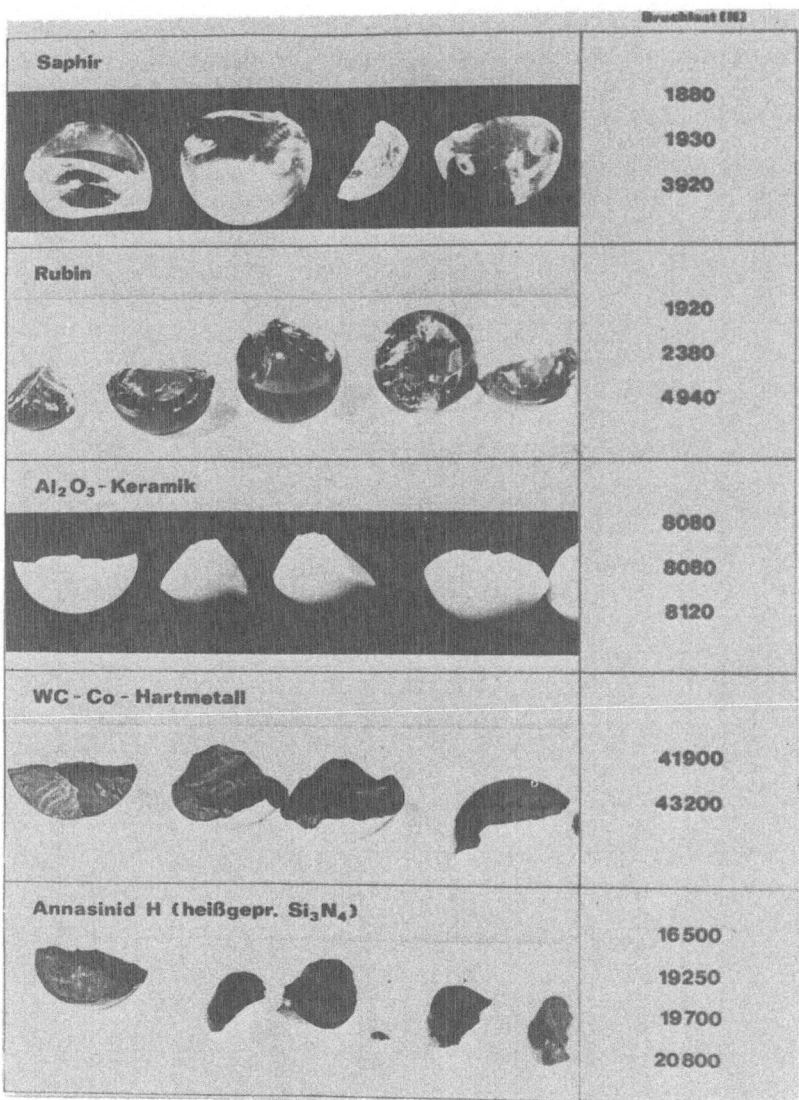

Fig. 2: Breaking load under compression of 8 mm balls of
different materials.

3.2.2 Thermal Behaviour

HS balls develop 10-20% less heat (19). This means that
cooling is easier and cheaper.

3.2.3 Surface Roughness

In order to obtain a high strength, a carefully polished surface is essential. Only one scratch on the surface decreases the strength by 40% (Fig. 3). The lifetime of a ball or roller bearing is extremely sensitive to the surface condition. The last polish must be carried out with 320 diamond grit, so the surface roughness is lower than 1-2 μm (17).

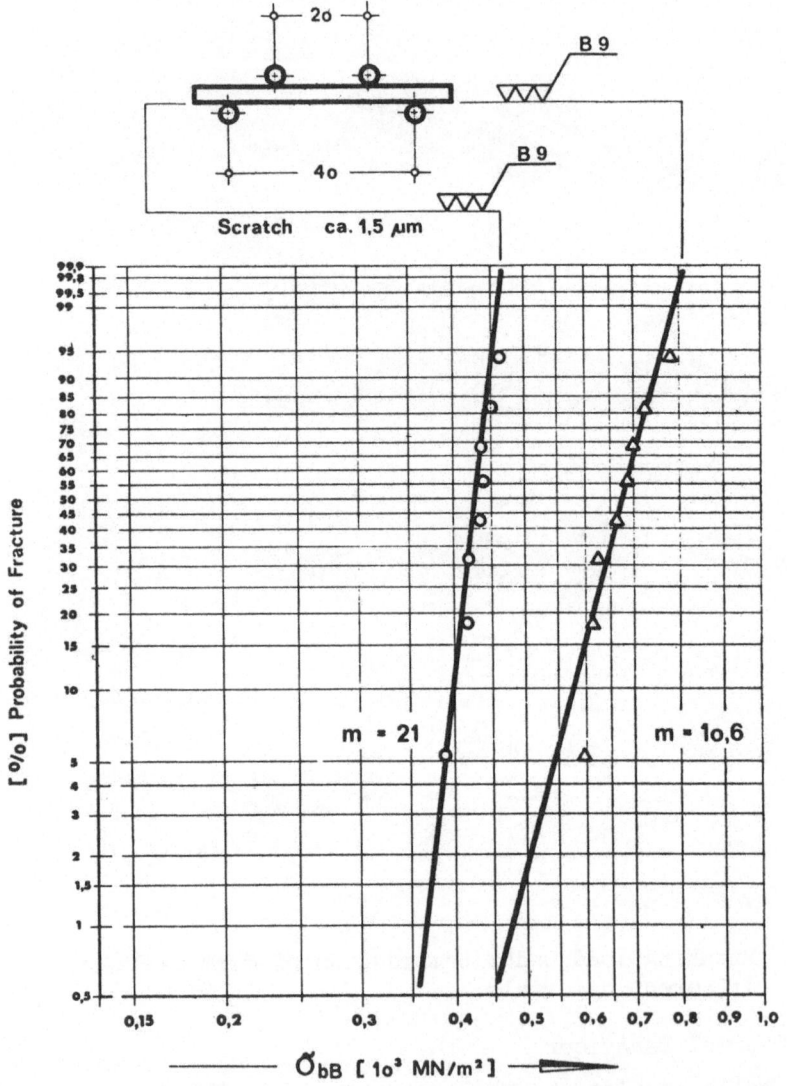

Fig. 3: Weibull-distribution of polished hot pressed silicon nitride (right line) and the same material scratched in the shown manner (left line).

3.2.4 Fatigue

HS does not fail by catastrophic breaking as one would expect with brittle ceramics. Spalling from the surface happens at those areas where the Hertzian stress is too high (14,15). These areas are where the surface is scratched due to faulty grinding, or when there are inhomogeneities, i.e. pores or inclusions, beneath the surface. This shows the importance of having a very homogeneous material with a fine-grained microstructure.

3.2.5 Lifetime

Tests have already proved the capability of ball and roller bearings made of HS. The most interesting results are summarised in Table 1. The HS races show an abrasion of only 0.6 μm after a 330 h test run.

Material roller=R ball =B	races	Hertzian stress (GN/mm^2)	rpm (min^{-1})	10^6 DN $(mm.min^{-1})$	Lifetime (h)	10^6 cycles
HS-R	steel	4.1	10,000		52-157	31-94
HS-R	steel	2.1	10,000	0.55	640	380
HS-R	HS	2.5	10,000	0.55	330	200
HS-B	steel	1.5	71,500	2.5	32	100

Table 1: Test results with HS ball/roller bearings from the literature (15,16,19)

4. CONCLUSIONS

All this shows that HS is a very hopeful material for bearing applications. The test proved its capability. Lubrication is no problem but the material can also work without lubrication. Two points are essential requirements with regard to the material: Firstly a homogeneous high strength HS with a good production reliability, and secondly grinding methods with a high degree of reliability. Both are already developed to a high extent: this means that balls, rollers and rings can be made with appropriate properties and shapes. What is to be done in the future? Economical methods are to be developed granting an acceptable price level in order to profit from the advantages of HS ceramics.

666

REFERENCES

1. C. Deville and F. Wöhler. Liebigs Annalen der Chemie und Pharmazie, 104 (1857) 256.
2. E. Gugel, A. Fickel and H. Kessel. Development in the Production of Hot-Pressed Silicon Nitride. Powder Met. Int., 6 (1974) 3, 136-140.
3. E.V. Zaretsky and W.J. Anderson. Rolling-control fatigue studies with four tool steels and a crystallized glass ceramic. J. Basic Eng. 83 (1961) 603-612.
4. K.M. Taylor, L.B. Sibley and J.C. Lawrence. Development of a ceramic rolling contact bearing for high temperature use. ASME 61, Lubs -12 (1961).
5. R.J. Parker, S.J. Grisaffe and E.V. Zaretsky. Rolling contact studies with four refractory materials to 2000°F. ASLE Trans. 8 (1965) 208-216.
6. C.W. Dee. Silicon Nitride: Tribological applications of a ceramic material. Tribology 3 (1970) 89-92.
7. D. Scott, J. Blackwell and P.J. McCullagh. Si_3N_4 as a rolling bearing material: A preliminary assessment. Wear 17 (1971) 73-82.
8. R.F. Coe, R.J. Lumby and M.F. Rearson. Some properties and applications of hot pressed Si_3N_4. Special Ceramics 5 (1970) 361-376.
9. D.J. Godfrey. Ceramics for high-temperature engineering? Proc. Brit. Ceram. Soc. 22 (1973) 1-25.
10. R.J. Parker and E.V. Zaretsky. Rolling-element fatigue life of silicon nitride balls. NASA Technical Memorandum (1972) TMX-68174.
11. J.H. Rumbarger and J.D. Dunfee. Survey of the state-of-the-art for the design of high speed rolling element bearings in gas turbine engines, Vol. I, Franklin Inst. Res. Lab., Philadelphia, Pa., U.S.A., NTIS, AD 705127 (1969).
12. Y.P. Chin and H. Dalal. Lubricant Interaction with Silicon Nitride in Rolling Contact Applications. Ceramics for High-Performance Applications, 2 (1974) 589-607.
13. H.M. Dalal. Bewertung des heissgepressten Si_3N_4 als Kugellagermaterial. ASLE Trans. 18 (1975) 3, 211-219.
14. H.R. Baumgartner. Evaluation of roller bearings containing hot-pressed silicon nitride rolling elements. Ceramics for High-Performance Applications, 2 (1974) 713-727.
15. D.V. Sundberg. Ceramic roller bearings for high speed and high temperature applications. SAE 740241 (1974).
16. H.R. Baumgartner, W.M. Wheildon. Rollkontakt-Müdigkeit von heissgepresstem Si_3N_4 im Hinblick auf die Oberflächenbehandlung. Surfaces and Interfaces Glass & Ceramics (1974) 179-192.
17. L.K. Heung, P.J. Gielisse and T.J. Rockett. Friction and wear modes in SiC and Si_3N_4. Vortrag Jahrestagung Amer. Ceram. Soc. 1974.

18. J.C. Sirca, J.E. Krysiak, P.R. Eklund and R. Ruk. Friction and wear characteristics of selected ceramics. Amer. Ceram. Soc. Bull. 53 (1974) 581-582.
19. J.M. Reddecliff. Silicon nitride ball bearing demonstration test. Report NTIS AD-A012526 (1975).

DISCUSSION (Lumby, Pick)

Clarke: Are the performance figures you quote for the bearings dependent on the batch of silicon nitride used? What is the variability in performance figures?

Gugel: The performance is very sensitive to the quality. Best performances are achieved with a homogeneous, very fine grained, material.

Klomp: The coefficient of friction data given are probably room temperature values. Can you give the temperature dependence of the coefficient of friction?

Gugel: Yes, the data given are room temperature values. The co-efficient of friction at 400°C is about 0.4 for the unlubricated sliding condition. At temperatures exceeding 700°C the oxidation of the Si_3N_4 lets the coefficient of friction rise to an un-acceptable level.

Davidge: Could you give an estimate of the price for hot pressed silicon nitride balls?

Gugel: The price depends on the size of the ball. Today you can buy a 1 mm ball for less than DM 0.50, but a 12 mm ball will cost about DM 100.

Popper: What are the relative costs of material, grinding and polishing? Would a fabrication method giving a rough sphere reduce the costs substantially?

Gugel: The relative cost of material, grinding and polishing of a sphere to an accuracy of $\pm$ 0.2 µm, is roughly 1:1:1. It should be mentioned that this is dependent on the geometry of the pre-shape and on the size of the ball. So the fabrication cost of a rough sphere with a surface finish of about 50 µm is about 30% less.

Moulson: The current price (100 DM) quoted for a 12 mm ball in HPSN is certainly very high. But as pointed out this will fall if production gets on the way. In this context it is worth recalling that it is not so many years ago (~ 7) when a 150 mm long 'Lucalox' alumina tube cost £5 (~ 20 DM) - this cost has fallen dramatically as the manufacture of the high pressure Na-vapour lamps has built up. The current price of a 150 mm Lucalox-type tube is something more like 2 DM.

SURVEY AND CONCLUSIONS

CONCLUDING REMARKS

R.J. Brook

Department of Ceramics, The University of Leeds, U.K.

There are two features about the topic of nitrogen ceramics that may be noted in the context of a meeting such as the present one. The first of these is that the subject has a tendency to display two rival growth directions. Since our knowledge is incomplete about any one material, we must study it in depth if it is ever to be used in the exacting applications that are being proposed; conversely, with the vista opened up by the sialons, very many alternative compositions and systems become possible and the wish to explore the full breadth of this range opens an alternative avenue of exceptional promise. The second feature is that the subject is of great interest to the scientist and engineer alike; as a consequence, there will again be the tendency for the subject to grow in different modes depending on the motives of the individual concerned.

For these reasons the subject can be seen to lend itself to the more extended type of meeting allowed by the NATO ASI concept. Such Institutes can provide an occasion for several lines of development to be compared and for an exchange of ideas between those committed to different lines. An attempt can be made to assess the state of development of the subject as a whole and, following the identification of common needs, it may be possible to see more clearly the major objectives for the next stage of development of the subject. The success of the present Institute in achieving this result can best be judged from the papers in this volume; however, it may be helpful to give some comment on the state of the art in a more condensed form and a few impressions are given in the paragraphs of these concluding remarks.

The materials considered in the Institute programme can with few exceptions be classified under four main categories, namely the three types of silicon nitride (sintered, reaction-bonded, and hot-pressed), and the sialons. This classification is very imprecise, and it is indeed one of the dangers of the subject that one is tempted to forget that two people talking, for example, of hot pressed silicon nitride are not necessarily talking about the same thing. This difficulty makes the drawing of general conclusions somewhat tentative, particularly when structural studies of one material are for example linked with kinetic studies of another; it lends weight to the plea made at the meeting that in the reporting of data, the fullest possible characterisation of materials should be given, and further, that, in quoting results of work on commercial materials, the trade identification should not in itself be regarded as a sufficient description.

If, with these reservations, the classification is acceptable as a broad indication of groups of materials, one can summarise the state of development of each type as in the table. Here the open circles show for each material the experimental aspects that are requiring and usually receiving attention; the solid circles indicate those aspects where a substantial degree of understanding has been achieved.

	Sintered Si_3N_4	RBSN	HPSN	Sialons
Phase Identification and Phase Interactions	–	–	O	O
Processing:				
finding techniques	O	●	●	O
studying mechanisms		O	●	
refining processes			O	
Properties:				
measurement	O	●	●	O
control		O	O	

Thus HPSN appears the most developed and the best understood of the four; the basic patterns of its processing and properties are known, and it is now a question of refining these with the objective of further improving well-recognised properties. In contrast, understanding of the sialons is at a very early stage and, as the meeting has seen, some of the a priori assumptions made about their properties will need to be confirmed. Reaction-bonded material is in

an intermediate state, with much basic work on mechanisms and
structure-property relations needing to be completed, but also
with the promise of substantial property improvements to be gained
as a result of the acquired knowledge.

Across the four categories a number of common problems can be
identified and it is perhaps worth noting these as research areas
in need of attention. These include work on:

Phase diagrams - both to provide a basic framework for the
    study of sialons and to provide specific guidelines
    for the choice of hot pressing additives for Si_3N_4
    itself.

Diffusion coefficients - to provide guidance in the identif-
    ication of rate controlling processes in the fabrication
    and high temperature mechanical properties of all
    nitrogen ceramics.

Clean systems and powders - to avoid the confusion introduced
    by trace impurities into the interpretation of measure-
    ments. Kinetic studies of reaction-bonding have been
    much plagued by this problem which is of course also
    well known in work on oxides.

Achievement of uniform microstructures - so that the prop-
    erties of these materials can reach their full potential.
    There are indications that macroscopic flaws introduced
    during processing are setting property limits rather
    than the intrinsic properties of the materials them-
    selves.

Property measurements - particularly in the case of sialons.
    Such measurements must include parameters such as
    thermal conductivity even if they are less easily
    measured.

Complementary measurements - where structural and other
    studies are done on the same material so that the
    causal links between the two can be identified with
    greater confidence.

The emphasis of the meeting is generally placed on the more
scientific aspects of the nitrogen ceramics as is to be expected
since these aspects are the ones more easily presented and the
ones less inhibited by commercial restrictions. However, the
sessions devoted to applications have been notable for showing
the range of openings being explored and for showing the extent
of technical success already achieved in a number of these. A
problem requiring much attention in this connection concerns the
matter of scale-up; this will become crucial upon the success of
any large-scale application, and both the feasibility and cost of
this operation have been emphasised in a number of the contributions.

It will be clear that these notes can be no more than a very approximate sketch of one person's view of the meeting and one can but urge that the reader return to the many excellent papers included in the present volume. A final point that can be made, however, concerns the timing of the meeting; in retrospect it can be seen to have occurred at an opportune moment when results from major programmes are becoming available but when, at the same time, sufficient flexibility remains for one to believe that the papers and discussions at the meeting can be helpful to those concerned with future development of the subject. In this last connection, it is hoped that this book can be instrumental in bringing the full flavour of the meeting to all those who are now concerned with the nitrogen ceramics.

AUTHOR INDEX

SUBJECT INDEX

INDEX OF CRYSTAL STRUCTURES

INDEX OF SYSTEMS AND SUB-SYSTEMS